Lecture Notes in

The Lecture Notes in Physics

The series Lecture Notes in Physics (LNP), founded in 1969, reports new developments in physics research and teaching – quickly and informally, but with a high quality and the explicit aim to summarize and communicate current knowledge in an accessible way. Books published in this series are conceived as bridging material between advanced graduate textbooks and the forefront of research and to serve three purposes:

- to be a compact and modern up-to-date source of reference on a well-defined topic

- to serve as an accessible introduction to the field to postgraduate students and nonspecialist researchers from related areas

- to be a source of advanced teaching material for specialized seminars, courses and schools

Both monographs and multi-author volumes will be considered for publication. Edited volumes should, however, consist of a very limited number of contributions only. Proceedings will not be considered for LNP.

Volumes published in LNP are disseminated both in print and in electronic formats, the electronic archive being available at springerlink.com. The series content is indexed, abstracted and referenced by many abstracting and information services, bibliographic networks, subscription agencies, library networks, and consortia.

Proposals should be sent to a member of the Editorial Board, or directly to the managing editor at Springer:

Christian Caron
Springer Heidelberg
Physics Editorial Department I
Tiergartenstrasse 17
69121 Heidelberg / Germany
christian.caron@springer.com

J.G. Muga
R. Sala Mayato
Í.L. Egusquiza (Eds.)

Time in Quantum Mechanics

Second Edition

 Springer

Editors

J.G. Muga
Departamento de Química-Física
Facultad de Ciencia y Tecnologia
Universidad del País Vasco
Apdo 644
48080 Bilbao, Spain
jg.muga@ehu.es

Í.L. Egusquiza
Física Teorikoaren Salla
Zientzia eta Teknologia
Euskal Herriko Unibertsitatea
644 PK.
Bilbao, Spain
inigo.egusquiza@ehu.es

R. Sala Mayato
Universidad de La Laguna
Departamento de Física Fundamental II
38203 La Laguna (S/C de Tenerife), Spain
rsala@ull.es

J.G. Muga, R. Sala Mayato and Í.L. Egusquiza (Eds.), *Time in Quantum Mechanics*, Lect. Notes Phys. 734 (Springer, Berlin Heidelberg 2008), DOI 10.1007/978-3-540-73473-4

This first edition of this book was originally published as Vol. *m 72* in the series *Lecture Notes in Physics*

ISSN 0075-8450

ISBN 978-3-642-09256-5   e-ISBN 978-3-540-73473-4

Springer is a part of Springer Science+Business Media
springer.com
© Springer-Verlag Berlin Heidelberg 2008
Softcover reprint of the hardcover 2nd edition 2008

Cover design: eStudio Calamar S.L., F. Steinen-Broo, Pau/Girona, Spain

Preface to the Second Edition

Time and quantum mechanics are, by and unto themselves, words of such force and attraction that the first edition of this book quite rapidly went out of print. The idea of bringing out a second edition became, thus, more compelling as time (indeed) flew on. Among the different possiblities (a wholly new text, a mere reprinting) we finally settled on a middle way, namely that each of the contributors has decided to what extent their respective chapters needed updating. One of the reasons for this decision is that the field has indeed been evolving, and the increase in editions of the TQM workshops held in La Laguna and Bilbao has provided us with quite a number of possible contributors for preparing, in addition to this second edition, a second volume with entirely new chapters. But more important is that the results and texts of the authors have withstood the passage of time and have weathered very well its ravages. All chapters have been updated, however, in order to include new results and bibliography.

We hope that this new edition meets with as much success as its predecessor and, as we write this, that in the future we can provide the interested reader with more contributions of other participants in our TQM workshops.

We are grateful to R. F. Snider and G. N. Flemming for pointing out several typos and errors of the first edition. We also acknowledge additional support from Ministerio de Educación y Ciencia (Grants BFM2003-01003, FIS2006-10268-C03-01 and FIS2004/05687), Gobierno Autónomo de Canarias (PI2004/025), and UPV-EHU (Grant 00039.310-15968/2004).

Bilbao - La Laguna,
May 2007

J. Gonzalo Muga
Rafael Sala Mayato
Iñigo L. Egusquiza

Preface to the First Edition

Time and *quantum mechanics* have, each of them separately, captivated scientists and laymen alike, as shown by the abundance of popular publications on "time" or on the many quantum mysteries or paradoxes. We too have been seduced by these two topics, and in particular by their combination. Indeed, the treatment of time in quantum mechanics is one of the important and challenging open questions in the foundations of quantum theory.

This book describes the problems, and the attempts and achievements in defining, formalizing and measuring different time quantities in quantum theory, such as the parametric (clock) time, tunneling times, decay times, dwell times, delay times, arrival times, or jump times. The theoretical analysis of several of these quantities has been controversial and is still subject to debate. For example, there are literally hundreds of research papers on the *tunneling time*. In fact, the standard recipe to link the observables and the formalism does not seem to apply, at least in an obvious manner, to time observables. This has posed the challenge of extending the domain of ordinary quantum mechanics.

The difficulties in dealing with time in quantum theory were made explicit very early on, most clearly by Pauli in his famous "theorem," which seemed to impose a serious limitation on the possibility of formulating time as a quantum observable and which has hindered the investigation of time in quantum mechanics for many years. Another disturbing historical landmark is the discovery of quantum Zeno's effect, a paradox that arises when attempting to find an algorithm for computing time probabilities by means of frequently repeated measurements: in the continuum limit these measurements "freeze" the occurrence of events. More recently, however, in the last 15 years, there has been much interest in overcoming these difficulties. Researchers from atomic, molecular, and optical sciences, or from mesoscopic, high-energy, and mathematical physics, have converged to study different time quantities in quantum mechanics. In addition, modern laser technology and the ability to manipulate atomic and molecular motions and the internal state of quantum systems allow nowadays the experimental realization of some of the questions on time in quantum mechanics. As a matter of fact, a number of time observables are already routinely measured in laboratories, for example arrival times in time-of-flight experiments, but the theoretical foundation of these measurements is still being discussed.

The book reflects a good number of these recent trends, but it is not an encyclopedic attempt to cover all the many different open questions, even entire fields, where time plays an important role, frequently a puzzling one, in quantum theory. It is not possible to cover all these subjects in a comprehensive manner unless the treatment becomes very superficial, so we abandoned that objective. While many of these topics are strongly related to the ones presented here, others are basically decoupled. We have thus preferred to make a more compact selection of topics based on the workshops on "Time in Quantum Mechanics" (TQM) in La Laguna and Bilbao. Our aim is to edit further volumes containing new aspects of quantum time not treated in the present one with additional material from old and new TQM workshops.

Most authors are or have been involved in the definition or study of "characteristic times" or "time observables" in quantum mechanics, and more specifically on tunneling and/or arrival times, which have been central subjects in the TQM workshop series. Even so, this book goes clearly beyond these two seed subjects. Thorny issues such as the relations between quantum mechanics and the world of classical events, the theory of measurement, hidden variable theories, time–energy uncertainty principles, superluminal effects, or extensions of the standard formalism are frequently ingredients of this research, as demonstrated in several chapters. It became evident to us that these topics, surrounding the central ones, had to be addressed too, in order to get a better handle on the original problems. Understanding the mysteries of time in quantum mechanics is thus inextricably linked to understanding quantum mechanics itself.

The chapters that follow are reviews that may serve both as an introductory guide for the non-initiated and as a useful tool for the expert. We have essentially allowed full freedom to each contributing author in choosing their presentation and emphasis. This has the major advantage of freshness and, in this manner, the actual state of the issue is much better presented: in this field, there is a host of diverse approaches, tools, languages, notations, nomenclature ... and results. Nonetheless, there has been progress, and consensus in some topics has been achieved. This consensus is also well reflected in the following chapters, as the reader will soon discover. The main disadvantage is of course that the presentation is not fully unified, and that notation and nomenclature are occasionally divergent. In particular, terms such as "tunneling," "traversal," "delay," or "arrival" times are used in different ways in different chapters.

Finally, we would like to acknowledge our coworkers D. Alonso, A.D. Baute, S. Brouard, M. Buttiker, J.A. Damborenea, V. Delgado, C.R. Leavens, D. Macías, J.P. Palao, A. Pérez, D. Sokolovski, R.F. Snider, and G.W. Wei for all their support in timing quantum mechanics and for so many good times with them. We are also indebted to the funding agencies of the TQM work-shop series: Universidad de La Laguna, Ministerio de Educación y Cultura, Ministerio de Ciencia y Tecnología, European Union (through the Canadian–European Research Initiative on Nanostructures, CERION), and the Basque Government.

Bilbao - La Laguna,   J. Gonzalo Muga
August 2001   Rafael Sala Mayato
                                                          Iñigo L. Egusquiza

Contents

1

Introduction

J. Gonzalo Muga[1], Rafael Sala Mayato[2], and Iñigo L. Egusquiza[3]

[1] Departamento de Química-Física, Universidad del País Vasco, Apdo 644. Bilbao, Spain
jg.muga@ehu.es

[2] Departamento de Física Fundamental II, Universidad de La Laguna, Tenerife, Spain
rsala@ull.es

[3] Department of Theoretical Physics and History of Science, University of the Basque Country, P.O. Box 644, 48080 Bilbao, Spain
wtpegegi@lg.ehu.es

This chapter is a historical sketch of "time in quantum mechanics." We have adopted a "physicist's approach" to history, summarizing some of the contributions underlying the topics and concepts treated in the book in rough chronological order. The account is complemented with comments on later or recent repercussions of the original works, giving certain prominence to some of our own contributions. The time span of this "story" is almost 100 years, so necessarily many of the comments and references will be minimal, and even entire fields are absent, in particular those not related to the topics addressed in the subsequent chapters.

1.1 Role of Time in the Early Days of Quantum Theory

In 1913 Bohr suggested that the interaction of radiation and atoms occurred by means of instantaneous transitions, "quantum jumps," among the allowed atomic orbits [1].[1] The jumps were accompanied by the absorption or the emission of radiation, whose frequency corresponded to the energy difference of the stationary orbits, $\nu = \Delta E/h$. However, no mechanism for the timing of these transitions was provided. Soon Rutherford pointed out to Bohr that this lacuna was a "grave difficulty" of his theory, and Slater noted a contradiction between the assumed instantaneous character of the jumps and the observed narrow widths of the spectral lines. There was some interest in discovering an

---

[1] The main source for this section is a book by Mara Beller [2]. An interesting review on time in quantum mechanics during the period 1925–1933 may be found in [3].

J. G. Muga et al.: *Introduction*, Lect. Notes Phys. **734**, 1–30 (2008)
DOI 10.1007/978-3-540-73473-4_1   © Springer-Verlag Berlin Heidelberg 2008

explanatory mechanism, but none was found. This led Einstein to propose a statistical theory for the transition probabilities in 1917.

Heisenberg and the Göttingen group tried to solve the problems of the old quantum theory by creating a "truly discontinuous" theory, matrix mechanics, in which unobservables, in particular the orbits, would be eliminated and in which visualizable models based on a space–time continuum would play no role at all. However, the timing of events did not quite fit in this scheme, as evidenced by a letter from Pauli to Bohr in 1925 [4]:

> "In the new theory, all physically observable quantities still do not really occur. Absent, namely, are the time instants of transition processes, which are certainly in principle observable (as for example, are the instants of the emission of photoelectrons). It is now my firm conviction that a really satisfying physical theory must not only involve no unobservable quantities, but must also connect all observable quantities with each other. Also, I remain convinced that the concept of 'probability' should not occur in the fundamental laws of a satisfying physical theory."

He also speculated that perhaps time could be defined through the concept of energy and wondered about the meaning of a time duration.

Heisenberg answered a few days later [5]:[2]

> "Your problem of 'duration' plays a fundamental role, and I've thought over several matters for domestic use. First, I believe that one can distinguish between a 'coarse' and a 'fine' duration. When, as in the new theory, a point in space has no longer a fixed place, or when this place is still only defined formally and symbolically, then the same is true also of the time-point of an event. But there is always given a rough duration, as also a rough place in space: with our geometric picture we shall still be able to achieve a rough picture of the phenomena. I think it is possible that this rough description is perhaps the only one we may ask for in the formalism."

Heisenberg's words announce some aspects of his uncertainty paper, but things were not quite mature for that step as yet. The original agenda of matrix theorists was quite radical; they doubted that the "position of the electron in time" could be given any meaning. This opinion goes even beyond the impossibility of an accurate definition of conjugate variables, a statement that was formulated later. However, reacting to the success of the visualizable and continuous wave mechanics of Schrödinger, matrix theorists retreated eventually from the extreme original program. Even though Schrödinger had shown the formal equivalence between matrix and wave mechanics for bound systems, an interpretational war began, and the two approaches were for a while

---

[2] Just one week later, Heisenberg, in a letter to Einstein, wondered if the times of transition should be regarded as observable or not [6].

competing, each claiming a superior or more fundamental status. Matrix theorists reemphasized the importance of the discontinuities, and thus of the quantum jumps, since they were not evident in the continuum wave theory.

1.1.1 Born's Collision Papers

Born's aim in his two 1926 collision papers was to harmonize the jumps and the wave picture [7, 8]. For an electron–atom collision, he interpreted the squared modulus of the stationary wave function coefficients at infinite distances as probabilities for the electron "to be thrown" into a given direction for a given atomic state. The theory would therefore not provide the actual state of the atom after the collision, but rather the probability of a certain event, identified by Born as a quantum jump. In this description, however, the wave function considered was stationary, so the question of the timing of the event was not really addressed.

Two early reactions to Born's proposal were due to Pauli and Jordan: Pauli extended the probabilistic interpretation to the position of the electron and opened the way for the return of Heisenberg to visualizable space–time concepts. According to Jordan, the probability interpretation of Born changed the status of transition probabilities from primary to secondary concepts, calculable now from the more elementary probabilities of the stationary states through their time variation [9]. He believed at that time that these were probabilities to *be* in the states, not "to be found." However, the distinction is important with regard to a consistent "transition rate interpretation" of time derivatives. In general, it is not possible to write the exact quantum mechanical equation for the derivative of a "population" P_a as a rate equation

$$\frac{dP_a}{dt} = \sum_b (W_{ab} P_b - W_{ba} P_a) \tag{1.1}$$

in terms of transition rates W, since the interference (nondiagonal) terms of the density matrix also matter. In other words, one cannot visualize the ensemble as being divided into different subensembles until a measurement is performed, and therefore the probabilities under consideration are not probabilities to *be* in the state, as Jordan thought. Note, however, that in open systems or systems repeatedly measured, an equation such as (1.1) and a genuine transition rate interpretation may be justified. More on this in Sect. 1.6.2 below.

1.1.2 Pauli's First Encyclopedia Article

In 1926, Pauli still acknowledged in his first Encyclopedia article that he had doubts about the role of time, along the lines of his letter to Bohr [10]. He emphasized that Einstein's probabilistic treatment of absorption and emission was mute about the times of transition. Did this fact indicate a fundamental causality, or was it due to the incompleteness of the theory? *"This is very much debated, yet still an unsolved issue,"* he concluded. He also insisted on the problem of the duration of the jump, suggesting that perhaps the precision limit of the time of transition was of the same order of magnitude as the period of the light emitted, but admitted that he could not offer a more precise analysis. The duration question is still being studied nowadays and is discussed in Chap. 4 by L. S. Schulman.

1.1.3 Uncertainty Relations

Heisenberg's 1927 paper is in part a response to Pauli's questions [11]. It was largely based on the discussion of several thought experiments where the disturbance of the measuring device was treated in terms of classical particle space–time pictures and quantum jumping, possibly to counterattack Schrödinger's wave picture.

Heisenberg insisted on the instantaneous quantum jump concept,

> *"... the time of transitions or 'quantum jumps' must be as concrete and determinable as, say, energies in stationary states."*

However, in one of the idealized experiments discussed, he argued that the imprecision within which the instant of transition is specifiable is given by Δt in $\Delta E \Delta t \sim h$, where ΔE is the change of energy in the quantum jump.

Without denying the fundamental importance of Heisenberg's uncertainty paper, his introduction of the "principle of indeterminacy" is rather unsatisfactory and imprecise for most modern commentators. After more than seven decades, many of the issues involved have been clarified for position and momentum (see e.g. a detailed analysis by M. Appleby where distinction is made between retrodictive errors, predictive errors, and disturbances, which lead to six uncertainty-like relations [12]), but similar clarity has not yet been achieved for time and energy.

In a "note added in proof," Heisenberg already acknowledges the deficiencies, pointed out by Bohr, of some of the arguments given to that date. Apart from that, several expressions, concepts or formulae were ambiguous or undefined. In particular, he presented the "familiar equation" $\mathbf{E}t - \mathbf{t}E = -i\hbar$ (these are his words and notation) without further comment on the nature of the mysterious time matrix $\mathbf{t}$. Perhaps the explanation of the assumed "familiarity" is that there was nothing exceptional in the conjugate character of the time span and frequency width of a signal. The novelty in Heisenberg's work is to associate Δt with the breadth of imprecision in evaluation of a point in time and to extend the uncertainty product relations to the realm of particle concepts such as position and momentum.

In Bohr's 1928 paper in Nature [13], a reelaboration of his Como lecture, he also discussed the uncertainty product relations for position–momentum and time–energy, but Δt and Δx became the extension of the wave packet in time or space, whereas $\Delta \nu$ and $\Delta \sigma$ were introduced as the wave packet frequency and wave number widths ($\sigma = 1/\lambda$=inverse of the wave length). Bohr wrote, "in the most favourable case",

$$\Delta t \Delta \nu = \Delta x \Delta \sigma = 1 \qquad (1.2)$$

as "well-known relations from the theory of optical instruments." Using the simple Planck–Einstein–de Broglie formulae $E = h\nu$ and $\lambda = h/p$, he obtained

$$\Delta t \Delta E = \Delta x \Delta p = h , \qquad (1.3)$$

which would determine the "highest possible accuracy in the definition of the energy and the momentum of the individuals associated with the wave field."

We see that already in the first papers on the quantum time–energy uncertainty principle by the founding fathers, there is a clear disparity of meanings in Δt and ΔE symbols. The seed had been planted for a long history of confusion and controversy on the time–energy relations. This is the subject of Chap. 3, so we shall skip further references to them, save for a brief mention of the contributions by Mandelstam and Tamm in 1945 [14] and Aharonov and Bohm in 1961 [15].

Other physicists, such as Campbell and Sentfleben, contributed to the early discussions on the role of time in quantum theory. The interested reader may find additional material in [2, 3].

Much of the initial hesitation and questioning faded away after the claims that the theory was final, following the rapid discoveries of the late twenties, became generally accepted. The Copenhagen–Göttingen group closed ranks and promoted the idea that the interpretative issues were essentially solved. In this vein, Bohr's paper in Nature was supposed to have eventually put an end to the contradiction between quantum jumps and the continuous theory, something that we fail to see nowadays. In particular, if we to try extract a recipe for the time distribution of the jumps from it, we would be in for a difficult ride. Anyhow, it was a ripe time to apply the theory to many different systems and experiments, and the applications were so successful that any and all discording voices had little echo for decades.

1.2 The Thirties and Forties

1.2.1 Pauli's "Theorem"

An important landmark in the history of time in quantum mechanics is a footnote of Pauli's second Encyclopedia article of 1933 [16], reedited with minor changes in 1958 [17], in which the argument runs as follows: if there existed a self-adjoint time operator $\widehat{T}$ canonically conjugate to the Hamiltonian,

$$[\widehat{H}, \widehat{T}] = i\hbar , \qquad (1.4)$$

the application of the unitary operator $\exp(iE_1\widehat{T}/\hbar)$ to the energy eigenstate $|E\rangle$ would produce a new energy eigenstate with energy eigenvalue $E - E_1$, so that the spectrum of E would necessarily extend continuously over the range $[-\infty, \infty]$. In principle, this precludes the existence of a self-adjoint time operator for systems where the spectrum of the Hamiltonian is bounded, semibounded, or discrete, i.e., for most of the systems of physical interest [16, 17, 18, 19, 20, 21]. Pauli's conclusion was that

> "...the introduction of an operator $\widehat{T}$ must fundamentally be abandoned...."

In a modern reading of Pauli's footnote, it should be noted, first of all, that it was presented as a formal argument and not, as frequently quoted, as a "theorem." No mathematical rigor was used in the original formulation, in particular no attention was paid to the domains of the operators involved. This point has been emphasized by Galapon, who has shown that it is consistent to assume a bounded, self-adjoint time operator conjugate to a Hamiltonian with an unbounded, semibounded, or finitely countable point spectrum [21, 22, 23, 24, 25]. In any case, there is no need to impose the requirement of self-adjointness in a formulation of quantum mechanics that links observables with positive operator-valued measures (POVMs), that is to say, with nonorthogonal resolutions of the identity (see [26] and Chap. 3 for an introduction to POVMs, and more specifically Chap. 10 for their usefulness in understanding times of arrival). For a different way out in the context of a theory of irreversible evolution, see [27] and references therein.

The traditional reading of Pauli's argument has been, however, quite different. It has been generally regarded as a no-go theorem for the possibility to describe time observables within quantum theory and as the basis for the lapidary sentence "time is only a parameter in quantum mechanics."

Time has of course a parametric aspect in the classical or quantum equations of motion. But quite often we are interested in time as an observable, namely, in the instants when certain events occur or in the durations of processes. Given the central role of the observable concept in quantum theory, it is indeed strange, even absurd, that the theoretical treatment of time observables has been essentially ruled out for so many years as a consequence of Pauli's footnote, without too much questioning. The weight of authority may have played some role, and other possible reasons will be pointed out below, see Sect. 1.4.3.

1.2.2 Von Neumann

One year before Pauli's article, von Neumann had published his "Mathematical Foundations of Quantum Mechanics" [28] where he pinpoints the "chief weakness of quantum mechanics" as its nonrelativistic character, which distinguishes t (without a corresponding operator) from the three space coordinates x, y, z represented by operators. In recent times, Hilgevoord has criticized this view, arguing that von Neumann confuses the position of a particle with the coordinates of a point in space [29, 3]. The partners of the parametric time t are the three coordinates of three-dimensional space, and none of these quantities is quantized. This does not preclude, according to Hilgevoord, the existence of true time observables.

In the same book, von Neumann formalizes the process of measurement by means of a dynamical discontinuity, a "reduction" or a "collapse" of the wave function,

$$|\psi\rangle = \sum c_i |\psi_i\rangle \to |\psi_i\rangle , \tag{1.5}$$

which has thereafter been frequently identified as a quantum jump, even though it is not exactly the quantum jump of Bohr's theory.

1.2.3 Mandelstam–Tamm

Mandelstam and Tamm produced observable-dependent uncertainty relations that refer to the variations in time of their expectation values [14]. They are possibly the most widely accepted and less controversial time–energy uncertainty relations nowadays. Note, however, that they do not refer to time of occurrence of events but to characteristic times of average expectation values. Consider a generic observable $\widehat{A}$. Then we have

$$\tau_A \Delta E \geq \hbar/2 ,$$

where

$$\tau_A = \frac{\Delta A}{d\langle\widehat{A}\rangle/dt} .$$

In words, τ_A is the characteristic time pertaining to $\widehat{A}$; it is the typical time interval required for there to be a substantial modification (in the scale given by ΔA) of the value of $\langle\widehat{A}\rangle$. Here $(\Delta A)^2$ and $(\Delta E)^2$ are the variances for the observable $\widehat{A}$ and for the energy, respectively (see Chap. 3 for details). For a mathematical analysis pointing out some limitations of the Mandelstamm and Tamm relation, see [30].

1.3 The Fifties

1.3.1 Bohm's Book on Quantum Theory

Twenty years after the magic late twenties, there had been no substantial change in the foundations of quantum theory, and the Copenhagen interpretation was rather well rounded and accepted. Possibly the best account of the standard interpretation is Bohm's 1951 "Quantum Theory" [31]. A central subject of this text is the potentiality concept and the transition from potential to actual. According to the standard interpretation, a quantum probability does not mean that a given fraction of the ensemble has a particular value of an observable before measurement; instead it represents a physical tendency to realize that particular value. In other words, a physical system has the potential to reveal a particular fact in a proper interaction (in particular in a measurement). Notice that this view is much influenced by developments in epistemology and the philosophical aspects of probability, of which an important (later) exponent is Popper [32]. It is also required to address experiments on individual systems.

However, also from the orthodox view, facts or events are not described by quantum theory. The latter, without an appeal to classical concepts, would have no meaning. This classical level is necessary to explain the preparation and the results of any experiment, for example, to say without ambiguity that a dial is in a given definite position:

> *"It is characteristic of the classical domain that within it exist objects, phenomena, and events that are distinct and well-defined and that exhibit reliable and reproducible properties ... It is this aspect of the world that is most readily described in terms of our customary scientific language*
>
> *..quantum theory presupposes the classical level... it does not deduce classical concepts as limiting cases of quantum concepts."* [31]

Nonetheless, these classical definite properties and events must be consistent with the quantum description via Born's probability interpretation; additionally, a definite result at the classical level implies a modification of the corresponding potentialities at the quantum level. Classical definiteness and quantum potentialities would, in this manner, complement each other in providing a complete description of the system as a whole:

> *"The continuously changing potentialities and the discontinuous forms in which these potentialities may be realized are, in fact, opposing, but complementary, properties of the electron, each of which expresses an equally important aspect of the electron's behaviour."* [31]

In other words, confronted with the difficulty to explain events, or facts, within quantum theory itself, Bohr, and Bohm as the standard bearer of the orthodox interpretation in [31], essentially give up: *"the large scale behaviour of a system is not completely expressible in terms of concepts appropriate at the small scale level."* They cannot find definite facts or events in quantum mechanics so, accepting their existence, they simply require a certain consistency between quantum theory and the macroscopic level. The events would occur through discontinuous jumps, which are *not* contained in the quantum equations:

> *"One of the most puzzling features of quantum processes is the transition of a system from one discrete energy level to another. Throughout the process of transition, the potentialities associated with the electron change in a continuous way, but the energy changes discontinuously."* [31]

If they are not contained in the quantum equations, how can we predict the time distribution for these events?

This lack of an explicit description of events has been considered by some physicists as a major problem of the theory. Some have tried to modify or complement quantum mechanics to include events and/or discontinuous jumps explicitly, see Sects. 1.6.1 and 1.6.2 below.

1.3.2 Bohm's "Causal" Theory

It is ironic that the same person who wrote the best account of the Copenhagen interpretation also created the best-known hidden variable theory. Bohm's causal theory, which attributes a precise value to the position and the momentum of the particle, solves trivially, in a sense, the timing of events and durations. However, the durations or events timed in Bohm's theory are, in general, "hidden" since we do not see the postulated trajectories.

For many the main problem with this theory though is that, according to the standard claim, it is impossible to determine if it is true or not,[3] although the assumed equivalence of the predictions of the causal theory and the standard approach have been recently questioned and debated for time correlations and double-slit experiments [38, 39, 40, 41, 42, 43, 44, 45, 46, 47, 48].

In any case, the theory has had and has quite a number of devoted followers, see [49, 50, 51] for an account of this viewpoint. Several authors [52, 53, 54, 55, 56, 57] have advocated the use of Bohm's approach to investigate tunneling or arrival times. The merits of this proposal are explained in Chap. 5 by C. R. Leavens.

1.3.3 Are There Quantum Jumps?

Schrödinger hated "quantum jumps" all along, and in 1952 he wrote a paper criticizing once again "the modern analog of epicycles" [58]. He believed that gradually changing amplitudes and wave packets would prove enough to explain even those experiments where the probability interpretation of his waves in terms of localized particles and jumps seemed most natural (the Wilson chamber, for example), all the while accepting that the language of quantum jumps could be a convenient shorthand for describing a wide range of phenomena. Nonetheless, in his view, this attribute of convenience should not make us think that quantum jumps could be taken literally.

He also criticized the arbitrariness of the choice of energy levels when two microsystems interact, the transition probability concept, as evaluated from time derivatives, and the interpretation of stationary scattering wave functions in terms of quantum jumps (*"Where anything happens, we are not facing pure energy states"*).

---

[3] A similar comment is valid for the stochastic interpretations [33, 34, 35, 36, 37].

Born answered that

> "*any of us theoretical physicists would use equivalent mathematical methods, and our prediction for the experimental verification would be practically the same.*" [59]

so the question was regarded by him more as a matter of philosophy than real life physics.

Indeed, at that time, unlike today, experiments with individual systems (one atom or one molecule) were out of question, so that there was no urgency to consider the times of occurrence of single events as really observable quantities. This lack of urgency is well represented in the following excerpt from Schrödinger:

> "*We never experiment with just one electron or atom or (small) molecule. In thought experiments we sometimes assume that we do, this invariably entails ridiculous consequences... In the first place it is fair to state that we are not experimenting with single particles any more that we can raise Ichthyosauria in the zoo.*" [58]

In modern times, the idea that the wave function describes reality and that it does not merely determine probabilities for real events is defended by a number of approaches, such as Bohmian mechanics and others. For Zeh [60], in particular, the discontinuous events (quantum jumps, collapses), and those particle-like aspects that seem to occur during measurements, would only be apparent discontinuities in time in the form of smooth, though very fast, processes of decoherence induced by the environment. In order to explain the observation of only one definite individual outcome among the possibilities represented in the density matrix, Zeh proposes splitting or branching of the observer's identity, following a variant of Everetts many-worlds interpretation [61].

As mentioned in a footnote above, quantum jumps are incorporated into some extensions and modifications of quantum mechanics, which we shall enumerate later in Sect. 1.6.1.

1.3.4 Time-of-Flight (TOF) Experiments

Time-of-flight experiments began in the 1940s and the 1950s. Quite obviously they measure a time quantity that regards quantum objects (atoms and molecules), for which the translational motion has nonetheless been almost invariably interpreted in classical terms. Due to the masses, velocities, and distances involved, this classical analysis of the arrival times is generally justified for most TOF experiments, and it did not trigger the need for a quantum theoretical treatment. The quantum nature of the translational motion in TOF experiments has, however, become evident more recently by means of interference effects among different time (rather than spatial) slits [62], and by the atomic realization of the "diffraction in time" predicted by Moshinsky in 1952 [63], see also Sect. 3.6.3.

1.3.5 Quantum Clocks and Their Limitations

A clock may be defined as a system with a dynamical variable, say a pointer position, having a simple time dependence, so that the value of the parametric time t may be obtained from it. It was soon realized by Salecker and Wigner [64] that the quantum nature of the pointer observable would imply some uncertainty product relation and limit the accuracy. In later theoretical applications of simple clock models (by Peres [65] and many others) the clock is coupled to a system to measure, as a stopwatch, the duration of a certain process or the time of occurrence of some events. Similarly, accuracy limitations may be expected, as discussed by Aharonov and coworkers [66] (see Chap. 8 for further discussion of this topic and quantum clocks in general).

The use of clocks for measuring time observables also poses the problem of connecting defined "ideal" time quantities (i.e., based solely on imposing certain properties and on the system state, without explicitly considering extra degrees of freedom) with the operationally defined quantities that emerge from the coupled system-clock models or from actual experiments [67].

1.3.6 Atomic Clocks

Independently of the works on model quantum clocks described before, actual atomic clocks have undergone an impressive development reaching fractional frequency uncertainties below 10^{-15} [68]. Modern navigation systems, measurement of distances and fundamental constants, mobile telephones and communications, digital television, and many other industrial, commercial, and scientific activities need the accuracy provided by atomic clocks and motivate a continuous effort to improve them. Louis Essen, at the National Physical Laboratory in the UK, constructed in 1955 the first atomic clock, which performed better than the Earth's rotation or the best pendulum or quartz clocks; it used a microwave transition of the caesium-133 atom to calibrate the frequency of an external quartz oscillator. Once the oscillator frequency is fixed and stable, the clock sets the time by counting the number of oscillations. The idea of using atomic transitions to define a standard of frequency, and thus a clock, traces back to I. Rabi in 1944; another important early contribution was the invention by N. Ramsey of the "separated oscillating fields" interferometry, in which the atoms cross two field regions and spend a time in free flight between them. Increasing this time narrows the fringes of the interference pattern for the probability to excite the atom, which leads to a more accurate lock of the external oscillator frequency to the atomic transition frequency [69]. The most advanced atomic clocks nowadays use a variant of this principle in a vertical (fountain) configuration.

Rapid progress in national laboratories led to the definition of the second in 1967 as 9,192,631,770 cycles of the radiation corresponding to the transition between two hyperfine energy levels of the ground state of the caesium-133 atom.

Quantum mechanics is a crucial ingredient of modern time and frequency standards, first of all, because the basic physical process is a quantum transition. Recent developments make quantum mechanics even more relevant because of a tendency to use colder and colder atoms to narrow the resonance peak. (For an alternative, see [70].) This implies the need to consider the atomic motion quantally, in free flight settings (taking into account recoil effects [71] and quantum reflection from the fields [72]) or in traps, where the quantization of vibrational levels provides a way to circumvent the Doppler effect with tight confinement (Lamb–Dicke effect). In addition, other quantum effects have to be taken into account: the "quantum projection noise" [73], in particular, is due to the statistical character of quantum mechanical laws so that a finite number of experiments to determine the population of excited atoms provide averages with fluctuations around the "true average" corresponding to infinitely many experiments. Quantum entanglement among the atoms has also been proposed to improve the accuracy [74, 75]. We refer the reader to any of the numerous recent books and reviews on time and frequency standards, see e.g. [76, 77] for further information on this fascinating topic.

1.4 The Sixties

We have picked up four influential contributions in the 1960s by Smith, Aharonov and Bohm, Allcock, and Aharonov–Bergmann–Lebowitz.

1.4.1 Smith, 1960

Quite surprisingly, there was no standard operator treatment for quantum lifetimes until Smith's work in 1960 [78]. His complaint about the lack of a formal theory for the lifetime is to some extent still valid 40 years later for other time quantities,

> "It is surprising that the current mathematical apparatus of quantum mechanics does not include a simple representation for so eminently observable a quantity as the lifetime of metastable entities ... Unlike other dynamical variables, for which corresponding operators are available, the lifetime is usually computed by various indirect devices." [78]

Eisenbud and Wigner were precursors of Smith's work by interpreting the energy derivative of the scattering phase shift as a time delay [79, 80]. Since then there has been a considerable interest among quantum scattering theorists to define and study lifetimes, delay times, and other characteristic times. Chapter 2 reflects this trend.

1.4.2 Aharonov–Bohm, 1961

A paper by Aharonov and Bohm in 1961 [15] is an interesting example of the difficulties associated with the time–energy uncertainty principle. At that time it was generally believed, as we may still find in some textbooks, that a measurement of energy must have an uncertainty

$$\delta E \geq h/\delta t ,\tag{1.6}$$

where δt is the time duration of the measurement. This had been justified with examples by such heavyweights as Landau and Peierls or Fock and Krylov. However, Aharonov and Bohm provided counterexamples and showed that the energy can be measured with arbitrary accuracy in an arbitrarily short time.

As part of their analysis they considered a free particle's motion as a clock to measure time and introduced a time operator

$$\widehat{T}_{AB} = \frac{m}{2} \left(\widehat{y}\frac{1}{\widehat{p}} + \frac{1}{\widehat{p}}\widehat{y} \right)\tag{1.7}$$

by simple symmetrization of the classical expression for the time when a particle with momentum p, initially at the origin, passes point y, $t = my/p$. With a sign change this becomes the time of arrival at the origin for a particle that at time $t = 0$ is at point y with momentum p. The corresponding operator and eigenfunctions have been studied in detail [20, 81]. The operator is maximally symmetric, which is the nearest best thing to a self-adjoint operator; it satisfies a conjugate relation with the Hamiltonian and therefore implies an "ordinary" uncertainty relation. Other aspects of this construction, its relation to Kijowski's distribution (Sect. 1.5.1), and possible generalizations are considered in Chap. 10.

1.4.3 Allcock, 1969

In a series of three papers published in 1969, Allcock tried to introduce a particular time observable, the time of arrival, in the quantum formalism [18, 82, 83]. His conclusion was a negative one, in line with the pessimism of Pauli. We have analyzed Allcock's work in detail along the years and have found flaws in each set of his detailed arguments, confirming a preliminary intuition of Wigner [84]. These arguments can be classified in three main groups:

1. Pauli's "theorem".
2. Quantum mechanics with sources.
3. Using complex potentials as detector models.

Pauli's argument is easily sidestepped with the aid of POVMs, as discussed earlier and in more detailed manner in Chap. 10. We shall next briefly comment on the remaining two items.

Quantum Mechanics with Sources

Allcock defined a "source" by imposing the following condition on the wave function (in one dimension)

$$\langle x|\psi(t)\rangle = 0 , \quad t < 0 , x > 0 , \tag{1.8}$$

in other words, by confining the wave in the left axis up to $t = 0$. He also showed that the solution of the Schrödinger equation subject to (1.8) is given by

$$\langle x|\psi(t)\rangle = \frac{1}{h^{1/2}} \int_{-\infty}^{\infty} dE \, e^{i(2mE)^{1/2}x/\hbar} e^{-iEt/\hbar} \chi(E), \quad x > 0 , t > 0 , \tag{1.9}$$

with

$$\chi(E) = \frac{1}{h^{1/2}} \int_{-\infty}^{\infty} dt \, \langle x = 0|\psi(t)\rangle e^{iEt/\hbar} . \tag{1.10}$$

Note that $\chi(E)$ is not a standard energy amplitude but the Fourier transform with respect to time of the wave function at the source point. Since (1.9) includes negative energies, Allcock thought that Pauli's "theorem" could perhaps be avoided in this way. He also realized that the *final* (i.e., after infinite time) total norm to the right of a certain point X is given by the integral

$$P(x > X)_{t\to\infty} = \lim_{t\to\infty} \int_X^{\infty} dx |\langle x|\psi(t)\rangle|^2 = \int_0^{\infty} dE \left(\frac{2E}{m}\right)^{1/2} |\chi(E)|^2 \tag{1.11}$$

(note the lower limit in the energy integral) and identified this probability, as in classical mechanics, with the total arrival probability. He then, using independent arguments, derived an arrival-time distribution in contradiction with this equation because it had contributions from the negative energies too, so he concluded *"unequivocally that an ideal concept of arrival time cannot be established for the problem with sources $(-\infty < E < \infty)$."*

This question has been reexamined recently [85]. It turns out that one can express the source solution (1.9) as a standard initial value problem by using contour deformation in the complex momentum plane. In particular, for free motion on the full line, the negative-energy contribution is exactly equal to the contribution of negative momenta in the standard integral expression of $\psi(t)$ over plane waves

$$\langle x|\psi(t)\rangle = \int_{-\infty}^{\infty} dp \, \langle x|p\rangle e^{-iEt/\hbar} \langle p|\psi(0)\rangle . \tag{1.12}$$

In classical mechanics the particles with negative momentum cannot reach a point with positive x and t when the ensemble is initially confined within the left half line. By contrast, the quantum mechanical wave at $x > 0$ and $t > 0$ is affected by negative momentum components [86]. Thus Allcock's classical assumption amounts to incorrectly neglecting this quantum contribution, as already noted in [87].

Complex Potentials

Allcock tried to reinforce his formal arguments with specific models for detection and chose a complex, imaginary potential step

$$V(x) = -iV_0\Theta(x) \ . \tag{1.13}$$

He found that when V_0 is very large the particle is not absorbed (i.e., detected) but reflected. In the opposite limit, when $V_0 \to 0$ the particle is absorbed, but in a very large length, so the resolution to determine where or when the particles arrive is very poor.

We know nowadays that Allcock's results are not general. It is possible to construct, using inverse scattering techniques, potentials that absorb essentially the full wave packet in a very short spatial interval [88, 89, 90, 91]. For plane waves this interval can be made arbitrarily small. For wave packets the length can be small but not zero because of a peculiar property of quantum perfect absorbers; they exactly reproduce the behaviour of the wave function defined in the absence of the absorber. Therefore, they must accumulate and give back norm to reproduce the "backflow effect" [92, 93], by which wave functions without negative momentum components have negative current density at certain time and position intervals [94].

1.4.4 Aharonov–Bergmann–Lebowitz, 1964

In quantum mechanics the results of measurements in the future are only partially constrained by the results of measurements in the past. Thus, the concept of quantum state, when defined by the results of measurements in the past only, is time-asymmetric. The two-state vector formalism of quantum mechanics (TSVF), originated in a seminal work of Aharonov, Bergmann, and Lebowitz (ABL) [95], removes this asymmetry. In this approach, a system at a given time t is described completely by a *two-state vector*

$$\langle\Phi| \ |\Psi\rangle \ , \tag{1.14}$$

which consists of a quantum state $|\Psi\rangle$ defined by the results of measurements performed on the system in the past relative to the time t and of a backward-evolving quantum state $\langle\Phi|$ defined by the results of measurements performed on this system after the time t.

An important and at times controversial application of the TSVF has been the *weak measurement theory* [96], which allows to see that systems described by some two-state vectors can affect another system at time t in a very peculiar way. This has led to the discovery of numerous bizarre effects. All these questions are treated in detail by Yakir Aharonov and Lev Vaidman in Chap. 13.[4]

---

[4] See also other references to weak measurements in Sects. 1.6.3, 7.4, 5.6 and in Chap. 11.

1.5 The Seventies: the Zeno Effect, TOA Distributions, POVMs

1.5.1 Kijowski's Time-of-Arrival Distribution

Kijowski noted that four essential characteristics uniquely determine the classical arrival-time distribution $\Pi(t; f)$ at a point X (Π is a function of time of arrival t and a functional of the phase-space distribution f) [97]. He then obtained a quantum distribution Π_K by demanding these properties in the quantum case (the distribution itself had been obtained previously by Allcock [18, 82, 83]). The quantum distribution has been later associated [81, 98, 99, 20] with the POVM of the Aharonov–Bohm maximally symmetric [100] time-of-arrival operator [15], which has a degenerate set of improper "eigenstates" $|t, \alpha\rangle$ (with the $\alpha = \pm$ associated with left or right arrivals, and t the eigenvalue) that provide a nonorthogonal resolution of the identity. This time-of-arrival distribution,

$$\Pi_K(t; \psi(t_0)) = \sum_\alpha |\langle t, \alpha | \psi(t_0) \rangle|^2 \,, \tag{1.15}$$

is obtained in a "predictive manner" from the wave function at t_0. Note that t is the time span from t_0 to the arrival instant. Alternatively, due to the "covariance" of the distribution, we may use $\Pi_K(t; \psi(t_0)) = \Pi_K(0; \psi(t + t_0))$. In words, the arrival-time distribution may be computed "on the spot" for each time instant by evaluating the overlaps between the wave function and the "crossing states" corresponding to the arrival instant $t = 0$. This is a useful change of perspective that has made possible to generalize Kijowski's distribution for interaction potentials [101] or for multiparticle systems [102], something that could not have been achieved by means of the original derivation or from the free motion time-of-arrival operator. We will reformulate these points in Chap. 10.

Kijowski himself considered a relativistic version of the arrival-time distribution [97]. This has been retaken by several authors in later times, see e.g. [103].

The status of the distribution $\Pi_K(t)$ should be understood as completely parallel to other distributions, such as $|\psi(x)|^2$ for positions. That is, they are not predictive for the actual location, be it in time or space, of a single realization, but they are predictive, in an ideal manner, for the distribution of measurements. This character of being ideal means that the proper apparatus must be found to realize it! In this vein, a recent series of articles that model the measurement of the first spontaneous fluorescence photon emitted by an atom excited by a localized laser beam culminates in the description of an operational procedure that provides Kijowski's distribution in a well-defined limit [104, 105, 67, 106].

1.5.2 Zeno Effect

The suppression of transitions due to frequently repeated, first-kind (instantaneous and collapsing) measurements in the limit of infinite frequency was already noticed by Allcock for the arrival time, i.e., for the transition from left to right subspaces in the real line [18]. The effect was later generalized and examined in a mathematically rigorous way by Misra and Sudarshan [107], who named it "Zeno's paradox in quantum theory" [108]. Even though they paid attention mostly to the decay of unstable particles, these authors sought, in general, a *"trustworthy algorithm for the probability that the particle makes a transition from a preassigned subspace of states to the orthogonal subspace sometime during a given period of time"*.

They considered that a "natural interpretation" of a continuous observation in this period is a sequence of instantaneous collapses into the original subspace, followed by normal Schrödinger evolution. They then showed that this seemingly natural approach prevents any transition in the limit $\delta t \to 0$. Misra and Sudarshan considered this answer physically unacceptable and concluded that the *"completeness of quantum theory must remain in doubt until a proper algorithm is found for the above probabilities."*

Misra and Sudarshan speculated on physical grounds that a hypothetical operator-based algorithm should have a number of properties that essentially coincide with those of generalized resolutions of the identity or POVMs. They even "predicted" the existence of a unnecessarily self-adjoint time operator as the first moment of the POVM. It thus appears that the program sketched by these authors has already been implemented for the free-motion arrival time. As a matter of fact, Kijowski had formulated his distribution 2 years before, but the explicit connection with POVMs and the time-of-arrival operator is much more recent [81, 98, 99, 20, 109]. It does not seem, however, that finding an optimal POVM ("optimal" in some suitably defined sense) for a specific time quantity is a simple matter. In other words, the generalization of Kijowski's distribution for other characteristic times is not necessarily straightforward [20].

The Zeno effect is still much debated nowadays and subjected to experimental scrutiny too. For modern reviews on the Zeno effect, see [110, 111]. Some recent works study its physical origin [112] (as a consequence of the initial-state reconstruction and short-time deviations from exponential decay, see Chap. 2), and its disappearance by increasing the distance between the detector and the decaying system [113].

1.5.3 Times of Occurrence in Quantum Mechanics and POVMs

We have already mentioned several times the POVMs and the corresponding nonorthogonal resolutions of the identity. Helstrom [114] and Holevo [115] became aware of the possibilities of these concepts to describe time observables. Srinivas and Vijalakshmi formulated the POVM program quite explicitly in this regard [109] *"there is no single unique time observable, but actually a whole class of time of occurrence observables – one associated with each observable event that could occur. In fact some of the confusion on the notion of time in quantum theory is mainly due to the fact that it is not appreciated that there is nothing like the time observable, but a class of time of occurrence observables associated with each physical system. The main objective of the theory should be to provide a mathematical characterization of such time of occurrence observables so that one arrives at a prescription for calculating say, the probability that an event occurs in a given interval of time."*

They thus advocate the use of POVMs based on the theory of continuous measurements and subject to the general requirements of causality (the results up to time t should not depend on later times) and time-translation invariance of the theory (or "time covariance").

1.6 Some Recent Trends

1.6.1 Theories that Include Events in Quantum Mechanics

According to the dictionary, "event" is anything that "takes place," "happens," or "occurs." The event for a classical system could be simply defined as *"the realization of some specified value of a variable at a given instant"*.

In classical mechanics the questions: "When does an event occur?" (e.g., when does a particle moving in one dimension cross the point X?) and "What is the probability distribution of the corresponding instants?" (e.g., the time-of-arrival distribution at X) are no more problematic than say "What is the position of the particle at $t = 0$?" This does not mean they are of the same nature, though. There are some differences that may become important in the quantization process:

- For a given t, only one x is possible in the trajectory, whereas for a given x, there may be several crossing instants $t_1, t_2, \ldots$.
- The particle must be somewhere at time t, whereas the particle may or may not arrive at the selected point X.

In spite of these differences, one can define time distributions, in general, just for a subensemble of the original ensemble of particles, by ordering the crossings as first arrivals, nth arrivals, or last arrivals. It is also possible to define time distributions for all arrivals, irrespective of their order. For a more detailed discussion, see [87].

However, things are not so simple in quantum mechanics. The solutions of the standard equations of quantum mechanics are continuous and probabilistic, and, as discussed in Sect. 1.3.1, they do not contain the abrupt events explicitly. Are events somehow implicit in the formalism, in the same way that classical statistical densities smooth out the individual occurrences of the classical events? The answer seems to be negative in general: the theorems by Kochen–Specker [116] and Bell [117] discourage a direct "naive realist" statistical interpretation of quantum theory.[5]

It might be possible to include events explicitly by modifying quantum theory without too much violence to the well-established results. This is the case for example of the *spontaneous collapse* theory of Ghirardi, Rimini, and Weber [119], and related approaches such as the "event-enhanced quantum theory" (EEQT) of Jadczyk and Blanchard [120, 121]. The modification of Schrödinger's equation by adding a random fluctuating collapse has been examined from very many different points of view. For a review, see [122].

Another approach, quite different in spirit and techniques, aspires to include events as part of the formalism without any modification of the dynamics. This is the approach taken by the followers of the "quantum histories" or "consistent histories" program [123, 124, 125, 126, 127, 128, 129], which is reviewed in Chap. 6. The basic idea of consistent histories is to decompose the density operator $\rho(t)$ by dividing the total evolution into smaller intervals and using resolutions of the identity (normally, but not only, by projector decompositions; see also [130, 131, 132] for POVM decompositions) at the intermediate times. One then looks for cases where the interference terms without a classical event interpretation do not contribute. This would allow for a consistent assignment of probabilities to classical-like histories.

These ideas were applied by Yamada and Takagi [133, 134, 135] to the arrival time. Interference between amplitudes associated with Feynman paths that contact or does not contact the arrival point X within a specified finite time interval does not vanish; so, according to the premises of this interpretation, probabilities cannot be defined for the events of arriving or not arriving at X within such a time interval (there are, however, some exceptional cases [136, 133]). Later, Halliwell and Zafiris [137] pointed out that coupling the particle to a thermal bath allows the definition of coupling-dependent, arrival-time probabilities.

In any case, if the quantum mechanical formalism is regarded as a pool of potentialities to be realized by appropriate measurements or interactions, the nonvanishing interferences just described are not a major stumbling block [136, 87]. They tell us that the arrival of the particles, left alone, is not classical-like, but they do not preclude the fact that proper interactions (measurements) lead to arrival-time distributions that can be modeled theoretically. It is moreover possible to select one of them as being optimal in some specified sense, as it occurs with Kijowski's distribution.

1.6.2 The Resurrection of Quantum Jumps

Apart from Bohm's book, standard texts on quantum mechanics have kept basically silent about quantum jumps (in the old sense). This is true even

---

[5] Only "contextual" and "nonlocal" hidden variable theories would be allowed, see e.g. [118].

for the first monographs written by Pauli or Dirac. However, the modern advances in the manipulation of single atoms have fostered new theoretical techniques, in particular "quantum trajectory" or "quantum jump" approaches [138, 139, 140, 141], that solve the master equation for an open system, typically an atom interacting with laser fields and a background bath, by means of many individual histories where a series of photon detections are represented by wave-function jumps. These "trajectories" provide the smooth master equation solution when averaged and make the statistical (smooth) and individual (abrupt) descriptions compatible. The latter is directly comparable with the experimental records of successive photon detections for individual atoms. They can explain, for example, the statistics of dark periods of individual atoms whose ground states are coupled to a metastable and a rapidly decaying state by two lasers. These modern quantum jumps are not interpreted, e.g., by Cook [142], as intrinsic properties of the isolated system, but as sudden counting events of systems subject to frequent measurements. These measurements would effectively prevent the development of superpositions (by repeated collapses) [142] and would lead to dynamics governed by rate equations rather than by Schrödinger's equation. The Zeno effect is always a menace for the occurrence of the jumps although, Cook argues, it is much more difficult to realize when detecting spontaneously emitted photons. Stenholm reasons similarly that an external intervention is necessary (caused by an observer or not) to interpret the time evolution in terms of a stochastic sequences of instantaneous transitions [143], stressing that the justification of the necessary master equations is not always an easy task. The connection between consistent histories and quantum jump techniques has been discussed by Brun [144].

"Quantum jump" techniques have been applied recently to simulate a time-of-flight experiment for individual atoms, see [104, 105, 67, 106] and Sect. 8.5.3. The operational time-of-arrival distribution corresponds in this case to the detection of the first spontaneously emitted photons after the excitation induced by a laser. It is possible, by subtracting the detection delay for the atom at rest, to obtain exactly the quantum current density from the distribution of first photons, in the limit of a large Einstein coefficient [104, 105]. With some additional manipulations consisting of a previous filtering of the initial state to compensate for detection losses, it is also possible to obtain Kijowski's distribution [67, 106].

1.6.3 Tunneling Times

The tunneling time is an example of a duration type of quantity. Actually the word "tunneling" here is a red herring since the same conceptual problem arises without tunneling, even for free motion. The actual question is:

"How long does a particle take to traverse a spatial region?"

It is true, however, that tunneling enhances some of the puzzling aspects of the possible answers. There are hundreds of papers devoted to tunneling or traversal times. The modern stage of the tunneling-time conundrum started with a letter by Büttiker and Landauer in 1982 [145]. They set a "traversal time" as the time characterizing the transition between sudden and adiabatic regimes in oscillating potential barriers. Very many different proposals appeared subsequently in the literature throughout the 1980s, with each author tending to defend their own time as "the good one." The controversy has kept going all through the 1990s, moving however in the direction to a certain degree of consensus: that there is no single quantity that contains the whole and only truth about timing the particle's traversal, see e.g. [146]. The tunneling time is in fact one example of quantization of a classical quantity that involves products of noncommuting observables: a unique, classical question corresponds to different quantum versions. Comprehensive theories [147] and experiments [148] show explicitly several of these times.

For previous review articles see [149, 150, 151, 152, 153, 154, 155, 146]. In this book Chaps. 2, 5, 7, 8, 11, and 12 review extensively various aspects of the timing of particle traversal through a given region, and the contributions of different authors. Here, we shall introduce the subject by summarizing an "umbrella theory," the BSM approach,[6] which has the merit of describing many different proposals of tunneling times as particular cases of a single formal framework [147], providing a unifying perspective, useful to relate, classify, and study the properties of different times. It dissolves the puzzling fact that many different times have been or may be defined, even though not all proposals may be included as particular cases, in particular the times derived from Bohm's interpretation, cf. Chap. 5.

The BSM Approach

In classical mechanics the transit times for a region of space can be easily defined from the trajectories. If the region contains a potential barrier and the particles are sent toward the barrier at time $t = 0$, the ensemble of classical particles can be divided into particles that will eventually pass the barrier and into particles that will not pass the barrier. This means that the average *dwell time* can be separated into two contributions

$$\tau_D = P_T \tau_T + P_R \tau_R , \qquad (1.16)$$

where P_T and P_R are the probabilities for transmission and reflection, and $\tau_{T,R}$ are the average dwell times for the transmission and reflection subensembles. In the question "how long does a transmitted particle take to cross the barrier" there are, apart from time itself, two observables implied: (1) "being at the barrier region," $\widehat{D} = \int_a^b dx\, |x\rangle\langle x|$, which is a projector ($\widehat{D}^2 = \widehat{D}$) and (2) "being transmitted,"

$$\widehat{T} = \int_0^\infty dp\, |p^-\rangle\langle p^-| . \qquad (1.17)$$

---

[6] It was termed "systematic projector approach" in the original paper.

The states $|p^-\rangle$ are scattering eigenstates of the Hamiltonian, formed by a combination of incident plane waves, suitably chosen to produce the outgoing plane wave $|p\rangle$, see Chap. 2. "Being reflected" corresponds to the complementary projector $\widehat{R}$, such that $\widehat{T} + \widehat{R} = 1$. The operator $\widehat{D}$, however, does not commute with $\widehat{T}$ or $\widehat{R}$. Thus, the quantum mechanical dwell time[7]

$$\tau_D = \int_0^\infty dt \, \langle \psi(t) | \widehat{D} | \psi(t) \rangle \tag{1.18}$$

can be decomposed in many different ways depending on the decomposition chosen for $\widehat{D}$ [147]. Using $\widehat{T} + \widehat{R} = 1$ and $\widehat{D}\widehat{D} = \widehat{D}$ we may write, for example,

$$\widehat{D} = \begin{cases} (\widehat{T} + \widehat{R})\widehat{D} = \widehat{T}\widehat{D} + \widehat{R}\widehat{D}, \\ (\widehat{T} + \widehat{R})\widehat{D}(\widehat{T} + \widehat{R}), \\ \widehat{D}(\widehat{T} + \widehat{R})\widehat{D} = \widehat{D}\widehat{T}\widehat{D} + \widehat{D}\widehat{R}\widehat{D}, \\ \frac{1}{2}\left([\widehat{T}, \widehat{D}]_+ + [\widehat{T}, \widehat{D}]_- + [\widehat{R}, \widehat{D}]_+ + [\widehat{R}, \widehat{D}]_-\right). \\ \cdots \end{cases} \tag{1.19}$$

These are some of the simplest decompositions, but there are infinitely many possibilities. The first decomposition leads to a separation of the dwell time into reflected and transmitted parts, as in the classical expression (1.16). However, the operators $\widehat{T}\widehat{D}$ and $\widehat{R}\widehat{D}$ are not Hermitian and therefore the corresponding times turn out to be complex quantities

$$\tau_T^{TD} = \frac{1}{P_T} \int_0^\infty dt \, \langle \psi(t) | \widehat{T}\widehat{D} | \psi(t) \rangle \,, \tag{1.20}$$

$$\tau_R^{RD} = \frac{1}{P_R} \int_0^\infty dt \, \langle \psi(t) | \widehat{R}\widehat{D} | \psi(t) \rangle \,. \tag{1.21}$$

The stationary wave versions, see Sect. 8.4 for more details, were obtained by Leavens and Aers [156] combining two time scales of the Larmor clock and by Sokolovski and Baskin, using a classification of paths in the path integral formulation of the wave function [157, 158, 159].[8] For opaque barriers, the moduli of the complex times coincide with the traversal time obtained by Büttiker and Landauer [145].

In the second decomposition there are two terms, $\widehat{T}\widehat{D}\widehat{T}$ and $\widehat{R}\widehat{D}\widehat{R}$, with a simple classical-like interpretation; they lead to dwell times for the "to-be-transmitted" and "to-be-reflected" components of the wave function [162, 163, 164]. Note, however, the presence of interference terms, such as $\widehat{T}\widehat{D}\widehat{R}$,

---

[7] See Chaps. 2, 7, and 11 for discussions on the dwell time concept in quantum mechanics.

[8] A *different* complex time had been defined before by Pollak and Miller [160] and Pollak [161], with real and imaginary parts corresponding to the time averages of the real and imaginary parts of the quantal microcanonical flux–flux correlation function.

without a classical counterpart. The interference terms are avoided in the third decomposition by first imposing particle presence in the selected region and then separating into to-be-transmitted and to-be-reflected components. Is this "THE" answer? The "good" resolution? While this decomposition leads to (1.16) and avoids interferences, it is not free from nonclassical features: the resulting times are not additive, because of the presence of two $\widehat{D}$ operators, so that the time for going from a to c is not the sum of the times for going from a to b plus the times for going from b to c. Finally, the last decomposition shown is based on separating the Hermitian and anti-Hermitian parts of $\widehat{T}\widehat{D}$ and $\widehat{R}\widehat{D}$. Interestingly enough, it may be associated with Larmor clock or weak measurements, where a weak magnetic field is used to rotate the spin of the electron crossing the region [165, 166], see Chaps. 7, 8, 11, and other applications of the Larmor clock concept in Chap. 9. The amount of precession should be an indication of the traversal time. This sounds like quite a good idea, but unfortunately the resulting time may be negative... Is this necessarily a bad thing? It is if one is looking for classical-like quantities and nothing else; it is not if one simply wants to know how nature behaves in different quantum scenarios corresponding to the same classical question.

As in other joint measurement problems involving noncommuting observables, the quantum multiplicity described above is linked to the lack of a proper *joint probability distribution* for being in the barrier and being eventually transmitted, but it is possible to define "marginal" and "conditional" probability distributions

$$P_D \equiv \langle \psi(t)|\widehat{D}|\psi(t)\rangle , \tag{1.22}$$

$$P_T \equiv \langle \psi(t)|\widehat{T}|\psi(t)\rangle , \tag{1.23}$$

$$P_{D|T} \equiv \langle \psi(t)|\widehat{T}\widehat{D}\widehat{T}|\psi(t)\rangle / P_T , \tag{1.24}$$

$$P_{T|D} \equiv \langle \psi(t)|\widehat{D}\widehat{T}\widehat{D}|\psi(t)\rangle / P_D , \tag{1.25}$$

and some of these decompositions have a simple interpretation in terms of them. In particular, the transmission times derived from the second and third decompositions are

$$\tau_T^{TDT} = \int_0^\infty P_{D|T}\, dt, \tag{1.26}$$

$$\tau_T^{DTD} = \frac{1}{P_T} \int_0^\infty P_D P_{T|D}\, dt . \tag{1.27}$$

Notice that the following equality, which holds for a classical ensemble of particles, is no longer valid in quantum mechanics:

$$P_{D|T} P_T = P_D P_{T|D} . \tag{1.28}$$

Relation (1.28), when it holds (i.e., in the classical case), implies the equality $\tau_T^{TDT} = \tau_T^{DTD}$ – this is of course not generically true in quantum mechanics.

Operational Models, Superluminality, Hartman Effect

At a less abstract level there are also *operational* models that define traversal times associated with specific idealized measurements, e.g. by simulating two-detector experiments [167] with one detector before and one detector after the barrier [168].

It is possible that these models lead to quantities defined in the abstract framework described above, as happens in the case of the Larmor clock. For other experimental setups the connections between abstract or "ideal" results and operational models may of course be more involved.

Other works have focused on describing the temporal behaviour of the wave function and its main features, such as peaks or fronts. A striking phenomenon is that in tunneling conditions the peak of the transmitted wave packet may appear at the rear barrier edge even before the extrapolated peak of the incident wave packet arrives at the front barrier edge. Related to this situation is the "Hartman effect" [169]: the time of appearance of the transmitted peak does not grow with the barrier width, up to a critical point where the components above the barrier maximum start to dominate [147, 170], see Sect. 2.4.1. Also, a sligtly absorbing medium may cause, for the appropriate initial state, the simultaneous arrival of the wave packet peak at different locations [171, 172]. Similar effects have been experimentally verified for photons, or microwaves [173, 174, 175, 176], and have led to considerable attention, even in the mass media. A violation of relativistic causality is, however, not implied [177, 178]. For further details, see Chaps. 2, 5, 7, 12, and 11.

1.7 Discussion

Understanding time in quantum mechanics is in fact intimately linked to understanding quantum mechanics itself, in particular, the transition between the potentialities described by the formalism and actual events. However, we are far from a consensus on how, and even if, this transition takes place. Different solutions proposed for this theoretical lacuna amount to different answers for several of the mysteries related to time in quantum mechanics. We have paid particular attention to work done on "tunneling times," and "arrival times," two topics that have been controversial throughout the past two decades. Some results have been firmly established though, and much progress has been achieved in avoiding a number of stumbling blocks. For example, Pauli's argument against the existence of a self-adjoint time operator in quantum mechanics is not a problem when realizing that observables are not necessarily linked to self-adjoint operators; similarly, difficulties pointed out by other authors, most prominently by Allcock, have been overcome; the multiplicity of quantum answers obtained for some unique classical questions (such as the traversal time) has been also well understood and formalized with a compact systematic theory. Of course different theoretical or experimental conditions may select one of them for a specific application. For the arrival time, in particular, distributions, which are "optimal" with respect to a number of classical constraints, have been identified.

All in all, we have reached a much better understanding of the phenomenology and underlying theoretical issues concerning time observables in quantum mechanics. Much remains to be done, and many issues are still contended, as the diverse character of the contributions to this volume will make clear. Nonetheless, we are convinced that the patient reader will have ample proof in the following chapters of the improvement in our understanding time in quantum mechanics that has taken place in the last decades. And we hope that the same reader will in turn contribute and improve on this work.

Acknowledgments

We are grateful to C. R. Leavens and A. Steinberg for reading and commenting on the manuscript. We acknowledge support by Ministerio de Educación y Cultura (Grants PB97-1482 and AEN99-0315), The University of the Basque Country (Grants UPV 063.310-EB187/98 and 9/UPV 00039.310-13507/2001), CERION, MCYT (BFM2000-0816-C03-03), and the Basque Government (PI-1999-28).

References

1. N. Bohr: Philos. Mag. **26**, 1 (1913)
2. M. Beller: *Quantum Dialogue* (University of Chicago Press, Chicago, IL, 1999)
3. J. Hilgevoord: Stud. History Philos. Mod. Phys. **36**, 29 (2005)
4. A. Hermann, K.v. Meyen, V.F. Weisskopf, eds.: *Wolfgang Pauli, Scientific Correspondence with Bohr, Einstein, Heisenberg* (Springer, Berlin, 1979) Letter of Pauli to Bohr, November 17, 1925.
5. See [4], Letter of Heisenberg to Pauli, November 24, 1925
6. Heisenberg to Einstein, November 30, 1925, Einstein Archive, Hebrew University of Jerusalem
7. M. Born: *Problems of Atomic Dynamics* (MIT Press, Cambridge, MA, 1926)
8. M. Born: Z. Phys. **37**, 863 (1926)
9. P. Jordan: Naturwissenschaften **37**, 614 (1927)
10. W. Pauli: in *Handbuch der Physik* (Springer, Berlin, 1926), Vol. 23, pp. 1–278
11. W. Heisenberg: Z. Physik **43**, 172 (1927). Translated in *Quantum Theory and Measurement*, ed. by J. A. Wheeler, W. H. Zurek (Princeton University Press, Princeton, NJ, 1983)
12. D.M. Appleby: J. Phys. A **31**, 6419 (1998)
13. N. Bohr: Nature **121** (suppl.), 580 (1928)
14. L. Mandelstam, I. Tamm: J. Phys. (USSR) **9**, 249 (1945)
15. Y. Aharonov, D. Bohm: Phys. Rev. **122**, 1649 (1961)
16. W. Pauli: in *Handbuch der Physik* (Springer, Berlin, 1933), Vol. 24, pp. 83–272

17. W. Pauli: in *Encyclopedia of Physics*, ed. by S. Flugge (Springer, Berlin, 1958), Vol. V/1, p. 60
18. G.R. Allcock: Ann. Phys. (N.Y.) **53**, 253 (1969)
19. V. Delgado, J.G. Muga: Phys. Rev. **A56**, 3425 (1997)
20. I.L. Egusquiza, J.G. Muga: Phys. Rev. **A61**, 012104 (2000). See also erratum, Phys. Rev. A **61** (2000) 059901(E)
21. E.A. Galapon: Proc. R. Soc. A **458**, 451 (2002)
22. E.A. Galapon: Proc. R. Soc. A **487**, 2671 (2002)
23. E.A. Galapon, R. Caballar, R.T. Bahague: Phys. Rev. Lett. **93**, 180406 (2004)
24. E.A. Galapon, F. Delgado, J.G. Muga, I. Egusquiza: Phys. Rev. A **72**, 042107 (2005)
25. E.A. Galapon, R.F. Caballar, R.T. Bahague: Phys. Rev. A **72**, 062107 (2005)
26. P. Busch, M. Grabowski, P. Lahti: *Operational Quantum Physics* (Springer–Verlag, Berlin, 1997)
27. B. Misra: Found. Phys. **25**, 1087 (1995)
28. J. von Neumann: *Mathematical Foundations of Quantum Mechanics* (Princeton University Press, Princeton, NJ, 1955)
29. J. Hilgevoord: Am. J. Phys. **64**, 1451 (1996)
30. J.E. Gray, A. Vogt: J. Math. Phys. **46**, 052108 (2005)
31. D. Bohm: *Quantum Theory* (Prentice-Hall, Upper Saddle River, NJ, 1951)
32. K.R. Popper: Br. J. Philos. Sci. **10**, 25 (1959)
33. E. Nelson: Phys. Rev. **150**, 1079 (1966)
34. D. Bohm, J.P. Vigier: Phys. Rev. **96**, 208 (1954)
35. D. Bohm, B.J. Hiley: Phys. Rep. **172**, 93 (1989)
36. M.Q. Chen, M.S. Wang: Phys. Lett. A **149**, 441 (1990)
37. K. Imafaku, I. Ohba, Y. Yamanaka: Phys. Lett. A **204**, 329 (1995)
38. P. Ghose, A. S. Majumdar, S. Guha, J. Sau: Phys. Lett. A **290**, 205 (2001)
39. M. Golshani, O. Akhavan: J. Phys. A **34**, 5259 (2001)
40. W. Struyve, W. D. Baere, J. D. Neve, S. D. Weirdt: J. Phys. A. **36**, 1525 (2003)
41. E. Guay, L. Marchildon: J. Phys. A **36**, 5617 (2003)
42. O. Akhavan: Thesis. `quant-ph/0402141`
43. G. Brida, E. Cagliero, G. Falzetta, M. Genovese, M. Gramegna, C. Novero: J. Phys. B **35**, 4751 (2002)
44. G. Brida, E. Cagliero, G. Falzetta, M. Genovese, M. Gramegna, E. Predazzi: Phys. Rev. A **68**, 033803 (2003)
45. O. Akhavan, M. Golshani: J. Phys. B **37**, 3777 (2004)
46. G. Brida, E. Cagliero, M. Genovese, M. Gramegna: J. Phys. B **37**, 3781 (2004)
47. M. Genovese: Phys. Rep. **413**, 319 (2005)
48. X. Oriols: Phys. Rev. A **71**, 017801 (2005)
49. P.R. Holland: *The Quantum Theory of Motion* (Cambridge University Press, Cambridge, MA, 1993)
50. D. Bohm, B. Hiley: *The Undivided Universe: An Ontological Interpretation of Quantum Mechanics* (Routledge, London, 1993)
51. J.R. Barker: Semicond. Sci. Technol. **9**, 911 (1994)
52. C.R. Leavens: Phys. Rev. A **58**, 840 (1998)
53. M. Daumer: in *Bohmian Mechanics and Quantum Theory: An Appraisal*, ed. by J.T. Cushing, A. Fine, S. Goldstein (Kluwer, Dordrecht, 1996), pp. 87–98
54. X. Oriols, F. Martín, J. Suñé: Phys. Rev. A **54**, 2594 (1996)
55. Y. Nogami, F.M. Toyama, W. van Dijk: Phys. Lett. **A270**, 279 (2000)

56. S. Kreidl, G. Grubl, H.G. Embacher: J. Phys. A **36**, 8851 (2003)
57. M. Ruggenthaler, G. Grubl, S. Kreidl: J. Phys. A **38**, 8441 (2005)
58. E. Schrödinger: Br. J. Philos. Sci. **3**, 109 (1952)
59. M. Born: Br. J. Philos. Sci. **4** (1953)
60. H.D. Zeh: Phys. Lett. A **174**, 189 (1993)
61. H. Everett: Rev. Mod. Phys. **29**, 454 (1957)
62. P. Szriftgiser, D. Guéry-Odelin, M. Arndt, J. Dalibard: Phys. Rev. Lett. **77**, 4 (1996)
63. M. Moshinsky: Phys. Rev. **88**, 625 (1952)
64. H. Salecker, E.P. Wigner: Phys. Rev. **109**, 571 (1958)
65. A. Peres: Am. J. Phys. **48**, 552 (1980)
66. Y. Aharonov, J. Oppenheim, S. Popescu, B. Reznik, W.G. Unruh: Phys. Rev. A **57**, 4130 (1998)
67. G.C. Hegerfeldt, D. Seidel, J. G. Muga: Phys. Rev. A **68**, 022111 (2003)
68. W. Oskay et al: Phys. Rev. Lett. **97**, 020801 (2006)
69. N.F. Ramsey: Phys. Rev. **78**, 695 (1950); *Molecular Beams* (Clarendon Press, Oxford, 1956)
70. D. Seidel, J. G. Muga: Phys. Rev. A **75**, 023811 (2007)
71. C.J. Bordé: Metrologia **39**, 435 (2002)
72. D. Seidel, J.G. Muga: Eur. Phys. J. D **41**, 71 (2007)
73. W.M. Itano et al: Phys. Rev. A **47**, 3554 (1993)
74. D.J. Wineland, J.J. Bollinger, W.M. Itano, D.J. Heinzen: Phys. Rev. A **50**, 67 (1994)
75. S.F. Huelga, C. Macchiavello, T. Pellizari, A.K. Ekert, M.B. Plenio, J.I. Cirac, Phy. Rev. Lett. **79**, 3865 (1997)
76. Fifty years of atomic time-keeping: 1955 to 2005, special issue of Metrologia **42**, S1-S153 (2005)
77. C. Audoin, B. Guinot: *The Measurement of Time: Time, Frequency and the Atomic Clock* (Cambridge University Press, Cambridge, MA, 2001)
78. F.T. Smith: Phys. Rev. **118**, 349 (1960)
79. L. Eisenbud: Ph.D. thesis, Princeton Univ. (1948). Unpublished
80. E.P. Wigner: Phys. Rev. **98**, 145 (1955)
81. J.G. Muga, C.R. Leavens, J.P. Palao: Phys. Rev. A **58**, 4336 (1998)
82. G.R. Allcock: Ann. Phys. (N.Y.) **53**, 286 (1969)
83. G.R. Allcock: Ann. Phys. (N.Y.) **53**, 311 (1969)
84. E.P. Wigner: in *Aspects of Quantum Theory*, ed. by A. Salam, E.P. Wigner (Cambridge University Press, London, 1972)
85. A.D. Baute, I.L. Egusquiza, J.G. Muga: J. Phys. A **34**, 4289 (2000)
86. A. D. Baute, I. L. Egusquiza, J. G. Muga: Int. J. Theor. Phys. Group Theo. Nonlinear Opt. **8**, 1 (2002)
87. J.G. Muga, C. R. Leavens: Phys. Rep. **338**, 353 (2000)
88. J.G. Muga, S. Brouard, D. Macías: Ann. Phys. (N.Y.) **240**, 351 (1995)
89. J.P. Palao, J.G. Muga, R. Sala: Phys. Rev. Lett. **80**, 5469 (1998)
90. J.P. Palao: *Teoría cuántica de colisiones con potenciales complejos en una dimensión*. Ph.D. thesis, Universidad de La Laguna (1999)
91. J.G. Muga, J.P. Palao, B. Navarro, I.L. Egusquiza: Phys. Rep. **395**, 357 (2004)
92. A.J. Bracken, G.F. Melloy: J. Phys. A: Math. Gen. **27**, 2197 (1994)
93. G.F. Melloy, A.J. Bracken: Ann. Phys. (Leipzig) **7**, 726 (1998)
94. J.G. Muga, J.P. Palao, C.R. Leavens: Phys. Lett. A **253**, 21 (1999)

95. Y. Aharonov, P.G. Bergmann, J.L. Lebowitz: Phys. Rev. B **134**, 1410 (1964)
96. Y. Aharonov, D. Albert, A. Casher, L. Vaidman: Phys. Lett. A **124**, 199 (1987)
97. J. Kijowski: Rept. Math. Phys. **6**, 361 (1974)
98. R. Giannitrapani: Int. J. Theor. Phys. **36**, 1575 (1997)
99. P. Busch, M. Grabowski, P.J. Lahti: Phys. Lett. A **191**, 357 (1994)
100. N.I. Akhiezer, I.M. Glazman: *Theory of Linear Operators in Hilbert Space* (Dover, New York, 1963)
101. A.D. Baute, I.L. Egusquiza, J.G. Muga: Phys. Rev. A **64**, 012501 (2001)
102. A.D. Baute, I.L. Egusquiza, J.G. Muga: Phys. Rev. A **65**, 032114 (2002)
103. J. León: J. Phys. A **30**, 4791 (1997)
104. J.A. Damborenea, I.L. Egusquiza, G.C. Hegerfeldt, J.G. Muga: Phys. Rev. A **66**, 052104 (2002)
105. J.A. Damborenea, I.L. Egusquiza, G.C. Hegerfeldt, J.G. Muga: J. Phys. B: At. Mol. Opt. Phys. **36**, 2657 (2003)
106. G.C. Hegerfeldt, D. Seidel, J.G. Muga, B. Navarro: Phys. Rev. A **70**, 012110 (2004)
107. B. Misra, E.C.G. Sudarshan: J. Math. Phys. **18**, 756 (1977)
108. D. Home, M.A.B. Whitaker: Ann. Phys. (N.Y.) **258**, 237 (1997)
109. M.D. Srinivas, R. Vijayalakshmi: Pramana **16**, 173 (1981)
110. P. Facchi, S. Pascazio: Prog. Opt. **42**, 147 (2001), Chap. 3
111. K. Koshino, A. Shimizu, Phys. Rep. **412**, 191 (2005)
112. J. G. Muga, F. Delgado, A. del Campo, G. García-Calderó n, Phys. Rev. A **73**, 052112 (2006)
113. F. Delgado, J. G. Muga, G. García-Calderó n, Phys. Rev. A **74**, 052112 (2006)
114. C.W. Helstrom: Int. J. Theor. Phys. **11**, 357 (1974)
115. A.S. Holevo: Rep. Math. Phys. **13**, 379 (1978)
116. S. Kochen, E. P. Specker: J. Math. Mech. **17**, 59 (1967)
117. J.S. Bell: Rev. Mod. Phys. **38**, 447 (1966)
118. R.I.G. Hughes: *The Structure and Interpretation of Quantum Mechanics* (Harvard University Press, Cambridge, MA, 1989)
119. G.C. Ghirardi, A. Rimini, T. Weber: Phys. Rev. D **34**, 470 (1986)
120. P. Blanchard, A. Jadczyk: Helv. Phys. Acta **69**, 613 (1996)
121. P. Blanchard, A. Jadczyk: in *Open Systems and Measurement in Relativistic Quantum Theory*, ed. by H.P. Breuer, F. Petruccione (Springer Verlag, Berlin, 1999). quant-ph/9812081
122. P. Pearle: in *Open Systems and Measurement in Relativistic Quantum Theory*, ed. by F. Petruccione, H.P. Breuer (Springer Verlag, Berlin, 1999). quant-ph/9901077
123. R.B. Griffiths: J. Stat. Phys. **36**, 219 (1984)
124. M. Gell-Mann, J.B. Hartle: in *Third International Symposium on the Foundations of Quantum Mechanics in the Light of New Technology*, ed. by S. Kobayashi, H. Ezawa, Y. Murayama, S. Nomura (Physical Society of Japan, Tokyo, 1990)
125. M. Gell-Mann, J.B. Hartle: in *25th International Conference on High Energy Physics*, ed. by K. K. Phua, Y. Yamaguchi (World Scientific, Singapore, 1991)
126. R. Omnès: Rev. Mod. Phys. **64**, 339 (1992)
127. R. Omnès: *The Interpretation of Quantum Mechanics* (Princeton University Press, Princeton, NJ, 1994)
128. J.B. Hartle: Phys. Rev. D **37**, 2818 (1988)

129. J.B. Hartle: Phys. Rev. D **44**, 3173 (1991)
130. O. Rudolph: Int. J. Theor. Phys. **35**, 1581 (1996)
131. O. Rudolph: J. Math. Phys. **37**, 5368 (1996)
132. A. Kent: Phys. Scripta **T76**, 78 (1998)
133. N. Yamada, S. Takagi: Prog. Theor. Phys. **85**, 985 (1991)
134. N. Yamada, S. Takagi: Prog. Theor. Phys. **86**, 599 (1991)
135. N. Yamada, S. Takagi: Prog. Theor. Phys. **87**, 77 (1992)
136. I.L. Egusquiza, J.G. Muga: Phys. Rev. A **62**, 032103 (2000)
137. J.J. Halliwell, E. Zafiris: Phys. Rev. D **57**, 3351 (1998)
138. J. Dalibard, Y. Castin, K. Molmer: Phys. Rev. Lett. **68**, 580 (1992)
139. H. Carmichael: *An Open Systems Approach to Quantum Optics* (Springer-Verlag Berlin, 1993)
140. G.C. Hegerfeldt: Phys. Rev. A **47**, 449 (1993)
141. M.B. Plenio, P.L. Knight: Rev. Mod. Phys. **70**, 101 (1998)
142. R.J. Cook: Prog. Opt. **28**, 361 (1990)
143. S. Stenholm: Phys. Scr. **47**, 724 (1993)
144. T.A. Brun: Phys. Rev. Lett. **78**, 1833 (1997)
145. M. Büttiker, R. Landauer: Phys. Rev. Lett. **49**, 1739 (1982)
146. R.Y. Chiao: Tunneling times and superluminality: a tutorial (1998). quant-ph/9811019
147. S. Brouard, R. Sala Mayato, J.G. Muga: Phys. Rev. A **49**, 4312 (1994)
148. P. Balcou, L. Dutriaux: Phys. Rev. Lett. **78**, 851 (1997)
149. E.H. Hauge, J.A. Stovneng: Rev. Mod. Phys. **61**, 917 (1989)
150. M. Büttiker: in *Electronic Properties of Multilayers and Low-dimensional Semiconductor Structures*, ed. by J.C. et al. (Plenum, New York, 1990), p. 297
151. R. Landauer: Ber. Bunsenges. Phys. Chem. **95**, 404 (1991)
152. C.R. Leavens, G.C. Aers: in *Scanning Tunneling Microscopy and Related Techniques*, ed. by R. J. Behm, N. García, H. Rohrer (Kluwer, Dordrecht, 1990)
153. M. Jonson: 'Tunneling times in quantum mechanical tunneling'. In: *Quantum Transport in Semiconductors*, ed. by D.K. Ferry, C. Jacoboni (Plenum, New York, 1991)
154. V.S. Olkhovsky, E. Recami: Phys. Rep. **214**, 339 (1992)
155. R. Landauer, Th. Martin: Rev. Mod. Phys. **66**, 217 (1994)
156. C.R. Leavens, G.C. Aers: Solid State Commun. **63**, 1101 (1987)
157. D. Sokolovski, L.M. Baskin: Phys. Rev. A **36**, 4604 (1987)
158. D. Sokolovski, J.N.L. Connor: Phys. Rev. A **42**, 6512 (1990)
159. D. Sokolovski, J.N.L. Connor: Phys. Rev. A **47**, 4677 (1993)
160. E. Pollak, W.H. Miller: Phys. Rev. Lett. **53**, 115 (1984)
161. E. Pollak: J. Chem. Phys. **83**, 1111 (1985)
162. J.G. Muga, S. Brouard, R. Sala: Phys. Lett. A **167**, 24 (1992)
163. B.A. van Tiggelen, A. Tip, A. Lagendijk: J. Phys. A **26**, 1731 (1993)
164. S. Brouard, R. Sala, J.G. Muga: Europhys. Lett. **22**, 159 (1993)
165. Y. Aharonov, L. Vaidman: Phys. Rev. A **41**, 11 (1990)
166. A.M. Steinberg: Phys. Rev. Lett. **74**, 2405 (1995)
167. C. Grupen: *Particle Detectors* (Cambridge University Press, Cambridge, MA, 1996)
168. J.P. Palao, J.G. Muga, S. Brouard, A. Jadczyk: Phys. Lett. **A233**, 227 (1997)
169. T.E. Hartman: J. Appl. Phys. **33**, 3427 (1962)
170. V. Delgado, J.G. Muga: Ann. Phys. (N.Y.) **248**, 122 (1996)

171. F. Delgado, J. G. Muga, A. Ruschhaupt: Phys. Rev. A **69**, 022106 (2004)
172. A. Ruschhaupt, J. G. Muga: Phys. Rev. Lett. **93**, 020403 (2004)
173. A.M. Steinberg, P.G. Kwiat, R.Y. Chiao: Phys. Rev. Lett. **71**, 708 (1993)
174. A. Enders, G. Nimtz: J. Phys. I France **3**, 1089 (1993)
175. A. Ranfagni, P. Fabeni, G.P. Pazzi, D. Mugnai: Phys. Rev. E **48**, 1453 (1993)
176. P. Mittelstädt, G. Nimtz, eds.: *Workshop on Superluminal(?) Velocities*, Ann. Phys. Leipzig. Vol. 7 (1998)
177. R.Y. Chiao, A.M. Steinberg: in *Progress in Optics XXXVII*, ed. by E. Wolf (Elsevier, Amsterdam, 1997), pp. 347–406
178. G. Hegerfeldt: 'Particle localization and the notion of Einstein causality'. In: *Extensions of Quantum Theory*, ed. by A. Horzela, E. Kapuscik (Apeiron, Mintreal, 2001), pp. 9–16. quant-ph/0109044.

2

Characteristic Times in One-Dimensional Scattering

J. Gonzalo Muga

Departamento de Química-Física, Universidad del País Vasco, Apdo 644. Bilbao, Spain
jg.muga@ehu.es

2.1 Introduction

Quantum scattering theory deals with collisions, namely, interactions that are essentially localized in time and space. This means that the interaction potential must vanish rapidly enough in coordinate space, so that the wave packet tends to free-motion incoming and outgoing asymptotic states before and after the interaction is effective. The scope of scattering theory also includes "half-collisions" or "decay processes" where the stage before the collision is ignored, i.e., the evolution of the system is considered only from the interaction region.

This chapter reviews various quantities that have been proposed in scattering theory to characterize the temporal aspects of collision. A quantum wave packet collision with a potential barrier in one dimension (1D) is fully described by the evolution of the wave function $\psi(x,t)$ from the incoming to the outgoing asymptotic states. However, the whole information contained in $\psi(x,t)$ is hardly required. A few well-chosen quantities are often enough to provide a fair picture of the dynamics. In particular, one of these elementary parameters is the transmission probability P_T, but to describe the time dependence we also need to quantify the duration of the collision, the arrival time at a detector, the decay time of an unstable state, the asymptotic behavior at short and large times, or response times, such as the time required to "charge" a well or to achieve stationary conditions when a source is turned on.

In spite of the inherent time dependence of collisions, the treatises on quantum mechanics or scattering theory concentrate on solutions of the time-independent Schrödinger equation. This is in part because many scattering experiments to obtain cross sections are performed in quasi-stationary conditions and also because the stationary scattering states form a basis to analyze the actual time-dependent collision. In many cases wave packet scattering is relegated to justify the cavalier obtention by stationary methods of cross-section expressions and occasionally to discuss resonance lifetimes. Another widespread limitation of textbooks is the exclusive interest in the final results of the collision at asymptotic distances and times, which has been generally

J. G. Muga: *Characteristic Times in One-Dimensional Scattering*, Lect. Notes Phys. **734**, 31–72 (2008)
DOI 10.1007/978-3-540-73473-4_2 © Springer-Verlag Berlin Heidelberg 2008

justified because "the midst of the collision cannot be observed." However, while it is true that in many collision experiments only the asymptotic results are observed, modern experiments with femtosecond laser pulses or other techniques known as "spectroscopy of the transition state" do probe the structure and the evolution of the collision complex [1]. Also, in quantum kinetic theory of gases, accurate treatments must abandon the "completed collision" approximation and use a nonasymptotic description, e.g. in terms of Möller wave operators instead of S matrices, as in the Waldmann–Snider equation and its generalizations for moderately dense gases [2].

The theory has to adapt to these new trends by paying more attention to the temporal description of the collisions. Even if we restrict ourselves to asymptotic aspects, the cross section does not contain the whole information available in a scattering process, since it is only proportional to the *modulus* of the S-matrix elements. Information on the phase is available from delay times with respect to free motion. In fact, the full collision process and not just the asymptotic regimes should be understood to control or to modify the products. This has motivated a recent trend of theoretical and experimental work to investigate the details of the interaction region and the transient phenomena.

In this chapter we restrict ourselves to 1D scattering. Many physical systems can be described in 1D: the application of the effective mass approximation to layered semiconductor structures leads to effective 1D systems [3]; some surface phenomena are described by 1D models [4]; chemical reactions can in certain conditions be modeled by effective 1D potentials [5]; and atomic motion may be 1D in nanometric confining waveguides [6, 7]. Moreover, the simplicity of 1D models has made them valuable as pedagogical and research tools. They facilitate testing hypotheses, new ideas, approximation methods and theories without unnecessary and costly complications. For the same reasons they are frequently used to examine fundamental questions of quantum mechanics. In particular, the time quantities treated in this book, such as tunneling or arrival times, have in most cases been examined in 1D models. Many results for 1D are inspired by results previously obtained in 3D, although the direct translation is not always trivial or possible. This is because 3D collisions with spherically symmetric potentials are described on the half line, by decomposition into partial waves, whereas 1D collisions involve the full line and a doubly degenerate spectrum.

The chapter is organized as follows: Section 2.2 provides a minimal overview of formal 1D scattering theory. The treatment is "formal" because no mathematically rigorous proofs are given. Instead, we summarize the operator structure of the theory and the results needed to define characteristic times later on. For a more rigorous mathematical presentation, see e.g. [8]. Sections 2.3, 2.4, and 2.5 are devoted to the dwell time, the delay time, and decay times (the exponential decay and its deviations), respectively. Quantities related to the tunneling time conundrum are scattered in several parts of the book. In this chapter, Sects. 2.4.1 and 2.4.4 discuss the Hartman effect and negative delays, while Sect. 2.6 discusses the role of the Büttiker–Landauer "traversal time" in the time dependence of evanescent waves. A detailed discussion of the arrival times is left for Chaps. 5, 6, and 10.

2.2 Scattering Theory in 1D

2.2.1 Basic Premises and Notation

Let $\widehat{H} = \widehat{H}_0 + \widehat{V}$ be the Hamiltonian operator for a single particle in 1D, where

$$\widehat{H}_0 = \frac{\widehat{p}^2}{2m} \tag{2.1}$$

is the kinetic energy operator in terms of the momentum operator $\widehat{p}$, and $\widehat{V}$ is a "local" potential operator with coordinate representation

$$\langle x|\widehat{V}|x'\rangle = \delta(x - x')V(x) . \tag{2.2}$$

$V(x)$ must vanish for large values of $|x|$ so that the Möller operators, defined below, exist.[1] This may certainly be accomplished by finite-range potentials, but spatial decays with infinite tails are also possible.

The plane waves $|p\rangle$ with coordinate representation given by

$$\langle x|p\rangle = h^{-1/2}e^{ixp/\hbar} \tag{2.3}$$

are improper eigenstates[2] of $\widehat{p}$ and $\widehat{H}_0$, normalized according to Dirac's delta function,

$$\langle p|p'\rangle = \delta(p - p') . \tag{2.4}$$

Closure relations (or resolutions of the unit operator $\widehat{1}$) may therefore be written in momentum or coordinate representation as

$$\widehat{1} = \int_{-\infty}^{\infty} dx\, |x\rangle\langle x| = \int_{-\infty}^{\infty} dp\, |p\rangle\langle p| . \tag{2.5}$$

2.2.2 Basic Abstract and Parameterized Operators

The state vector of the particle at time t is denoted as $|\psi(t)\rangle$ or simply as $\psi(t)$. We shall only deal with potentials such that at large times in the past and future certain states ψ, the scattering states, tend (in a strong sense) to freely moving asymptotic states ϕ_{in} and ϕ_{out}, respectively,

$$\psi(t) \to \phi_{\text{in}}(t), \quad t \to -\infty , \tag{2.6}$$
$$\psi(t) \to \phi_{\text{out}}(t), \quad t \to \infty . \tag{2.7}$$

The central objects in the scattering theory are the *abstract* Möller operators. They link the asymptotic states with ψ,

---

[1] We shall not deal here with "step" potentials with different asymptotic levels on both sides. For a detailed treatment of this case, see [9, 10] and references therein.

[2] That is, not in the Hilbert space of square integrable states.

$$\psi(t) = \widehat{\Omega}_+ \phi_{\text{in}}(t) , \tag{2.8}$$

$$\psi(t) = \widehat{\Omega}_- \phi_{\text{out}}(t) . \tag{2.9}$$

Another important operator is

$$\widehat{S} = \widehat{\Omega}_-^\dagger \widehat{\Omega}_+ , \tag{2.10}$$

which links the two asymptotes,

$$\phi_{\text{out}}(t) = \widehat{S} \phi_{\text{in}}(t) . \tag{2.11}$$

It is also convenient to introduce the auxiliary "transition" operators $\widehat{T}_\pm$ as

$$\widehat{T}_\pm = \widehat{V} \widehat{\Omega}_\pm . \tag{2.12}$$

The explicit definition of the Möller operators is given by infinite time (strong) limits,

$$\widehat{\Omega}_\pm = \lim_{t \to \mp\infty} e^{i\widehat{H}t/\hbar} e^{-i\widehat{H}_0 t/\hbar} . \tag{2.13}$$

The domain of these operators is the Hilbert space of square integrable states, although it is very useful to consider an extension that can be applied on plane waves and allows us to work in a momentum representation. To this end let us first define the *parameterized* operators

$$\widehat{\Omega}(z) = \widehat{1} + \widehat{G}_0(z)\widehat{T}(z) , \tag{2.14}$$

$$\widehat{T}(z) = \widehat{V} + \widehat{V}\widehat{G}(z)\widehat{V} , \tag{2.15}$$

where z is a complex variable with dimensions of energy, and $\widehat{G}(z) = (z - \widehat{H})^{-1}$ and $\widehat{G}_0(z) = (z - \widehat{H}_0)^{-1}$ are the *resolvents* of $\widehat{H}$ and $\widehat{H}_0$. $\widehat{T}(z)$, $\widehat{V}$, $\widehat{G}(z)$ and $\widehat{G}_0(z)$ are also related by

$$\widehat{G}(z) = \widehat{G}_0(z) + \widehat{G}_0(z)\widehat{T}(z)\widehat{G}_0(z) , \tag{2.16}$$

$$\widehat{T}(z)\widehat{G}_0(z) = \widehat{V}\widehat{G}(z) . \tag{2.17}$$

We shall see that the matrix elements of the resolvents in coordinate representation are singular on the real positive axis and at poles on the negative real axis (bound states). Further singularities may occur by analytical continuation on the second energy sheet.

Note that the operators of scattering theory have *abstract* or *parameterized* versions [11]. Confusion may arise if they are not properly distinguished. The relation between abstract and parameterized operators is found by acting with (2.13) on a square integrable state. The resulting infinite time limits can be substituted by the following limits, see e.g., [12]:

$$\widehat{\Omega}_+ = \lim_{\varepsilon \to 0+} \varepsilon \int_{-\infty}^0 dt \, e^{\varepsilon t} e^{i\widehat{H}t/\hbar} e^{-i\widehat{H}_0 t/\hbar} , \tag{2.18}$$

$$\widehat{\Omega}_- = \lim_{\varepsilon \to 0+} \varepsilon \int_0^\infty dt \, e^{-\varepsilon t} e^{i\widehat{H}t/\hbar} e^{-i\widehat{H}_0 t/\hbar} . \tag{2.19}$$

Integrating, and introducing a closure relation in momentum,

$$\widehat{\Omega}_{\pm} = \int_{-\infty}^{\infty} dp\, \widehat{\Omega}(E_p \pm i0)|p\rangle\langle p| , \qquad (2.20)$$

$$\widehat{T}_{\pm} = \int_{-\infty}^{\infty} dp\, \widehat{T}(E_p \pm i0)|p\rangle\langle p| . \qquad (2.21)$$

The action of these operators on plane waves is now well defined. In particular, the improper eigenvectors of $\widehat{H}$ are obtained by acting with the parameterized Möller operators on the plane waves,

$$|p^{\pm}\rangle = \widehat{\Omega}(E_p \pm i0)|p\rangle = |p\rangle + \frac{1}{E_p \pm i0 - \widehat{H}_0}\widehat{T}(E_p \pm i0)|p\rangle , \qquad (2.22)$$

where $E_p = p^2/(2m)$ is the energy of the plane wave and of the corresponding eigenstate of $\widehat{H}$. This is the Lippmann–Schwinger integral equation for the states $|p^{\pm}\rangle$, which are composed of a "free" plane wave and a "scattering" wave. To evaluate the coordinate representation and the asymptotic behavior of the states at large distances, the matrix elements of the free-motion resolvent are required,

$$\langle x|\frac{1}{E_p \pm i0 - \widehat{H}_0}|x'\rangle = \mp\frac{im}{\hbar|p|} e^{\pm i|p||x-x'|/\hbar} . \qquad (2.23)$$

Equation (2.23) is obtained by introducing a resolution of unity in momentum representation and using contour integration in the complex momentum plane. Note that the two ways of approaching the real axis in (2.23), from below or from above, imply different boundary conditions at large $|x|$ for the two states in (2.22): the scattering wave of $|p^+\rangle$ is formed by outgoing plane waves moving off the potential region, whereas the scattering wave of $|p^-\rangle$ involves incoming plane waves toward the potential region.

Since the plane waves $|p\rangle$ form a complete set, the following resolutions of the operators $\widehat{\Omega}$, $\widehat{T}$, and S can be introduced:

$$\widehat{\Omega}_{\pm} = \int_{-\infty}^{\infty} dp\, |p^{\pm}\rangle\langle p| , \qquad (2.24)$$

$$\widehat{T}_{\pm} = V\int_{-\infty}^{\infty} dp\, |p^{\pm}\rangle\langle p| , \qquad (2.25)$$

$$\widehat{S} = \int_{-\infty}^{\infty} dp \int_{-\infty}^{\infty} dp'\, |p'\rangle\langle p'^-|p^+\rangle\langle p| . \qquad (2.26)$$

Strictly speaking, the operators in (2.24)–(2.26) are not identical to the ones in (2.10), (2.12), and (2.13) since the former may be applied on plane waves. However, when acting on Hilbert space states they are equivalent so that, to avoid a clumsy notation, the same symbols will be used. A momentum representation is therefore allowed for these operators, which in general involves distributions (generalized functions such as Dirac's delta or Cauchy's principal part).

For real potential functions $V(x)$ the norm is conserved throughout the collision, $\langle\phi_{\text{in}}|\phi_{\text{in}}\rangle = \langle\psi|\psi\rangle = \langle\phi_{\text{out}}|\phi_{\text{out}}\rangle$. This means that the Möller operators are isometric, i.e.,

$$\widehat{\Omega}_{\pm}^{\dagger}\widehat{\Omega}_{\pm} = \widehat{1} . \tag{2.27}$$

As a consequence,

$$\langle p^{\pm}|p'^{\pm}\rangle = \delta(p - p') . \tag{2.28}$$

In general the Möller operators are not unitary because the bound states are not in their range. Contrast this to the operator $\widehat{S}$: it conserves the norm too, but it is unitary because it maps the whole Hilbert space onto the whole Hilbert space,

$$\widehat{S}\widehat{S}^{\dagger} = \widehat{S}^{\dagger}\widehat{S} = \widehat{1} . \tag{2.29}$$

The *scattering states* ψ with incoming and outgoing asymptotes move far away from the potential so they are orthogonal to the bound states $\{|\Phi_j\rangle\}$ at large (positive or negative) times. Since the overlap amplitude $\langle\psi|\Phi_j\rangle = 0$ is independent of time, the space of bound states $\mathcal{B}$ is orthogonal to the scattering states, namely to the range of the Möller operators. We shall always assume that the ranges of the two Möller operators are equal to the subspace of scattering states $\mathcal{R}$ and that the whole Hilbert space is the direct sum of the subspaces spanned by scattering and bound states, $\mathcal{H} = \mathcal{R} \oplus \mathcal{B}$. This assumption is known as *asymptotic completeness*,

$$\widehat{\Omega}_{\pm}\widehat{\Omega}_{\pm}^{\dagger} = \widehat{1} - \widehat{\Lambda} = \int_{-\infty}^{\infty} dp\, |p^{\pm}\rangle\langle p^{\pm}| . \tag{2.30}$$

In this expression the "unitary deficiency" $\widehat{\Lambda}$ is the projector onto the subspace of bound states,

$$\widehat{\Lambda} = \sum_{j} |\Phi_j\rangle\langle\Phi_j| . \tag{2.31}$$

Taking matrix elements in (2.26), the momentum representation of $\widehat{S}$ is given by

$$\langle p|\widehat{S}|p'\rangle = \delta(p - p') - 2i\pi\delta(E_p - E_{p'})\langle p|\widehat{T}(E_p + i0)|p'\rangle . \tag{2.32}$$

The collision conserves the energy, which is, asymptotically, kinetic energy. That is why $\widehat{S}$ commutes with $\widehat{H}_0$ and its matrix elements are proportional to an energy delta function. It is quite useful to factor out this delta function to define an on-the-energy-shell $\mathbf{S}(E)$ matrix. Using

$$\delta(p - p') = \frac{|p|}{m}\delta(E_p - E'_p)\delta_{pp'} , \tag{2.33}$$

where $\delta_{pp'}$ is the Kronecker delta,

$$\delta_{pp'} = \begin{cases} 1 & \text{if } p = p' \\ 0 & \text{if } p \neq p' \end{cases}, \tag{2.34}$$

and defining the matrix elements of $\mathbf{S}$, $S_{\alpha\beta}$, by

$$\langle p|\widehat{S}|p'\rangle = |p|m^{-1}\delta(E_p - E'_p)S_{\text{sign(p)sign(p')}}(E_p), \tag{2.35}$$

one finds

$$S_{\text{sign(p)sign(p')}}(E_p) = \delta_{pp'} - \frac{2i\pi m}{|p|}\langle p|\widehat{T}(E_p + i0)|p'\rangle, \qquad |p| = |p'|. \tag{2.36}$$

The subscripts $\alpha, \beta = \pm$ in the matrix elements $S_{\alpha\beta}$, denote the two possible "channels," which correspond to positive $(+)$ or negative $(-)$ momentum. A difference between the 1D scattering on the full line $(-\infty < x < \infty)$ and the radial scattering on the half line $(0 < r < \infty)$ is that in the former, the **S** matrix is a unitary 2×2 matrix while in the later it is a complex number of unit modulus.

2.2.3 Symmetries

Time Reversal Invariance. This symmetry holds for real potentials. It implies

$$S_{\alpha\beta} = S_{-\beta-\alpha}. \tag{2.37}$$

Parity. Frequently the potential is symmetrical with respect to its central position. In that case,

$$S_{\alpha\beta} = S_{-\alpha-\beta}. \tag{2.38}$$

2.2.4 Eigenstates of $\widehat{H}$

The eigenstates of $\widehat{H}$ given by the Lippmann–Schwinger integral equations (2.22) behave asymptotically as a combination of two plane waves with positive and negative momenta. The factors multiplying these plane waves are the *reflection and transmission amplitudes* according to the following table for asymptotic, long-distance behaviour (assume for the time being that $p > 0$)

$$\frac{1}{\hbar^{1/2}} \begin{cases} \exp(ipx/\hbar) + R^l(p)\exp(-ipx/\hbar), & \text{if } x \sim -\infty \\ T^l(p)\exp(ipx/\hbar), & \text{if } x \sim \infty, \end{cases} \tag{2.39}$$

$$\frac{1}{\hbar^{1/2}} \begin{cases} T^r(p)\exp(-ipx/\hbar), & \text{if } x \sim -\infty \\ \exp(-ipx/\hbar) + R^r(p)\exp(ipx/\hbar), & \text{if } x \sim \infty. \end{cases} \tag{2.40}$$

For potentials of finite range that vanish outside $[a, b]$ these are in fact exact expressions for $x < a$ and $x > b$.[3]

If $p > 0$, the boundary conditions in (2.39) define the states $\langle x|p^+\rangle$ corresponding to an *incoming* plane wave from the left, $\langle x|p\rangle$, while the boundary conditions in (2.40) define the states $\langle x|(-p)^+\rangle$ corresponding to an *incoming* plane wave from the right, $\langle x|-p\rangle$. $T(p)$ and $R(p)$, with superscripts r or l for right or left incidence, are the transmission and reflection amplitudes. A wave packet peaked around a given $|p^+\rangle$ would be dominated by the plane wave $|p\rangle$ before the collision, whereas after the collision, there would be two packets, one reflected and one transmitted with probabilities $|R(p)|^2$ and $|T(p)|^2$, dominated by $|-p\rangle$ and $|p\rangle$, respectively (see e.g., [13]).

For $p < 0$, however, the states determined by (2.39) and (2.40) correspond, respectively, to $\langle x|p^-\rangle$, with *outgoing* plane wave $\langle x|p\rangle$, and $\langle x|(-p)^-\rangle$, with *outgoing* plane wave $\langle x|-p\rangle$. A wave packet formed around $|p^-\rangle$ would be close to a plane wave $|p\rangle$ only *after* the collision occurs. To form this peculiar outgoing state, the incoming asymptote must combine waves incident from both sides of the potential barrier. This may of course be difficult to implement in practice, but it does not preclude the usefulness of these states as basis functions, and in general for applications where some control or selection of the products of the collision is required.

The previous discussion should make clear that $T(p)$, for $p < 0$, is *not* a standard transmission amplitude, because it is *not* the amplitude of the transmitted plane wave of the state $|p^+\rangle$, $p < 0$. However, it analytically continues the standard transmission amplitude ($T(p)$ for $p > 0$) onto the $p < 0$ domain, so the term "transmission amplitude" will be used irrespective of the sign of p, even though the physical meaning is different for the two possible signs. Of course a similar analysis applies for the reflection amplitudes. According to our notational convention, positive arguments of the amplitudes always correspond to states $|p^+\rangle$, while negative momentum arguments correspond to $|p^-\rangle$ states.

2.2.5 Relation Between Scattering Amplitudes and Basic Operators

Comparing the asymptotic (large $|x|$) behavior of the states in (2.39) and (2.40) with the asymptotic behavior in (2.22), the amplitudes $R(p)$ and $T(p)$

---

[3] Occasionally one may find a different convention for the "transmission amplitude." For barriers of width d, some authors write the transmitted wave for left incidence, up to the normalization factor and disregarding the "l" superscript, as $T \exp(ip(x - d)/\hbar)$ (this is the case in particular of Chap. 12) instead of $T \exp(ipx/\hbar)$ as it is done here. Checking the convention used is of importance to interpret correctly energy or momentum derivatives of the "phase of T" since depending on the convention this phase may differ by a factor $pd/\hbar$. These derivatives enter into the definition of several characteristic times (delay time and phase time), see Sect. 2.4 below.

can be related to on-the-energy-shell elements of the transition matrix. We shall work out one case in detail: the scattering part of $\langle x|p^+\rangle$ for $p > 0$ and $x \to \infty$ is

$$\int_{-\infty}^{\infty} dx' \langle x|\widehat{G}_0(E_p + i0)|x'\rangle\langle x'|\widehat{T}(E_p + i0)|p\rangle$$

$$\sim -\frac{2\pi mi}{h}\frac{e^{ipx/\hbar}}{p}\int_{-\infty}^{\infty} dx' e^{-ipx'/\hbar}\langle x'|\widehat{T}(E_p + i0)|p\rangle$$

$$= -\frac{2\pi mi}{p}\langle x|p\rangle\langle p|\widehat{T}_+|p\rangle .\tag{2.41}$$

Adding the free wave, $h^{-1/2}e^{ipx/\hbar}$, and comparing with (2.39), there results $T^l(p) = 1 - 2i\pi m\langle p|\widehat{T}_+|p\rangle/p$ for $p > 0$. The rest of the cases can be worked out similarly (because of time reversal invariance, $\langle p|\widehat{T}_\pm|p\rangle = \langle -p|\widehat{T}_\pm| - p\rangle$, and $T^r(p) = T^l(p)$; therefore the superscript for the transmission amplitude will be dropped hereafter):

$$T(p) = 1 - \frac{2i\pi m}{p}\langle p|\widehat{T}_{\text{sign}(p)}|p\rangle ,$$

$$R^l(p) = -\frac{2mi\pi}{p}\langle -p|\widehat{T}_{\text{sign}(p)}|p\rangle ,$$

$$R^r(p) = -\frac{2mi\pi}{p}\langle p|\widehat{T}_{\text{sign}(p)}| - p\rangle .\tag{2.42}$$

Some useful relations follow from (2.42),

$$[T(-p)]^* = T(p), \qquad p \text{ real} .\tag{2.43}$$
$$R^{r,l}(-p)^* = R^{r,l}(p), \qquad p \text{ real} .\tag{2.44}$$

From (2.36) and (2.42), the **S** matrix is given by

$$\mathbf{S}(p) \equiv \mathbf{S}(E) = \begin{pmatrix} T(p) & R^r(p) \\ R^l(p) & T(p) \end{pmatrix} , \qquad p > 0.\tag{2.45}$$

It is quite useful to consider **S** as a (matrix) function of p. In simple applications we only use $\mathbf{S}(p)$ with $p > 0$,[4] but in fact we may also define $\mathbf{S}(p)$ for $p < 0$ or even for complex p in terms of the analytical continuations of the amplitudes $T(p)$, $R^r(p)$, and $R^l(p)$. This extension will be discussed in Sect. 2.2.7.

---

[4] Keep in mind that $p > 0$ in the arguments of **S** or of the scattering amplitudes does not mean "incidence from the left." According to the sign convention described in Sect. 2.2.4, it means that the amplitudes correspond to states with outgoing scattering parts: $|p^+\rangle$ for left incidence and $| - p^+\rangle$ for right incidence.

2.2.6 The Diagonal S_d Matrix

The $\mathbf{S}$ matrix (2.45) has been obtained from the momentum representation of S using plane waves incident from one side, $|\pm p\rangle$, but other on-shell matrices may be defined in terms of a different basis formed by combinations of $|\pm p\rangle$. Of particular interest is the set $|u_j\rangle$, $j = 0, 1$, that provides a diagonal matrix,

$$\mathbf{S_d}(p) = \begin{pmatrix} S_0(p) & 0 \\ 0 & S_1(p) \end{pmatrix} . \tag{2.46}$$

Unitarity implies that $|S_j| = 1$, so the matrix elements may be written in terms of real eigenphase shifts δ_j, $S_j = e^{2i\delta_j}$. The $|u_j\rangle$ are not mixed by the collision; these incident states produce an outgoing combination equal to the incident one, except for a phase factor. The diagonal $\mathbf{S_d}$ matrix is most advantageous for parity invariant potentials, since the linear combinations become simply even and odd wave functions,

$$|u_0\rangle = 2^{-1/2}(|p\rangle + |-p\rangle) , \tag{2.47}$$
$$|u_1\rangle = 2^{-1/2}(|p\rangle - |-p\rangle) . \tag{2.48}$$

From the asymptotic behavior of $|u_j^+\rangle = \Omega_+|u_j\rangle$ and $|\pm p^+\rangle$ we may relate reflection and transmission amplitudes for even potentials to the eigenphase shifts,

$$R(p) = 2^{-1}\left(e^{2i\delta_0} - e^{2i\delta_1}\right) , \tag{2.49}$$
$$T(p) = 2^{-1}\left(e^{2i\delta_0} + e^{2i\delta_1}\right) . \tag{2.50}$$

(Equation (2.50) is in fact valid for arbitrary potentials.) The boundary conditions for the states $|u_j^+\rangle$ are

$$\lim_{x\to-\infty} \langle x|u_0^+\rangle = e^{i\delta_0}\left(\frac{2}{h}\right)^{1/2} \cos(-px/\hbar + \delta_0) ,$$

$$\lim_{x\to\infty} \langle x|u_0^+\rangle = e^{i\delta_0}\left(\frac{2}{h}\right)^{1/2} \cos(px/\hbar + \delta_0) ,$$

$$\lim_{x\to-\infty} \langle x|u_1^+\rangle = ie^{i\delta_1}\left(\frac{2}{h}\right)^{1/2} \sin(px/\hbar - \delta_1) ,$$

$$\lim_{x\to\infty} \langle x|u_1^+\rangle = ie^{i\delta_1}\left(\frac{2}{h}\right)^{1/2} \sin(px/\hbar + \delta_1) . \tag{2.51}$$

It will be convenient for later manipulations to drop the constant complex phase factors and define real eigenfunctions of $\widehat{H}$ as

$$\langle x|\psi_0\rangle = e^{-i\delta_0}\langle x|u_0+\rangle ,$$
$$\langle x|\psi_1\rangle = -ie^{-i\delta_1}\langle x|u_1+\rangle . \tag{2.52}$$

2.2.7 Complex Momentum

The properties of $T(p)$ as a function of the complex momentum p are of importance for many applications [8]. Let the potential function $V(x)$ be such that

$$\int_{-\infty}^{\infty} dx\, |V(x)|(1+x^2) < \infty \ . \tag{2.53}$$

Then $T(p)$ is meromorphic in $\mathrm{Im}\, p > 0$ with a finite number n_b of simple poles $i\beta_1, i\beta_2, ..., i\beta_n,\ \beta_j > 0$ on the imaginary axis. The numbers $-\beta_j^2/(2m)$ are the eigenvalues of H. Moreover,

$$T(p) = 1 + O(1/p) \quad \text{as} \quad |p| \to \infty, \ \mathrm{Im}\, p \geq 0 \ , \tag{2.54}$$

and there can only be a zero at the real axis, at $p = 0$,

$$|T(p)| > 0 \quad \mathrm{Im}\, p \geq 0, \ p \neq 0 \ . \tag{2.55}$$

In the generic case $T(0) = 0$, and

$$T(p) = \gamma p + o(p), \ \gamma \neq 0, \ \text{as} \ p \to 0, \ \mathrm{Im}\, p \geq 0 \ . \tag{2.56}$$

Since $T(p)$ is meromorphic and it does not have zeros in the upper plane, the integral

$$\frac{1}{2\pi i} \int_{\mathcal{A}} dp\, \frac{d\ln T(p)}{dp} = -n_b \tag{2.57}$$

along the contour $\mathcal{A}$ consisting of $[-R, -\epsilon]$, $[\epsilon, R]$, a semicircle of radius ϵ around the origin, and a large semicircle of radius R in the upper half plane, provides, according to a theorem of complex plane integration, the number of zeros (none in this case) minus the number of poles of $T(p)$ enclosed (the bound states). The integral may also be evaluated using (2.43), (2.54), and (2.56); this gives $2i\Phi_T(R) - 2i\Phi_T(\epsilon) - i\pi$, where $\Phi_T(p)$ is the phase of T,

$$T(p) = |T(p)| \exp(i\Phi_T) \ . \tag{2.58}$$

Combining the two results,

$$\Phi_T(0) - \Phi_T(\infty) = \pi(n_b - 1/2) \ , \tag{2.59}$$

which is *Levinson's theorem* for the case $T(p = 0) = 0$. Otherwise, there is no $-i\pi$ contribution from the small semicircle and the phase difference becomes just πn_b. The convention followed is that $\Phi_T(\infty) = 0$, so the theorem establishes the value of $\Phi_T(0)$.

The possibility to analytically continue $T(p)$ to the lower half plane will depend on the potential considered [14]. Here we shall assume that the continuation can be performed (this is the case for example for potentials of finite range) and discuss the properties that these continuations must obey. From $T^\dagger(z) = T(z^*)$ and the relations (2.42) we find

$$(R^{r,l}(p))^* = R^{r,l}(-p^*) \,, \tag{2.60}$$

$$(T(p))^* = T(-p^*) \,, \tag{2.61}$$

so that if there is a pole of $T(p)$ in the fourth quadrant at $p_R - ip_I$ ($p_R, p_I > 0$), there must also be a pole in the third quadrant at $-p_R - ip_I$. For an isolated pole, and if p_I is small, the phase of $T(p)$ along the positive real line will increase rapidly by π. From (2.50) we see that poles of $T(p)$ are generally poles of S_0 or of S_1. Since $|T(p)| = |\cos(\delta_0 - \delta_1)|$, if the resonance eigenphase shift also jumps by π, while the other one remains approximately constant, the transmission probability along the real axis will pass across a maximum (1) or a minimum (0) or both, depending on the initial phase difference of the two eigenphase shifts. The above simplified picture will be blurred if the resonances are very close to each other, or the pole is far from the real line.

2.2.8 Unitarity and its Consequences

The unitarity of the collision $\mathbf{S}$ matrix, $\mathbf{SS}^\dagger = \mathbf{S}^\dagger\mathbf{S} = \mathbf{1}$, reflects the conservation of norm in the collision. It provides two relations:
From the diagonal elements

$$|T(p)|^2 + |R^{r,l}(p)|^2 = 1 \,, \tag{2.62}$$

and from nondiagonal ones

$$T(p)[R^l(p)]^* + [T(p)]^* R^r(p) = 0, \quad p \text{ real} \,. \tag{2.63}$$

Equation (2.63) leads to a relation for the phases,

$$2\Phi_T - \Phi_{R^r} - \Phi_{R^l} = (2n+1)\pi, \quad n = 0, \pm 1, \pm 2, \dots \,, \tag{2.64}$$

where, as in (2.58),

$$R^{r,l}(p) = |R^{r,l}(p)|e^{i\Phi_R^{r,l}(p)} \,. \tag{2.65}$$

2.3 A Measure of the Collision Duration: The Dwell Time

In classical mechanics the quantity

$$\tau_D(a, b; t_1, t_2)_{classical} = \int_{t_1}^{t_2} dt \int_a^b dx \, \varrho(x, t) \,, \tag{2.66}$$

where $\varrho(x, t)$ is the probability density of an ensemble of independent particles, is the average over the ensemble of the time that each particle trajectory spends between a and b within the time window $[t_1, t_2]$ [15]. In other words, this is an average "dwell" or "sojourn" time in the selected space–time region.[5]

Its formal quantum mechanical counterpart is

$$\tau_D(a, b; t_1, t_2; \psi) = \int_{t_1}^{t_2} dt \int_a^b dx \, |\psi(x, t)|^2 \, . \tag{2.67}$$

In principle the coordinates a, and $b > a$, and the instants t_1 and $t_2 > t_1$ are arbitrary but most often a and b are chosen so that $V(x)$ is zero or negligible for $x < a$ and $x > b$. Hereafter t_1 will be, by default, $-\infty$, or occasionally 0, an initial preparation time, and $t_2 = \infty$.

In spite of the formal similarity of the classical and quantum expressions, the interpretation of (2.67) as a "mean time" spent in the region $[a, b]$, $[t_1, t_2]$ by quantum particles is not straightforward, since in the standard interpretation of the quantum mechanical formalism there are no trajectories and therefore there is no obvious way to assign a time (duration) of presence to a given member of the ensemble of particles associated with the quantum state. There are however several arguments that provide (2.67) by extending to the quantum case the classical dwell time, e.g. via Feynman path integrals [20], causal or Bohm trajectories [21], or as an expectation value of a hermitian sojourn time operator [22], see also Chap. 7 for an interpretation in terms of weak measurements, and the discussion of Sect. 11.3.1. Irrespective of a hypothetical statistical interpretation of the dwell time in terms of individual members of the ensemble, the dwell time is at the very least a characteristic quantity of the ensemble represented by the state ψ that quantifies the duration of the wave packet collision. In fact, the dwell time is considered an important parameter in high-speed applications of mesoscopic semiconductor structures [23].

τ_D can be written in several ways, in particular as

$$\tau_D = \tau_D(a, b; -\infty, \infty) = \int_{-\infty}^{\infty} dt \, P_{ab}(t) = \langle \psi(t = 0) | \widehat{T}_D | \psi(t = 0) \rangle \, , \tag{2.68}$$

where $P_{ab}(t) = \int_a^b dx \, \varrho(x, t)$, $\widehat{T}_D$ is the *sojourn time operator*,

$$\widehat{T}_D = \int_{-\infty}^{\infty} dt \, e^{i\widehat{H}t/\hbar} \widehat{D}(a, b) e^{-i\widehat{H}t/\hbar} \, , \tag{2.69}$$

and $\widehat{D}(a, b)$ is the projector onto the selected space region,

---

[5] The concept of "dwell time" for a finite space region in the stationary regime is due to Büttiker [16]. Previously, integrals of the form (2.66) had been used to define time delays by comparing the free motion to that with a scattering center and taking the limit of infinite volume, see e.g. [17]. For further review of early contributions to the "sojourn time" concept, see [18], and for their relation to the time delay see [19].

$$\widehat{D}(a, b) = \int_a^b dx \, |x\rangle\langle x| \, .$$
(2.70)

An experimental determination of the dwell time may be carried out by monitoring the time evolution of the probability inside the selected spatial region [24]. This is admittedly an indirect route, where the first moment of $\widehat{T}_D$, τ_D, is obtained without having measured individual dwell times for the members of the ensemble. It remains to be seen if second and higher moments of $\widehat{T}_D$ may be associated with some simple operational procedure.

Let us now find other useful expressions for the dwell time. Integrating the continuity equation over x between a and b, and over time between $-\infty$ and t, P_{ab} takes the form

$$P_{ab}(t) = \int_{-\infty}^t dt' \, [J(a, t') - J(b, t')] = \int_{-\infty}^t dt' \, \Delta J(a, b, t') \, ,$$
(2.71)

where $J(x, t')$ is the current density, $\Delta J(a, b) = J(a) - J(b)$, and the boundary condition $P_{ab}(-\infty) = 0$ has been assumed. Substituting (2.71) into (2.68), one finds

$$\tau_D = \int_{-\infty}^\infty dt \int_{-\infty}^t dt' \, \Delta J(t') = \int_{-\infty}^\infty dt \int_{-\infty}^\infty dt' \, \mathcal{H}(t - t') f(t')$$
(2.72)

$$= \lim_{t'' \to \infty} \int_{-\infty}^{t''} dt' \, (t'' - t') \, \Delta J(t') = \lim_{t'' \to \infty} \left[t'' P_{ab}(t'') - \int_{-\infty}^{t''} dt' \, t' \, \Delta J(t') \right] \, .$$

Unless $P_{ab}(t)$ decays faster than t^{-1}, the dwell time will diverge. The existence of a potential function leads generically to an asymptotic decay $\sim t^{-3}$, as discussed in Sect. 2.5.3. However, for free motion the dwell time will diverge unless the momentum wave function vanishes at $p = 0$, because of the dependence $\sim t^{1/2}$ of the free-motion propagator, see (2.141) below and the related discussion. In terms of the sojourn time operator (2.69) for $\widehat{H}_0$, the possible divergence is due to a $|p|^{-1}$ factor,

$$\widehat{T}_{D,H_0} = \sum_{\alpha=\pm} \int_{-\infty}^\infty dp \, \frac{mh}{|p|} |p\rangle\langle p|\widehat{D}|\alpha p\rangle\langle \alpha p| \, .$$
(2.73)

In this and the following sections we shall limit ourselves in general to incoming asymptotes in the positive momentum channel $(+)$ that vanish at $p = 0$, so that the dwell time for free motion does exist. This will allow us to compare dwell times with and without potential and to define delay times. These states, with a bounded support in momentum space, have necessarily a Fourier transform in coordinate space that can only vanish at some set of points of measure zero. But this is not a problem since the total probability for positive positions tends to zero as $t \to -\infty$,

$$\lim_{t \to -\infty} \int_a^\infty dx \, |\langle x|\phi_{\text{in}}(t)\rangle|^2 = \int_{-\infty}^0 dp \, |\langle p|\phi_{\text{in}}(0)\rangle|^2$$
(2.74)

for any a and any ϕ_{in} [25].

Assuming that $tP_{ab}(t) \to 0$ as $t \to \infty$, the dwell time (2.68) takes the local form

$$\tau_D(a, b) = \int_{-\infty}^{\infty} dt' \, [J(b, t') - J(a, t')] \, t' \, . \tag{2.75}$$

Other expression for states incident in the positive momentum channel may be obtained by using resolutions of the identity in terms of the states $|p^+\rangle$,

$$\tau_D(a, b; \psi) = \int_0^{\infty} dp \, |\langle p | \phi_{\text{in}}(0) \rangle|^2 \tau_D(p) \, , \tag{2.76}$$

where

$$\tau_D(a, b; p) \equiv \frac{\int_a^b dx \, |\langle x | p^+ \rangle|^2}{p/mh} \, , \tag{2.77}$$

which suggests the interpretation of $\tau_D(a, b; p)$ as a dwell time for particles of definite momentum p [16].

Suppose now that $a < 0$ and $b > 0$ are both far from the barrier region, before and after the barrier, respectively, so that the first passage of the wave packet across a can be described accurately in terms of the free-motion asymptote ϕ_{in}, while the passage of the transmitted and reflected wave packets can be evaluated with the asymptotic expressions:

$$\psi_T(b, t) = \frac{1}{\sqrt{h}} \int_0^{\infty} dp \, \langle p | \phi_{\text{in}}(0) \rangle \, T(p) \, e^{i(pb - Et)/\hbar} \, , \tag{2.78}$$

$$\psi_R(a, t) = \frac{1}{\sqrt{h}} \int_0^{\infty} dp \, \langle p | \phi_{\text{in}}(0) \rangle \, R(p) \, e^{-i(pa + Et)/\hbar} \, . \tag{2.79}$$

(For a potential with support between 0 and d, b could be taken at the very barrier edge, $b = d$, but a cannot be 0 because of the strong interference between the incident and reflected parts. $|a|$ should be much greater than the incident wave packet width in order to distinguish clearly the entrance passage from the reflected one.) Then,

$$\int_{-\infty}^{\infty} dt' \, J_T(b, t') = \int_0^{\infty} dp \, |T(p)|^2 |\langle p | \phi_{\text{in}}(0) \rangle|^2 = P_T \, ,$$

$$\int_{-\infty}^{\infty} dt' \, J_I(a, t') = \int_0^{\infty} dp \, |\langle p | \phi_{\text{in}}(0) \rangle|^2 = 1 \, ,$$

$$\int_{-\infty}^{\infty} dt' \, J_R(a, t') = -\int_0^{\infty} dp \, |R(p)|^2 |\langle p | \phi_{\text{in}}(0) \rangle|^2 = -P_R \, , \tag{2.80}$$

where the subscripts I, T, and R in J_I, J_T, and J_R mean that ϕ_{in}, ψ_T, and ψ_R have been used to calculate the fluxes. One can then write (2.75) as

$$\tau_D = P_T \langle t \rangle_b^{\text{out}} - \langle t \rangle_a^{\text{in}} + P_R \langle t \rangle_a^{\text{out}} \, , \tag{2.81}$$

where

$$\langle t \rangle_b^{\text{out}} \equiv \frac{\int_{-\infty}^{\infty} dt' \, J_T(b, t') \, t'}{\int_{-\infty}^{\infty} dt' \, J_T(b, t')} \, , \qquad (2.82)$$

$$\langle t \rangle_a^{\text{in}} \equiv \int_{-\infty}^{\infty} dt' \, J_I(a, t') \, t' \, , \qquad (2.83)$$

$$\langle t \rangle_a^{\text{out}} \equiv \frac{-\int_{-\infty}^{\infty} dt' \, J_R(a, t') \, t'}{\int_{-\infty}^{\infty} dt' \, |J_R(a, t')|} \, . \qquad (2.84)$$

In each case the "average passage instant" is obtained by properly normalizing the fluxes. One may rightly wonder whether the notation and terminology used (as average passage times) are justified. The "averages" are taken over the current density J, a quantity that is not definite positive even for an incident wave packet without negative momentum components [26, 27, 28]. It turns out, however, that the above "averages" over J are equal to averages over a positively defined arrival-time distribution (Kijowski's arrival-time distribution) [29], as will be discussed in Chap. 10. Models of detectors based on complex non-hermitian potentials also lead to these average times, delayed only by the small (dwell) time that the particle spends in the detector before being detected [30]. In the next section we shall relate these times to the "phase times."

Finally, note that (2.81) could be, and has been, used to partition the dwell time into transmission and reflection components [15, 31], see also the closely related approach of Olkhovsky and Recami [32]. The main drawback is that the defined entrance average instant is common for both contributions, see [33, 34], which is not correct in the classical ensemble limit, and may lead to negative transmission times [35] even in the classical case [33]. A two-detector model avoided this problem by assigning different entrance instants for each member of the ensemble [34]. The distinction between the dwell time and its components was first done by Büttiker, [16] and raised some controversy. As summarized in Chap. 1, Muga, Brouard, and Sala have emphasized the multiplicity of possible quantum partitionings versus the uniqueness of the classical case and developed a systematic theory to generate partitionings with the correct classical limit. Some of these include interference terms that cannot be assigned to transmission or reflection but to both of them [15]. For arguments in favor of one particular partitioning based on weak measurements, see Chap. 11 and [36].

2.4 Importance of the Phases: Time Delays

If the **S** matrix is known or simply one of the amplitudes R^l or R^r is given as a function of momentum and there are no bound states, necessary and sufficient conditions are known for a unique potential to exist, and there are well-established construction procedures [37, 8]. However a knowledge of the probabilities is not enough to determine the amplitudes. The phases are associated with observable time-dependent properties.

Consider a wave packet impinging from the left on a barrier potential located near $x = 0$. The exact barrier position is not important for our present purposes: two typical choices for $x = 0$ are the center of a symmetrical barrier or the left edge of a finite-range potential. Let us take as before the spatial interval $[a, b]$ well outside the barrier, so that there is a clear separation between incoming and reflected passages.

Since the incoming state is in the positive momentum channel,

$$\langle x|\phi_{\text{in}}(t)\rangle = \int_0^\infty dp\, \langle x|p\rangle\langle p|\phi_{\text{in}}(0)\rangle e^{-iEt/\hbar} , \tag{2.85}$$

applying the Möller operator $\widehat{\Omega}_+$ one obtains

$$\langle x|\psi(t)\rangle = \int_0^\infty dp\, \langle x|p^+\rangle\langle p|\phi_{\text{in}}(0)\rangle e^{-iEt/\hbar} . \tag{2.86}$$

(This relation is exact. If the zero of time is taken well before the wave packet interacts significantly with the barrier, one could also substitute $\langle p|\phi_{\text{in}}(0)\rangle \to \langle p|\psi(0)\rangle$ without introducing any significant error.)

Substituting (2.85), (2.78), and (2.79) in the time averages (2.82)–(2.84) and using the standard expression for the current density,

$$J(x,t) = \frac{\hbar}{m} \text{Im}\left(\psi(x,t)^* \frac{\partial\psi(x,t)}{\partial x}\right) , \tag{2.87}$$

the derivative of an energy Dirac's delta may be identified and then used to perform one of the momentum integrals. The results are

$$\langle t\rangle_b^{out} = \frac{1}{P_T} \int_0^\infty dp\, |\langle p|\phi_{\text{in}}(0)\rangle|^2 \,|T(p)|^2 \frac{m}{p}\,[b - x_0 + \hbar\Phi'_T(p)] , \tag{2.88}$$

$$\langle t\rangle_a^{out} = \frac{1}{P_R} \int_0^\infty dp\, |\langle p|\phi_{\text{in}}(0)\rangle|^2 \,|R(p)|^2 \frac{m}{p}\,[-a - x_0 + \hbar\Phi'_R(p)] , \tag{2.89}$$

$$\langle t\rangle_a^{in} = \int_0^\infty dp\, |\langle p|\phi_{\text{in}}(0)\rangle|^2 \frac{m}{p}\,[a - x_0] , \tag{2.90}$$

where the prime means derivative with respect to p, and

$$x_0 \equiv \hbar\, \text{Im}\,(\langle\phi_{\text{in}}(0)|p\rangle'/\langle\phi_{\text{in}}(0)|p\rangle) . \tag{2.91}$$

These results do not require to assume a narrow packet in momentum representation.

The quantity

$$\tau_T^{Ph}(x_0, b; p) \equiv m\,[b - x_0 + \hbar\Phi'_T(p)]\,/p \tag{2.92}$$

in the integrand of (2.88) consists of the time that a classical free particle with mass m and momentum p would spend from x_0 to b, plus the *time delay* $m\hbar\Phi'_T(p)/p$. Similarly, the term in brackets in (2.89),

$$\tau_R^{Ph}(x_0, a; p) \equiv m\left[-a - x_0 + \hbar\Phi'_R(p)\right]/p \,, \tag{2.93}$$

is the time spent by a classical particle that travels freely from x_0 to $x = 0$, where its momentum is instantly reversed, and from $x = 0$ to a, plus a delay contribution. It is to be noted that unless $a = -b$ the reference time associated with classical free motion is different in the transmission and reflection cases. We shall see a consequence of this disparity in Sect. 2.4.2 when calculating average delays.

Formally we may use (2.92) and (2.93) to *define* "phase times" for arbitrary values of a, b, and x_0. In particular, for a finite-range barrier between $x = 0$ and d let us define

$$\tau_T^{Ph}(0, d; p) = \frac{md}{p} + \frac{m\hbar}{p}\Phi'_T(p) \tag{2.94}$$

by subtracting from $\tau_T^{Ph}(x_0, d; p)$ the classical flight time between x_0 and 0, $-mx_0/p$. These "extrapolated phase times" for traversal should not be over-interpreted as actual traversal times [38, 39]. Not only because, as pointed out in Chap. 1, there is no unique traversal time, but also because a wave packet peaked around p is very broad in coordinate representation, so it is severely deformed before the hypothetical "entrance" instant $t_{ent} = |x_0|m/p$, and at $x = 0$ there is an important interference effect between incident and reflected components. The wave functions ϕ_{in} and ψ_R used to calculate the fluxes J_I and J_R do not faithfully represent the actual wave, so that the average instants (2.89) and (2.90) lose their physical meaning as average detection times.

2.4.1 The Hartman Effect

Relation (2.88) is suitable for examining the "Hartman effect" [40, 41, 38, 32, 31]. Hartman [40] studied the evolution of a wave packet with momentum distribution centered around p_c, colliding with a rectangular barrier of height $V_0 > p_c^2/(2m)$ and width d. He found three regions according to the value of d. For large barrier widths (opaque barrier conditions), the phase time associated with p, under the barrier, goes to a constant, $\tau_T^{Ph}(x_0, d; p) = 2m/(p\kappa) - x_0 m/p$, independent of d, where

$$\kappa = [2m(V_0 - E)]^{1/2}/\hbar \,. \tag{2.95}$$

When transmission is dominated by momentum components below the barrier, the transmitted wave packet seems to traverse the potential region in a time interval independent of d. This is the "Hartman effect," which, as Winful has recently pointed out, may be viewed as a result of a saturation of the integrated probability density and correspondingly of the dwell time, with increasing barrier width [42, 43]. If d is increased further, plane waves with momentum above the barrier height dominate the transmission, and classical behavior results, i.e., time grows linearly with d. Finally, for small barrier widths, Hartman defined a "thin barrier region" where the phase time depends generally on d.

To be more specific, let us consider the initial Gaussian wave packet

$$\langle x|\phi_{in}(0)\rangle = \left[\frac{1}{2\pi\delta^2}\right]^{1/4} \exp\left[ip_c x/\hbar - (x - x_c)^2/(4\delta^2)\right], \qquad (2.96)$$

of average momentum $p_c = \hbar k_c$ and spatial width (square root of the variance) δ. Here x_0 becomes equal the wave packet center x_c. The initial momentum distribution is a Gaussian distribution with variance $\sigma^2 = [\hbar/(2\delta)]^2$. We assume that $p_c \gg \sigma^2$ so that the truncation at $p = 0$ in (2.88) is not significant. For an energy distribution peaked around $E_c < V_0$ the following results can be drawn [31]:

If $\kappa_c d \equiv \sqrt{2m(V_0 - E_c)}d/\hbar \gg 1$, $\langle t \rangle_d^{out}$ does not vary appreciably when d increases, thus showing the Hartman effect. When d is sufficiently large, the components of the wave packet under the barrier are so strongly depressed by $|T(p)|^2$ that higher momenta start to dominate, and $\langle t \rangle_d^{out}$ grows almost linearly, as one expects classically. As δ is increased, larger values of d are needed to pass from the first regime to the second one. An estimation of the value of d, which gives the transition between Hartman effect and quasiclassical behavior, can be obtained for each value of δ by equating the factor $|T(p)|^2 |\langle|\phi_{in}(p)\rangle|^2$ for $p = p_c$ and for $p = p_r$, where p_r is the momentum of the first resonance above the barrier. This leads to the relation

$$\delta = \frac{\hbar\sqrt{-\ln|T(p_c)|}}{|p_r - p_c|} \approx \frac{\hbar\sqrt{\kappa_c d}}{|p_r - p_c|}, \qquad (2.97)$$

between δ and d, which clearly separates quantum and quasiclassical behavior. Also, for fixed δ, the transition is sharper at larger δ as a consequence of the narrower momentum distribution.

We have already warned the reader against a naive overinterpretation of the extrapolated phase time $\tau_T^{Ph}(0, d; p)$, which becomes $2m/(p\kappa)$ for the barrier traversal in the Hartman effect, mainly because of the strong deformation of the broad incident wave packet. We could try to avoid the interpretational pitfalls of this quantity and instead look at the time $\langle t \rangle_d^{out}$ for a wave packet initially localized near the edge of the barrier, and with a small spatial width compared to the barrier length d. In this way one may identify the entrance time and the preparation instant with a tolerable small uncertainty. However, Low and Mende [44] speculated and then Delgado and Muga [45] have shown that this localization leads to the dominance of over-the-barrier components. Similar conclusions are drawn from a two-detector model (one before and one

after the barrier) when the detector before the barrier localizes the particle
into a small spatial width compared to d [34].

2.4.2 The Lifetime and Delay Time Matrices

The four delay times corresponding to reflection and transmission for right
and left incidence form the *delay time matrix* introduced by Eisenbud in his
thesis [46],

$$\Delta t_{\alpha\beta} = \text{Re}\left[-i\hbar \frac{1}{S_{\alpha\beta}} \frac{dS_{\alpha\beta}}{dE}\right] . \tag{2.98}$$

The matrix element $\Delta t_{\alpha\beta}$ is the delay time in the appearance of the peak
outgoing signal in channel β, after the injection of a pulse narrowly peaked
in momentum in channel α. The "delay" may in fact become negative as discussed already. These delay times have been traditionally obtained by means
of the "stationary phase approximation." Let us rewrite the transmitted wave
function as

$$\langle x|\psi_T(t)\rangle = h^{-1/2} \int_0^\infty dp\, e^{ixp/\hbar - iE_p t/\hbar + i\Phi_T} \langle p|\phi_{\text{in}}(0)\rangle |T(p)| . \tag{2.99}$$

If the initial state is narrowly peaked around p_0, the integral will be appreciably different from zero only if the phase of the exponential function is
stationary near $p = p_0$. This implies a "spatial delay" with respect to the
free-motion wave packet,

$$\Delta x = \hbar \frac{d\Phi_T}{dp}\bigg|_{p=p_0} , \tag{2.100}$$

and a corresponding "time delay",

$$\Delta t_{++}(p_0) = \frac{\hbar m}{p_0} \frac{d\Phi_T}{dp}\bigg|_{p=p_0} . \tag{2.101}$$

The time delays are also related to the on-the-energy-shell lifetime matrix of
Smith [47],

$$\mathbf{Q}(E) = i\hbar \mathbf{S}(E) \frac{d\mathbf{S}(E)^\dagger}{dE} . \tag{2.102}$$

$\mathbf{S}$ is unitary, so $\mathbf{Q}$ is Hermitian. Thus the diagonal matrix elements of $\mathbf{Q}$ are
real and take the form

$$Q_{\alpha\alpha} = \sum_\beta |S_{\alpha\beta}|^2 \Delta t_{\alpha\beta} . \tag{2.103}$$

Since the particle has a probability $|S_{\alpha\beta}|^2$ to emerge in the channel β, $Q_{\alpha\alpha}$ is
the average delay experienced by the particle injected in channel α.

We shall now relate the $\mathbf{Q}$ matrix with the "wave packet lifetime," defined as the difference between dwell times with and without potential [47, 48],

$$\langle Q \rangle \equiv \tau_{D,\psi} - \tau_{D,\phi_{\text{in}}} . \tag{2.104}$$

As before, the incidence is in the positive momentum channel. $\tau_{D,\psi}$ is given by (2.81) whereas the dwell time for free motion is

$$\tau_{D,\phi_{\text{in}}} = \langle t \rangle_{b,\phi_{\text{in}}}^{out} - \langle t \rangle_{a,\phi_{\text{in}}}^{in} = \int_0^\infty dp \, |\langle p|\phi_{\text{in}}(0)\rangle|^2 \frac{m}{p}[b-a] , \tag{2.105}$$

where, similarly to (2.90),

$$\langle t \rangle_{b,\phi_{\text{in}}}^{out} = \int_0^\infty dp \, |\langle p|\phi_{\text{in}}(0)\rangle|^2 \frac{m}{p}[b-x_0] . \tag{2.106}$$

Since, by hypothesis, $\langle t \rangle_{a,\psi}^{in} = \langle t \rangle_{a,\phi_{\text{in}}}^{in}$, $\langle Q \rangle$ takes the form

$$\begin{aligned}
\langle Q \rangle &= \int_0^\infty dt \int_a^b dx \left(|\langle x|\psi(t)\rangle|^2 - |\langle x|\phi_{\text{in}}(t)\rangle|^2 \right) \\
&= P_T[\langle t \rangle_{b,\psi}^{out} - \langle t \rangle_{b,\phi_{\text{in}}}^{out}] + P_R[\langle t \rangle_{a,\psi}^{out} - \langle t \rangle_{b,\phi_{\text{in}}}^{out}] . \tag{2.107}
\end{aligned}$$

Substituting all the integral expressions obtained for the passage times, and writing $c = -a - b$,

$$\langle Q \rangle = \hbar \int_0^\infty dp \, \frac{m}{p} |\langle p|\phi_{\text{in}}(0)\rangle|^2 \left[\Phi_T' |T(p)|^2 + \left(\Phi_R' + \frac{c}{\hbar} \right) |R(p)|^2 \right] . \tag{2.108}$$

Note the term proportional to c in the reflection part. It arises because of the mismatch between the free-motion reference times used to define the reflection and transmission time delays when $c \neq 0$. Choosing $c = 0$, $\langle Q \rangle$ represents the weighted momentum average of the mean delay for each momentum,[6]

$$\langle Q \rangle = \int_0^\infty dp \, |\langle p|\phi_{\text{in}}(0)\rangle|^2 Q(E)_{++} . \tag{2.109}$$

The eigenvalues of $\mathbf{Q}$ have been used as good indicators of resonances [50], see Sect. 2.4.3 below, and may be interpreted for symmetrical potentials as the delays associated with symmetrical or antisymmetrical bilateral incidence [49]. However, their operational interpretation in terms of individual measurements is puzzling. An asymptotic measurement of the arrival time at b in the transmission side could be done in principle for one of the identically prepared systems represented by the wave packet. Because of the coordinate spread of the wave packet, however, there is a large uncertainty in the time that

---

[6] Additional oscillatory terms, see e.g. [39, 49], appear when the no-interference condition between the reflected and incident wave packets is not imposed.

the *same* particle enters the region $[a, b]$. If a detector is placed at a before the collision occurs, the entrance time can be determined, but in general either the particle is destroyed or its behavior afterwards is modified by the measurement. We are thus faced with an intrinsic difficulty to measure *individual* delays. This means that, at variance with other quantum mechanical averages that are interpreted as averages of the eigenvalues measured for the individual members of the ensemble, the operational meaning of (2.109) does not require to assign a lifetime to a given particle. It depends on the average times defined in (2.82)–(2.84), which are measurable, at least in principle, by the time-of-flight technique (another operational procedure making use of particle absorption along the chosen interval has been described by Golub et al. [51]). This peculiarity of the delay time was already noted by Goldrich and Wigner [52]. A consequence is that the ordinary quantum fluctuations around the average value are not operationally meaningful. Instead, the relevant fluctuations refer to variations of the average values themselves, corresponding to **S** matrix (or Hamiltonian) ensembles [53].

The trace of (2.102) in the on-shell space is related to the change in density of states $\Delta\rho(E) \equiv \mathrm{Tr}[\delta(E-H) - \delta(E-H_0)]$, which is a fundamental quantity to characterize the continuous spectrum [54] according to the "spectral theorem" (the 3D elastic and multichannel versions of the spectral theorem have been extensively discussed and proven rigorously [55]),

$$\Delta\rho(E) = -\pi^{-1}\mathrm{Im}\,\mathrm{Tr}[\widehat{G}(E+i0) - \widehat{G}_0(E+i0)]$$
$$= \frac{1}{h}\sum_\alpha Q(E)_\alpha = \pi^{-1}\frac{d\Phi_T(E)}{dE}. \tag{2.110}$$

The second equality (spectral theorem) follows from a result of Dashen, Ma, and Bernstein [56]. To obtain the final expression, (2.62) and (2.64) [57] have been used; see [58] for an alternative derivation consisting in evaluating $\Delta\rho$ for a finite system and then going to infinity. Note that the maxima of the trace of **Q** may be used to identify resonance energies and widths [59]. For further relations between the density of states and the dwell time, see [60, 61, 62]. Chapter 9 discusses the concept of local density of states and its relation to the Larmor clock and transport properties.

2.4.3 Breit–Wigner Resonances

The simplest model of resonance behavior is the Breit–Wigner model for an isolated resonance,

$$\mathbf{S}(E) = 1 - \frac{i\mathbf{A}}{E - E_0 + i\Gamma/2}. \tag{2.111}$$

By imposing unitarity to **S** and assuming that **A** and the resonance parameters E_0 and Γ are independent of E, it follows that $\mathbf{A} = \mathbf{A}^\dagger$ and

$$\mathbf{A}^2 = \Gamma\mathbf{A}. \tag{2.112}$$

This means that the matrix $\mathbf{A}$ factorizes as $A_{\alpha\beta} = \gamma_\alpha \gamma_\beta^*$ and that it is proportional to a projector matrix $\mathbf{P} = \mathbf{A}/\Gamma$ with eigenvalues 1 and 0. Thus, (2.112) takes the form

$$\Gamma = \sum_\alpha |\gamma_\alpha|^2 \ . \tag{2.113}$$

The corresponding $\mathbf{Q}$ matrix may now be written as

$$\mathbf{Q} = \mathbf{P} q_m \ , \tag{2.114}$$

with eigenvalues q_m and zero, where

$$q_m = \frac{\hbar \Gamma}{(E - E_0)^2 + \Gamma^2/4} \tag{2.115}$$

is the maximum value allowed for a diagonal element of $\mathbf{Q}$. The Breit–Wigner model for $\mathbf{S}$ and $\mathbf{Q}$ can be generalized in various ways, in particular to account for multiple overlapping resonances [53].

2.4.4 Negative Delays

In partial wave analysis of 3D collisions with spherical potentials, the time delay has been used mainly as a way to characterize resonance scattering. One of the standard definitions of a resonance is a jump by π in the eigenphases of the $\mathbf{S}$ matrix. In 1D collisions the time delay has also been used frequently to characterize (non-resonant) tunneling, where it may become negative. In fact the different delay signs associated with the two types of effects, resonances and tunneling, are not independent. In 3D it was soon understood by Wigner [63] that the increases and decreases of the phase should balance each other. Since Levinson's theorem imposes a fixed phase difference from $p = 0$ to ∞, there must be intervals of negative delay to compensate for the phase increases associated with the resonances. A similar analysis applies in 1D to the transmission amplitude. In Fig. 2.1, the phase of the transmission amplitude for a square barrier is shown versus p for different values of the barrier width d. As d increases, the scattering resonances "above the barrier" $p > p_0 = (2mV_0)^{1/2}$ become more dense and are defined better because of the approach of the resonance poles in the fourth complex momentum quadrant to the real axis. The corresponding increases of the phase are compensated by a more and more negative delay in the tunneling region.

Negative delays also arise if a pole of $T(p)$ crosses the real axis upwards, when varying the interaction strength, to become a loosely bound state in the positive imaginary axis. Levinson's theorem, see (2.59), then imposes a sudden jump in the phase $\Phi_T(0)$ that must be compensated by a strong negative slope. This effect is more important near threshold, i.e., when the pole is very close to the real axis [64]. Similar effects have been described for nonbound state poles in complex potential scattering [65].

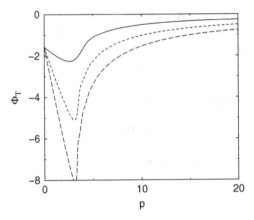

Fig. 2.1. Phase of the transmission amplitude versus momentum for a square barrier of "height" $V_0 = 5$ and for three different widths, $d = 1$ (*solid line*), 2 (*short dashed line*), and 3 (*long dashed line*). $m = 1$ (all quantities in atomic units)

Wigner also found a bound for the negative (partial wave) delay time of a potential of finite radius. Whereas positive delays can be arbitrarily large, negative delays are restricted by "causality conditions" [66]. Some back-of-the-envelope causality arguments may, however, be misleading. For example, assume a barrier of length d, and let a coincide with the left edge and b with the right edge. If the total time $\tau_T^{Ph}(0, d)$ is to be positive, the delay "cannot be more negative than the reference free time,"

$$\Delta t_{++} > -\frac{md}{p} ,$$ (2.116)

see e.g. [67]. In fact this bound may be violated, in particular at low energy in the proximity of a loosely bound state. This should not surprise the reader after our repeated warnings against an overinterpretation of the extrapolated time $\tau_T^{Ph}(0, d)$. The flaw in the argument is the assumption of positivity of τ_T^{Ph}. Nevertheless, rigorous bounds have been established by Wigner himself and various authors in 3D collisions, see [68, 66] for review. In 1D collisions, the following bound holds for even potentials with finite support between $-b$ and b [64, 69]:

$$\Delta t_{++} \geq \frac{m}{p} \left\{ -2b - \frac{\hbar}{2p} [\sin(2pb/\hbar + 2\delta_0) - \sin(2pb/\hbar + 2\delta_1)] \right\}$$

$$\geq \frac{m}{p} \left(-d - \frac{\hbar}{p} \right) .$$ (2.117)

This may be proven by using the even and odd eigenfunctions $\langle x|\psi_j\rangle$ introduced in (2.52), in particular the fact that $\int_{-b}^{b} dx \, \psi_j^2 > 0$. We start by calculating the logarithmic derivative of $\langle x|\psi_0\rangle$ at $x = b$ from the known expression for the outer region, see (2.51),

$$L_b \equiv \frac{d\langle x|\psi_0\rangle/dx}{\langle x|\psi_0\rangle}\bigg|_{x=b} = -\frac{p}{\hbar}\tan(pb/\hbar + \delta_0) . \tag{2.118}$$

Taking the derivative of L_b with respect to p,

$$\frac{d\delta_0}{dp} = -\left\{\frac{\hbar}{p}\frac{dL_b}{dp}\cos^2(pb/\hbar + \delta_0) + \frac{1}{2p}\sin[2(pb/\hbar + \delta_0)] + \frac{b}{\hbar}\right\} . \tag{2.119}$$

The first term on the right-hand side may also be written as

$$\frac{h\hbar}{2m}[\langle x|\psi_0\rangle_E\langle x|\psi_0\rangle_x - \langle x|\psi_0\rangle\langle x|\psi_0\rangle_{E,x}](x = b) , \tag{2.120}$$

where the subscripts E and x are shorthand notation for the derivatives with respect to E and x. Repeating the same operations for $x = -b$ one finds that

$$[\langle x|\psi_0\rangle_E\langle x|\psi_0\rangle_x - \langle x|\psi_0\rangle\langle x|\psi_0\rangle_{E,x}](x = b)$$
$$= -[\langle x|\psi_0\rangle_E\langle x|\psi_0\rangle_x - \langle x|\psi_0\rangle\langle x|\psi_0\rangle_{E,x}](x = -b) . \tag{2.121}$$

We shall now prove that this is a positive quantity. Taking the derivative of the stationary Schrödinger equation with respect to energy one obtains the identity for real eigenfunctions of $\widehat{H}$ [47]

$$\langle x|\psi\rangle^2 = -\frac{\hbar^2}{2m}\frac{\partial}{\partial x}\left(\langle x|\psi\rangle\langle x|\psi\rangle_{E,x} - \langle x|\psi\rangle_E\langle x|\psi\rangle_x\right) , \tag{2.122}$$

so that, using (2.121),

$$\int_{-b}^{b} dx \,\langle x|\psi_0\rangle^2 = \frac{\hbar^2}{m}\left(\langle x|\psi_0\rangle_E\langle x|\psi_0\rangle_x - \langle x|\psi_0\rangle\langle x|\psi_0\rangle_{E,x}\right)(x = b) . \tag{2.123}$$

Carrying out similar manipulations for the odd wave function $\langle x|\psi_1\rangle$ and using $\Phi_T = \delta_0 + \delta_1$ and (2.101), (2.117) is found as a consequence of the positivity of the probability to find the particle in the barrier region.

According to this bound the negative delay may be arbitrarily large for small enough momenta and may diverge at $p = 0$, as it occurs when a bound state appears when making the potential more attractive [64]. For the square barrier, which does not have bound states, the time advancement of the Hartman effect is less important, and it is actually bound by (2.116). Thus, whereas the experiments looking for anomalously large traversal velocities ("superluminal effects") have been frequently based on evanescent conditions in square barriers (tunneling), square wells with the proper depth may in fact lead to much larger advancement effects at threshold energies [70].

2.5 Time Dependence of Survival Probability: Exponential Decay and Deviations

The decay of unstable quantum states is an ubiquitous process in virtually all fields of physics and energy ranges, from particle and nuclear physics to condensed matter, or atomic and molecular science. The exponential decay, by far the most common type, is surrounded by deviations at short and long times [71, 72]. In fact other deviations and decay functions are also possible by a proper choice of initial state [73], but they may require in general rather artificial and complicated preparations except in some peculiar systems in which the exponential decay may be totally absent for small ratios between the resonance energy and its width [74]. We shall, however, concentrate here on the standard and much more common case of an exponential decay at intermediate times, surrounded by short-time and long-time deviations. The short-time deviations have been much discussed, in particular in connection with the Zeno effect [75, 76, 77] and the anti-Zeno effect [78, 79, 80, 81]. Experimental observations of short- [82, 83] and long-time deviations [84] are very recent. A difficulty in the experimental verification of long-time deviations has been the weakness of the decaying signal [85], and also the measurement itself may suppress the initial state reconstruction [72, 86], which is ultimately responsible for the deviations. The short- and long-time deviations may indeed be distinguished because of the different role played by the initial-state reconstruction [86]: The long-time decay can be attributed to a wave that was, in a classical-like, probabilistic sense, fully outside the initial state or the inner region at intermediate times, i.e., to a completely "regenerated" or reconstructed state, whereas the decay during the exponential regime is due to a nonregenerated wave. At short times a small quantum interference between regenerated and nonregenerated paths is responsible for the deviation from the exponential decay. We may thus conclude that state reconstruction is a "consistent history" for long-time deviations but not for short-time ones, see [86] for a full discussion of these aspects. In this section we shall concentrate, instead, on identifying possible dependences on time of the deviations in simple one-particle 1D systems following [87, 88]. Similar techniques have been applied to study the decay of a more complicated multiparticle system, a Tonks–Girardeau gas [89], in [90].

The quantum mechanical decay of unstable states can be described in different ways [91, 92]. In many theoretical works the emphasis has been on justifying the approximately valid exponential decay law. A possible treatment for the survival amplitude $A(t, \psi) \equiv \langle \psi(0) | \psi(t) \rangle$ decomposes the state ψ by the usual resolution into proper and improper eigenstates of the Hamiltonian $\widehat{H}$, corresponding to bound and continuum states. Even though it contains all the information, this is not convenient in general either for calculation purposes or for rationalizing the decay behavior in a simple manner, except in favorable circumstances where the integral is easily approximated and parameterized, e.g. for isolated resonances and particular initial states. An ideal description would handle arbitrarily complex initial states and potentials in simple terms, and allow for an understanding of both the dominant exponential decay and the deviations from it. Much progress in this direction has been achieved by representing $A(t, \psi)$ as a discrete sum over resonant terms [93, 87]. The discretization allows a clear identification and separation of the physically dominant contributions, different terms being important for different time regimes.

The survival amplitude $A(t, \psi) = \langle \psi(0) | \psi(t) \rangle$ requires the diagonal matrix elements of the unitary evolution operator $e^{-i\hat{H}t/\hbar}$. When this operator is expressed in terms of the resolvent, $A(t, \psi)$ takes the form

$$A(t, \psi) = \langle \psi | e^{-i\hat{H}t/\hbar} | \psi \rangle$$

$$= \frac{i}{2\pi m} \int_C dq \, q \langle \psi | \frac{e^{-izt/\hbar}}{z - \hat{H}} | \psi \rangle = \frac{i}{2\pi} \int_C dq \, e^{-izt/\hbar} M(q) \,, \quad (2.124)$$

where $z = q^2/2m$ is a complex energy and the contour C goes from $-\infty$ to $+\infty$ passing above all the singularities of the resolvent due to the spectrum of $\hat{H}$ (discrete poles for bound states and the natural boundary of the real axis for the continuum) and

$$M(q) \equiv \frac{q}{m} \langle \psi | \frac{1}{z - \hat{H}} | \psi \rangle \,. \quad (2.125)$$

The survival probability is to be calculated as $\mathcal{S}(t, \psi) = |A(t, \psi)|^2$.

2.5.1 Predicted Time Behavior

The function $M(q)$ is evaluated in the upper half q-plane and then analytically continued into the lower half plane. Provided that the continuation exists, $M(q)$ has in general a set of *core* singularities, depending only on the potential, and possibly other *structural* state-dependent singularities. It is then useful to deform the original integration contour to the diagonal $\mathcal{D}$ of the second and fourth quadrants of the q-plane. This provides both physical insight by identifying the most relevant time dependence (exponential decay) of the survival and a calculational advantage for the remainder, since for $t > 0$ the exponential $e^{-izt/\hbar} = e^{-iq^2t/(2m\hbar)}$ is a real Gaussian on this diagonal.

Let us assume that a pole expansion of the form

$$M(q) = \sum_k \frac{a_k}{(q - q_k)} \quad (2.126)$$

is possible (higher-order poles can be treated in a similar fashion). Here $k = 1, 2, 3 \cdots$, indexes the poles. On deforming the q integration from contour C to $\mathcal{D}$, the residues of the poles q_k crossed in the fourth quadrant on carrying out this deformation provide contributions to $A(t)$ that decay exponentially with time, whereas the residues are purely oscillatory for poles in the upper half plane (bound states),

$$E_k(t) = a_k e^{-iq_k^2 t/(2m\hbar)} = a_k e^{-u_k^2} \,, \quad (2.127)$$

where

$$u \equiv q/f, \quad f \equiv (1-i)\sqrt{(m\hbar/t)} \tag{2.128}$$

becomes real along the diagonal $\mathcal{D}$. Independently of providing or not providing a residue, all poles contribute because of the integral along the diagonal. Each pole contribution is expressed in terms of the w function, see [94] or Appendix, as

$$D_k(t) = -\frac{a_k}{2}\mathrm{sign}(\mathrm{Im}u_k)\, w[\mathrm{sign}(\mathrm{Im}u_k)u_k] \; . \tag{2.129}$$

The exponential term may be added to this contribution to give the compact result [94],

$$A(t) = \sum_k [E_k(t) + D_k(t)] = \sum_k \frac{1}{2}a_k w(-u_k) \; . \tag{2.130}$$

(It is understood that $E_k(t) = 0$ for poles in the lower half plane that have not been crossed when deforming the contour.) The second expression is very useful for studying the short-time behavior, but the first one has the advantage of separating explicitly the exponential decay, E_k, from the "correction" D_k, which is given in terms of the known entire function w parameterized by the pole position and time. Numerical values and asymptotic properties of this function for small or large times are easy to calculate.

The above treatment may be extended for an $M(q)$ that includes an entire function in addition to the pole expansion. This would add to the w functions the integral along $\mathcal{D}$ of the entire function times a real Gaussian.

2.5.2 Short Time Behavior

The short time behavior of the quantum survival probability is easily analyzed in terms of the above formalism, which allows to classify several possible non exponential dependences.

Many authors have described a short time t^2 dependence of the *decay probability* $P_{\mathrm{decay}} \equiv 1 - \mathcal{S}$, provided the mean energy and the second energy moment of these states exist, see in particular the work related to the "quantum Zeno paradox" [75, 95]. Less attention has been paid to the short time behavior if these conditions are not fulfilled. A formal treatment and examples by Moshinsky and coworkers suggest a $t^{1/2}$ dependence of the decay probability at short times [96, 97]. We shall clarify how these two seemingly different claims can be compatible and describe other possible dependences.

The Taylor series (2.167) of the w functions in (2.130) gives a series in powers of $t^{1/2}$,

$$A(t) = \sum_k \frac{a_k}{2} \sum_{n=0}^{\infty} \frac{[2^{-1}q_k(1-i)(t/m\hbar)^{1/2}]^n}{\Gamma(\frac{n}{2}+1)} \; . \tag{2.131}$$

This suggests a short time $t^{1/2}$ dependence of the decay probability, as claimed by Moshinsky and coworkers [96, 97]. On the other hand, the formal series based on expanding the evolution operator,

$$A(t, \psi) = \langle \psi | e^{-i\widehat{H}t/\hbar} | \psi \rangle = 1 - \frac{it}{\hbar} \langle \psi | \widehat{H} | \psi \rangle - \frac{t^2}{2\hbar^2} \langle \psi | \widehat{H}^2 | \psi \rangle + \cdots, \quad (2.132)$$

provides a t^2 dependence,

$$P_{\text{decay}} = \frac{t^2}{\hbar^2} (\langle \psi | \widehat{H}^2 | \psi \rangle - \langle \psi | \widehat{H} | \psi \rangle^2) + \cdots. \quad (2.133)$$

However, the expectation values of $\widehat{H}$ and/or higher powers of $\widehat{H}$ may not exist. Several behaviors are possible depending on the existence of these moments. The question of the physical realizability of Hilbert space states with infinite first or second energy moments is subject to debate [19]. We shall leave this debate aside and determine the possible implications on the short-time behavior.

Consider the first two derivatives of A at time $t = 0$ first from (2.132) and then by assuming a general short time dependence of the form $A \sim 1 + b t^c$, where b and c are finite constants,

$$\frac{dA}{dt}\bigg|_{t=0} = \frac{-i}{\hbar} \langle \psi | \widehat{H} | \psi \rangle = b c t^{c-1}\big|_{t=0}, \quad (2.134)$$

$$\frac{dA^2}{dt^2}\bigg|_{t=0} = -\frac{1}{\hbar^2} \langle \psi | \widehat{H}^2 | \psi \rangle = b c (c-1) t^{c-2}\big|_{t=0}. \quad (2.135)$$

If the mean energy of the initial state does not exist, a $t^{1/2}$ dependence of the decay probability is possible, see examples in [88] and [97].

If the mean energy is finite so that $d\mathcal{S}/dt|_{t=0} = 2\text{Re}(dA/dt|_{t=0}) = 0$, then $c \geq 1$. This rules out a $t^{1/2}$ dependence of A since a $t^{1/2}$ dependence implies an infinite time derivative of A at $t = 0$. The corresponding coefficient for $t^{1/2}$ in (2.131) must vanish by compensation between the different pole contributions.

The second derivative is only finite at time zero if $c \geq 2$. This means that if the first energy moment exists but not the second, a dependence t^c where $1 \leq c < 2$ is possible for A (and for the decay probability), in particular $t^{3/2}$. (The coefficient for $t^{1/2}$ must also vanish in this case.) Otherwise, one can expect that the series (2.132) will be effective at short times for states with finite moments $\langle \psi | \widehat{H}^n | \psi \rangle$ leading to a t^2 behavior. Examples where $t^{3/2}$ and t^2 dominate the short-time behavior of P_{decay} are provided in [88].

2.5.3 Large-Time Behavior

At first sight the asymptotic expansion of the w function for $t \sim \infty$ in the correction term to the exponential decay suggests a long-time dependence of the survival probability as t^{-1}, but in fact the general behavior is t^{-3} because of the cancellation of all the t^{-1} contributions. Due to the exponential $e^{-izt/\hbar}$ in (2.124), the large t behavior is dominated by the region around the origin. The origin is actually a saddle point for the steepest descent path for this exponential factor that crosses the origin along the diagonal $\mathcal{D}$ of the second and fourth quadrants. By introducing u and f variables as in (2.128) the exponential becomes e^{-u^2} and u remains real along the steepest descent path.

The resolvent matrix element $\langle\psi|(z-\widehat{H})^{-1}|\psi\rangle$, which is defined for $\mathrm{Im}\,q > 0$ (first energy sheet), has to be analytically continued into the lower-half q-plane (or second sheet of the complex z plane) to allow for this type of analysis, which will be valid in particular for finite-range potentials. Provided that the analytically continued function is analytical at the origin it has a Taylor series expansion

$$\langle\psi|(z - \widehat{H})^{-1}|\psi\rangle = a_0 + a_1 q + a_2 q^2 + \dots \qquad (2.136)$$

with coefficients a_i depending on ψ. But because of the (odd) q factor in (2.125), the first term, a_0, does not contribute to the integral (2.124). The asymptotic formula for the survival amplitude comes therefore from the second term and takes the form

$$\langle\psi|e^{-i\widehat{H}t/\hbar}|\psi\rangle \sim \frac{i}{2m\pi}a_1 f^3 \int_{-\infty}^{\infty} du\, u^2 e^{-u^2} = \frac{1-i}{2m\sqrt{\pi}}a_1\left(\frac{m\hbar}{t}\right)^{3/2}. \qquad (2.137)$$

This formal result depends on the validity of (2.136) and on the assumption that no additional contributions due to the deformation of the contour are to be considered asymptotically. In general, the analytically continued matrix elements of the resolvent will have poles in the lower-half q-plane that may be crossed when deforming the contour, but these can only yield contributions that decay *exponentially* with time, so they are negligible at long times.

A similar analysis may be performed for the propagator (no bound states) [98]

$$\langle x|e^{-i\widehat{H}t/\hbar}|x'\rangle = \frac{i}{2\pi}\int_C dq\, I(q)e^{-izt/\hbar}, \qquad (2.138)$$

$$I(q) = \frac{q}{m}\langle x|\frac{1}{z - \widehat{H}}|x'\rangle, \qquad (2.139)$$

substituting $M(q)$ by $I(q)$. Quite generally, $I(q)$ vanishes at $q = 0$, and a $t^{3/2}$ dependence results. An exception is free motion on the full line, where

$$\langle x|\frac{1}{z - \widehat{H}_0}|x'\rangle = \frac{-im}{q\hbar}e^{i|x-x'|q/\hbar}, \qquad (2.140)$$

so that $I(0) = -i/\hbar \neq 0$. As a consequence, the asymptotic behavior of the probability density for free motion on the full line is generically t^{-1}. This is an important case in which (2.136) is *not* satisfied. Explicitly, by carrying out the integral in (2.138), the well-known propagator

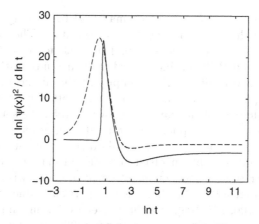

Fig. 2.2. $d\ln|\langle x|\psi(t)\rangle|^2/d\ln t$ versus $d\ln t$ for two different wave packets: one of them vanishes at $p = 0$, $\langle p|\psi(0)\rangle = C(1 - e^{-\alpha p^2/\hbar^2})e^{-\delta^2(p-p_0)^2/\hbar^2-ipx_0/\hbar}\Theta(p)$ (*solid line*), and the other one is a Gaussian wave packet, $\langle p|\psi(0)\rangle = C'e^{-\delta^2(p-p_0)^2/\hbar^2-ipx_0/\hbar}$ (*dashed line*). C and C' are normalization constants; the parameters are $p_0 = 1$, $x_0 = -10$, $\alpha = 0.5$, $\delta = 1$, $x = 0$, and $m = 1$ (all quantities in atomic units). Note the asymptotic dependences of the probability densities: t^{-3} and t^{-1}, respectively

$$\langle x|e^{-i\widehat{H}_0 t/\hbar}|x'\rangle = \left(\frac{m}{iht}\right)^{1/2} e^{im(x-x')^2/2\hbar t} \tag{2.141}$$

is obtained. A t^{-1} behavior will also occur exceptionally when the potential allows for a zero energy pole of the resolvent.

The free-motion probability density may decay faster than t^{-1} when the momentum amplitude $\langle p|\psi\rangle$ vanishes at $p = 0$, so that the q^{-1} singularity is canceled, see Fig. 2.2. The exceptional cases of decay slower than t^{-1} have been studied by Unnikrishnan [99].

2.6 Other Characteristic Times of Wave Propagation

In the previous section we have seen that contour deformation techniques in the complex plane allow us to single out contributions to the survival amplitude from resonance poles. In general, the integral that provides the time-dependent wave function may involve other critical points, "structural" poles, saddle points, or branch points, that determine the transient and the asymptotic behavior of the wave propagation. It is frequently possible to write explicit expressions or asymptotic expansions for the contributions of these critical points. In simple cases the effect of (the dominant term of) one of the critical points provides already a good approximation and a simple picture emerges, where characteristic times or velocities for the arrival of the main signal may be identified. Also typical is the transition from the dominance of one critical point to another, which may lead to a change in qualitative behavior and to a characteristic time for the transition. The pioneering work in this direction is due to Stevens [100, 101], who followed the techniques that Sommerfeld and Brillouin introduced in their study of the propagation of light in dispersive media [102]. Examples of the application of a square barrier and a separable potential to quantum scattering may be found in [103] and [104]. Here we shall examine, following [105], the somewhat simplified case corresponding to a point source producing evanescent waves. This is not a "scattering problem" in the standard sense, but it illustrates quite clearly methods and concepts involved in more conventional scattering problems and in other time-dependent quantum phenomena where a stationary state is reached after a transient behavior, in particular due to a sudden potential switching [106, 107, 108], or a shutter removal as in "diffraction in time phenomena" after Moshinsky's pioneering work [96], see [109, 110] and references therein. Similar techniques to the ones exemplified below have been used to determine characteristic times of matter-wave pulses [109, 110], permanent particle trapping, resonance build-up [106, 107], also including interparticle interactions [111], and of expansions of a Tonks–Girardeau gas [112]. Indeed, the attainment of ultracold temperatures by laser cooling makes the discussion of quantum effects for the atomic translational motion relevant and timely.

In order to summarize essential aspects of the time dependence of wave phenomena a number of characteristic velocities or times have been traditionally defined. (We will see that some of them coincide with times associated with critical points.) The *phase velocity*, ω/k, is the velocity of constant phase points in the stationary wave (assume $k > 0$ for the time being)

$$e^{ikx - i\omega t} . \tag{2.142}$$

The boundary conditions, the superposition principle, and the *dispersion relation* $\omega = \omega(k)$ between the frequency ω and the wave number k determine the time evolution of the waves in a given medium. If a "group" is formed by superposition of stationary waves around a particular ω, it propagates with the *group velocity* $d\omega/dk$. In *dispersive media* (where the group velocity depends on ω), the group velocity can be smaller (normal dispersion) or greater (anomalous dispersion) than the phase velocity. It was soon understood that both these velocities could be greater than c for the propagation of light; Sommerfeld and Brillouin [102], studying the fields that result from an input-step function-modulated signal in a single Lorentz resonance medium, introduced other useful velocities, such as the velocity of the very first wave front (equal to c) or the *signal velocity* for the propagation of the main front of the wave.

The above description is, however, problematic for *evanescent waves*, characterized by imaginary wave numbers instead of the real wave numbers of propagating waves. The role played by the imaginary part of the group velocity $d\omega/dk$ and the possible definition of a signal velocity in the evanescent case have been much discussed. Assume that a source is placed at $x = 0$ and emits with frequency ω_0 from $t = 0$ on. If ω_0 is above the *cutoff frequency* of the medium (the one that makes $k = 0$) a somewhat distorted but recognizable front propagates with the velocity corresponding to ω_0. For the dimensionless Schrödinger equation,

$$i\frac{\partial \psi}{\partial t} = -\frac{\partial^2 \psi}{\partial x^2} + \psi \,, \tag{2.143}$$

the dispersion relation takes the form

$$\omega = 1 + k^2 \,, \tag{2.144}$$

and the signal propagation velocity for the main front is equal to the group velocity, $v_p = (d\omega/dk)_{\omega_0} = 2(\omega_0 - 1)^{1/2}$. In other words, at some distance x from the source, the amplitude behaves, in first approximation, as

$$\psi(x,t) \approx e^{-i\omega_0 t} e^{+ik_0 x} \Theta(t - x/v_p) \,, \tag{2.145}$$

where $k_0 = (\omega_0 - 1)^{1/2}$ is the wave number related to ω_0 by the dispersion relation and Θ is the Heaviside (step) function. In the evanescent case, $\omega_0 < 1$, a preliminary analysis by Stevens [100, 101, 113], following the contour deformation techniques used by Brillouin and Sommerfeld, suggested that a main front, moving now with velocity $v_m = 2(1 - \omega_0)^{1/2} = \mathrm{Im}(d\omega/dk)_{\omega_0}$, and attenuated exponentially by $\exp(\kappa_0 x)$, where

$$\kappa_0 = (1 - \omega_0)^{1/2} \,, \tag{2.146}$$

could be also identified,

$$\psi(x,t) \approx e^{-i\omega_0 t} e^{-\kappa_0 x} \Theta(t - x/v_m) \,. \tag{2.147}$$

The contour for the integral defining the field evolution was deformed along the steepest descent path from the saddle point, and the main front (2.147) was associated with a residue due to the crossing of a pole at $i\kappa_0$ by the steepest descent path.

The result seemed to be supported by a different approximate analysis of Moretti based on the exact solution [113] and by the fact that the time of arrival of the evanescent front, $\tau = x/v_m$, had been found independently by Büttiker and Landauer [114, 16] as a characteristic *traversal time* for tunneling using rather different criteria (semiclassical arguments, the rotation of the electron spin in a weak magnetic field, and the transition from adiabatic to sudden regimes in an oscillating potential barrier).

However, more accurate studies of the point source problem and other boundary conditions have shown that the contribution from the saddle point (due to frequency components above or at the frequency cutoff created by the sharp onset of the source emission) and possibly from other critical points (e.g., resonance poles when a square barrier is located in front of the source [103]) are generally dominant at τ, so that no sign of the ω_0 front is seen in the total wave density at that instant, see [115, 116, 117, 118, 103, 105] and Sect. 2.6.1.

Büttiker and Thomas reconsidered the signal sent out by a source that has a sharp onset in time [119]. They proposed two approaches to enhance the monochromatic fronts compared to the forerunners due to the saddle. First, the dominance of the high-frequency forerunners could be avoided if the source is frequency limited such that all frequencies of the source are within the evanescent case. Of course this makes the onset of the signal unsharp. A second option is not to limit the source but to frequency limit the detection. We can chose a detector that is tuned to the frequency of the source and that responds when the monochromatic front arrives.

These two proposals and the sharp onset case were later implemented and examined in detail by Muga and Büttiker [105]. For a source with a sharp onset, they found that the traversal time τ plays a basic and unexpected role in the transient regime. For strongly attenuating conditions in the WKB (Wentzel-Kramers-Brillouin) limit, the traversal time governs the appearance of the first main peak of the forerunner. In contrast, the transition from the forerunner to an asymptotic regime that is dominated by the monochromatic signal of the source is given by an exponentially long time, see more details in Sect. 2.6.1 below. If the source is frequency band limited such that it switches on gradually but still fast compared to the traversal time, the situation remains much the same as for the sharp source, except that now the transition from the transient regime to the stationary regime occurs much faster, but still on an exponentially long timescale. This changes if we permit the source to be switched on a timescale comparable to or larger than the traversal time for tunneling. Clearly, in this case a precise definition of the traversal time is not possible. But for such a source the transition from the transient regime to the asymptotic regime is now determined by the traversal time. Much the same picture emerges if we limit the detector instead of the source. Muga and Büttiker model the detector response by means of a "spectrogram," a time–frequency representation of the wave function at a fixed point. As long as the frequency window of the detector is made sharp enough to determine the traversal time with accuracy, the detector response is dominated by the uppermost frequencies. In contrast, if the frequency window of the detector is made so narrow that the possible uncertainty in the determination of the traversal time is on the order of the traversal time itself, the detector sees a crossover from the transient regime to the monochromatic asymptotic regime at a time determined by the traversal time.

Possibly, the fact that we cannot determine the traversal time with an accuracy better than the traversal time itself tells us something fundamental about the tunneling time problem and is not a property of the two particular methods investigated.

2.6.1 Role of the Traversal Time for a Source with a Sharp Onset

We shall obtain exact and approximate expressions of the time-dependent wave function for $x > 0$ and $t > 0$ corresponding to the Schrödinger equation (2.143) and the "source boundary condition"

$$\psi(x = 0, t) = e^{-i\omega_0 t} \Theta(t) , \qquad (2.148)$$

in the evanescent case $\omega_0 < 1$. (A discussion of the physical meaning of "source boundary conditions" as compared to standard "initial value" conditions may be found in [120].) The solution may be constructed from its Fourier transform as

$$\psi(x, t) = -\frac{e^{-it}}{2\pi i} \int_{\Gamma_+} dk \left[\frac{1}{k + i\kappa_0} + \frac{1}{k - i\kappa_0} \right] e^{ikx - ik^2 t} , \qquad (2.149)$$

where the contour Γ_+ goes from $-\infty$ to ∞ passing above the pole at $i\kappa_0$, and κ_0 is given by (2.146). The contour can be deformed along the steepest descent path from the saddle at $k_s = x/2t$, the straight line

$$k_I = -k_R + x/2t , \qquad (2.150)$$

(k_R and k_I are the real and imaginary parts of k.) plus a small circle around the pole at $i\kappa_0$ after it has been crossed by the steepest descent path, for fixed x, at the critical time

$$\tau = \frac{x}{2\kappa_0} . \qquad (2.151)$$

This procedure allows to recognize two w functions [94] one for each integral,

$$\psi(x, t) = \frac{1}{2} e^{-it + ik_s^2 t} [w(-u_0') + w(-u_0'')] . \qquad (2.152)$$

Here,

$$u_0' = \frac{1 + i}{2^{1/2}} t^{1/2} \kappa_0 \left(-i - \frac{\tau}{t} \right) , \qquad (2.153)$$

$$u_0'' = \frac{1 + i}{2^{1/2}} t^{1/2} \kappa_0 \left(i - \frac{\tau}{t} \right) .$$

It is clear from the exact result, (2.152) and (2.153), that τ is an important parameter that appears naturally in the w-function arguments, and determines with κ_0 the global properties of the solution. Its detailed role will be discussed next.

The simplest approximation for $\psi(x, t)$ for times before τ is to retain the dominant contribution of the saddle by putting $k = k_s$ in the denominators of (2.149) and integrating along the steepest descent path,

$$\psi_s(x, t) = \frac{e^{-it + ik_s^2 t}}{2i\pi^{1/2}} \left(\frac{1}{u_0'} + \frac{1}{u_0''} \right) . \qquad (2.154)$$

The average local instantaneous frequency for this saddle contribution is equal to the frequency of the saddle point [105],

$$\omega_s \equiv 1 + x^2/4t^2 . \qquad (2.155)$$

After the crossing of the pole $i\kappa_0$ by the steepest descent path at $t = \tau$ the residue

$$\psi_0(x,t) = e^{-i\omega_0 t} e^{-\kappa_0 x} \Theta(t - \tau) \tag{2.156}$$

has to be added to (2.154),

$$\psi(x,t) \approx \psi_s(x,t) + \psi_0(x,t) . \tag{2.157}$$

The solution given by (2.156) describes a monochromatic front that carries the signal into the evanescent medium. The conditions of validity of this approximation can be determined by examining the asymptotic series of the $w(z)$ functions in (2.152) for large $|z|$, see Appendix. In fact (2.157) is obtained from the dominant terms of these expansions. Large values of $|z|$ are obtained with large values of κ_0, t, or x, and also when $t \to 0$. Within the conditions that make the saddle approximation valid, the contribution of the pole is negligible. To see this more precisely let us examine the ratio between the moduli of the two contributions,

$$R(t) \equiv \frac{|\psi_0|}{|\psi_s|} = \frac{2\pi^{1/2}}{x} e^{-\kappa_0 x} t^{3/2} (x^2/4t^2 + \kappa_0^2) . \tag{2.158}$$

Its value at τ is an exponentially small quantity,

$$R(t = \tau) = e^{-\kappa_0 x} (2\pi\kappa_0 x)^{1/2} . \tag{2.159}$$

In summary, for the source with a sharp onset described here, the monochromatic front is not visible when the approximation (2.157) remains valid around $t = \tau$. A complementary analysis is carried out in Chap. 12.

However, two very important observable features of the wave can be extracted easily from (2.157). The first one is the arrival of the *transient front*, characterized by its maximum density at $t_f \equiv \tau/3^{1/2}$. (The peak time is exactly τ if it is evaluated with respect to x for t fixed [121].) This time is, surprisingly, on the order of τ, but the wave front that arrives does not oscillate with the pole frequency ω_0, but with the saddle point frequency ω_s, i.e., it does not tunnel.

The second observable feature that we can extract from (2.157) is the timescale for the attainment of the stationary regime, or equivalently, the duration t_{tr} of the transient regime dominated by the saddle before the pole dominates. t_{tr} can be identified formally as the time where the saddle and pole contributions are equal, $R = 1$. Because of (2.159) we shall assume $\tau \ll t_{tr}$ to obtain the explicit result

$$t_{tr} \approx \left(\frac{x e^{\kappa_0 x}}{2\kappa_0^2 \pi^{1/2}} \right)^{2/3} . \tag{2.160}$$

Finally, when $x\kappa_0$ is small ($\lesssim 1$), the saddle approximation describes correctly the very short-time initial growth, but fails around τ because the pole is within the width of the Gaussian centered at the saddle point. The pole cancels part of the Gaussian contribution so that the bump predicted by ψ_s at $\tau/3^{1/2}$ is not seen in this regime. In fact, the forerunner is dominated in this case by under-the-barrier components [122], and its characteristic arrival time is inversely proportional to the difference between the potential energy and the incidence energy, a tunneling timescale different from both the phase time and the Büttiker–Landauer time [123, 122]. A generalization of the results of this subsection for relativistic equations may be found in [124].

2.6.2 Ultrafast propagation in absorbing media

If a small imaginary potential is added to the setup described in Sect. 2.6.1, an interesting effect occurs [125, 126]: the temporal peak arrives at different locations simultaneously. The arrival time corresponds to the lifetime of the particle in the medium from the instant when the point source with a sharp onset is turned on. The simultaneous arrival due to absorption, unlike the Hartman effect, occurs for carrier frequencies under or above the cutoff, and for arbitrarily large distances. It also holds in a relativistic generalization but limited by causality. A possible physical realization has been proposed by illuminating a two-level atom with a detuned laser [125]. The effect is also found within a broad spatial range in an absorbing waveguide when the source emits (more realistic) smoothed pulses instead of a perfectly sharp step signal. The optimal carrier frequency is barely below the cutoff but, at variance with other "ultrafast" wave phenomena based on anomalous dispersion in absorbing media [127, 128, 129, 130], which depend on the dominance of the carrier (central) frequency associated with faster than light, infinite, or negative group velocities, the ubiquitous peak is, at each position, dominated by the saddle point contributions above the cutoff frequency. It is thus a fundamentally different phenomenon. In the case described by the Schrödinger equation [125] it is closer in nature to the over-the-barrier, saddle-dominated peak that arrives at the Büttiker–Landauer time in a nonabsorbing medium [105, 121], and in fact it tends to it continuously when the absorption vanishes. However, that peak "moves" with a semiclassical tunneling velocity whereas in the absorbing medium it appears everywhere simultaneously within the domain of the effect.

The task of sending information to arrive at different receivers simultaneously is different from the question of superluminal velocities because the information always arrives subluminally [131, 132], and it will be possible in principle to send information faster to a single fixed receiver than with the present effect.

Acknowledgements

This chapter is largely based on work done with many collaborators over the years: S. Brouard, M. Büttiker, A. del Campo, J. A. Damborenea, F. Delgado, V. Delgado, I. Egusquiza, G. García-Calderón, P. Palao, M. G. Raizen,

68   J. G. Muga

A. Ruschhaupt, R. Sala, R. F. Snider, G. Wei, D. Wardlaw, and J. Villavicencio. Thanks to them all. I also acknowledge support by the Basque Government, MEC, and UPV-EHU.

Appendix: Properties of w Functions

The w function is an entire function defined in terms of the complementary error function as [94]

$$w(z) = e^{-z^2}\text{erfc}(-iz) .\tag{2.161}$$

$w(z)$ is frequently recognized by its integral expression

$$w(z) = \frac{1}{i\pi} \int_{\Gamma_-} \frac{e^{-u^2}}{u - z} du ,\tag{2.162}$$

where Γ_- goes from $-\infty$ to ∞ passing below the pole at z. For $\text{Im} z > 0$ this corresponds to an integral along the real axis. For $\text{Im} z < 0$ the contribution of the residue has to be added, and for $\text{Im} z = 0$ the integral becomes the principal part contribution along the real axis plus half the residue. From (2.162) two important properties are deduced:

$$w(-z) = 2e^{-z^2} - w(z)\tag{2.163}$$

and

$$w(z^*) = [w(-z)]^* .\tag{2.164}$$

To obtain an asymptotic series as $z \to \infty$ for $\text{Im} z > 0$ one may expand $(u - z)^{-1}$ around the origin (the radius of convergence is the distance from the origin to the pole, $|z|$) and integrate term by term. This provides

$$w(z) \sim \frac{i}{\sqrt{\pi}\, z} \left[1 + \sum_{m=1}^{\infty} \frac{1 \cdot 3 \cdot ... \cdot (2m-1)}{(2z^2)^m} \right] \quad \text{Im} z > 0 ,\tag{2.165}$$

which is a uniform expansion in the sector $\text{Im} z > 0$. For the sector $\text{Im} z < 0$, (2.163) gives

$$w(z) \sim \frac{i}{\sqrt{\pi}\, z} \left[1 + \sum_{m=1}^{\infty} \frac{1 \cdot 3 \cdot ... \cdot (2m-1)}{(2z^2)^m} \right] + 2e^{-z^2} , \quad \text{Im} z < 0 .\tag{2.166}$$

If z is in one of the bisectors then $-z^2$ is purely imaginary and the exponential becomes dominant. But right at the crossing of the real axis, $\text{Im} z = 0$, the exponential term is of order $o(z^{-n})$, (all n), so that (2.165) and (2.166) are asymptotically equivalent as $|z| \to \infty$.

$w(z)$ has the series expansion

$$w(z) = \sum_{0}^{\infty} \frac{(iz)^n}{\Gamma(\frac{n}{2}+1)} \ . \tag{2.167}$$

The w function is a particular case of the Moshinsky function [96], which can be regarded as "the basic propagator for a Schrödinger transient mode" [133].

References

1. A.H. Zewail: *Femtochemistry* (World Scientific, Singapore 1994)
2. R.F. Snider: J. Stat. Phys. **61**, 443 (1990)
3. G. Bastard: *Wave Mechanics Applied to Semiconductor Heterostructures* (Les Editions de Physique, Paris 1988)
4. Y.B. By, S. Efrima: Phys. Rev. B **28**, 4126 (1983)
5. R.D. Levine: *Quantum Mechanics of Molecular Rate Processes* (Oxford University Press, Oxford 1969)
6. R. Folman, P. Krüger, J. Schmiedmayer, J. Denschlag, C. Hankel: Adv. At. Mol. Opt. Phys. **48**, 263 (2002)
7. M. Olshanii, Phys. Rev. Lett. **81**, 938 (1998)
8. P. Deift, E. Trubowitz: Commun. Pure Appl. Math. **32**, 121 (1979)
9. A.D. Baute, I.L. Egusquiza, J.G. Muga: Int. J. Theor. Phys. Group Theor. Nonlinear. Opt **8**, 1 (2002). quant-ph/0007079
10. A.D. Baute, I.L. Egusquiza, J.G. Muga: J. Phys. A **34**, 5341 (2001)
11. R.F. Snider: J. Chem. Phys. **88**, 6438 (1988)
12. J.R. Tayor: *Scattering Theory* (John Wiley, New York 1972)
13. B. Diu: Eur. Phys. Lett. **1**, 231 (1980)
14. M.S. Marinov, B. Segev: J. Phys. A: Math. Gen **29**, 2839 (1996)
15. J.G. Muga, S. Brouard, R. Sala: Phys. Lett. A **167**, 24 (1992)
16. M. Büttiker: Phys. Rev. B. **27**, 6178 (1983)
17. M.L. Goldberger, K.M. Watson: *Collision Theory* (Krieger, Huntington 1975)
18. H. Ekstein, A.J.F. Siegert: Ann. Phys. **68**, 509 (1971)
19. P. Exner: *Open Quantum Systems and Feynman Integrals* (Reidel, Dordrecht 1985)
20. D. Sokolovski, J.N.L. Connor: Phys. Rev. A **44**, 1500 (1992)
21. C.R. Leavens, G.C. Aers: in *Scanning Tunneling Microscopy III.* ed. by R. Wiesendanger, H.J. Günterodt (Springer, Berlin 1993), pp. 105–140
22. W. Jaworski, D. Wardlaw: Phys. Rev. A **40**, 6210 (1989)
23. H. Mizuta, T. Tanoue: *The Physics y Applications de Resonant Tunnelling Diodes* (Cambridge University Press, Cambridge 1995)
24. M. Tsuchiya, T. Matsusue, H. Sakaki: Phys. Rev. Lett. **59**, 2356 (1987)
25. W. Jaworski, D.M. Wardlaw: Phys. Rev. A **37**, 2843 (1987)
26. G.R. Allcock: Ann. Phys. **53**, 311 (1969)
27. A.J. Bracken, G.F. Melloy: J. Phys. A **27**, 2197 (1994)
28. J.G. Muga, J.P. Palao, C.R. Leavens: Phys. Lett. A **253**, 21 (1999)
29. J.G. Muga, C.R. Leavens: Phys. Rep. **338**, 353 (2000)
30. J.G. Muga, S. Brouard, D. Macías: Ann. Phys. (N.Y.) **240**, 351 (1995)

31. S. Brouard, R. Sala, J. G. Muga: Phys. Rev. A **49**, 4312 (1994)
32. V.S. Olkhovsky, E. Recami: Phys. Rep. **214**, 339 (1992)
33. V. Delgado, S. Brouard, J.G. Muga: Solid State Commun. **94**, 979 (1995)
34. J.P. Palao, J.G. Muga, S. Brouard, A. Jadcyk: Phys. Lett. A **233**, 227 (1997)
35. R. Leavens: Physics Lett. A, **178**, 27 (1993)
36. A.M. Steinberg: Phys. Rev. Lett. **74**, 2405 (1995); Phys. Rev. A **52**, 32 (1995)
37. R.G. Newton: J. Math. Phys. **21**, 493 (1980)
38. C.R. Leavens, G.C. Aers: Phys. Rev. B **39**, 1202 (1989)
39. E.H. Hauge, J.A. Stovneng: Rev. Mod. Phys. **61**, 917 (1989)
40. T.E. Hartman: J. Appl. Phys. **33**, 3427 (1962)
41. J.R. Fletcher: J. Phys. C: Solid State Phys. **18**, L55 (1985)
42. H. G. Winful: Phys. Rev. Lett. **91**, 260401 (2003)
43. H. G. Winful: Phys. Rep. **436**, 1 (2006)
44. F.E. Low, P.F. Mende: Ann. Phys. (NY) **210**, 380 (1991)
45. V. Delgado, J.G. Muga: Ann. Phys. (NY) **248**, 122 (1996)
46. L. Eisenbud: Dissertation, Princeton University, June 1948 (unpublished)
47. F.T. Smith: Phys. Rev **118**, 349 (1960)
48. M. Sassoli de Bianchi: Helv. Phys. Acta **66**, 361 (1993)
49. H.M. Nussenzveig: Phys. Rev. **62**, 042107 (2000)
50. A. Kuppermann, J. Kaye: J. Phys. Chem. **85**, 1969 (1981)
51. R. Golub, S. Felber, R. Gähler, E. Gutsmiedl: Phys. Lett. A **148**, 27 (1990)
52. F. Goldrich, E.P. Wigner: in *Magic Without Magic: John Archibald Wheeler.* ed. by J. R. Klauder (W. H. Freeman, San Francisco, CA, 1972), p. 147
53. J.G. Muga, D.M. Wardlaw: Phys. Rev. E **51**, 5377 (1995)
54. L.N. Pandey, D. Sahu, T.F. George: Appl. Phys. Lett. **56**, 277 (1990)
55. J.M. Jauch, K.B. Sinha, B.N. Misra: Helv. Phys. Acta **45**, 398 (1972); T.Y. Tsang, T.A. Osborn: Nucl. Phys. A **247**, 43 (1975); P. Brumer, D.E. Fitz, D. Wardlaw: J. Chem. Phys. **72**, 386 (1980)
56. R. Dashen, S. Ma, H.J. Bernstein: Phys. Rev. **187**, 345 (1969)
57. Y. Avishai, Y.B. Band: Phys. Rev. B **32**, 2674 (1985)
58. W. Trzeciakowski, M. Gurioli: J. Phys. Condens. Matter **5**, 105 (1993); **5**, 1701 (1993)
59. H.W. Jang, S.E. Choi, J.C. Light: J. Chem. Phys. **100**, 4188 (1994)
60. V. Gasparian, M. Pollak: Phys. Rev. B **47**, 2038 (1993)
61. G. Iannaccone: Phys. Rev. 51 **51**, 4727 (1995)
62. G. Iannaccone, B. Pellegrini: Phys. Rev. B **53**, 2020 (1996)
63. E.P. Wigner: Phys. Rev. **98**, 145 (1955)
64. W. van Dijk, K.A. Kiers: Am. J. Phys. **60**, 520 (1992)
65. J.G. Muga, J.P. Palao: Ann. Phys. **7**, 671 (1998)
66. H.M. Nussenzveig: *Causality and Dispersion Relations* (Academic Press, New York 1972)
67. A. Galindo, P. Pascual: *Quantum Mechanics I,II* (Springer-Verlag, Berlin 1990,1991)
68. Ph.A. Martin: Acta Phys. Aust. Suppl. **23**, 159 (1981)
69. M. Sassoli de Bianchi: J. Math. Phys. **35**, 2719 (1994)
70. J.G. Muga, I.L. Egusquiza, J.A. Damborenea, F. Delgado: Phys. Rev. A **66**, 04442115 (2002)
71. L.A. Khalfin: Zurn. Eksp. Teor. Fiz. **33**, 1371 (1957), English translation: Sov. Phys. JETP **6**, 1053 (1958)

72. L. Fonda, G.C. Ghirardi: Il Nuovo Cimento **7A**, 180 (1972)
73. M.L. Goldberger, K.M. Watson: Phys. Rev. **136**, B1472 (1964)
74. T. Jittoh, S. Matsumoto, J. Sato, Y. Sato, K. Takeda: Phys. Rev. A **71**, 012109 (2005)
75. B. Misra, E.C.G. Sudarshan: J. Math. Phys. **18**, 756 (1977)
76. C.B. Chiu, E.C.G. Sudarshan, B. Misra: Phys. Rev. D **16**, 520 (1977)
77. J.J. Sakurai: *Modern Quantum Mechanics* (Addison- Wesley, Reading, MA 1994)
78. W.C. Schieve, L.P. Horwitz, J. Levitan: Phys. Lett. A **136**, 264 (1989)
79. A.G. Kofman, G. Kurizki: Phys. Rev. A **54**, R3750 (1996)
80. A.G. Kofman, G. Kurizki: Nature (London) **405**, 546 (2000)
81. P. Facchi, H. Nakazato, S. Pascazio: Phys. Rev. Lett. **86**, 2699 (2001)
82. S.R. Wilkinson, C.F. Bharucha, M.C. Fischer, K.W. Madison, P.R. Morrow, Q. Niu, B. Sundaram, M.G. Raizen: Nature (London) **387**, 575 (1997)
83. M.C. Fischer, B. Gutiérrez-Medina, M.G. Raizen: Phys. Rev. Lett. **87**, 040402 (2001)
84. C. Rothe, S.I. Hintschich, A.P. Monkman: Phys. Rev. Lett. **96**, 163601 (2006)
85. G. García-Calderón, V. Riquer, R. Romo: J. Phys. A **34**, 4155 (2001)
86. J.G. Muga, F. Delgado, A. del Campo, G. García-Calderón: Phys. Rev. A **73**, 052112 (2006)
87. J.G. Muga, G.W. Wei, R.F. Snider: Ann. Phys. **252**, 336 (1996)
88. J.G. Muga, R.F. Snider, G.W. Wei: Europhys. Lett. **35**, 247 (1996)
89. M. Girardeau: J. Math. Phys. **1**, 516 (1960)
90. A. del Campo, F. Delgado, G. García Calderón, J.G. Muga, M. G. Raizen: Phys. Rev. A **74**, 013605 (2006)
91. V.F. Weiskopf, B.P. Wigner: Z. Phys. **63**, 54 (1930); **65**, 18 (1930)
92. H. Jakobovits, Y. Rothschild, J. Lecitan: Am. J. Phys. **63**, 439 (1995)
93. G. García Calderón, J.L. Mateos, M. Moshinsky: Phys. Rev. Lett. **74**, 337 (1995)
94. M. Abramowitz, I.A. Stegun: *Handbook of Mathematical Functions* (Dover, New York 1972); The function w is a particular case of Moshinsky's function, see [96] and H. M. Nussenzveig: in *Symmetries in Physics*, ed. by A. Frank, K. B. Wolf (Springer, Berlin, Heidelberg 1992), p. 293
95. A. Peres: Am. J. Phys. **48**, 931 (1980); W.M.Itano, J.D. Heinzen, J.J. Bollinger, D.J. Winely: Phys. Rev. A **41**, 2295 (1990); K. Urbanowski: Phys. Rev. A **50**, 2847 (1994)
96. M. Moshinsky: Phys. Rev. **84**, 525 (1951)
97. G. García Calderón, G. Loyola, M. Moshinsky: in *Symmetries in Physics*, ed. by A. Frank and K. B. Wolf (Springer, Berlin, Heidelberg 1992), p. 273.
98. J.G. Muga, V. Delgado, R.F. Snider: Phys. Rev. B **52**, 16381 (1995)
99. K. Unnikrishnan: Am. J. Phys. **66**, 632 (1998)
100. K.W.H. Stevens: Eur. J. Phys. **1**, 98 (1989)
101. K.W.H. Stevens: J. Phys. C: Solid State Phys. **16**, 3649 (1983)
102. L. Brillouin: *Wave Propagation and Group Velocity* (Academic Press, New York 1960)
103. S. Brouard, J.G. Muga: Phys. Rev. A **54**, 3055 (1996)
104. J.G. Muga, J.P. Palao: J. Phys. A **31**, 9519 (1998)
105. J.G. Muga, M. Büttiker: Phys. Rev. A **62**, 023808 (2000)
106. F. Delgado, H. Cruz, J. G. Muga: J. Phys. A **35**, 10377 (2002)

107. F. Delgado, J.G. Muga, D.G. Austing, G. García-Calderón: J. Appl. Phys. **97**, 013705 (2005)
108. A. Ruschhaupt, F. Delgado, J. G. Muga: J. Phys. B **38**, 2665 (2005)
109. A. Del Campo, J.G. Muga: J. Phys. A **38**, 9803 (2005)
110. A. del Campo, J.G. Muga: J. Phys. A **39**, 5897 (2006)
111. F. Delgado, J.G. Muga, H. Cruz, D. Luis, D.G. Austing: Phys. Rev. B **72**, 195318 (2005)
112. A. del Campo, J.G. Muga: Europhys. Lett. **74**, 965 (2006)
113. P. Moretti: Phys. Scr. **45**, 18 (1992)
114. M. Büttiker, R. Landauer: Phys. Rev. Lett. **49**, 1739 (1982)
115. A. Ranfagni, D. Mugnai, P. Fabeni, P. Pazzi: Phys. Scr. **42**, 508 (1990)
116. A. Ranfagni, D. Mugnai, A. Agresti: Phys. Lett. A **158**, 161 (1991)
117. N. Teranishi, A.M. Kriman, D.K. Ferry: Superlattices and Microstructures **3**, 509 (1987)
118. A.P. Jauho, M. Jonson: Superlattices and Microstructures **6**, 303 (1989)
119. M. Büttiker, H. Thomas: Ann. Phys. (Leipzig) **7**, 602 (1998); Superlattices Microstruct. **23**, 781 (1998)
120. A.D. Baute, I.L. Egusquiza, J.G. Muga: J. Phys. A **34**, 4289 (2001)
121. J. Villavicencio, R. Romo, S. Sosa y Silva: Phys. Rev. A. **66**, 042110 (2002)
122. G. García Calderón, J. Villavicencio, F. Delgado, J. G. Muga: Phys. Rev. A **66**, 042119 (2002)
123. G. García-Calderón, J. Villavicencio: Phys. Rev. A **64**, 012107 (2001)
124. F. Delgado, J.G. Muga, A. Ruschhaupt, G. García-Calderón, J. Villavicencio: Phys. Rev. A. **68**, 032101 (2003)
125. F. Delgado, J.G. Muga, A. Ruschhaupt: Phys. Rev. A **69**, 022106 (2004)
126. A. Ruschhaupt, J. G. Muga: Phys. Rev. Lett. **93**, 020403 (2004)
127. C.G.B. Garrett, D.E. McCumber: Phys. Rev. A **1**, 305 (1970)
128. S. Chu, S. Wong; Phys. Rev. Lett. **48**, 738 (1982)
129. M.A.I. Talukder, Y. Amigishi, M. Tomita: Phys. Rev. Lett. **86**, 3546 (2001)
130. E. Rosenthal, B. Segev: Phys. Rev. **65**, 032110 (2002)
131. K. Wynne: Opt. Commun. **209**, 85 (2002)
132. M. D. Stenner, D. J. Gauthier, M. A. Neifeld: Nature **425**, 695 (2003)
133. H.M. Nussenzveig: in *Symmetries in Physics*. ed. by A. Frank, K. B. Wolf (Springer, Berlin, Heidelberg 1992), p. 293

3

The Time–Energy Uncertainty Relation

Paul Busch

Department of Mathematics, University of York, York, UK

3.1 Introduction

The time–energy uncertainty relation

$$\Delta T \, \Delta E \geq \frac{1}{2}\hbar \qquad (3.1)$$

has been a controversial issue since the advent of quantum theory, with respect to appropriate formalisation, validity, and possible meanings. Already the first formulations due to Bohr, Heisenberg, Pauli, and Schrödinger are very different, as are the interpretations of the terms used. A comprehensive account of the development of this subject up to the 1980s is provided by a combination of the reviews of Jammer [1], Bauer and Mello [2], and Busch [3, 4]. More recent reviews are concerned with different specific aspects of the subject: [5, 6, 7]. The purpose of this chapter is to show that different types of time–energy uncertainty relation can indeed be deduced in specific contexts, but that there is no unique universal relation that could stand on equal footing with the position–momentum uncertainty relation. To this end, we will survey the various formulations of a time–energy uncertainty relation, with a brief assessment of their validity, and along the way we will indicate some new developments that emerged since the 1990s (Sects. 3.3, 3.4, and 3.6). In view of the existing reviews, references to older work will be restricted to a few key sources. A distinction of three aspects of time in quantum theory introduced in [3] will serve as a guide for a systematic classification of the different approaches (Sect. 3.2).

3.2 The Three-fold Role of Time in Quantum Theory

The conundrum of the time–energy uncertainty relation is related to an ambiguity concerning the role of time in quantum theory. In the first place, time is identified as the parameter entering the Schrödinger equation and measured

P. Busch: *The Time–Energy Uncertainty Relation*, Lect. Notes Phys. **734**, 73–105 (2008)
DOI 10.1007/978-3-540-73473-4_3   © Springer-Verlag Berlin Heidelberg 2008

by an external, detached laboratory clock. This aspect will be referred to as *pragmatic, laboratory,* or *external time.* By contrast, time as *dynamical* or *intrinsic time* is defined through the dynamical behaviour of the quantum objects themselves. Finally, time can also be considered as an observable – here called *observable time* or *event time.* These three aspects of time in quantum theory will be explained in some more detail.

3.2.1 External Time

The description of every experiment is based on a spatio-temporal coordinatisation of the relevant pieces of equipment. For example, one will specify the relative distances and orientations of particle sources and detectors, as well as control the times at which external fields are switched on and off, or record the times at which a detector fires. Such *external time* measurements are carried out with clocks that are not dynamically connected with the objects studied in the experiment. The resulting data are used to specify parameters in the theoretical model describing the physical system, such as the instant or the duration of its preparation, or the time period between the preparation and the instant at which a measurement of, say, position is performed, or the duration of a certain measurement coupling applied.

External time is sharply defined at all scales relevant to a given experiment. Hence there is no scope for an uncertainty interpretation with respect to external time. However, it has been argued that the duration of an energy measurement limits the accuracy of its outcomes. According to an alternative proposal, the energy of an object is uncertain, or indeterminate, during a period of preparation or measurement, since this involves interactions. These two types of conjectured relations will be scrutinised in Sects. 3.3.1 and 3.3.2.

3.2.2 Intrinsic Time

As a physical magnitude, time is defined and measured in terms of physical systems undergoing changes, such as the straight line motion of a free particle, the periodic circular motion of a clock dial, or the oscillations of atoms in an atomic clock. In accordance with this observation, it can be said that every *dynamical variable* of a physical system marks the passage of time, as well as give an (at least approximate) quantitative measure of the length of the time interval between two events. Hence every non-stationary observable A of a quantum system constitutes its own characteristic time $\tau_\varphi(A)$ within which its mean value changes significantly (φ being any initial state). For example, if $A = Q$, the position of a particle, then $\tau_\varphi(Q)$ could be defined as the time it takes for the bulk of the wave packet associated with a state vector φ to shift by a distance equal to the width of the packet. Or for a projection P, $\tau_\varphi(P)$ could be the length of the greatest time interval for which the probability $\langle \varphi_t | P \varphi_t \rangle \geq 1 - \varepsilon$. Here $\varphi_t = e^{-itH/\hbar}\varphi$ is the state at time t in the Schrödinger picture. Further concrete examples of characteristic times are the time delay in scattering theory, the dwell time in tunnelling, or the lifetime of an unstable state (cf. Chap. 2).

Considering of time as an entity intrinsic to the dynamical behaviour of a physical system entails a variety of time–energy uncertainty relations in which ΔT is given by a characteristic time $\tau_\varphi(A)$ associated with some dynamical variable A. On the other hand, the study of dynamics often involves experimental questions about the time of an event, the time difference between events, or the duration of a process associated with the object system. This raises the quest for the treatment of time as an observable.

3.2.3 Observable Time

A standard experimental question in the study of decaying systems is about the temporal distribution of the decay events over an ensemble. More precisely, rather than the instant of decay one will be measuring the *time of arrival* of the decay products in a detector. A related question is that about the *time of flight* of a particle. Attempts to represent these *time observables* in terms of appropriate operators have been hampered by Pauli's theorem [8] (cf. Chap. 1), according to which the semi-boundedness of any Hamiltonian H precludes the existence of a self-adjoint operator T acting as a generator of a unitary group representation of translations in the energy spectrum. In fact, the covariance relation

$$e^{ihT/\hbar} H e^{-ihT/\hbar} = H + hI \, , \tag{3.2}$$

valid for all $h \in \mathbb{R}$, immediately entails that the spectrum of H should be $\mathbb{R}$. If the covariance was satisfied, it would entail the Heisenberg canonical commutation relation, valid in a dense domain,

$$[H, T] = iI \, , \tag{3.3}$$

so that a shift generator T would be canonically conjugate to the energy, with ensuing observable time–energy uncertainty relation for any state ρ,

$$\Delta_\rho T \, \Delta_\rho H \geq \frac{\hbar}{2} \, . \tag{3.4}$$

In his classic paper on the uncertainty relation, Heisenberg [9] posited a time operator T conjugate to the Hamiltonian H and gave the canonical commutation relation and uncertainty relation, without any comment on the formal or conceptual problematics.

It should be noted, however, that the Heisenberg relation is weaker than the covariance relation; hence it is possible that the former can be satisfied even when the latter cannot. We shall refer to operators conjugate to a given Hamiltonian as *canonical time operators*. For example, for the harmonic oscillator Hamiltonian there do exist self-adjoint canonical time operators T.

In other cases, such as the free particle, symmetric operators have been constructed, which are conjugate to the Hamiltonian, but which are not self-adjoint and do not admit self-adjoint extensions.

No general method seems to exist by which one could decide which Hamiltonians do admit canonical, self-adjoint *time* operators. Moreover, even in cases where such time operators do not exist, there may still be relevant experimental questions about the time of the occurrence of an event. It is therefore appropriate to consider the approach to defining observables in terms of the totality of statistics, i.e., in terms of positive operator-valued measures (POVMs). All standard observables represented as self-adjoint operators are subsumed under this general concept as special cases by virtue of their associated projection-valued spectral measures. The theory of POVMs as representatives of quantum observables and the ensuing measurement theory are developed in [10], including a comprehensive review of relevant literature. In Sect. 3.6 we will consider examples of POVMs describing time observables and elucidate the scope of an uncertainty relation for observable time and Hamiltonian.

In Sect. 3 we will also address the important question of interpretation of time uncertainties. The uncertainty of the decay time has always been quoted as the prime example of the fundamental indeterminacy of the time of occurrence of a quantum event. Yet the question remains as to whether such an indeterminacy interpretation is inevitable or whether the time uncertainty is just a matter of subjective ignorance.

3.3 Relation Between External Time and Energy Spread

One of the earliest proposed versions of the time–energy uncertainty relation $\Delta T \, \Delta E \gtrsim h$ identifies the quantity ΔT not as an *uncertainty* but as the *duration* of the measurement of energy. The quantity ΔE has been interpreted in two ways: either as the range within which an uncontrollable change of the energy of the object must occur due to the measurement (starting with a state in which the energy was more or less well defined) or as the resolution of the measurement of energy. On the latter interpretation, if the energy measurement is repeatable, the energy measurement resolution ΔE is also reflected in the uncertainty of the energy in the outgoing state φ of the object system, i.e., it is approximately equal to the root of the variance of the Hamiltonian, $\Delta H = \left(\langle \varphi | H^2 \varphi \rangle - \langle \varphi | H \varphi \rangle^2 \right)^{1/2}$.

The original arguments were rather informal, and this has given rise to long controversies, leading eventually to precise quantum mechanical models on which a decision could be based. Prominent players in this debate were Bohr, Heisenberg, and Pauli versus Einstein, with their qualitative discussions of *Gedanken* experiments; Landau and Peierls, Fock and Krylov, Aharonov and Bohm, Kraus, Vorontsov, and Stenholm (for a detailed account, cf. [4]).

The conclusion maintained here is that an uncertainty relation between external time duration and energy spread is not universally valid. It may hold for certain types of Hamiltonians, but it turns out wrong in some cases. A counter example was first provided by an energy measurement model due to Aharonov and Bohm [11]. The debate about the validity of this argument suffered from a lack of precise definitions of measurement resolution and reproducibility of outcomes. This difficulty can be overcome by recasting the model in the language of modern measurement theory using positive operator valued-measures. This analysis [4] will be reviewed and elaborated next.

3.3.1 Aharonov–Bohm Energy Measurement Model

We consider a system of two particles in one dimension, one particle being the object and the other serving as a probe for a measurement of momentum. The total Hamiltonian is given by

$$H = \frac{P_x^2}{2m} + \frac{P_y^2}{2M} + Y P_x g(t) \; ,$$

where (X, P_x) and the (Y, P_y) are the position and the momentum observables of the object and probe, respectively, and m, M are their masses. The interaction term produces a coupling between the object momentum P_x to be measured and the momentum P_y of the probe as the read-out observable. The function $g(t)$ serves to specify the duration and the strength of the interaction as follows:

$$g(t) = \begin{cases} g_0 & \text{if } 0 \le t \le \Delta t \; , \\ 0 & \text{otherwise} \; . \end{cases}$$

The Heisenberg equations for the positions and momenta read

$$\dot{X} = \tfrac{1}{m} P_x + Y g(t) \; , \quad \dot{P}_x = 0,$$

$$\dot{Y} = \tfrac{1}{M} P_y \; , \qquad\qquad \dot{P}_y = -P_x g(t) \; .$$

This is solved as follows:

$$P_x = P_x^0 \; , \quad P_y = P_y^0 - P_x^0 \, g_0 \, \Delta t \; , \quad \text{for } t \ge \Delta t \; .$$

The kinetic energy of the object before and after the interaction is given by one and the same operator:

$$H_0 = \frac{m}{2} \dot{X}^2 = \frac{P_x^2}{2m} \; .$$

Thus, the value of kinetic energy H_0 can be obtained by determining the momentum P_x in this measurement. During the interaction period the kinetic energy $\frac{m}{2} \dot{X}^2$ varies but the first moments before and after the measurement are the same. This is an indication of a *reproducible* energy measurement. Following Aharonov and Bohm, one could argue that achieving a given resolution Δp_x requires the change of deflection of the probe $\Delta \left(P_y - P_y^0 \right)$ due to a shift of the value of P_x of magnitude Δp_x to be greater than the initial uncertainty of the probe momentum, ΔP_y^0. This yields the following threshold condition:

$$\Delta p_x \, g_0 \, \Delta t \cong \Delta P_y^0 \, .$$

By making g_0 large enough, "both Δt and Δp_x can be made arbitrarily small for a given ΔP_y^0" [11].

This is the core of Aharonov and Bohm's refutation of the external time–energy uncertainty relation: the energy measurement can be made in an arbitrarily short time and yet be reproducible and arbitrarily accurate.

It is instructive to reformulate the whole argument within the Schrödinger picture, as this will allow us to find the POVMs for momentum and kinetic energy associated with the relevant measurement statistics. The property of reproducibility presupposes a notion of initially relatively sharp values of the measured observable. We take the defining condition for this to be the following: the uncertainty of the final probe momentum is approximately equal to the initial uncertainty. Let $\Phi = \varphi \otimes \phi$ be the total Heisenberg state of the object (φ) plus probe (ϕ). The final probe momentum variance is found to be

$$\left(\Delta_\Phi P_y \right)^2 = \left(\Delta_\phi P_y^0 \right)^2 + g_0^2 \, \Delta t^2 \left(\Delta_\varphi P_x \right)^2 \, .$$

Sharpness of the object momentum corresponds to the last term being negligible.

First we calculate the probability of obtaining a value P_y in an interval S. The corresponding spectral projection will be denoted $E^{P_y}(S)$. The following condition determines the POVM of the measured *unsharp* momentum observable of the object:

$$\langle \Phi_{\Delta t} | I \otimes E^{P_y}(S) \, \Phi_{\Delta t} \rangle = \langle \varphi | A(S) \, \varphi \rangle \quad \text{for all } \varphi \, ,$$

where $\Phi_{\Delta t} = \exp\left(-i\Delta t \, H/\hbar \right) \varphi \otimes \phi$ is the total state immediately after the interaction period, i.e.,

$$\Phi_{\Delta t} \left(p_x, p_y \right) = e^{-ip_x^2 \, \Delta t/2m\hbar - i\gamma(p_x, p_y, \Delta t)/\hbar} \, \varphi \left(p_x \right) \, \phi \left(p_y + p_x g_0 \Delta t \right) \, ,$$

$$\gamma \left(p_x, p_y, \Delta t \right) = \frac{1}{6M} \, p_x^2 \, g_0 \, \Delta t^3 + \frac{1}{2M} \, p_x \, p_y \, g_0 \, \Delta t^2 + \frac{1}{2M} \, p_y^2 \, \Delta t \, .$$

One obtains:

$$A(S) = E_f^{P_x} \left(-\frac{S}{g_0 \Delta t} \right) \, ,$$

which is an *unsharp momentum* observable ($*$ denoting convolution),

$$E_f^{P_x}(R) = \chi_R * f(P_x) = \int_{\mathbb{R}} dp\, f(p)\, E^{P_x}(R+p) \; , \tag{3.5}$$

$$f(p) = g_0\, \Delta t\, |\phi(p\, g_0\, \Delta t)|^2 \; . \tag{3.6}$$

Due to the properties of the convolution it is straightforward to verify that these positive operators form a POVM, i.e., (countable) additivity over disjoint sets and normalisation $E_f^{P_x}(\mathbb{R}) = I$ are satisfied. It is thus seen that the resolution of the measurement, described by the confidence distribution f, is determined by the initial probe state as well as the interaction parameter $g_0 \Delta t$. In fact, a measure of the inaccuracy is given by the width of the distribution f, which can be characterised (for suitable probe states ϕ) by the variance:

$$(\Delta p_x)^2 = \mathrm{Var}_f(p) = \left(\frac{1}{g_0 \Delta t}\right)^2 \mathrm{Var}_\phi(P_y) \; . \tag{3.7}$$

It is clear that increasing the parameter $g_0 \Delta t$ leads to a more and more sharply peaked function f. This is to say that the inaccuracy of the momentum measurement, given by the width Δf of f, can be arbitrarily increased for any fixed value of the duration Δt. The same will be seen to be true for the inaccuracy of the measured values of energy inferred from this momentum measurement. This disproves the *inaccuracy* version of the external time–energy uncertainty relation where ΔE is taken to be the energy measurement inaccuracy.

In order to assess the reproducibility properties of the measurement, we need to investigate the state change of the object due to the measurement. The final object state ρ_R conditional upon an outcome p_x in R is determined via the following relation: for all states φ and all object operators a,

$$\mathrm{tr}\,[a\, \rho_R] = \langle \Phi_{\Delta t}|a \otimes E^{P_y}(-R g_0 \Delta t)\, \Phi_{\Delta t}\rangle \; .$$

One obtains:

$$\rho_R = \int_R dp'_x\, A_{p'_x}\, |\varphi\rangle\langle\varphi|\, A_{p'_x}^*$$

where the operators $A_{p'_x}$ act as

$$(A_{p'_x}\varphi)(p_x) = (g_0 \Delta t)^{1/2}\, e^{-ip_x^2 \Delta t/2m\hbar}\, e^{-i\gamma(p_x, -p'_x g_0 \Delta t, \Delta t)/\hbar} \times$$
$$\times \phi((p_x - p'_x)\, g_0 \Delta t)\, \varphi(p_x) \; .$$

The momentum distribution is (up to normalisation):

$$\langle p_x|\rho_R|p_x\rangle = \int_R dp'_x\, |(A_{p'_x}\varphi)(p_x)|^2 = \chi_X * f(p_x)\, |\varphi(p_x)|^2 \; .$$

If $|\varphi(p_x)|^2$ is sharply peaked at p_x^0, in the sense that

$$|\phi((p_x - p'_x)\, g_0 \Delta t)|^2\, |\varphi(p_x)|^2 \cong |\phi((p_x^0 - p'_x)\, g_0 \Delta t)|^2\, |\varphi(p_x)|^2 \; ,$$

then one has

$$\langle p_x | \rho_R | p_x \rangle \cong \chi_R * f \left(p_x^0 \right) \, |\varphi \left(p_x \right)|^2 \ . \tag{3.8}$$

Hence if φ is such a *near-eigenstate* of P_x, then the conditional final state has practically the same sharply peaked momentum distribution. In other words, the present model practically preserves near-eigenstates. It follows indeed that the measurement allows one to determine the kinetic energy with negligible disturbance of any pre-existing (approximately sharp) value. Thus the *disturbance* version of the purported external time–energy uncertainty relation is ruled out.

We show next in which sense the above momentum measurement scheme serves as a measurement of kinetic energy. In fact the relation $H_0 = P_x^2/2m$ translates into the following functional relationship between the spectral measures of H_0 and P_x: we have

$$H_0 = \frac{P_x^2}{2m} = \int_{-\infty}^{+\infty} \frac{p^2}{2m} E^{P_x} \left(dp \right) = \int_0^{+\infty} e \, E^{H_0} \left(de \right) \ ,$$

and so

$$E^{H_0} \left(Z \right) = E^{P_x} \left(h^{-1} \left(Z \right) \right), \quad Z \subseteq \mathbb{R}^+, \quad h \left(p \right) = \frac{p^2}{2m} \ .$$

This suggests that in the above unsharp momentum measurement, one should record such subsets R of the momentum spectrum, which are images of some $Z \subseteq \mathbb{R}^+$ under the map h^{-1}. This leads to the following positive operators that constitute a POVM on $\mathbb{R}^+$:

$$E_f^{H_0} \left(Z \right) := E_f^{P_x} \left(h^{-1} \left(Z \right) \right) = \int_R f \left(p \right) E^{P_x} \left(h^{-1} \left(Z \right) + p \right) dp \ .$$

Let us assume the confidence function f is inversion symmetric, $f \left(-p \right) = f \left(p \right)$. Then, since the set $h^{-1} \left(Z \right)$ is inversion symmetric, the convolution $\chi_{h^{-1}(Z)} * f$ also shares this property. Hence the positive operators $E_f^{H_0} \left(Z \right)$ are actually functions of H_0 and constitute a smearing of the spectral measure of H_0:

$$E_f^{H_0} \left(Z \right) = \chi_{h^{-1}(Z)} * f \left(P_x \right) = \chi_{h^{-1}(Z)} * f \left((2mH_0)^{1/2} \right)$$

$$= \int_Z \left(\frac{m}{2e} \right)^{1/2} f \left((2mH_0)^{1/2} - (2me)^{1/2} \right) de \ . \tag{3.9}$$

This is a corroboration of the fact that the unsharp momentum measurement constitutes an unsharp measurement of energy. The expected readings and their variances are obtained as follows:

$$\langle p^n \rangle_f = \int_{\mathbb{R}} p^n f \left(p \right) dp = \left(\frac{1}{g_0 \Delta t} \right)^n \langle P_y^n \rangle_\phi \ ,$$

then

$$\langle H_0 \rangle_{\varphi,f} = \langle \varphi | \int_0^\infty e \, E_f^{H_0} (de) \, \varphi \rangle = \langle H_0 \rangle_\varphi + \left(\frac{1}{g_0 \Delta t} \right)^2 \left\langle \frac{P_y^2}{2m} \right\rangle_\phi , \qquad (3.10)$$

and

$$\mathrm{Var}_{\varphi,f} (H_0) = \langle H_0^2 \rangle_{\varphi,f} - \left(\langle H_0 \rangle_{\varphi,f} \right)^2$$
$$= \mathrm{Var}_\varphi (H_0) + \left(\frac{1}{g_0 \Delta t} \right)^4 \mathrm{Var}_\phi \left(\frac{P_y^2}{2m} \right) + 4 \left(\frac{1}{g_0 \Delta t} \right)^2 \langle H_0 \rangle_\varphi \left\langle \frac{P_y^2}{2m} \right\rangle_\phi .$$
$$(3.11)$$

There is a distortion of the expected values towards slightly larger values, and the energy measurement inaccuracy is measured by the last two terms in the last equation. Both the distortion and the accuracy can be made arbitrarily small by choosing a suitably large coupling parameter g_0, although it must be noted that the inaccuracy depends on the value of the object energy.

We conclude, therefore, in agreement with Aharonov and Bohm, that a reproducible energy measurement is possible with arbitrary accuracy and in arbitrarily short time.

However, very recently Aharonov and Reznik [12] have taken up the issue again, considering this time–energy measurements carried out from *within* the system. In this situation the conclusion is that due to a back reaction of the energy measurement on the internal clock, an accuracy δE requires the duration τ_0, measured internally, to be limited by the uncertainty relation

$$\tau_0 \, \delta E \geq \hbar . \qquad (3.12)$$

What is actually shown in the analysis of [12] is that the clock rate is uncertain and hence the duration has an *uncertainty* $\Delta \tau_0 \geq \hbar / \delta E$. This conclusion is in accordance with the quantum clock uncertainty relation, which will be presented in Sect. 3.5.

3.3.2 Relation Between Preparation Time and Energy

An uncertainty relation for the indeterminacy of the energy of a system and the duration of an external perturbation has been proposed and accepted as valid even by opponents to the external time–energy relation (cf. the review of Bauer and Mello [2]). The duration of the perturbation is defined *dynamically* as the approximate time period during which the interaction energy is non-negligible. Hence this type of time–energy uncertainty relation is best classified as one associated with dynamical time, although in a measurement context the duration of interaction is fixed with reference to a laboratory clock. A particular instance of this type of uncertainty relation occurs in the preparation of a quantum system: the interaction with the preparation devices can be regarded as an external perturbation so that one may note

$$T_{\mathrm{prep}} \Delta E \gtrsim \hbar \,, \tag{3.13}$$

where T_{prep} denotes the duration of the preparation (perturbation) and ΔE is some suitable measure of the width of the energy distribution, such as those introduced in Sect. 3.4.

This preparation time relation has been deduced by Moshinsky [13] in an exactly soluble potential model of the preparation of a particle by means of a slit with a shutter that is opened during a time interval T_{prep}. This time period determines the width of the Bohr–Wigner time of passage distribution (cf. Section 3.4.3, (3.26)), whereas the energy uncertainty ΔE is given by the width of the energy distribution of the outgoing particle, given a sharp initial energy E_0:

$$\mathfrak{p}\left(E : E_0, T_{\mathrm{prep}}\right) \propto E^{1/2} \frac{\sin^2\left((E - E_0)\, T/2\hbar\right)}{(E - E_0)^2} \,.$$

Similar distributions are known to arise for the short-time energy distribution of a decaying state as well as in first-order perturbation theory. We conclude that it is impossible to simultaneously prepare a sharp energy and a sharp time of passage. This is an indication of the complementarity of event time and energy.

A relation of the form (3.13) was derived in a somewhat different context by Partovi and Blankenbecler [14]; they showed that the most likely state compatible with the probability distributions of the position of a free particle measured at two times with separation T has an energy dispersion that must satisfy (3.13). These authors interpret the time interval T between the two measurement as the duration of a multi-time measurement whose aim it is to estimate the state that gives rise to the statistical data obtained.

3.4 Relations Involving Intrinsic Time

In this section we review different ways of quantifying measures of times that are intrinsic to the system and its evolution.

3.4.1 Mandelstam–Tamm Relation

A wide class of measures of intrinsic times has been provided by Mandelstam and Tamm [15]. An elegant formulation of the ensuing universal *dynamical* or *intrinsic* time–energy uncertainty relations was given in the textbook of Messiah. Let A be a non-stationary observable. Combining the Heisenberg equation of motion for A,

$$i\hbar \frac{dA}{dt} = AH - HA \,, \tag{3.14}$$

with the general uncertainty relation,

$$\Delta_\rho A \, \Delta_\rho H \geq \frac{1}{2} |\langle AH - HA \rangle_\rho| \; , \tag{3.15}$$

and introducing the characteristic time

$$\tau_\rho (A) = \frac{\Delta_\rho A}{\left| \frac{d}{dt} \langle A \rangle_\rho \right|} \tag{3.16}$$

(whenever the denominator is non-zero), one obtains the inequality

$$\tau_\rho (A) \, \Delta_\rho H \geq \frac{1}{2} \hbar \; . \tag{3.17}$$

Here we have used the notation $\langle X \rangle_\rho = \mathrm{tr}\,[\rho X]$, $(\Delta_\rho X)^2 = \langle X^2 \rangle_\rho - \langle X \rangle_\rho^2$.

As an illustration we consider the case of a free particle. Let $A = Q$ be the particle position and let ρ be a pure state represented by a unit vector φ. Assume the momentum P is fairly sharply defined in that state, i.e., $\Delta_\varphi P \ll |\langle P \rangle_\varphi|$. Now the time derivative of position is the velocity, $d\langle Q \rangle_\varphi / dt = \langle P \rangle_\varphi / m = \langle V \rangle_\varphi$, so we have

$$\tau_\varphi (Q) = \frac{\Delta_\varphi Q}{|\langle V \rangle_\varphi|} \; . \tag{3.18}$$

From the Schrödinger equation for a free particle we have

$$(\Delta_\varphi Q)^2 = (\Delta_\varphi Q\,(0))^2 + (\Delta_\varphi V)^2 \, t^2 + \{\langle Q\,(0)\,V + VQ\,(0)\rangle_\varphi - 2\langle V \rangle_\varphi \langle Q\,(0)\rangle_\varphi\} t \; .$$

Using the uncertainty relation in the general form

$$(\Delta_\varphi Q)^2 \, (\Delta_\varphi P)^2 \geq \frac{1}{4} |\langle Q\,(0)\,P - PQ\,(0)\rangle_\varphi|^2$$

$$+ \frac{1}{4} \{\langle Q\,(0)\,V + VQ\,(0)\rangle_\varphi - 2\langle V \rangle_\varphi \langle Q\,(0)\rangle_\varphi\}^2 \; ,$$

we find the estimate

$$(\Delta_\varphi Q)^2 \leq (\Delta_\varphi Q\,(0) + t\,\Delta_\varphi V)^2 \; .$$

Putting $t = \tau_\varphi (Q)$, this gives

$$\Delta_\varphi Q \leq \Delta_\varphi Q\,(0) \left[1 + \frac{\Delta_\varphi P}{|\langle P \rangle_\varphi|} \right]^{1/2} \cong \Delta_\varphi Q\,(0) \; .$$

This estimate follows from the assumption of small variance for P, and this corresponds to the limiting case of slow wave packet spreading. Thus the characteristic time $\tau_\varphi (Q)$ is indeed seen to be the period of time it takes the wave packet to propagate by a distance equal to its width. It can also be said that this is the approximate time for the packet to pass a fixed point in space. Insofar as the position of the particle is indeterminate within approximately $\Delta_\varphi Q\,(0)$ one may be tempted to interpret this characteristic time as the indeterminacy of the time of passage. The event 'particle passes a point x_0' has an appreciable probability only within a period of duration $\tau_\varphi (Q)$.

3.4.2 Lifetime of a Property

Let P be a projection, $U_t = \exp(-itH/\hbar)$, ψ_0 be a unit vector representing the state of a quantum system. We consider the function

$$\mathfrak{p}(t) = \langle \psi_0 | U_t^{-1} P U_t \psi_0 \rangle. \tag{3.19}$$

The Mandelstam–Tamm relation yields

$$\left| \frac{d\mathfrak{p}}{dt} \right| \leq \frac{2}{\hbar} \Delta_{\psi_0} H \left[\mathfrak{p}(1-\mathfrak{p}) \right]^{1/2}.$$

Integration of this inequality with the initial condition $\mathfrak{p}(0) = 1$ yields

$$\mathfrak{p}(t) \geq \cos^2\left(t\,\Delta_{\psi_0} H/\hbar\right), \quad 0 \leq t \leq \frac{\pi}{2} \frac{\hbar}{\Delta_{\psi_0} H} \equiv t_0. \tag{3.20}$$

The initial condition means that the property P was actually in the state ψ_0 at time $t = 0$. One may define the lifetime τ_P of the property P by means of the condition $\mathfrak{p}(\tau_P) = \frac{1}{2}$. Hence one obtains the uncertainty relation

$$\tau_P \Delta_{\psi_0} H \geq \frac{\pi \hbar}{4}. \tag{3.21}$$

This relation was derived by Mandelstam and Tamm for the special case of $P = |\psi_0\rangle\langle\psi_0|$.

There are alternative approaches of defining the lifetime of a state and obtaining an energy–time uncertainty relation for the lifetime. For example, Grabowski [16] defines

$$\tau_0 = \int_0^\infty \mathfrak{p}(t)\, dt, \tag{3.22}$$

which yields

$$\tau_0 \Delta_{\psi_0} H \geq \frac{\hbar}{2}, \tag{3.23}$$

provided the Hamiltonian has no singular continuous spectrum.

The variance of H may be infinite in many situations, so that the above relations are of limited use. We will review below a variety of approaches based on alternative measures of the width of the energy distribution in a state ψ_0.

3.4.3 Bohr–Wigner Uncertainty Relation

Fourier analysis gives 'uncertainty' relations for any wave propagation phenomenon in that it gives a reciprocal relationship between the widths of the spatial/temporal wave pattern on one hand and the wave number/frequency distributions on the other. On the basis of this classical wave analogy, Bohr [17] proposed a time–energy uncertainty relation, which appeared to assume the same status as the corresponding position/momentum relation,

$$\Delta t \,\Delta E \gtrsim h, \quad \Delta x \,\Delta p \gtrsim h \,. \tag{3.24}$$

Hilgevoord [5] presents a careful discussion of the sense in which a treatment of time and energy variables on equal footing to position and momentum variables is justified.

A more formal approach in this spirit was pursued by Wigner [18], who considered a positive temporal distribution function associated with the wave function ψ of a particle:

$$\mathfrak{p}_{x_0}(t) = |f(t)|^2, \quad f(t) = \psi(x_0, t) \,. \tag{3.25}$$

In the limit $\Delta_\psi P \ll |\langle P\rangle_\psi|$, the width of this distribution is of the order of $\tau_\psi(Q)$. The quantity ΔE measures the width of the Fourier transform $\tilde{f}$ of f. This method can be extended to other types of characteristic times. Define

$$f(t) = \langle\varphi|\psi_t\rangle, \quad \tilde{f}(E) = (2\pi)^{-1} \int_{-\infty}^{\infty} f(t)\, e^{itE/\hbar}\, dt \,, \tag{3.26}$$

and the moments (providing the denominators are finite)

$$\langle t^n\rangle_f = \frac{\int_0^\infty |f(t)|^2\, t^n\, dt}{\int_0^\infty |f(t)|^2\, dt}, \quad \langle E^n\rangle_{\tilde{f}} = \frac{\int_0^\infty \left|\tilde{f}(E)\right|^2 E^n\, dE}{\int_0^\infty \left|\tilde{f}(E)\right|^2\, dE}. \tag{3.27}$$

The previous case considered by Bohr is formally included by replacing $|\varphi\rangle$ with $|x_0\rangle$, an improper position eigenstate. One obtains an uncertainty relation for the variances $(\Delta_f t)^2 = \langle t^2\rangle_f - \langle t\rangle_f^2$, $\left(\Delta_{\tilde{f}} E\right)^2 = \langle E^2\rangle_{\tilde{f}} - \langle E\rangle_{\tilde{f}}^2$:

$$\Delta_f t \,\Delta_{\tilde{f}} E \geq \frac{\hbar}{2} \,. \tag{3.28}$$

It must be noted that neither of the distributions $|f(t)|^2$ and $\left|\tilde{f}(E)\right|^2$ is normalised, nor will they always be normalisable. Moreover, their operational meaning is not immediately obvious. The following is a possible, albeit indirect, way of associating these distributions with physical measurements.

Assume the state ψ is prepared at time $t = 0$, and that at time $t > 0$ a repeatable measurement of energy is made and found to give a value in a small interval Z of width δE and centre E_0, after which a measurement of the property $P_\varphi = |\varphi\rangle\langle\varphi|$ is made. We calculate the probability for this sequence of events, under the assumption that H has a non-degenerate spectrum with improper eigenstates $|E\rangle$:

$$\begin{aligned}
\mathfrak{p} &= \mathfrak{p}_\psi\left(E^H(Z), P_\varphi\right) = \mathrm{tr}\left[P_\varphi\, E^H(Z)\, e^{-itH/\hbar}\, |\psi\rangle\langle\psi|\, e^{itH/\hbar}\, E^H(Z)\, P_\varphi\right] \\
&= \int_Z dE \int_Z dE'\, \langle\varphi|E\rangle\, \langle E'|\varphi\rangle\, \langle\psi|E'\rangle\, \langle E|\psi\rangle\, e^{-it(E-E')/\hbar}.
\end{aligned} \tag{3.29}$$

Assuming that Z is sufficiently small so that the functions $\langle E|\psi\rangle$ and $\langle E|\varphi\rangle$ are practically constant within Z, we have:

$$\mathfrak{p} \cong |\langle E_0|\varphi\rangle|^2 \, |\langle E_0|\psi\rangle|^2 \, (\delta E)^2 \cong \left|\tilde{f}\,(E_0)\right|^2 (\delta E)^2 \,. \tag{3.30}$$

As an illustration we reproduce the standard formulas for the exponential decay law. This is known to hold in an intermediate time range, while deviations must occur for short as well as long times, see Sects. 2.5.2 and 2.5.3. The Mandelstam–Tamm relation for the lifetime of a property already indicates that the short-time behaviour of the survival probability is a power law $1 - \mathfrak{p} \propto t^2$.

For H with non-degenerate spectrum, one has

$$f\,(t) = \langle\psi_0|\psi_t\rangle = \int_{-\infty}^{\infty} e^{-itE/\hbar}\,\tilde{f}\,(E)\,dE$$

$$\cong \exp\left(-\,|t|\,(\Gamma/2\hbar) - itE_0/\hbar\right), \tag{3.31}$$

$$\tilde{f}\,(E) = |\langle E|\psi_0\rangle|^2 \cong \frac{1}{\pi}\,\frac{\Gamma/2}{(E - E_0)^2 + (\Gamma/2)^2}\,. \tag{3.32}$$

The Lorentzian distribution $\tilde{f}\,(E)$ has no finite variance, hence as an alternative measure of the energy spread one usually takes the full width at half-height, $\delta E = \Gamma$. The lifetime τ of the state ψ_0 is defined via

$$\mathfrak{p}\,(\tau) = e^{-\tau\Gamma/\hbar} = 1/e\,, \tag{3.33}$$

so that one obtains the famous lifetime–linewidth relation

$$\tau\,\Gamma = \hbar\,. \tag{3.34}$$

One can also use the Wigner measures that are

$$\Delta_f t = \frac{\hbar}{2\Gamma} = \sqrt{2\tau}\,, \quad \Delta_{\tilde{f}}E = \Gamma/2\,. \tag{3.35}$$

It must be noted that here the relevant distribution is $\left|\tilde{f}\,(E)\right|^2 = |\langle E|\psi_0\rangle|^4$. Hence we have

$$\Delta_f t\,\Delta_{\tilde{f}}E = \frac{\sqrt{2}}{2}\,\hbar\,. \tag{3.36}$$

A novel application of a Wigner-type uncertainty relation has been proposed recently [19], which identifies $\tilde{f}(E)$ as the energy amplitude of a state, in which case the associated $|f\,(t)|^2$ is found to coincide with the time-of-arrival distribution due to Kijowski [20].

Another approach to defining a formal probability distribution for time based on the statistics of measurements of a time dependent observable A was attempted by Partovi and Blankenbecler [14]. This approach presupposes that the time dependence of the expectation $A(t) := \text{tr}[\rho(t)A]$ is strictly monotonic. It seems that the scheme of a proof of a time–energy uncertainty relation for the dispersion of the ensuing time distribution provided in [14] gives tangible results essentially when the (self-adjoint) operator A satisfies the canonical commutation relation with the Hamiltonian, which is known to be possible only in very special cases.

3.4.4 Further Relations Involving Intrinsic Time

In more realistic models of decaying systems, the measures of spread introduced in Sect. 3.4.3 turn out inadequate. Bauer and Mello [2] have studied alternative measures with a wider scope of applications. For example, they define a concept of *equivalent width*, given by

$$W(\phi) = (\phi(x_0))^{-1} \int_{-\infty}^{\infty} \phi(x)\, dx \tag{3.37}$$

whenever the right-hand side is well defined. They then prove that the following relation holds:

$$W(\phi)\, W\left(\tilde{\phi}\right) = 2\pi\hbar . \tag{3.38}$$

In the case of a decaying state,

$$\tilde{\phi}(E) = \left|\tilde{f}(E)\right|^2 = |\langle E|\psi_0\rangle|^4 ,$$

so that the inverse Fourier transform turns out to be the autocorrelation function of f:

$$\phi(t) = \int_{-\infty}^{\infty} e^{-itE/\hbar}\tilde{\phi}(E)\, dE = \frac{1}{2\pi\hbar} \int_{-\infty}^{\infty} \overline{f}(t')\, f(t+t')\, dt' = \overline{f} \times f(t) .$$

On proving the inequality $\left|W\left(\overline{f} \times f\right)\right| \leq W(|f| \times |f|)$, one obtains a time–energy uncertainty relation for equivalent widths:

$$W(|f| \times |f|)\, W\left(\left|\tilde{f}\right|^2\right) \geq 2\pi\hbar . \tag{3.39}$$

If the exponential decay formulas are inserted and the constant $x_0 = t = 0$ (for f), and $E = E_0$ (for $\tilde{f}$), then one obtains equality in the above relation.

It is interesting to observe that the autocorrelation function describes coherence in time. This is a useful measure of the *fine structure* of the temporal distribution function $\mathfrak{p}(t) = |f(t)|^2$.

A different approach to describing width and fine structure was taken by Hilgevoord and Uffink (cf. the review of Hilgevoord [5, 6]), who adopted the concepts of *overall width* and *translation width* from the theory of signal analysis as follows. Let χ be a square-integrable function, normalised to unity, and $\tilde{\chi}$ its Fourier transform. The *overall width* $\mathfrak{W}\left(|\chi|^2, \alpha\right)$ of the distribution $|\chi|^2$ is defined as the width of the smallest time interval $\mathfrak{f}$ such that

$$\int_{\mathfrak{f}} |\chi(t)|^2 \, dt = \alpha .$$

Then the following relation holds:

$$\mathfrak{W}\left(|\chi|^2, \alpha\right) \mathfrak{W}\left(|\tilde{\chi}|^2, \alpha\right) \geq C(\alpha) , \quad \text{for} \quad \alpha > \frac{1}{2} , \qquad (3.40)$$

with a constant $C(\alpha)$ independent of χ. This yields an energy–time uncertainty relation in the spirit of the Wigner relation (3.28) if we put $\chi(t) = f(t) = \langle \varphi | \psi_t \rangle$, $\tilde{\chi}(E) = \tilde{f}(E)$; in the case of $\varphi = \psi_0$ and H having a non-degenerate spectrum, then $\tilde{f}(E) = |\langle E | \psi_0 \rangle|^2$.

For the analysis of interference experiments, a relation between the overall width of the energy distribution and the translation width of the temporal distribution has proved enormously useful. The *translation width* $\mathfrak{w}(f, \rho)$ is defined as the smallest number t for which

$$|f(t)| = |\langle \psi_0 | \psi_t \rangle| = 1 - \rho .$$

Then observing that $\tilde{f}(E) = |\langle E | \psi_0 \rangle|^2$, Hilgevoord and Uffink [21] show:

$$\mathfrak{w}(f, \rho) \, \mathfrak{W}\left(\tilde{f}, \alpha\right) \geq 2\hbar \arccos\left(\frac{2 - \alpha - \rho}{\alpha}\right) , \quad \text{for} \quad \rho \geq 2(1 - \alpha) . \qquad (3.41)$$

The lifetime–linewidth relation is recovered for any decaying state by putting $T_{1/2} = \mathfrak{w}\left(f, \sqrt{1/2}\right)$, $\alpha = 0.9$, which yields [6]

$$T_{1/2} \, \mathfrak{W}\left(\tilde{f}, 0.9\right) \geq 0.9\hbar . \qquad (3.42)$$

An interesting connection between the Mandelstam–Tamm relation and the Hilgevoord–Uffink relation is pointed out in [22].

With this example we conclude our survey of intrinsic time–energy relations, without any claim to completeness. For example, a number of rigorous results on the rate with which an evolving state 'passes through' a reference subspace are reported by Pfeifer and Frohlich [23, 7]. We also recommend the recent reviews of Hilgevoord [5, 6] as a lucid didactic account demonstrating the importance of the translation width–overall width uncertainty relation in substantiating Bohr's rebuttal of Einstein's attempts to achieve simultaneous sharp determinations of complementary quantities.

3.5 Quantum Clock

The constituents of real rods and clocks and other measuring devices are elementary particles, atoms and molecules, which are subject to the laws of quantum mechanics. Hence it is natural to investigate the effect of the quantum nature of measuring instruments. This thought has played a leading role in the early debates between Einstein and the other founders of quantum mechanics. By taking into account quantum features of the experimental setup, Bohr was able to refute Einstein's *Gedanken* experiments that were aimed at beating quantum limitations of joint measurements of position and momentum, or time and energy. Later Wigner exhibited limitations of space–time measurements due to the quantum nature of test particles, and it was in this context that he introduced the idea of a *quantum clock* [24, 25], see Chap. 8.

The issue of quantum clocks belongs, in a sense, to the realm of the theory of time measurements: time is being measured by means of observing the dynamical behaviour of a quantum system. However, the ensuing uncertainty relations are clearly of the intrinsic-time type, and the theory of quantum clocks is actually based on the theory of repeated measurements, or monitoring, of a non-stationary quantum-nondemolition variable. By contrast, time as an observable is recorded in experiments in which typically a detector waits to be triggered by the occurrence of some *event*, such as a particle hitting a scintillation screen. The latter type of *event time* measurement will be discussed in Sect. 3.6.

The Salecker–Wigner quantum clock has experienced renewed interest in recent years in three areas of research: investigations on the detectability of the quantum nature of space–time on length scales far larger than the Planck length (e.g. [26, 27]); studies of tunnelling times (e.g. [28]) and superluminal photon propagation through evanescent media [29]; and quantum information approaches for optimising quantum clock resolution [30] and synchronisation via non-local entangled systems [31]. All of these questions and proposals are subject to ongoing controversial scrutiny, so that it is too early to attempt an assessment. Instead we will be content with a brief outline of the principal features of a quantum clock and explain the relevance of the intrinsic time–energy uncertainty relation in this context.

A quantum clock is characterised as a system that, in the course of its time evolution, passes through a sequence of distinguishable states $\psi_1, \psi_2, \ldots$ at (laboratory) times $t_1, t_2, \ldots$. In order to be distinguishable as clock pointer positions, neighbouring states ψ_k, ψ_{k+1} must be (at least nearly) orthogonal. Under this assumption, the time resolution defined by this system is $\delta t = t_{k+1} - t_k$. It is known that a non-stationary state that runs through n orthogonal states in a period T must be a superposition of at least n energy eigenstates. For a harmonic oscillator with frequency ω and period $T = 2\pi/\omega$, the state $\psi_1 = \sum_{k=1}^{n} \varphi_k/\sqrt{n}$ will turn into ψ_2 perpendicular to ψ_1 if $\delta t = T/n$. It follows that the mean energy must be of the order $\hbar/\delta t = \hbar n/T$.

If one considers the mean position of a wave packet as the clock pointer, then according to the relevant Mandelstam–Tamm relation and the constraint $\delta t > \tau_{\psi_1}(Q)$ on the resolution, one obtains

$$\delta t \geq \frac{\hbar}{2\Delta_{\psi_1} H} \; .$$

These examples illustrate the fact that the rate of change of a property of the system decreases with increasing sharpness of the prepared energy. In the limit of an energy eigenstate, all quantities will have time-independent distributions and expectation values, hence *nothing happens*.

Another requirement to be imposed on a system to ensure its functioning as a quantum clock is that its pointer can be read in a non-disturbing way. This can be achieved for suitable families of pointer states, such as coherent states for the harmonic oscillator, which admit non-demolition measurements. The relevant theory of quantum-nondemolition measurements for continuous variables is developed in [32].

The quantum clock time–energy uncertainty relation can be derived in a very general way from the intrinsic-time uncertainty relations reviewed above. In order to achieve a time resolution δt, pairs of successive pointer states $\psi_1 = \psi_t$, $\psi_2 = \psi_{t+\delta t}$ need to be orthogonal: $\mathfrak{p}(\delta t) = |\langle \psi_t | \psi_{t+\delta t} \rangle|^2 = 0$. The relation (3.20) implies

$$\delta t \geq t_0 = \frac{\pi\hbar}{2\Delta_{\psi_1} H} \; . \tag{3.43}$$

As noted before, the variance is not always a good measure of the width of the energy distribution. A more stringent condition on the clock resolution can be obtained by application of the Hilgevoord–Uffink relation (3.41) between temporal translation width and overall energy width. If the clock is a periodic system, the resolution δt is given by the period divided by the number of pairwise orthogonal states, $\delta t = T/n$. This entails that the state ψ_1 has to have a translation width of the order of at most δt. Hence (3.41) yields

$$\delta t \geq \mathfrak{w}(f,\rho) \geq \frac{2\hbar \arccos\left((2-\alpha-\rho)/\alpha\right)}{\mathfrak{W}\left(\tilde{f},\alpha\right)} \; .$$

For a quantum clock, ρ should be close to unity. Taking $\rho = 1$ requires $\alpha \geq \frac{1}{2}$, and we have

$$\delta t \geq \frac{2\hbar \arccos\left((1-\alpha)/\alpha\right)}{\mathfrak{W}\left(\tilde{f},\alpha\right)} \equiv \hbar C(\alpha) , \quad \frac{1}{2} \leq \alpha \leq 1 \; .$$

Since both the enumerator and the denominator are increasing functions of α, and since the quotient $C(\alpha)$ is 0 both at $\alpha = \frac{1}{2}$ (as $\arccos 1 = 0$) and at $\alpha = 1$ (as $\mathfrak{W}\left(\tilde{f}, 1\right) = \infty$), it follows that there must be a value α_0 where $C(\alpha)$ is maximal. The inequality for the clock resolution must still hold at this point:

$$\delta t \geq \hbar C\left(\alpha_0\right) .\tag{3.44}$$

A universal quantum clock uncertainty relation in this spirit was proposed by this author [3] and independently by Hilgevoord and Uffink [33].

3.6 Relations Based on Time Observables

Let us recall the motivation for considering time as a quantum observable. First, there do exist a variety of experiments in which times of events are recorded, where these events occur at randomly distributed instants as monitored by means of laboratory clocks. The appropriate mathematical tool for the representation of these temporal statistics is that of a POVM over the time domain, which will be explained in Sect. 3.6.1. As an illustration of intrinsic time preparation and measurement inaccuracies, we will briefly review the famous Einstein photon box experiment in Sect. 3.6.2 . Secondly, having acknowledged the possible role of time as a random variable, the next question that arises concerns the nature of the randomness: for example, is the instant of decay of an unstable particle *truly* indeterminate, as would be appropriate to a quantum observable, or is it determined by some possibly hidden mechanism, albeit unpredictable? We shall argue in Sect. 3.6.3 that an indeterminacy interpretation is appropriate in the light of temporal interference experiments.

3.6.1 Event Time Observables

A measurement of an ordinary quantum observable is typically devised so as to provide an outcome at a specified instant of time. Often one aims at achieving the *impulsive measurement* limit where the duration of the interaction between object and probe is negligible, so that it makes sense to speak of an (approximate) instant of the measurement.

By contrast, event time measurements are extended in time, with sensitive detectors waiting to be triggered. The experimenter has no control over the time instant at which the detectors fire. This very instant constitutes the outcome of such a measurement.

Wigner [18] epitomises the distinction between these two types of measurements in terms of the localisation of particles. The first type of measurement amounts to measuring the position at a particular time. This will answer the question: *'Where is the particle – now?'* The second type of measurement corresponds to a determination of the instant of time at which the particle passes a particular point in space, thus answering the question: *'When is the particle – here?'*

Following [3], we explain the term *event* to refer to the (approximate) actuality of a property, in the sense that the probability for this property to occur is equal to (or close to) unity. The event to be observed in the above time of passage experiment is the approximate localisation of the particle at the given space point. We note that the Mandelstam–Tamm parameter $\tau_\rho(Q)$ seems to give an indication of the *indeterminacy* of the time of passage, owing to the indeterminacy of position in the state ρ.

With the exception of the photodetection theory [34, 35], a theory of *event time* measurements is very much in its initial stages. In the 1990s, interest in the theory of *time-of-arrival* measurements has grown significantly and ensuing results are reviewed in other chapters of this book. Here we focus on the formal representation of event time observables in terms of POVMs.

Suppose a detection experiment is repeated many times until a sufficiently large statistical distribution of times is obtained. A quantum mechanical account of the statistics will have to provide probabilities for the event times to lie within intervals Z of the time domain. Such probabilities should be expressed as expectation values of operators associated with each set Z, i.e., $p_\rho(Z) = \mathrm{tr}\,[\rho\,F_{Z_0}(Z)]$. These probabilities should be approximately equal to the observed frequencies. Here Z_0 denotes an interval that represents the time domain specified in the experiment in question. If the measurement can be thought of as being extended from the infinite past to the infinite future, one would have $Z_0 = \mathbb{R}$.

Due to the positivity of the numbers $p_\rho(Z)$ for all states ρ, the operators $F_{Z_0}(Z)$ will be positive. Similarly, since $p_\rho(Z) \leq 1$, we have $F_{Z_0}(Z) \leq I$. Finally, the (countable) additivity of probability measures entails the (countable) additivity of the $F_{Z_0}(Z)$ for disjoint families of sets Z_k, i.e. $F_{Z_0}(\cup Z_k) = \sum_k F_{Z_0}(Z_k)$. Taken together, these properties ensure that the family of $F_{Z_0}(Z)$ constitutes a (not necessarily normalised) POVM over Z_0. Due to the nature of time measurements, one anticipates that certain events will never occur (i.e., for no state ρ), so that indeed it may happen that $p_\rho(Z_0) < 1$, or $F_{Z_0}(Z_0) < I$.

Every observable can be characterised by its transformation behaviour under the fundamental space–time transformations. In particular, time observables will transform covariantly under time translations:

$$U_t F_{Z_0}(Z)\,U_t^{-1} = F_{Z_0-t}(Z-t)\ . \tag{3.45}$$

Properties of such time observables and specific examples (mainly in the context of decay observation) are studied in detail by Srinivas and Vijayalakshmi [35]. Detection times are axiomatically characterised as *screen observables* through further transformation covariance relations in work due to Werner [36].

Assuming that first and second moments for the POVM F_{Z_0} are defined on a dense domain, one can introduce a unique maximally symmetric (generally not self-adjoint) time operator

$$T = \int_{Z_0} t\,F_{Z_0}(dt)\ .$$

We put $\bar{t} = \mathrm{tr}\,[\rho \cdot T]$, then the temporal variance is defined as

$$(\Delta_\rho T)^2 = \frac{\int_{Z_0} \left(t - \bar{t}\right)^2 \mathrm{tr}\left[\rho \, F_{Z_0}\left(dt\right)\right]}{\mathrm{tr}\left[\rho \, F_{Z_0}\left(Z_0\right)\right]} \, . \tag{3.46}$$

The uncertainty relation (3.1) then follows for an event time observable and energy if the observation period $Z_0 = \mathbb{R}$,

$$\Delta_\rho T \, \Delta_\rho H \geq \frac{\hbar}{2} \, . \tag{3.47}$$

For an event time POVM with a finite interval Z_0, this relation is not generally valid.

It is still true, as it was in 1990 [3], that a systematic quantum theory of time measurements is lacking but will be necessary for an operational understanding of event time POVMs. The following examples may serve as guidance for the development of a better intuition about time observables and measurements.

Freely Falling Particle

For the Hamiltonian

$$H_g = \frac{P^2}{2m} - mgQ \, , \tag{3.48}$$

one easily verifies that the following self-adjoint operator T_g is canonically conjugate to H,

$$T_g = -\frac{1}{mg}P \, . \tag{3.49}$$

In fact this choice is suggested by the dynamical behaviour of the system: solving the Heisenberg equation of motion gives $P\left(t\right) = P - mgt\, I$, where $P\left(0\right) = P$. Time is measured dynamically as the linear increase of momentum. In this case even the Weyl relations are satisfied:

$$e^{itH/\hbar}\,e^{ihT/\hbar} = e^{-ith/\hbar}\,e^{ihT/\hbar}\,e^{itH/\hbar} \, . \tag{3.50}$$

As a further consequence, T_g and H act as generators of energy and time shifts, respectively, in the sense of the covariance relations

$$e^{ihT/\hbar}\,H\,e^{-ihT/\hbar} = H + hI \, , \tag{3.51}$$
$$e^{itH/\hbar}\,T\,e^{-itH/\hbar} = T - tI \, . \tag{3.52}$$

The associated time POVM is indeed a projection-valued measure, namely, the spectral measure

$$E^{T_g}\left(Z\right) = E^P\left(-mgZ\right) \, .$$

Both the covariance relations and the Weyl relation imply the Heisenberg canonical commutation relation and hence the uncertainty relation (3.47).

It must be noted that the present Hamiltonian is unbounded, its spectrum being absolutely continuous and extending over the whole real line. Thus the obstruction due to Pauli's theorem does not apply.

Oscillator Time

We now consider the Hamiltonian (putting $m = \hbar = 1$)

$$H_{\text{osc}} = \frac{1}{2} \left(P^2 + Q^2 \right) . \tag{3.53}$$

The spectrum consists of non-negative, equidistant values, so that there is no unitary shift group, hence no self-adjoint operator T satisfying the Weyl relation (3.50) can exist. Nevertheless, classical reasoning suggests the existence of a phase-like quantity that transforms covariantly (modulo 2π) under the time evolution group. This leads to the introduction of a time POVM and hence a periodic time variable proportional to the phase.

Introduce the ladder operator $a = \frac{1}{2}(Q + iP)$, which gives the number operator $N = a^* a$, with eigenvalues $n = 0, 1, 2, \ldots$ and eigenvectors $|n\rangle$. Then $H = N + \frac{1}{2}I$. For $t \in [0, 2\pi]$, we introduce the formal, non-normalisable vectors $|t\rangle = \sum_n e^{int} |n\rangle$, then we define

$$F_{\text{osc}}(Z) = (2\pi)^{-1} \int_Z dt \, |t\rangle\langle t| = \sum_{n,m \geq 0} (2\pi)^{-1} \int_Z e^{i(n-m)t} \, dt \, |n\rangle\langle m| .$$

It is easily verified that this defines a normalised, shift-covariant (mod 2π) POVM.

This oscillator-time POVM yields a whole family of self-adjoint operators canonically conjugate to H_{osc}: first define

$$T_{\text{osc}}^{(0)} = \int_0^{2\pi} t \, F_{\text{osc}}(dt) = \sum_{m \neq n \geq 0} \frac{1}{i(n-m)} |n\rangle\langle m| + \pi I .$$

This operator was first constructed as a self-adjoint solution of the canonical commutation relation (3.3), thus refuting a widespread erroneous reading of Pauli's theorem. Consequently, this operator does satisfy the uncertainty relation (3.47) in a dense domain (certainly not containing the energy eigenstates). Strangely enough, this aspect of the interesting papers of Garrison and Wong [37] and Galindo [38] has been widely ignored, while the fact as such is repeatedly being rediscovered in recent years. Next we calculate the time shifts of this operator,

$$T_{\text{osc}}^{(t)} = e^{itH} T_{\text{osc}}^{(0)} e^{-itH} = T_{\text{osc}}^{(0)} - tI + 2\pi F_{\text{osc}}([0, t]) .$$

Here we are facing a covariant family of non-commuting, self-adjoint operators, all of which satisfy the canonical commutation relation with $H = H_{\text{osc}}$. The non-commutativity corresponds to the fact that the phase quasi-eigenvectors $|t\rangle$ are mutually non-orthogonal, so that F_{osc} itself turns out to be a non-commutative POVM.

Here we have given just one example of a covariant oscillator-time (phase) POVM. There are in fact an infinite variety of such phase POVMs associated with H_{osc}. First significant steps towards a systematic account and operational analysis of covariant oscillator phase POVMs have been recently undertaken by Lahti and Pellonpää [39].

We note that a similar construction to the present one is possible for a finite quantum system with a spin Hamiltonian

$$H_{\text{spin}} = \beta s_3 \,,$$

where s_3 is the z component of the spin of a spin-s system. However, in this case a canonical commutation relation and the Heisenberg uncertainty relation are not valid.

Time POVMs vs Time Operators?

The preceding example shows in a striking way that observables may be more appropriately represented by means of a POVM instead of just a self-adjoint or symmetric operator: not only does the latter merely give the first moments of the experimental statistics, but, as seen here, there may exist a high degree of non-uniqueness in the choice of even a self-adjoint operator as a representative of an observable (here the phase, or oscillator time). An approach to defining event time observables taking into account the characteristic covariance may help to remove these ambiguities.

Nevertheless, for specific systems for which the physics of time measurements is well understood, the construction of canonical time operators may be sufficient and adequate.

By providing some mathematical qualifications on Pauli's claims concerning self-adjoint time operators canonically conjugate to the Hamiltonian of a physical system, Galapon [40, 41] made room for the construction of such canonical time operators for certain positive Hamiltonians with non-empty point spectrum. This was recently followed with a fresh approach to the time of arrival operator for a free particle in [42, 43, 44]; see also Chapter 10. In the next example we provide some general considerations on the search for covariant POVMs corresponding to the time of arrival.

Free-Particle Time Observables

Seemingly obvious candidates of a time operator conjugate to the free-particle Hamiltonian,

$$H_{\text{free}} = \frac{P^2}{2m} \,,$$

are given by suitably symmetrised expressions for the time-of-arrival variable suggested by classical reasoning, for example:

$$-\frac{1}{2} m \left(Q P^{-1} + P^{-1} Q \right) \qquad \text{or} \qquad - m P^{-1/2} Q P^{-1/2} \,.$$

While these expressions formally satisfy the canonical commutation relation, they are *not self-adjoint* but only symmetric (on suitably defined dense domains on which they actually coincide, see Sect. 10.4), and they do not possess a self-adjoint extensions. Hence this intuitive approach does not lead to a time observable in the usual sense of a self-adjoint operator conjugate to the free Hamiltonian. For a long time, this observation has been interpreted by many researchers as implying that time is not an observable in quantum mechanics. But this view does not take into account the fact that there are detection experiments that record the time of arrival of a particle, or more precisely, the time when the detector fires. The statistics of such measurements are appropriately described as probability distributions using suitable POVMs. For the present case of a free particle there do indeed exist time-shift covariant, normalised POVMs. An example is given by the following:

$$\langle\varphi|F_{\text{free}}(Z)\varphi\rangle = (2\pi)^{-1} \int_Z dt \left\{ \left| \int_0^\infty dp \sqrt{p/m\hbar} \, \exp\left(itp^2/m\hbar\right) \tilde{\varphi}(p) \right|^2 + \right.$$
$$\left. + \left| \int_{-\infty}^0 dp \sqrt{-p/m\hbar} \, \exp\left(itp^2/m\hbar\right) \tilde{\varphi}(p) \right|^2 \right\}.$$

Early explicit constructions of such POVM time observables and more general *screen* observables can be found in [45] and [36]. More recently, the question of constructing time-of-flight observables as covariant POVMs has been intensely studied; this development is reviewed in Chap. 10.

Time POVM associated with an effect.

The question of defining a time observable for any given type of event was investigated by Brunetti and Fredenhagen [46] who were able to define a time translation covariant POVM associated with a positive operator representing the event in question (an *effect*, in the terminology of Ludwig [47]). These authors also derived a new lower bound for the time uncertainty for covariant event time POVMs on the time domain $\mathbb{R}$, [48]:

$$\Delta_\rho T \geq \frac{d}{\langle H \rangle_\rho} \tag{3.54}$$

Using their approach, Brunetti and Fredenhagen were able to rederive the time delay operator of scattering theory. This work has inspired new model investigations on the theory of time measurements [49, 50].

In order to illustrate Brunetti and Fredenhagen's approach, we construct a simple example of a covariant time POVM associated with a Hamiltonian H with simple bounded, absolutely continuous spectrum $[0, 2\pi]$. One can think of a particle moving in one spatial dimension, with its momentum confined to the interval $[0, p_0]$, where $p_0^2/2m = 2\pi$.

Let $\mathcal{H}$ be the Hilbert space $L^2(0, 2\pi)$ in which H acts as the multiplication operator $H\psi(h) = h\psi(h)$. We choose a shift-covariant family of unit vectors

φ_t, $t \in \mathbb{R}$, as follows (putting $\hbar = 1$): $\varphi_t(h) = e^{iht}/\sqrt{2\pi}$. We can then define a time-shift covariant POVM via

$$P(X) := \int_X |\varphi_t\rangle\langle\varphi_t| \, dt, \quad X \in \mathcal{B}(\mathbb{R}). \tag{3.55}$$

The normalization $P(\mathbb{R}) = I$ can be verified by considering the integral

$$\int_{\mathbb{R}} \langle\psi|\varphi_t\rangle\langle\varphi_t|\xi\rangle \, dt.$$

for any $\psi, \xi \in \mathcal{H}$, and showing that its value is $\langle\psi|\xi\rangle$. This follows readily by observing that the function $t \mapsto \langle\varphi_t|\xi\rangle =: \hat{\xi}(t)$ is the Fourier-Plancherel transform $\hat{\xi} =: \mathcal{F}\xi$ of $\xi \in \mathcal{H}$. Note that $\hat{\xi} \in L^2(\mathbb{R})$, and that $\mathcal{F}(\mathcal{H})$ is a proper closed subspace of $L^2(\mathbb{R})$. Thus, we find that for $\psi \in \mathcal{H}$,

$$\mathcal{F}P(X)\mathcal{F}^{-1}\hat{\psi}(t) = \chi_X(t)\hat{\psi}(t),$$

which corresponds to the Naimark extension of the POVM P to a spectral measure on $L^2(\mathbb{R})$.

We are now in a position to compare the time observable (3.55) with the general construction of Brunetti and Fredenhagen in [46]. Given a bounded positive operator A, they consider the positive operator measure, defined first on intervals J via

$$B(J) := \int_J e^{itH} A e^{-itH} \, dt.$$

They then show that in certain circumstances this can be turned into a normalized POVM on a suitable closed subspace (provided this is not the null space). In the present case of the POVM (3.55), we see that the operator corresponding to A can be identified with the one-dimensional projection operator $|\varphi_0\rangle\langle\varphi_0|$. In that case the normalization condition is already satisfied, and $B(J) = P(J)$ holds on $\mathcal{H}$. The POVM P corresponds to a measurement of the time that the system spends (loosely speaking) in the state ϕ_0.

A formal time operator is obtained from the first moment operator of the POVM P:

$$T\psi(h) = \int_{\mathbb{R}} t \, \varphi_t(h)\langle\varphi_t|\psi\rangle \, dt = -i\frac{d}{dh}\psi(h). \tag{3.56}$$

This is well defined for functions $\psi \in L^2(0, 2\pi)$ which are absolutely continuous and such that the derivative $\psi' \in L^2(0, 2\pi)$. In order for this operator to be symmetric, the domain must be further restricted by appropriate boundary conditions. It is well known that the condition $\psi(2\pi) = c\psi(0)$ makes $-id/dh$ a self-adjoint operator $T^{(c)}$ for any c of modulus 1. Each such $T^{(c)}$ is a self-adjoint extension of the differential operator understood as a symmetric operator $T^{(0)}$ with the boundary condition $\psi(0) = \psi(2\pi) = 0$. Note that the spectrum of $T^{(c)}$ is $\mathbb{Z}$, with eigenvectors $e^{i\arg(c)H/2\pi}\varphi_m$, where $\varphi_m(h) = e^{imh}/\sqrt{2\pi}$, $m \in \mathbb{Z}$.

The covariance relation

$$e^{i\tau H} T e^{-i\tau H} = T - \tau I$$

is found to be satisfied for $T^{(0)}$ but not for any of its self-adjoint extensions since $e^{i\tau H} T^{(c)} e^{-i\tau H}) = T^{(c')}$ with $c' = e^{i2\pi\tau} c$. In accordance with this, the canonical commutation relation between the Hamiltonian and the time operator is obtained only on the domain of $T^{(0)}$; therefore, the uncertainty relation (3.47) holds on this dense subspace, with the variance of the time distribution being defined via (3.46). Since the spectrum of H is a bounded interval of length $\lambda(H) = 2\pi$, there is an absolute bound to the temporal variance in any state ρ:

$$\Delta_\rho T \geq \frac{\hbar}{2\lambda(H)}. \tag{3.57}$$

These examples show that for a variety of Hamiltonians, event time observables can be defined as time-shift covariant POVMs, the form of which is inferred by the aid of classical intuition or with reference to a class of experimental situations. Where the first moment operator satisfies a canonical commutation relation with the Hamiltonian on a dense domain, the observable-time energy uncertainty relation will follow. Whether or not this is the case depends on the nature of the spectrum of the Hamiltonian and the time domain [35].

We conclude this brief survey of the problem of time-covariant POVMs with the following pointers to some interesting related developments.

A connection between time observables represented by POVMs and irreversible dynamics has been explored by Amann and Atmanspacher [51].

Finally, there have been several studies of the representation of event time observables in terms of POVMs in the wider context of relativistic quantum mechanics and quantum gravity [52, 53, 54, 55, 56, 57]. It is too early and beyond the scope of the present chapter to give a conclusive review of these recent and ongoing developments.

3.6.2 Einstein's Photon Box

A comprehensive theory of event time measurements is missing to date, so that a first step towards an understanding of time as an observable seems to be to carry out case studies. Here we will revisit briefly the *Gedanken* experiment proposed by Einstein. In this experiment, a photon is allowed to escape from a box through a hole, which is closed and opened temporarily by a shutter. The opening time period is determined by a clock, which is part of the box system. Einstein argued that it should be possible to determine the energy of the outgoing photon by weighing the box before and after the opening period. Thus it would seem that one can obtain an arbitrarily sharp value for the energy of the photon, while at the same time the time period of preparation or emission of the outgoing photon could be made as short as one would wish, by setting the clock mechanism appropriately. This conclusion would contradict the preparation time–energy uncertainty relation (3.13).

Bohr's rebuttal [58] was based on the observation that the accuracy of the weighing process is limited by the indeterminacy of the box momentum, which in turn limits the unsharpness of position by virtue of the Heisenberg uncertainty relation for the box position and momentum. But an uncertainty in the box position entails an uncertainty in the rate of the clock, as a consequence of the equivalence principle. All this taken together, the accuracy of the determination of the photon energy and the uncertainty of the opening time do satisfy the uncertainty relation (3.1).

Bohr's informal way of reasoning has given rise to a host of attempts, by some, to make the argument more precise (or even more comprehensible) or, by others, to refute it in defence of Einstein. In fact if Bohr's were the only way of arguing, the consistency of non-relativistic quantum mechanics (replacing the photon with a (gas) particle) would appear to depend on the theory of relativity. Hence several authors have considered different methods of measuring the photon energy.

In his review of 1990, the present author has offered an argument that makes no assumptions concerning the method of measurement and is simply based on a version of quantum clock uncertainty relation. This argument goes as follows. If the photon energy is to be determined with an inaccuracy δE from the difference of box energies before and after the opening period, then these energies must be well defined within δE, i.e., the box energy uncertainty ΔE must satisfy $\Delta E \leq \delta E$. Then the clock uncertainty relation, either in the Mandelstam–Tamm form (3.43) or the Hilgevoord–Uffink form (3.44), allows us to conclude that the box system needs at least a time $t_0 \cong \hbar/\Delta E$ in order to evolve from the initial 'shutter-closed' state to the (orthogonal!) 'shutter-open' state (and back). During this transition time t_0 it is *objectively indeterminate* whether the shutter is open or closed. Accordingly, the time interval within which the photon can pass the shutter is also indeterminate by an amount $\Delta T = t_0$. We thus arrive precisely at Bohr's relation

$$\Delta T\, \delta E \cong \hbar\,. \tag{3.58}$$

It seems satisfying that this derivation works without advocating the box position–momentum uncertainty relation; instead it refers directly to the quantum dynamical features of the box. Without going into an analysis of the energy transfer between box and photon, it seems plausible that the energy measurement uncertainty δE of the box, which corresponds to an uncertainty of the box energy, will give rise to an uncertainty of the energy of the escaping photon. Similarly, the uncertainty in the shutter opening time gives a measure of the uncertainty of the time of passage of the photon through the hole. Hence the box uncertainty relation admits the following interpretation also: it is impossible to determine the energy and time of passage of a particle with accuracies better than those allowed by this uncertainty relation.

Thus the measurement uncertainty relation (3.58) accords with the dynamical Mandelstam–Tamm relation for the characteristic time $\tau_\rho(Q)$, (3.17), and thus with the preparation time–energy relation (3.13).

It is also interesting to note the close analogy between this experiment and the double-slit experiment where similar debates between Bohr and Einstein took place concerning the possibility of jointly determining the position and momentum of a particle. Time of passage and energy are complementary quantities in the same sense as position and momentum: the arrangements for determining time (position) and energy (momentum) are mutually exclusive. However, while these conclusions have been corroborated in the case of position and momentum with appropriate quantum mechanical joint measurement models (for details and a survey of this development, cf. [10]), a similarly comprehensive treatment for time and energy is as yet waiting to be carried out. Only very recently a first scheme of joint measurements of energy and time of arrival has been proposed [59] along the lines of the position–momentum measurement model due to Arthurs and Kelly.

3.6.3 Temporal Interference and Time Indeterminacy

In the preceding sections we have repeatedly referred to temporal indeterminacies of events such as the passage of a particle through a space region, and we have motivated this interpretation indirectly by invoking the quantum indeterminacies of the relevant dynamical properties. The analogy between the time–energy complementarity and the position–momentum complementarity that emerges in the context of the Einstein's photon box (a point strongly emphasised by Cook [60]) suggests, however, that it should be possible to obtain direct experimental evidence for the appropriateness of the indeterminacy interpretation of time uncertainties. In the case of position and momentum, the indeterminacy of the location of a particle passing through a screen with two slits is demonstrated by means of the interference pattern on the capture screen, which images the fine structure of the associated momentum amplitude function. As a simple model illustration, if the wave function of the particle at the location of the slit is given as

$$\psi_0(x) = \begin{cases} (4a)^{-1/2} & \text{if } A - a \leq |x| \leq A + a \,, \\ 0 & \text{elsewhere} \,, \end{cases}$$

then the momentum amplitude is given as the Fourier transform,

$$\tilde{\psi}_0(p) = 2\sqrt{a} \cos(Ap) \, \frac{\sin(ap)}{ap} \,.$$

If the slit width a is small compared to the distance between the slits then the factor $(\sin(ap)/ap)^2$ describes the slowly varying envelope of the momentum distribution while the factor $\cos^2(Ap)$ describes the rapid oscillations that constitute the interference pattern. If the path of the particle were known, one would have an incoherent mixture of two packets travelling through the slits, and no interference would appear. Hence the ignorance interpretation regarding the two paths is in conflict with the presence of the interference pattern, which is due to the coherent superposition of the two path states. In other words, the path is indeterminate, and it is *objectively undecided* through which slit the particle has passed.

In a similar way, if one were able to offer a particle a multiple *temporal* 'slit', then the indeterminacy of the time of passage would be reflected in an interference pattern in the associated energy distribution. As it turns out, experiments exhibiting such *diffraction in time*, or *temporal interference*, had already been carried out in the 1970s. In the experiment of Hauser, Neuwirth, and Thesen [61], a beam of Mössbauer quanta is emitted from excited ^{57}Fe nuclei, with a mean energy of $E_0 = 14.4$ keV and a lifetime $\tau = 141$ ns, and is sent through a slit that is periodically closed and opened by means of a fast-rotating chopper wheel. Then the energy distribution of the quanta is measured. The count rate is around 3000 events per second, so that on average there is about one photon within 2000 lifetimes passing the device. This suggests that one is observing interference of individual photons. We briefly sketch the analysis and interpretation proposed in [3].

The amplitude incident at the chopper,

$$f_0(t) = e^{-t/2\tau} e^{-i\omega_0 t}, \quad \omega_0 = E_0/\hbar, \quad t \geq 0,$$

is modulated into

$$f(t; t_0) = f_0(t) \, \chi(t; t_0).$$

Here the chopping function χ is equal to 1 for all $t > 0$, which fall into one of a family of equidistant intervals Z_k of equal length T_{open} distributed periodically, with period T_{chop}, over the whole real line. For all other values of t we have $\chi(t) = 0$. The time parameter t_0 indicates the difference between the zero point of the decay process and the beginning of a chopping period; its value is distributed uniformly over a chopping period if a large ensemble of events is observed.

The Fourier transform of f_0 reproduces the Lorentzian shape of (3.32). The energy distribution obtained behind the chopper should be given by the Fourier transform of $f(t; t_0)$,

$$\tilde{f}(\omega; t_0) = \int_{\mathbb{R}} dt \, f(t; t_0) \, e^{i\omega t} = \sum_k \int_{Z_k} dt \, f_0(t) \, e^{i\omega t} = \sum_k \tilde{f}_k(\omega; t_0).$$

Hence, the expected spectral intensity is

$$\mathfrak{I}(\omega; t_0) = \left| \tilde{f}(\omega; t_0) \right|^2 = \left| \sum_k \tilde{f}_k(\omega; t_0) \right|^2. \tag{3.59}$$

This corresponds to a coherent superposition of the temporal partial packets $f_k(t; t_0)$. The observed distribution is obtained by averaging $\mathfrak{I}(\omega; t_0)$ over one chopping period with respect to t_0,

$$\mathfrak{I}(\omega) = \frac{1}{T_{\text{chop}}} \int_0^{T_{\text{chop}}} dt_0\, \mathfrak{I}(\omega; t_0) \ . \tag{3.60}$$

Now, if one assumed the time window through which each photon passes to be *objectively* determined (albeit possibly unknown), then one would predict the t_0-average $\mathfrak{I}^{\text{ob}}(\omega)$ of the spectral distribution

$$\mathfrak{I}^{\text{ob}}(\omega; t_0) = \sum_k \left| \tilde{f}_k(\omega; t_0) \right|^2 \ . \tag{3.61}$$

A calculation yields that the shape of the distribution $\mathfrak{I}^{\text{ob}}(\omega)$ is very similar to a somewhat broadened Lorentzian curve, whereas $\mathfrak{I}(\omega)$ shows a sharp central peak and several distinguished, symmetric side peaks of much smaller amplitudes. The latter is in excellent agreement with the experimental spectral data.

The increase of the overall width of the spectral distribution can be seen as a consequence of the temporal fine structure introduced by the action of the chopper. Similarly, the fine structure of the spectral distribution is linked to the overall width of the temporal distribution: the latter is of the order of the lifetime, while the former is approximately equal to the undisturbed linewidth. This behaviour is in accordance with the Hilgevoord–Uffink relation (3.41) between the overall width and the translation width for a pair of Fourier-related distributions, which is thus found to be (at least qualitatively) confirmed.

We conclude that the spectral interference pattern exhibited in this experiment demonstrates the *non-objectivity*, or *indeterminacy* of the time of passage of the photon through the chopper. It is tempting to go one step further and claim that the time of the emission of the photon is equally indeterminate.

In 1986, time indeterminacies were demonstrated for material particles, in an observation of *quantum beats* in neutron interferometry by Badurek et al. [62]. Similar temporal diffraction experiments have been carried out in recent years with material particles, namely atoms [63] and neutrons [64]. The results obtained are in agreement with the time–energy uncertainty relation. The issue of the (non-)objectivity of event times has also been investigated from the perspective of Bell's inequalities. In a seminal paper of Franson [65], an interference experiment with time–energy entangled photons was proposed. Subsequent measurements by Brendel et al. [66] and Kwiat et al. [67] yielded observed fringe visibilities in accordance with quantum mechanical predictions and were significantly larger than allowed by a Bell inequality that follows from classical reasoning.

3.7 Conclusion

We summarise the main types of time–energy uncertainty relations

$$\Delta T \, \Delta E \gtrsim \hbar \tag{3.62}$$

and their range of validity depending on the interpretation of the quantities ΔT and ΔE:

(1) A relation involving *external time* is valid if ΔT is the *duration* of a perturbation or a preparation process and ΔE is the uncertainty of the energy in the system.

(2) There is *no* limitation to the duration of an energy measurement and the disturbance or inaccuracy of the measured energy.

(3) There are a variety of measures of *characteristic, intrinsic times*, with ensuing *universally* valid *dynamical time–energy uncertainty relations*, ΔE being a measure of the width of the energy distribution or its fine structure. This comprises the Bohr–Wigner, Mandelstam–Tamm, Bauer–Mello, and Hilgevoord–Uffink relations.

(4) *Event time observables* can be formally represented in terms of positive operator-valued measures over the relevant time domain. An *observable time–energy uncertainty relation*, with a constant positive lower bound for the product of inaccuracies, is *not universally* valid but will hold in specific cases, depending on the structure of the Hamiltonian and the time domain.

(5) Time measurements by means of *quantum clocks* are subject to a dynamical time–energy uncertainty relation, where the time resolution of the clock is bounded by the unsharpness of its energy, $\delta t \gtrsim \hbar/\Delta E$.

(6) Einstein's photon box experiment constitutes a demonstration of the *complementarity* of time of passage and energy: as a consequence of the quantum clock uncertainty relation, the inaccuracy δE in the determination of the energy of the escaping photon limits the uncertainty ΔT of the opening time of the shutter. This is in accordance with the *energy measurement* uncertainty relation based on *internal clocks* discovered recently by Aharonov and Reznik.

(7) Temporal diffraction experiments provide evidence for the *objective indeterminacy* of event time uncertainties such as time of passage.

Finally we have to recall that:

(8) A full-fledged quantum mechanical theory of time measurements is still waiting to be developed.

Acknowledgement

Work leading to this revised and expanded version was carried out during the author's stay at the Perimeter Institute, Waterloo, Canada. Hospitality and support by PI is gratefully acknowledged.

References

1. M. Jammer: *The Philosophy of Quantum Mechanics* (Wiley, New York 1974)
2. M. Bauer, P.A. Mello: Ann. Phys. (N.Y.) **111**, 38 (1978)
3. P. Busch: Found. Phys. **20**, 1 (1990)
4. P. Busch: Found. Phys. **20**, 33 (1990)
5. J. Hilgevoord: Am. J. Phys. **64**, 1451 (1996)
6. J. Hilgevoord: Am. J. Phys. **66**, 396 (1998)
7. P. Pfeifer, J. Frohlich: Rev. Mod. Phys. **67**, 759 (1995)
8. W. Pauli: 'Die allgemeinen Prinzipien der Wellenmechanik'. In: *Handbuch der Physik, 2nd ed., Vol 24.* ed. by H. Geiger, K. Scheel (Springer-Verlag, Berlin 1933); English translation: *General Principles of Quantum Mechanics* (Springer-Verlag, New York 1980)
9. W. Heisenberg: Z. Phys. **69**, 56 (1927)
10. P. Busch, M. Grabowski, P. Lahti: *Operational Quantum Physics* (Springer-Verlag, Berlin 1995, ^{2}1997)
11. Y. Aharonov, D. Bohm: Phys. Rev. **122**, 1649 (1961)
12. Y. Aharonov, B. Reznik: Phys. Rev. Lett. **84**, 1368 (2000)
13. M. Moshinsky: Am. J. Phys. **44**, 1037 (1976)
14. M.H. Partovi, R. Blankenbecler: Phys. Rev. Lett. **57**, 2887 (1986)
15. L. Mandelstam, I.G. Tamm: J. Phys. (USSR) **9**, 249 (1945)
16. M. Grabowski: Lett. Math. Phys. **8**, 455 (1984)
17. N. Bohr: Naturwissenschaften **16**, 245 (1928); Nature Suppl., April 14 (1928) p. 580
18. E.P. Wigner: 'On the time–energy uncertainty relation'. In: *Aspects of Quantum Theory*, ed. by A. Salam, E.P. Wigner (Cambridge University Press, Cambridge, MA 1972) pp. 237–247
19. A.D. Baute, R. Sala Mayato, J.P. Palao, J.G. Muga, I.L. Egusquiza: Phys. Rev. A **61**, 022118 (2000)
20. J. Kijowski: Rep. Math. Phys. **6**, 361 (1974)
21. J. Hilgevoord, J. Uffink: 'The mathematical expression of the uncertainty principle'. In: *Microphysical Reality and Quantum Formalism*, ed. by A. van der Merwe et al. (Kluwer, Dordrecht 1988) pp. 91–114
22. J. Uffink: Am. J. Phys. **61**, 935 (1993)
23. P. Pfeifer: Phys. Rev. Lett. **70**, 3365 (1993)
24. E.P. Wigner: Rev. Mod. Phys. **29**, 255 (1957)
25. H. Salecker, E.P. Wigner: Phys. Rev. **109**, 571 (1957)
26. R.J. Adler, I.M. Nemenman, J.M. Overduin, D.I. Santiago: Phys. Lett. B **477**, 424 (2000)
27. Y.J. Ng, H. van Dam: Phys. Lett. B **477**, 429 (2000)
28. C.R. Leavens, W.R. McKinnon: Phys. Lett. A **194**, 12 (1994)
29. Y. Japha, G. Kurizki: Phys. Rev. A **60**, 1811 (1999)
30. V. Buzek, R. Derka, S. Massar: Phys. Rev. Lett. **82**, 2207 (1999)
31. R. Josza, D.S. Abrams, J.P. Dowling, C.P. Williams: Phys. Rev. Lett. **85**, 2010 (2000)
32. P. Busch, P. Lahti: Ann. Physik (Leipzig) **47**, 369 (1990)
33. J. Hilgevoord, J. Uffink: Found. Phys. **21**, 323 (1991)
34. E.B. Davies: *Quantum Theory of Open Systems* (Academic Press, New York 1976)

35. M.D. Srinivas, R. Vijayalakshmi: Pramana **16**, 173 (1981)
36. R. Werner: J. Math. Phys. **27**, 793 (1986)
37. J.C. Garrison, J. Wong: J. Math. Phys. **11**, 2242 (1970)
38. A. Galindo: Lett. Math. Phys. **8**, 495 (1984)
39. P. Lahti, J.-P. Pellonpää: J. Math. Phys. **40**, 4688 (1999); **41**, 7352 (2000)
40. E.A. Galapon: Proc. R. Soc. Lond. A **458**, 451 (2002)
41. E.A. Galapon: Proc. R. Soc. Lond. A **458**, 2671 (2002)
42. E.A. Galapon, R.F. Caballar, R.T. Bahague: Phys. Rev. Lett. **93**, 180406 (2004)
43. E.A. Galapon, R.F. Caballar, R.T. Bahague: Phys. Rev. A **72** 062107 (2005)
44. E.A. Galapon, F. Delgado, J. G. Muga, I. Egusquiza: Phys. Rev. A **72** 042107 (2005)
45. A. Holevo: *Probabilistic and Statistical Aspects of Quantum Theory* (North-Holland, Amsterdam 1980)
46. R. Brunetti, K. Fredenhagen: Phys. Rev. A **66**, 044101 (2002)
47. G. Ludwig: *Foundations of Quantum Mechanics 1* (Springer–Verlag, Berlin 1983)
48. R. Brunetti, K. Fredenhagen: Rev. Math. Phys. **14**, 897 (2002)
49. J.A. Damborenea, I.L. Egusquiza, G.L. Hegerfeldt: Phys. Rev. A **66**, 052104 (2002)
50. G.L. Hegerfeldt, D. Seidel, J.G. Muga: Phys. Rev. A **68**, 022111 (2003)
51. A. Amann, H. Atmanspacher: Int. J. Theor. Phys. **37**, 629 (1998)
52. R. Giannitrapani: Int. J. Theor. Phys. **36**, 1575 (1997)
53. R. Giannitrapani: J. Math. Phys. **39**, 5180 (1998)
54. M. Toller: Phys. Rev. A **59** 960 (1999)
55. M. Toller: Int. J. Theor. Phys. **38** 2015 (1999)
56. S. Mazzucchi: J. Math. Phys. **42** 2477 (2001)
57. C. Rovelli: Phys. Rev. D **65**, 124013 (2002)
58. N. Bohr: 'Discussion with Einstein on epistemological problems in atomic physics'. In: *Albert Einstein: Philosopher–Scientist*, ed. by P.A. Schilpp (Open Court, LaSalle 1949) pp. 201–241
59. A.D. Baute, I.L. Egusquiza, J.G. Muga, R. Sala Mayato: Phys. Rev. A **61**, 052111 (2000)
60. L.F. Cook: Am. J. Phys. **48**, 142 (1980)
61. U. Hauser, W. Neuwirth, N. Thesen: Phys. Lett. A **49**, 57 (1974)
62. G. Badurek, H. Rauch, D. Tuppinger: Phys. Rev. A **34**, 2600 (1986)
63. P. Szriftgiser, D. GueryOdelin, M. Arndt, J. Dalibard: Phys. Rev. Lett. **77**, 4 (1996)
64. T. Hils, J. Felber, R. Gahler, W. Glaser, R. Golub, K. Habicht, P. Wille: Phys. Rev. A **58**, 4784 (1998)
65. J.D. Franson: Phys. Rev. Lett. **62**, 2205 (1989)
66. J. Brendel, E. Mohlers, W. Martienssen: Europhys. Lett. **10**, 575 (1992)
67. P.G. Kwiat, A.M. Steinberg, R.Y. Chiao: Phys. Rev. A **47**, R2472 (1993)

4

Jump Time and Passage Time: The Duration of a Quantum Transition

Lawrence S. Schulman

Physics Department, Clarkson University, Potsdam, NY 13699-5820, USA
schulman@clarkson.edu

4.1 Introduction

It is ironic that experimentally time is the most accurately measured physical quantity, while in quantum mechanics one must struggle to provide a definition of so practical a concept as time of arrival. Historically, one of the first temporal quantities analyzed in quantum mechanics was *lifetime*, a property of an unstable state. The theory of this quantity is satisfactory in two ways. First, with only the smallest of white lies, one predicts exponential decay, and generally this is what one sees. Second, at the quantitative level, one finds good agreement with a simply derived formula, the Fermi–Dirac Golden rule,

$$\Gamma = \frac{2\pi}{\hbar}\rho(E)|\langle f|H|i\rangle|^2 . \tag{4.1}$$

Equation (4.1) uses standard notation. Γ is the transition rate from an initial (unstable) state $|i\rangle$ to a final state $|f\rangle$. The transition occurs by means of a Hamiltonian H. The density of (final) states is ρ, evaluated at the (common) energy of the states $|i\rangle$ and $|f\rangle$. In terms of Γ, the lifetime is $\tau_{\mathrm{L}} = 1/\Gamma$.

The lifetime τ_{L} is not a property of any one atom (or whatever), but rather of an *ensemble* of like atoms. For much of the twentieth century this was sufficient. One was taught not to inquire too closely about the time evolution of an individual member of an ensemble. An exception to this informed neglect arose as technology allowed experimentalists to focus on transitions in individual atoms [1]. Although one can recast these phenomena in ensemble terms, the ensemble is typically conditioned on the fact of the ultimate decay of the system studied. But a similar extension of naive ensemble interpretations was already present in studies of *tunneling time*. The barrier penetration phenomenon of quantum mechanics was sufficiently provocative in its denial of classical notions that one sought places where conventional ideas *could* be applied, e.g., trying to assign a time of passage through the barrier. This subject has a long history and a collection of recent views can be found in [2]. Again, in principle, for barrier penetration one deals with ensembles, but if one

L. S. Schulman: *Jump Time and Passage Time: The Duration of a Quantum Transition*, Lect. Notes Phys. **734**, 107–128 (2008)
DOI 10.1007/978-3-540-73473-4_4   © Springer-Verlag Berlin Heidelberg 2008

measures passage time there would need to be conditioning on the fact of the transition, observations of individual transits, and a time interval measured for each. Our notation for tunneling time (without distinguishing among the many definitions) is τ_T.

The tunneling-time concept allowed further probes of the Copenhagen view of quantum mechanics. A decaying particle, for example, a nucleus in the Gamow model of alpha decay, was said to undergo a *quantum jump*. The idea (I guess) was that you could measure the particle in its initial state or in its final state. But getting from one to the other was a "jump." It took a measurement to distinguish one state from the other, putting the jump itself beyond the scope of quantum mechanics, or at least of ordinary unitary time evolution. However, if one could ascribe to the particle a trajectory under the barrier, along with a time during which the particle tunnels, then one has made the first steps in the analysis of this "jump." Assigning a trajectory is problematic [3], although several authors have used the Feynman path integral [4, 5, 6], acknowledging the limitation that the path contributions only add as amplitudes, not probabilities.

A different "quantum jump" was exhibited experimentally in the 1980s [7, 8, 9]. This involved an atomic transition for a *single* atom. There was not any "path under a barrier" and as indicated the notion of ensemble needed updating. In particular, one could no longer muddle the distinction between an abstract ensemble and the large number of atoms participating in decay experiments.

In these experiments [7, 8, 9] one monitored the atom closely, noting when it was in its excited state. The duration of its stay in the excited state was (quantumly) random, and repetition of the experiment gave statistics that could be used to evaluate the lifetime. But from the data it was evident that something else was happening—the famous or infamous jump—and that its timescale, if any could be defined for it, had little to do with lifetime.

The question that I raised [10] was whether one could say anything about the time interval that elapses between finding the atom in one state and finding it in the other. One does not need the drama of [7, 8, 9] to ask this question. Radium (^{226}Ra) has a half-life of about 1600 year and one can imagine putting a single such nucleus in an inert matrix and waiting to see an α-decay. (This is similar to the experiments cited, where seeing *nothing* meant that the system was still in its metastable state.) The interval between being in one state and being in the other is certainly brief. But is it, as early and perhaps loose interpretations of measurement theory would have it, instantaneous? Another example goes back to arguments for the quantized nature of light, as demonstrated by the photoelectric effect. The "instantaneous" appearance of electrons when ultraviolet light was turned on was a blow to classical interpretations [11]. But again, "instantaneous" is a matter of technology, and the bounds on this time interval were only about a nanosecond.

In this chapter I will define two times, each related to the question asked. But they are in general quantitatively different from one another. The times

Table 4.1. Characteristic times

Time	Name	Description
τ_L	Lifetime	Usual lifetime for decay, $\hbar/2\pi\rho(E)|\langle f|H|i\rangle|^2$
τ_Z	Zeno time	Inverse of energy spread, $\hbar/\sqrt{\langle\psi|(H-E_\psi)^2|\psi\rangle}$
τ_J	Jump time	τ_Z^2/τ_L
τ_T	Tunneling time	As in barrier penetration
τ_P	Passage time	Minimum time to go from a state to a $\perp$ one, $\pi\tau_Z/2$
τ_R	Response time	A property of monitoring apparatus
τ_{PM}	Pulse time	Interval between ideal pulsed measurements (cf. QZE)
τ_{Door}	Door time	Metaphorical

are called *jump time* and *passage time*. Roughly speaking, the first measures how long it takes for the transition process to get seriously underway and the second measures how long it takes to complete the process in a single exemplar. But it is better not to use too much verbal description. From the definitions below and from the applications, the relevance of each should emerge.

The jump time is designated by the symbol τ_J. In Sect. 4.2 I will motivate my definition and arrive at a quantitative expression. The considerations parallel arguments arising in the quantum Zeno effect (QZE). The formula for τ_J turns out to be the next simplest thing you could construct from the Hamiltonian after (4.1).

The passage time, designated τ_P, has a precise mathematical definition, although in a specific experimental situation it will depend on the apparatus as well as on the system undergoing the transition. It arises from a bound on the minimum time for a state to evolve (with given Hamiltonian) to a state orthogonal to itself.

Definitions are tested by what you can do with them, what they unify. I will show that τ_J arises in several contexts. It is a generalization of tunneling time (τ_T). It satisfies a kind of time–energy uncertainty relation. For certain transitions it establishes the experimentally observed timescale, although for atomic decays it is immeasurably short. Passage time is related to a theoretical bound found by Fleming [12], a bound not hampered by requiring notions of what a "measurement" is supposed to be. Although with respect to my own ideas on quantum measurement theory τ_P may prove more significant than τ_J, its dependence on the measuring apparatus limits its general applicability.

Finally, because of the many characteristic temporal quantities that will be defined here, I have included Table 4.1 for reference.

4.2 Jump Time

How long does it take to walk through a doorway? Call this time τ_{Door}. Consider the following experiment. A stream of people pass through a door, one at a time. From time to time, and without looking, I fire a marshmallow across the doorway. Anyone hit by a marshmallow must turn back. Assume the marshmallow crosses the doorway instantaneously. If I fire N times during a time interval of duration T, then I expect to turn back a fraction $N\tau_{\text{Door}}/T$ of the people. An experiment to measure τ_{Door} would consist of gradually increasing the marshmallow firing rate until no one can cross. The estimate for τ_{Door} would then be T/N. In other words, when my firing rate reaches $1/\tau_{\text{Door}}$ I stop the traffic. Without further refinements this measurement would not define τ_{Door} by better than a factor 2, i.e., it defines a *timescale*, rather than a precise time.

The same perspective motivates the definition of quantum jump time. The decay, or other quantum transition, corresponds to getting through the door. The process-terminating interruption is an "observation," a quantum measurement. As for tunneling time, the use of classical concepts means that the doorway analogy is incomplete.

We formalize the discussion: at intervals δt project onto the initial states, i.e., measure whether the system is still in its initial state. If these disturbances do not slow the decay, then δt is to be considered longer than the jump. On the other hand, if these projections do slow the decay, then they have reached its timescale. In this way I arrive at a context similar to that of the quantum Zeno effect [13].

Let the system begin in a state ψ and let the full Hamiltonian be H. After a time δt, ψ evolves to $\exp(-iH\delta t/\hbar)|\psi\rangle$. One checks for decay by applying $\langle\psi|$. The probability that it is still in ψ is $p(\delta t) = |\langle\psi|\exp(-iH\delta t/\hbar)|\psi\rangle|^2$. A short calculation shows that

$$p(\delta t) = 1 - \left(\frac{\delta t}{\tau_z}\right)^2 + \text{O}(\delta t^4)\,, \tag{4.2}$$

where

$$\tau_z^2 \equiv \frac{\hbar^2}{\langle\psi|(H - E_\psi)^2|\psi\rangle} \tag{4.3}$$

and $E_\psi \equiv \langle\psi|H|\psi\rangle$. I call τ_z the *"Zeno time,"* notwithstanding my lack of full concurrence with the classical allusion [14].

Remark: It is worth taking a second look at the derivation of (4.2), since the appearance of high-frequency terms in the off-diagonal matrix elements has exercised some authors [15]. Let

$$f(t) \equiv \langle\psi|\exp(-i(H - E_\psi)t/\hbar)|\psi\rangle\,.$$

First, assume that this function has at least three derivatives in $[0, t]$, so that in particular, besides E_ψ, $\langle\psi|H^2|\psi\rangle$ and $\langle\psi|H^3|\psi\rangle$ must be finite. Then by standard theorems, one can write $f(t) = 1 - (t/\tau_z)^2/2 + t^3\,\dddot{f}(t^*)$ for some t^* between 0 and t. Calculating $|f|^2$ (to get $p(t)$) shows the deviation from $1 - t^2/\tau_z^2$ to be no larger than $\text{O}(t^3)$. When a fourth derivative exists, $\text{Re}\,\dddot{f}(0) = 0$ implies (4.2).

Now suppose that many projections are made during a time t, carried out at intervals τ_{PM}. Then to leading order, at t, the probability of being in ψ is

$$p_{\mathrm{Interrupted}}(t) = \left[p(\tau_{\mathrm{PM}})\right]^{t/\tau_{\mathrm{PM}}} \approx \left[1 - \left(\frac{\tau_{\mathrm{PM}}}{t}\frac{t\tau_{\mathrm{PM}}}{\tau_{\mathrm{z}}^2}\right)\right]^{t/\tau_{\mathrm{PM}}} \approx \exp\left(-t\tau_{\mathrm{PM}}/\tau_{\mathrm{z}}^2\right) .$$
(4.4)

To define a jump time, we want to know whether this differs from standard decay. Without projections the probability for being in ψ is

$$p_{\mathrm{Uninterrupted}}(t) = \exp(-t/\tau_{\mathrm{L}})$$
(4.5)

with $\tau_{\mathrm{L}} \equiv 1/\Gamma$ the usual lifetime ("Γ" of (4.1)). Comparing (4.4) and (4.5), we see that the interrupted decay will be slower for $\tau_{\mathrm{PM}} < \tau_{\mathrm{z}}^2/\tau_{\mathrm{L}}$ [16]. We are thus led to define the "jump time" as the time for which the slowdown would begin to be significant, namely

$$\tau_{\mathrm{J}} \equiv \tau_{\mathrm{z}}^2/\tau_{\mathrm{L}} .$$
(4.6)

In words, τ_{J} is the time such that if one inspected a system's integrity at intervals of this duration, the decay would be slowed significantly [10, 17].

Remark: Because my goal is only to define a *timescale*, I do not attempt greater precision. For example, in (4.5), because of the initial quadratic dependence, one may want to change the extrapolated time-zero value. For our purposes, however, the normalization is irrelevant, since it is the decay *rate* whose equality fixes τ_{J}.

Remark: Recall that (4.1) uses the first moment of the Hamiltonian. The jump-time definition, (4.6), involves the second moment, in a way, the simplest step beyond minimal decay information.

4.3 Corroborations of the Definition

The usefulness of jump time will be demonstrated in a number of contexts: (1) comparison with tunneling time; (2) time–energy uncertainty principle; (3) reconciling continuous measurement with the QZE; and (4) experiments on the quadratic regime of decay.

4.3.1 Comparison with Tunneling Time

In [14] a simple example of quantum tunneling was studied in an effort to estimate τ_{z}. There is an interesting complication in this calculation, namely the dependence of τ_{z} on the initial state (ψ of (4.3)). This complication is the reflection of a recurrent problem: What is a metastable state? For τ_{L} this question is not critical, since by the time the exponential decay sets in, transients have disappeared. But now it is the transients we study. Our choice in [14] was to minimize the second moment of the Hamiltonian, hence to maximize τ_Z. With this approach we found, with fairly rough approximations, that

$$\tau_Z^2 = \tau_L \tau_T \ . \tag{4.7}$$

Comparing this to (4.6), it is seen that for this kind of transition, the tunneling time is the jump time.

4.3.2 Time–Energy Uncertainty Principle

An interpretation of τ_J in terms of bandwidth and uncertainty relations can be found by combining (4.3), for τ_Z, with (4.1), for lifetime, τ_L. After some manipulation one obtains $(|\psi\rangle = |i\rangle$, the initial state)

$$\tau_J = \frac{\tau_Z^2}{\tau_L} = \frac{1}{\int \frac{dE}{2\pi\hbar} \frac{\rho(E)}{\rho(E_\psi)} \frac{|\langle E|H - E_\psi|\psi\rangle|^2}{|\langle f|H|\psi\rangle|^2}} \ . \tag{4.8}$$

Because of the orthogonality of the initial and final states, one can insert a "$-E_\psi$" into the Golden rule matrix element. Thus the ratio

$$\frac{\rho(E)}{\rho(E_\psi)} \frac{|\langle E|H - E_\psi|\psi\rangle|^2}{|\langle f|H|\psi\rangle|^2} \tag{4.9}$$

is of order unity when E passes through E_ψ. As E moves away from E_ψ a variety of patterns is possible, depending on the specific physical situation. One scenario is for this ratio to become smaller, mainly because with increasing energy deviation, $|E\rangle$ becomes rather different from $|i\rangle$ [18]. In any case, this ratio, whose numerator incorporates transitions to all possible on-shell and off-shell states, measures the ability of the Hamiltonian H to move the system away from its initial state. One thus has a band of accessible transition states.

With this perspective, τ_J is (the inverse of) an integral over energies (or frequencies) of an order unity function describing the modulation of the lowest band of accessible states. It follows that τ_J is the inverse bandwidth for the transition. This is a completely reasonable conclusion: you would like to create a situation where the system's transition is sudden. Your success is governed by the frequencies available. The accessibility of those frequencies is the essence of the bandwidth. This makes the jump time a reflection of a kind of time–energy uncertainty relation. As such it is a statement of this relation that is consistent with the views expressed in [19].

4.3.3 Reconciling Continuous Measurement with the QZE

The sequence of infinitely rapid projections envisioned in the usual derivation of the quantum Zeno effect is hardly the way measurements actually take place. Mostly they could be described as "continuous," in the following sense. An apparatus monitors a system and when some particular event takes place it is triggered and reports that event. Before that report, the apparatus, by its silence, is telling you, "No, the event has not yet taken place." If this picture is true, then one should expect *all* decay to be suppressed, since the unwavering attention of the apparatus should act like a continuous check – effectively with a zero time interval between measurements – that no decay has taken place.

This problem was addressed some years ago by several authors [20, 21, 22, 23], some of whom also wished to dispense with the (perhaps metaphysical) traditional notion of "measurement" and instead include the apparatus as part of the quantum system. They found that adding apparatus-like terms to the Hamiltonian could stop or slow the decay.

In recent work [24], I found that the important criterion for determining which "continuous" measurements could affect decay (or any transition) was a comparison of two quantities: the response time of the apparatus and the jump time of the system being measured. The essential physical idea is that no measurement is "instantaneous" and any apparatus represents a sequence of physical processes, first getting the signal to the apparatus and then having the apparatus register that signal – the latter typically involves irreversible amplification. What I found was that when the response time of the apparatus, τ_R, was on the order of τ_J the decay would be hindered. In particular, an apparatus with response time τ_R had the same effect in slowing the decay as idealized pulsed measurements with pulse time $\tau_{PM} = 4\tau_R$. Moreover, from the development of Sect. 4.2 of the present article, τ_{PM} should be less than or equal to τ_J for there to be a significant effect. Consequently the same criterion should hold for τ_R. This prediction—including the factor 4—has recently been tested experimentally [48] and indeed it turns out that the response time of the apparatus plays the role of pulse time with respect to the QZE.

The demonstration proceeds by making a model of a decay plus an apparatus that "continuously" monitors that decay. The model Hamiltonian and wave function for the decay alone are

$$H = \begin{pmatrix} 0 & \Phi^\dagger \\ \Phi & \omega \end{pmatrix} \qquad \text{and} \qquad \psi = \begin{pmatrix} x \\ y \end{pmatrix}, \tag{4.10}$$

where $x \in \mathbf{C}$, Φ and y are complex column vectors of the same dimension, and ω is a diagonal matrix. The Schrödinger equation (with $\hbar = 1$) becomes

$$i\dot{x} = \Phi^\dagger y, \qquad i\dot{y} = \omega y + \Phi x. \tag{4.11}$$

One can derive the decay rate from (4.11) by assuming the time dependence $\exp[-i(E - i\Gamma/2)t]$ for both x and y. One obtains

$$E - i\frac{\Gamma}{2} = \Phi^\dagger \frac{1}{E - \omega - i\Gamma/2} \Phi \longrightarrow \int d\omega \frac{\rho(\omega)|\phi(\omega)|^2}{E - \omega - i\Gamma/2} \tag{4.12}$$

where the arrow indicates a continuum limit, ρ is the density of states, and ϕ the appropriate limit of Φ. The usual manipulations now give

$\Gamma = 2\pi\rho(0)|\phi(0)|^2$, the Fermi–Dirac golden rule. The Zeno time for the state with $x = 1$ (and $y = 0$) is simply $\tau_z = 1/\sqrt{\Phi^\dagger\Phi}$.

The Hamiltonian in (4.10) can be thought of as describing a two-level atom coupled to the electromagnetic field. For $\psi^\dagger = (x^*, 0)$ the atom is in the unstable state (call this level #1), while $\psi^\dagger = (0, y^\dagger)$ describes the decayed atom (in level #2) with photon(s) emitted. As a monitoring device we imagine another system coupled to the atom that allows the atom to decay once more (to atomic level #3), emitting one or more additional photons, providing sufficient decoherence for this to be considered a measurement. The coupling strength between levels 2 and 3 will be thought of as adjustable (perhaps some function of an external electric field). Such a model is embodied in the following Hamiltonian

$$H = \begin{pmatrix} 0 & \Phi^\dagger & 0 \\ \Phi & \omega & \Theta^\dagger \\ 0 & \Theta & W \end{pmatrix}. \tag{4.13}$$

The additional levels, $\{W\}$, can be thought of as the apparatus and Θ is the 2–3 coupling. We assume that the levels are numerous enough and so distributed that the transition induced by this coupling is effectively irreversible.

To see how the combined system behaves we make a substitution similar to that done above: all components of the wave function are given the time dependence $\exp(-izt)$. One obtains

$$z = \Phi^\dagger \frac{1}{z - \omega - \Theta^\dagger \frac{1}{z-W}\Theta}\Phi. \tag{4.14}$$

In the usual way (which was implicit above), $1/(z - W)$ is evaluated using the formula $1/(x \pm i\epsilon) = P(1/x) \mp i\pi\delta(x)$. We write the result as

$$\Theta^\dagger \frac{1}{z - W}\Theta = \Delta E - i\frac{\Gamma_\theta}{2}. \tag{4.15}$$

This formula uses the reasonable assumption that Θ does not depend on which photon was emitted in the 1–2 transition. Γ_θ is the essential descriptor of the apparatus, indicating the rate at which it takes the atom from level #2 to level #3. The inverse of Γ_θ is thus the response time of the apparatus, which we denote τ_R. Equation (4.15) is inserted in (4.14) to yield

$$z = \Phi^\dagger \frac{1}{z - \omega - \Delta E + \frac{i}{2\tau_R}}\Phi. \tag{4.16}$$

We next assume that the response time is *so* small that its inverse dominates the $z - \omega - \Delta E$ term in the denominator of (4.16). The imaginary part of z is thus a transition rate *away from the initial excited state, in the presence of the observing apparatus*. Writing $\text{Im } z = -\Gamma_{\text{effective}}/2$, (4.16) implies

$$\Gamma_{\text{effective}} = \frac{4\tau_R}{\tau_z^2} \tag{4.17}$$

(using $\Phi^\dagger\Phi\tau_z^2 = 1$, which is still true for the full H, including the apparatus). If ΔE is itself comparable to $1/\tau_R$ there is a slight modification of (4.17), reducing $\Gamma_{\text{effective}}$, but unless $\tau_R\Delta E \gg 1$ this does not change our qualitative conclusions.

The expression (4.17) is to be compared to the effective decay rate when under *pulsed* idealized observation, as conventionally described in the QZE. From our (4.4), this rate is $\tau_{\text{PM}}/\tau_z^2$. Comparing this with (4.17), we see that *the same degree of hindrance is obtained for an apparatus with response time* τ_R *and pulsed measurements (projections) at intervals* τ_{PM}, *provided*

$$\tau_R = \tau_{\text{PM}}/4 \ . \tag{4.18}$$

Moreover, as discussed in Sect. 4.2, neither interruption will slow the decay unless it is $\lesssim \tau_J$. As indicated, the relation (4.18) has been checked experimentally [48].

Remark: Once one deals with Hamiltonians and ordinary unitary evolution (rather than mysterious wave function "collapses") both for the "system" and for the "apparatus," another perspective is opened for understanding the hindering of decay because of continuous, rapid-response observation [25]. One starts with a system (with Hamiltonian (4.10)), which has a continuum into which to decay. Coupling a detector to this can be thought of as changing the spectral properties of the combined system. In particular what it can do is push the energy of the excited level and the continuum into which it decayed away from one another. Thus the halting of decay occurs because there are no longer levels that match (including the photon energy) the energy of the excited atom. This is discussed in [25] and [24]. A continuous version of the *anti*-QZE [16] has a corresponding explanation.

4.3.4 Experiments on the Quadratic Regime of Decay

Atomic and nuclear transitions take place quickly, putting the times discussed in this chapter out of reach of contemporary measurement for those systems. In [10, 17] I estimated that for atomic transitions $\tau_J \sim 10^{-20}$ s. However, there is a recent experiment [26] where the potential seen by the particles, including a barrier, has a distance scale of a few hundred nanometers. This experiment, a measurement of Landau–Zener tunneling, has (for us) two benefits: the timescales are *much* longer and the potential can be quickly modified.

The experiment [26] consists of putting ultracold Na atoms in opposing laser beams that have a relatively small frequency difference between them. As a result the potential seen by each atom is time-dependent. Going into the atom's accelerated frame, the potential can be written $V = V_0\cos(2k_L x) + aMx$ ("tilted washboard"), where a is the acceleration arising from the frequency mismatch. Initially a small value of a is given to get rid of atoms not caught in the potential, after which it is sharply increased, giving rise to the tunneling situation. It is then switched off in such a way that it is possible to deduce what fraction of the atoms has escaped from the potential. For long times this quantity dies exponentially with a timescale of 70 μs. However, for short times it is demonstrably *not* exponential – it begins with what appears to be zero slope, tilts a bit, and then after roughly 5–10 μs goes over to the exponential form.

In [27] I showed how one could get a back-of-the-envelope estimate of the duration of this transient period. Recall that my derivation of the jump time, τ_J, was essentially a play-off of the quadratic and exponential time dependencies (ignoring finer nuances of the decay curve). Hence it should provide an estimate of the duration of the transient period in the experiment just described.

To make this estimate it was not necessary to calculate either of the quantities τ_Z or τ_L. Instead I appealed to the interpretation of τ_J as inverse bandwidth, (4.8). Which states are accessible to the atom in this potential? In fact it is a periodic potential and the atom is initially in its lowest band. If it were not for the tilt, the states in this band would be eigenstates of the Hamiltonian. The tilt couples these states and makes the otherwise stable states unstable. I thus take the band of accessible states to be just the band of Bloch states. But the width of this band can be calculated from the period of the potential and the mass of the atom. The bandwidth is just

$$E_b = \frac{\hbar^2 K^2}{2M} \, , \tag{4.19}$$

where $M \simeq 23 M_p$ and the wave number is $K = 1/94$ nm [28]. We evaluate

$$\tau_J = \frac{\hbar}{E_b} = \frac{2M}{\hbar K^2} \simeq 6 \; \mu s \, . \tag{4.20}$$

Comparing this to Figs. 3 or 4 in [26], it can be seen that the agreement is excellent. In evaluating (4.20) there are *many* powers of 10, and I found it remarkable that they condense to *any* reasonable result, much less one that was close to the actual experiment [29].

Remark: The closeness of the evaluated time in (4.20) to the experimental result should be considered fortuitous. My estimate depends on the wavelength of the light and the mass of the particle. It does *not* explicitly depend on the strength of the potential or on the rate of acceleration, features that are known to affect the duration of the nonexponential decay.

4.4 Passage Time

4.4.1 Fleming's Bound and the Ersak Equation

Given a Hamiltonian H and a state ψ, define $U(t) \equiv \exp[-i(H - E_\psi)t/\hbar]$, with $E_\psi \equiv \langle \psi | H | \psi \rangle$. We define a quantity related to what Fleming [12] calls the *integrity* amplitude

$$f(t) \equiv \langle \psi | U(t) | \psi \rangle \, . \tag{4.21}$$

Next, the function ϕ_t is defined to be that portion of the evolute that is orthogonal to ψ,

$$U(t)|\psi\rangle = f(t)|\psi\rangle + |\phi_t\rangle \, , \tag{4.22}$$

with $\langle \psi | \phi_t \rangle = 0$. Successive application of $U(t)$ and $U(t')$ to ψ, followed by left multiplication by $\psi^\dagger$, leads to

$$f(t + t') = f(t)f(t') + \langle \psi | U(t') | \phi_t \rangle \, . \tag{4.23}$$

Using the variable $-t'$, the adjoint of (4.22) is

$$\langle \psi | U(t') = \langle \psi | f^*(-t') + \phi_{-t'} \, . \tag{4.24}$$

Multiply this equation on the right by $|\phi_t\rangle$ to yield $\langle \psi | U(t') | \phi_t \rangle = \langle \phi_{-t'} | \phi_t \rangle$. When this is substituted in (4.23), we get

$$f(t + t') = f(t')f(t) + \langle \phi_{-t'} | \phi_t \rangle \, . \tag{4.25}$$

Fleming calls this the Ersak equation. Take the derivative of (4.25) with respect to t', set t' to zero, and use the fact (from (4.21)) that $\dot{f}(0) = 0$ to yield

$$\dot{f}(t) = -\langle \dot{\phi}_0 | \phi_t \rangle \, . \tag{4.26}$$

From the derivative of (4.22) it is clear that

$$\langle \dot{\phi}_0 | \dot{\phi}_0 \rangle = \frac{1}{\hbar^2} \langle \psi | (H - E_\psi)^2 | \psi \rangle \equiv \frac{(\Delta H)^2}{\hbar^2} \, . \tag{4.27}$$

For convenience we write $f \equiv g \exp(i\gamma)$, with g real and nonnegative and γ real. We apply the Schwarz inequality to (4.26):

$$|\dot{f}| \leq \frac{\Delta H}{\hbar} \sqrt{1 - g(t)^2} \, . \tag{4.28}$$

Using $|\dot{f}|^2 = \dot{g}^2 + \dot{\gamma}^2 g^2$, we immediately have

$$|\dot{g}| \leq \frac{\Delta H}{\hbar} \sqrt{1 - g^2} \, . \tag{4.29}$$

Finally, letting $g \equiv \cos\theta$ provides a bound on $\dot{\theta}$, specifically, $|\dot{\theta}| \leq \frac{\Delta H}{\hbar}$. Since g starts at 1, θ starts at 0, and it follows that

$$\theta(t) \leq \frac{\Delta H}{\hbar} t \, . \tag{4.30}$$

This gives our desired bound. Recalling the definition of g, it shows that no state can become orthogonal to itself in less than $\pi\hbar/2\Delta H$. But this last quantity is just $\pi\tau_z/2$, in our earlier notation.

This result was derived by Fleming [12] and leads us to define the passage time, $\tau_P \equiv \pi\tau_z/2$.

To confirm that the bound can in fact be attained, let $H = \alpha^2\sigma_x$ ($\sigma =$ Pauli spin matrices) and $\psi = |+\rangle$. Then $\tau_z = \hbar/\alpha$, and the system turns over in $\pi\tau_z/2$. This example, however, does not clarify the relations among the many times that have been defined in this article. Because there is no exponential decay in this case, τ_L is not clearly defined. If one takes it to be the time to first extinction, then τ_z, τ_L, τ_P, and τ_J are all essentially the same.

The example of the last paragraph is realized in a real-world system: chiral molecules that can exist in two reflection-related isomers. In practice those molecules for which the transition time between isomers is anything but microscopic appear as one or the other isomer, never the symmetric superposition that is the system's true ground state. This and similar phenomena have been attributed [30, 31] to a manifestation of the QZE. The idea is that merely by virtue of being in solution the molecules are constantly buffeted about and "observed," or decohered. The timescale for this is the inverse of the energy split between the theoretical symmetric and asymmetric states of the isomers, which is expected to be extremely long (hence the decay is subject to interruptions on the timescale of collisions in solution). But as remarked in the last paragraph, this situation does not distinguish among the various characteristic times, since all are the same.

4.4.2 Implications of the Bound in Measurements

As just shown, no quantum system, under unitary evolution alone, can become orthogonal to itself in less than τ_P, where τ_P is, up to a trivial factor, what we have called τ_z. In particular, for a given state, ψ, and given Hamiltonian, H,

$$\tau_P = \frac{\pi\tau_z}{2} = \frac{\pi}{2}\frac{\hbar}{\sqrt{\langle\psi|(H - E_\psi)^2|\psi\rangle}} . \tag{4.31}$$

Moreover, we showed that for at least one system, possessing only two levels, the bound is actually attained.

In general measurements, however, the Fleming bound may have little to do with the time the system needs to complete its transition. Thus the Landau–Zener tunneling experiment shows transitions within the first microsecond, although the jump time is $\sim 5\,\mu s$. In this case, since the measured τ_L is $\sim 70\,\mu s$, the Zeno time would presumably be the algebraic mean, $\sim 20\,\mu s$. There is no doubt, from the inspection of the data, that many transitions occur well before τ_P. How can that be?

The answer is that proper use of the bound requires that the Hamiltonian of the measurement apparatus be included in the "H" of Sect. 4.4.1. In general this can involve enormous energies, much larger than those of the system measured (were it in isolation). Thus, for the *full* system, τ_z^{full} may be extremely short, in particular shorter than even τ_J of the isolated system. In the tunneling experiment [26], one has a *time-dependent* Hamiltonian, reflecting the fact that controlling the value of the acceleration, a, as a function of time, is an important part of the successful performance of that experiment. Thus during the time that the crossed beams are turned on at their maximum a, the wave function of the atom in the tunneling experiment is partly in the well, partly in the barrier, partly outside. The sudden change in the confining potential means that the apparatus is interacting directly with the system, leading to a large energy spread. This remark is related to the story told to students when they first encounter barrier penetration: if you check whether the particle "really" is in the barrier, you would introduce enough energy to overcome that barrier. (A change in τ_Z due to measurement was also seen in [25], but there the "apparatus" coupling stops the decay rather than facilitates it.)

What I now show is that for some kinds of measurement the Fleming bound provides direct physical information. Moreover, serious attention to this bound can provide an experimental test for my own theory of what takes place in a quantum measurement [32].

We again consider the "apparatus" of Sect. 4.3.3. The Hamiltonian is

$$H = \begin{pmatrix} 0 & \Phi^\dagger & 0 \\ \Phi & \omega & \Theta^\dagger \\ 0 & \Theta & W \end{pmatrix} , \tag{4.32}$$

where H is a $(1+N+M) \times (1+N+M)$-dimensional matrix; N is the dimension of the diagonal matrix ω; and M $(\gg N)$ the dimension of the diagonal matrix W. The states of the system are of the form $\psi^\dagger = (x^*, y^\dagger, z^\dagger)$, $x \in \mathbf{C}$, $y \in \mathbf{C}^N$ and $z \in \mathbf{C}^M$. The physical scenario is this. The normalized state ψ with $x = 1$ represents an undecayed atom; call its level #1. It is coupled, perhaps electromagnetically, to states with $y \neq 0$, $z = 0$, via the coupling terms Φ. The "y" states represent the atom in its decayed state (call it #2) plus one or more photons. Now it may happen that the atom can continue its decay to a third level, or perhaps by varying an external field that decay can be encouraged. Let the atom in that third level plus *all* emitted photons (from both steps) correspond to the various "z" levels. As in Sect. 4.3.3, this second transition involves considerable decoherence and provides the irreversibility and amplification characteristic of the measurement process. Thus the way the rest of the world knows that the system has decayed from level 1 to level 2 is realized through the coupling, Θ, and the states with $z \neq 0$.

The important point is that *for this kind of apparatus–system coupling, there is* no *change in* τ_Z. It is still $\hbar/\sqrt{\Phi^\dagger \Phi}$. The key is that the measurement works by coupling to the decay products, *not* to the original state [33], thus leaving τ_Z and τ_P unchanged. For such measurements, the Fleming bound does not allow the state to be completely out of its original level, nor to be completely in any other, for $t < \tau_P$.

4.5 Experimental Discrimination among Quantum Measurement Theories and "Special States"

4.5.1 Testing the Foundations

Suppose you had an apparatus of the type described in Sect. 4.4.2, i.e., one that couples only to decay products (cf. (4.32)). If this were a system for which τ_z is known, then one could say with confidence that unitary evolution alone cannot bring the wave function entirely to the decay states before τ_P. What are the implications of this according to the Copenhagen interpretation of quantum mechanics? Answer: none. You can still (for $t < \tau_P$) measure the system to be in the decayed state (presumably, using *this* measurement apparatus), and as usual the probability of doing so would be the absolute value squared of the amplitude in the decayed state – no need for this to be unity, just strictly positive.

By contrast, according to the explanation for the definiteness of quantum measurements that I have proposed [32], you would only get a definite measurement of the decay when the entire wave function has entered the Hilbert space of decayed states. I will not review these ideas here, and refer the reader either to the indicated book or, for a less complete version, to [34].

This allows an explicit experimental test of my theory. A system is put in an unstable state and then shielded from the environment, except for an apparatus monitoring its state *indirectly*, i.e., by checking for decay products. For this system (for which I do not have a specific physical proposal yet) you would need to calculate or bound τ_P. I then predict *no* decays before τ_P, whereas the Copenhagen interpretation imposes no such ban (despite some relative reduction if the system is still in the quadratic decay regime).

Although complete blocking of the environment can be difficult (cf. [32]), the quest for quantum computers has in recent years developed experimental tools for just this goal. I look forward to exploring this further.

4.5.2 Special States for Decay

The motivation for this subsection is explained in [32]. Briefly, in Sect. 4.5.1 I indicated that according to my ideas no decay could take place until $t \geq \tau_P$. But what if $t = 2\tau_P$? Would the system *then* decay, i.e., exit *completely* from the state $x = 1$, as my theory requires? From the Hamiltonian (4.32) it does not look that way. For moderate Θ (hence τ_R) one gets the usual exponential decay: on a scale of τ_L the wave function gradually passes out of its initial state. Since, generally, $\tau_L \gg \tau_P$ this implies that at $2\tau_P$ most of the wave function is still in the undecayed state. My explanation for the manifest observations of decay at short times (but $> \tau_P$) is that there are *special* states of the environment for which the decay *does* go to completion, despite the fact that for the vast majority of environmental states this does not happen. Why Nature chooses these "special" states is discussed in [32]. What I wish to show in the present article is a special state for decay in the model Hamiltonian (4.32), or in something close to it.

The physical environment is not represented in (4.32). The main environmental richness is in the initial state of the ambient photon field when the atom is still in its level-1, undecayed state. But this requires a cross product of available photon states with the $(1 + N + M)$-dimensional Hilbert space I have heretofore considered. Instead of this, I will simplify by incorporating the field-initial-condition information in Θ itself. This quantity, in the rotating wave approximation, is of the form $\Theta = \sum_k |3\rangle\langle 2| a_k^\dagger$, with $a_k^\dagger$ the photon creation operator. (Multiple photon creation is also allowed.) If the field of preexisting photons (before the decay) is well occupied, both a_k and $a_k^\dagger$ can be approximated by $\sqrt{n_k}$, with n_k the occupation number of the kth photon mode. This means that the features of the environment appear as particular values of the components of Θ.

I have already presented something like this in [35]. I assumed that the environment fluctuates near the atom, effectively modifying Φ. With a particular $\Phi(t)$ the decay *is* complete by τ_P. However, this demonstration required beliefs about what the field could accomplish, beliefs that I did not explicitly justify.

I will show next that with a purely fixed set of interactions (Θ and Φ) the system will rapidly go completely over to an orthogonal state. The demonstration will not quite produce a state that makes it in τ_P, just $\sqrt{2}$ times that, but this establishes the main point.

With this in mind we break the subspace $\{(0, 0, z)\}$ ($z \neq 0$) into two pieces. One piece consists of a particular set of N levels (one for each dimension in the space $\{(0, y^\dagger, 0)\}$ with $y \neq 0$). We assume that each of these has the same energy as one of the "y" levels. (Recall this is the total energy, atom plus photons, so these levels correspond to the atom dropping to level-3 and emitting a photon of just the 2–3 energy difference, of which there are many.) At the same time, we assume that the occupation numbers of those levels in the ambient field are just such as to make the coupling to the "y" level with energy ω_k equal to that same ω_k. The coupling of the remaining degrees of freedom I call $\widetilde{\Theta}$, and the energies $\widetilde{W}$. The Hamiltonian and the wave function take the form

$$H = \begin{pmatrix} 0 & \Phi^\dagger & 0 & 0 \\ \Phi & \omega & \omega & \widetilde{\Theta}^\dagger \\ 0 & \omega & \omega & 0 \\ 0 & \widetilde{\Theta} & 0 & \widetilde{W} \end{pmatrix}, \qquad \psi = \begin{pmatrix} x \\ y \\ \zeta \\ \tilde{z} \end{pmatrix}. \qquad (4.33)$$

Now when most matrix elements of Θ are moderate, the passage out of the initial Hilbert subspace of undecayed atomic states is slow, on the order of τ_L. If it can be demonstrated that by using only the restriction of H to its first $2N + 1$ dimensions one can get decay in a time on the order of τ_P, then the remaining couplings and levels ($\widetilde{\Theta}$, etc.) will be negligible on that timescale. Therefore I restrict attention to the first $2N + 1$ levels and study the Hamiltonian and states

$$\widehat{H} = \begin{pmatrix} 0 & \varPhi^\dagger & 0 \\ \varPhi & \omega & \omega \\ 0 & \omega & \omega \end{pmatrix} , \qquad \psi = \begin{pmatrix} x \\ y \\ \zeta \end{pmatrix} . \tag{4.34}$$

Two approaches will be used to analyze the dynamics. First give ψ an overall dependence $\exp(-iEt)$ (with $\hbar = 1$). By the same manipulations that led to (4.14), E is found to satisfy

$$E = \varPhi^\dagger \frac{E - \omega}{E^2 - 2\omega E} \varPhi = \frac{1}{2} \left\{ \frac{1}{E\tau_Z^2} + \varPhi^\dagger \frac{1}{E - 2\omega} \varPhi \right\} . \tag{4.35}$$

As in (4.15), this becomes

$$E^2 = \frac{1}{2\tau_Z^2} + \frac{E}{2} \left(\Delta E - \frac{i}{4} \varGamma \right) . \tag{4.36}$$

(The denominator "4" for $\varGamma$ arises from the 2ω in (4.35).) Generally both ΔE and $\varGamma$ (which is the usual decay rate) are much smaller than $1/\tau_Z$, so that to a good approximation

$$E \approx \frac{1}{\sqrt{2}\tau_Z} - \frac{i}{16} \varGamma \tag{4.37}$$

(where ΔE is ignored). This implies that with the initial condition $x = 1$ the behavior of x will be $\cos(t/\sqrt{2}\tau_Z)$ to a very good approximation. This in turn implies that x will hit zero when $t = (\pi/2)\tau_Z\sqrt{2}$. That value differs by a factor $\sqrt{2}$ from the optimum defined by Fleming's bound. The point though is that with a bit of manipulation of the environment the decay has been speeded up from a scale of τ_L to one of τ_P. (If the coupling, ω, in $\widehat{H}$ is changed to $\alpha\omega$ and the ζ energies set to $\alpha^2\omega$, then the time for reaching orthogonality becomes $\tau_P(1 + 1/\alpha^2)^{1/2}$, which can be made closer to τ_P by increasing α.)

This result can also be obtained by looking at the time-dependent equations generated by the Hamiltonian of (4.34). They are

$$i\dot{x} = \varPhi^\dagger y , \quad i\dot{y} = \varPhi x + \omega(y + \zeta) , \quad i\dot{\zeta} = \omega(y + \zeta) . \tag{4.38}$$

Add and subtract the second and third equations, integrate the equation for the difference, substitute back for y, and finally take the derivative with respect to t to obtain

$$\ddot{x}(t) + \frac{1}{2\tau_Z^2} x(t) = -\frac{1}{2} \frac{\partial}{\partial t} \int_0^t \varPhi^\dagger e^{-2i\omega(t-s)} \varPhi x(s) \, ds . \tag{4.39}$$

Define $K(u) \equiv \varPhi^\dagger \exp(-i\omega u)\varPhi$. This is an important kernel for studying decay properties. Thus for unobserved decay (4.11) implies $\dot{x} = -\int_0^t K(u)x(t-u) \, du$. Although the possibilities for K's behavior are wide, for moderate times it typically drops rapidly, so that a reasonable approximation is $K(u) \approx (\varGamma/2)\delta(u)$. The normalization can be checked by plugging into the equation just written for $\dot{x}$. In (4.39) we have $K(2u)$, so that with the δ-function approximation we obtain

$$\ddot{x}(t) + \frac{\Gamma}{8}\dot{x} + \frac{1}{2\tau_z^2}x(t) = 0 . \qquad (4.40)$$

For times less than τ_J the δ-function approximation is not applicable, but the Zeno time is generally much longer and is the scale now considered. With initial conditions $x(0) = 1$, $\dot{x} = 0$ (from (4.38)) it follows that, to lowest order in Γ,

$$x(t) = \cos(t/\sqrt{2}\tau_z)e^{-\Gamma t/16} , \qquad (4.41)$$

which agrees with our previous result. An amusing perspective on the early vanishing of x is as the ultimate anti-QZE [16].

To further confirm that the approximations work, I have included Fig. 4.1. This is a numerical calculation of a decay that, in the absence of apparatus-induced "specializing" effects, would show normal exponential decay. For this calculation it is assumed that the coupling enhancement in the apparatus arising from the extra photons in the particular modes k (the extra factors "$\sqrt{n_k}$" mentioned earlier) only lasts for a period τ_P, after which the coupling returns to normal.

This time dependence of Θ illustrates the fact that the "specialness" of the microscopic state includes timing. The added coupling due to the ambient field is indeed ambient and once the transition is complete things return to

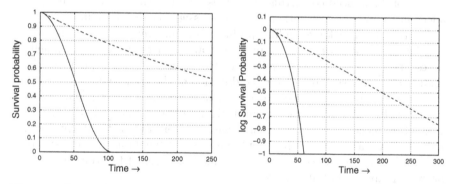

Fig. 4.1. Survival probability as a function of time, linear, and log plots. The *solid line* represents decay under the influence of the (matched photon) apparatus. The *dashed line* is ordinary decay, with no apparatus, and on the log plot shows appropriate linear decline. (In this system $\tau_Z \approx 48$ and a fit gives $\tau_L \approx 393$ yielding $\tau_J \approx 6$, which is too small for a deviation to be seen in this figure.) These are computer calculations of the survival probability with the Hamiltonians (4.10) and (4.34) (with transient coupling, as described in the text). From the analytic calculations, passage time should be $\sqrt{2}\tau_P \approx 106$, and as is evident the decay in the presence of the apparatus hits zero at a time close to this. The *dashed curve*, representing normal decay, is far from zero at this time. In this calculation $N = 101$, and continuation of the curve would eventually show quantum Poincaré recurrence

normal. If the reader is encountering my ideas for the first time and finds the choreography excessive, please be assured that the appearance of "unlikely" microscopic states has been addressed extensively. What is *likely* or *unlikely* is related to the thermodynamic arrow of time, and it is by exploring related foundational questions of statistical mechanics that I am able to argue for the plausibility of these ideas. If this has piqued your interest, see [32].

Remark: The states just exhibited are special states for quantum jumps. Another example, in which the environment plays an even more explicit role, is [36].

4.6 Discussion

Under the unitary evolution given by the formal mathematical structure of quantum mechanics, systems move gradually from state to state. For example, an unstable atom still has amplitude in its original state after many of its lifetimes. But in practice, which is to say in the lab, they go from being in one state to being in the other, seemingly instantaneously. This is the "quantum jump." Experiments that saw single-atom transitions [7, 8, 9] appear to confirm this perception. In the measurements, the system went from state to state in a time beneath the discrimination of the observers, whereas when the times spent in the unstable state were averaged, they recovered the lifetime of the atom.

The problem studied in this chapter is whether the "quantum jump" is indeed instantaneous or whether it could be assigned a duration, in theory and in experiment. The longstanding problem of tunneling time, in connection with barrier penetration, sets a precedent and is a guide. If that tunneling represents a process necessary for decay, then surely the associated time is a candidate (or a lower bound) for the duration of the transition.

Two characteristic times are defined here, *jump time* (τ_J) and *passage time* (τ_P). In general they are quantitatively different and it is the richness of quantum mechanics, as well as lingering questions about its interpretation, that allow two answers to what would be a well-defined question classically.

Both times use the *Zeno time*, τ_Z, defined in terms of the Hamiltonian of the system and its initial state as

$$\tau_Z \equiv \hbar \bigg/ \sqrt{\langle\psi|(H - E_\psi)^2|\psi\rangle} \qquad \text{with } E_\psi \equiv \langle\psi|H|\psi\rangle . \qquad (4.42)$$

The jump time, τ_J, takes what I consider to be a more traditional view and is defined in terms of the timescale needed to slow (à la the quantum Zeno effect) the decay. A "measurement" is an idealized projection leading to

$$\tau_J \equiv \tau_Z^2/\tau_L . \qquad (4.43)$$

This time shows up in several contexts. It is related to tunneling time [14], for those transitions where a physical barrier can be identified. Its inverse is the bandwidth of the Hamiltonian, in a kind of time–energy uncertainty principle that governs the ability of the system to change state. τ_{J} is also an indicator of the duration of the quadratic decay regime in both experiment [26] and in numerical calculations. (An illustration of the latter is Fig. 2 of [24].)

The jump time, τ_J, is motivated by considerations of the QZE. The usual formulation of that effect is in terms of pulsed measurements, say at (pulse) intervals τ_{PM}. One is then faced by the fact that apparently continuous measurements *do* allow decay. We resolve that paradox with a purely quantum calculation (no measurement theory "black magic") in which it is shown that the measurement apparatus *response time*; say τ_R, plays the role of pulse time, and for short enough τ_R observation will indeed inhibit decay. We obtain the relation $\tau_{PM} = 4\tau_R$ [24], a relation that has recently been confirmed experimentally [48].

The passage time, τ_P, arises from pure unitary evolution alone, sans interpretive steps. It is based on a bound [12] that shows that for *any* H and ψ the system cannot evolve to a state orthogonal to ψ in time less than τ_P, with

$$\tau_{\mathrm{P}} = \pi\tau_{\mathrm{Z}}/2 \ . \tag{4.44}$$

If you think of H as the Hamiltonian of the system alone, then it would appear that this bound has little to do with measurements. The "instantaneous" jump occurs outside the realm of unitary evolution (so they say) and could certainly happen faster than τ_P.

But I want to consider H to be the Hamiltonian of *the system and the measuring apparatus*. This is a view I have advocated for quite some time [37, 38, 34] and is the perspective taken by the many-worlds and decoherence interpretations of quantum mechanics.

But even among those who accept this view, there is still no consensus about the implication of τ_P for an actual measurement. In my theory [32] this bound implies that the apparatus could *not* detect a transition in less than τ_P, making this the ultimate transition time. (And in the present article an example was given of a special state that did manage the transition in close to τ_P.) Of course, one can still have detection in times less than the τ_P you would calculate using the system Hamiltonian alone, since the full passage time of system plus apparatus, $\tau_{\mathrm{z}}^{\mathrm{full}}$, is in general much shorter than the restricted one.

Finally, there is a particular kind of detection in which the presence of the apparatus does not change the passage time. This provides the possibility of an experimental test of my measurement theory. One of my current goals is to find a practical experimental setup in which this test can be made.

Acknowledgments

I am grateful to P. Facchi, E. Mihokova, D. Mozyrsky, and S. Pascazio for helpful discussions. This work is supported in part by the U.S. NSF under grants PHY 97 21459 and PHY 00 99471.

References

1. M.B. Plenio, P.L. Knight: The quantum-jump approach to dissipative dynamics in quantum optics, Rev. Mod. Phys. **70**, 101 (1998)
2. D. Mugnai, A. Ranfagni, L. S. Schulman: *Tunneling and its Implications* (World Scientific, Singapore 1997). Proc. Adriatico Research Conf., Trieste, Italy
3. N. Yamada: Speakable and unspeakable in the tunneling time problem, Phys. Rev. Lett. **83**, 3350 (1999)
4. L.S. Schulman, R.W. Ziolkowski: 'Path Integral Asymptotics in the Absence of Classical Paths'. In: *Path Integrals from meV to MeV*, ed. by V. Sa-yakanit et al. (World Scientific, Singapore 1989) pp. 253–278
5. D. Sokolovski, L.M. Baskin: Traversal time in quantum scattering, Phys. Rev. A **36**, 4604 (1987)
6. H.A. Fertig: Traversal-time distribution and the uncertainty principle in quantum tunneling, Phys. Rev. Lett. **65**, 2321 (1990)
7. J.C. Bergquist, R.G. Hulet, W.M. Itano, D.J. Wineland: Observation of quantum jumps in a single atom, Phys. Rev. Lett. **57**, 1699 (1986)
8. T. Sauter, W. Neuhauser, R. Blatt, P.E. Toschek: Observation of quantum jumps, Phys. Rev. Lett. **57**, 1696 (1986)
9. W. Nagourney, J. Sandberg, H. Dehmelt: Shelved optical electron amplifier: observation of quantum jumps, Phys. Rev. Lett. **56**, 2797 (1986)
10. L.S. Schulman: 'How quick is a quantum jump?', p. 121, in [2]
11. R.M. Eisberg: *Fundamentals of Modern Physics* (Wiley, New York 1961)
12. G.N. Fleming: A unitarity bound on the evolution of nonstationary states, Nuov. Cim. **16 A**, 232 (1973). Fleming cites [39] and [40] as partial sources.
13. For a review, see H. Nakazato, M. Namiki, S. Pascazio: Temporal behavior of quantum mechanical systems, Int. J. Mod. Phys. B **10**, 247 (1996)
14. L.S. Schulman, A. Ranfagni, D. Mugnai: Characteristic scales for dominated time evolution, Phys. Scr. **49**, 536 (1994)
15. A.G. Kofman, G. Kurizki: Acceleration of quantum decay processes by frequent observations, Nature **405**, 546 (2000)
16. Reversing the inequality suggests *faster* decay, and indeed this can happen [41, 42, 43, 44, 45, 46, 47, 15], although in the present work this feature arises from truncation of the series (4.2). Accelerated decay due to observation is called the *inverse quantum Zeno effect* and has been known as least as far back as the 1983 work of Lane [41]. A more general formulation was given by Pascazio and Facchi around 1996. They found that for typical electromagnetic coupling frequent measurement could probe regions of the spectrum where the coupling was stronger, and decay enhanced. Moreover, this effect can occur for projections that occur at times much longer than τ_J of (4.6). Similar general results are found in [15].
17. L.S. Schulman: Observational line broadening and the duration of a quantum jump, J. Phys. A **30**, L293 (1997)
18. Another scenario is for the ratio (4.9) to grow to a distant maximum. This is the situation for the inverse QZE, mentioned above [16].
19. J. Hilgevoord: The uncertainty principle for energy and time, Am. J. Phys. **64**, 1451 (1996)
20. A. Sudbery: The observation of decay, Ann. Phys. **157**, 512 (1984)
21. K. Kraus: Measuring processes in quantum mechanics I. Continuous observation and the watchdog effect, Found. Phys. **11**, 547 (1981)

22. A. Peres: Zeno paradox in quantum theory, Am. J. Phys. **48**, 931 (1980)
23. A. Peres: 'Continuous Monitoring of Quantum Systems'. In: *Information, Complexity and Control in Quantum Physics* (Springer, Berlin 1987) p. 235
24. L.S. Schulman: Continuous and pulsed observations in the quantum Zeno effect, Phys. Rev. A **57**, 1509 (1998)
25. E. Mihokova, S. Pascazio, L.S. Schulman: Hindered decay: quantum Zeno effect through electromagnetic field domination, Phys. Rev. A **56**, 25 (1997)
26. S.R. Wilkinson, C.F. Bharucha, M.C. Fischer, K.W. Madison, P.R. Morrow, Q. Niu, B. Sundaram, M.G. Raizen: Experimental evidence for non-exponential decay in quantum tunneling, Nature **387**, 575 (1997)
27. L.S. Schulman: Jump time in Landau–Zener tunneling, Phys. Rev. A **58**, 1595 (1998)
28. Q. Niu, M.G. Raizen: How Landau–Zener tunneling takes time, Phys. Rev. Lett. **80**, 3491 (1998)
29. Of course all those powers of 10 depend on the units used. Since E_b (of (4.19)) is a characteristic energy in this problem, the appearance of its associated time scale should not be unexpected. However, other time and energy scales are present. The height of the potential defining the bands is one such quantity (E_b uses *only* the laser wavelength) although in the actual experiment [26] it was not all that different ($h/V_0 \approx 20\mu s$). Lifetime (as opposed to jump time) is another characteristic time and is famous for having large dimensionless numbers that confound dimensional analysis alone (i.e., there is the well known sensitivity of $\exp(-\int \sqrt{2m(V-E)}dx/\hbar)$, and the "prefactor" is also subject to large variation). In the case of this experiment the lifetime is roughly 70 μs.
30. R.A. Harris, L. Stodolsky: On the time dependence of optical activity, J. Chem. Phys. **74**, 2145 (1981); J.A. Cina, R.A. Harris: Superpositions of handed wave functions, Science **267**, 832 (1995); R. Silbey, R.A. Harris: Tunneling of molecules in low-temperature media: an elementary description, J. Phys. Chem. **93**, 7062 (1989)
31. M. Simonius: Spontaneous symmetry breaking and blocking of metastable states, Phys. Rev. Lett. **40**, 980 (1978)
32. L.S. Schulman: *Time's Arrows and Quantum Measurement* (Cambridge University Press, Cambridge 1997)
33. The term "decay products" refers to states in Hilbert space. It is not the physical atom that is or is not a decay product – in fact, the atom in level 1, 2, or 3 appears in every state.
34. L.S. Schulman: Definite quantum measurements, Ann. Phys. **212**, 315 (1991)
35. L.S. Schulman: 'A Time Scale for Quantum Jumps'. In: *Macroscopic Quantum Tunneling and Coherence*, ed. by A. Barone, F. Petruccione, B. Ruggiero, P. Silvestrini (World Scientific, Singapore 1999)
36. L.S. Schulman, C.R. Doering, B. Gaveau: Linear decay in multi-level quantum systems, J. Phys. A **24**, 2053 (1991)
37. L.S. Schulman: Definite measurements and deterministic quantum evolution, Phys. Lett. A **102**, 396 (1984)
38. L.S. Schulman: Deterministic quantum evolution through modification of the hypotheses of statistical mechanics, J. Stat. Phys. **42**, 689 (1986)
39. I. Ersak: Sov. J. Nucl. Phys. **9**, 263 (1969)
40. F. Lurçat: Strongly decaying particles and relativistic invariance, Phys. Rev. **173**, 1461 (1968)

41. A.M. Lane: Decay at early times: larger or smaller than the Golden Rule?, Phys. Lett. A **99**, 359 (1983)

42. W.C. Schieve, L.P. Horwitz, J. Levitan, Numerical study of Zeno and anti-Zeno effects in a local potential model, Phys. Lett. A **136**, 264 (1989)

43. P. Facchi, S. Pascazio: Quantum Zeno effects with "pulsed" and "continuous" measurements, preprint, quant-ph/0101044

44. S. Pascazio: Quantum Zeno effect and inverse Zeno effect. In: *Quantum Interferometry*, ed. by F. De Martini et al. (VCH Publishing Group, Weinheim, 1996) p. 525

45. P. Facchi, S. Pascazio: Spontaneous emission and lifetime modification caused by an intense electromagnetic field, Phys. Rev. A **62**, 023804 (2000)

46. P. Facchi, H. Nakazato, S. Pascazio: From the quantum Zeno to the inverse quantum Zeno effect, quant-ph/0006094, Phys. Rev. Lett. **86**, 2699 (2001)

47. P. Facchi, S. Pascazio: Quantum Zeno and inverse quantum Zeno effects, Prog. Optics **42**, ed. by E. Wolf (Elsevier, Amsterdam 2001)

48. E.W. Streed, J. Mun, M. Boyd, G.K. Campbell, P. Medley, W. Ketterle, and D.E. Pritchard, Continous and Pulsed Quantum Zeno Effect, Phys. Rev. Lett. **97**, 260402 (2006)

Bohm Trajectory Approach to Timing Electrons

C. Richard Leavens

Institute for Microstructural Sciences, National Research Council of Canada, Ottawa, ON, Canada K1A 0R6

5.1 Introduction

The timing of quantum events and of the duration of quantum processes is an area of quantum theory that has been quite active for more than two decades. Despite much effort by many people using a wide-ranging variety of approaches based on conventional quantum mechanics, there are still controversial issues to be resolved and considerable work remaining to be done. Even the extent to which this is true is controversial. In the first edition of this book, Egusquiza, Muga, and Baute [1] concluded "the long prevalent view that there is no place for ideal time observables has been superseded" and "we do have the (admitted elementary) tools for dealing with the simplest ones without in any way distorting the standard framework of quantum mechanics." This contrasts with the recent statement of Dürr, Goldstein, and Zanghì [2] that "Time measurements ... are particularly embarrassing for the quantum formalism." Faced with such strong differences of opinion, it seems worthwhile to heed the advice of David Bohm that "there should be a kind of dialogue between different interpretations"[3]. This chapter was written with this attitude in mind. Arrival time at a spatial point – a simple example of an event time – and dwell, transmission, and reflection times for a finite spatial region – simple examples of time durations – are analyzed for Dirac and Schrödinger electrons using the "Bohm trajectory approach," i.e. within the framework of de Broglie's and Bohm's [2, 3, 4, 5, 6, 7, 8, 9] causal alternative to quantum mechanics. Then several approaches to timing quantum particles based on the conventional theory are discussed from this point of view.

The use of a trajectory approach to timing quantum particles is motivated in Sect. 5.2. In Sect. 5.3 the essentials of Bohm's ontological interpretation of quantum theory are sketched and then applied to timing electrons. Some conventional approaches are discussed from the perspective of Bohmian mechanics in Sect. 5.4. Spin-dependent arrival-time distributions for nonrelativistic electrons are considered in Sect. 5.5. A recent claim based on a gedanken

C. R. Leavens: *Bohm Trajectory Approach to Timing Electrons*, Lect. Notes Phys. **734**, 129–162 (2008)
DOI 10.1007/978-3-540-73473-4_5

protective measurement that Bohm trajectories are not real is addressed in Sect. 5.6. Concluding comments are made in Sect. 5.7.

5.2 Motivation for Using a Trajectory Approach

For an ensemble of quantum particles prepared in the initial state $\psi_0(x) \equiv \psi(x, t = 0)$, theoretical expressions [10, 11, 12, 13, 14, 15, 16] for the intrinsic (or ideal) distribution $\Pi(T; X; \psi_0)$ of their arrival times T at $x = X$ are usually written – for those that do arrive – as the sum of two terms $\Pi_+(T; X; \psi_0)$ and $\Pi_-(T; X; \psi_0)$, corresponding to arrivals from the left and from the right, respectively, with no left–right interference term. Think classically for the moment so that the possibility of such an interference term is not an issue, to say that a point particle arrives at $x = X$ at the time $t = T$ from the left (right) means that the particle is located at $x = X$ at $t = T$ and that it was in the region $x < X$ ($x > X$) at all times t in the interval $T - \Delta t \leq t < T$ for some nonzero Δt. This concept is a simple, clear, and meaningful one for a particle that moves along a continuous trajectory with a well-defined *finite* velocity at each instant of time. Returning to the quantum case, one can attempt to maintain this simplicity and clarity by adopting an approach involving continuous particle trajectories, either "virtual" or "real," and using a relativistic wave equation to avoid infinite velocities. For the special case of electrons, considered here, at least two such approaches come to mind. One is based on Feynman's path integral derivation [17] of the $1 + 1$ dimensional free-electron Dirac equation using a "checkerboard" model in which particles move along zigzag paths in space–time always at the vacuum speed of light c. It has recently been shown [18] that, when the finite correlation length for reversals of direction discovered by Jacobson and Schulman [19] is taken into account, this model leads to an arrival-time distribution for free motion in $1 + 1$ dimensions with no left–right interference term. The results for $\Pi_+(T; X)$ and $\Pi_-(T; X)$ are identical to those presented without derivation or discussion over 30 years ago by Wigner [20].[1] In another approach that comes to mind, the trajectory method based on the causal version of the de Broglie–Bohm ontological interpretation of relativistic quantum mechanics [3, 7, 8], the particle speed associated with a Dirac electron cannot exceed c and there is no left–right interference term in $\Pi(T; X)$. Both approaches are simple and clear but whether or not they are physically meaningful is open for discussion, particularly since the two arrival-time distributions are not identical. The first approach is based on the implicit assumption that the arrival-time distribution calculated using the virtual paths of the checkerboard model is identical to that for actual electrons, and the second approach is based on the explicit postulate that Bohm trajectories are real. It is the latter approach – the Bohm trajectory approach– as applied to electrons, that is the primary focus of this

---

[1] Wigner's expression for $\Pi_\pm$ is not restricted to the special case of free evolution.

chapter. This is developed in Sect. 5.3 after a quick review of the relevant essentials of Bohm's causal interpretation of quantum theory.

5.3 Bohm's Ontological Interpretation of Quantum Theory

5.3.1 Brief Introduction

Much of the work on characteristic times for quantum particles in terms of (assumed) real trajectories, as opposed to virtual paths such as Feynman's, has been carried out within the framework of Bohm's ontological interpretation of quantum mechanics [2, 3, 4, 5, 6, 7, 8, 9]. In the causal version of the theory, tailored to the single-particle problem of interest here, it is postulated that an electron propagating in a potential $V(r, t)$ is an actual point-like particle *and* an accompanying wave $\psi(r, t)$, which probes the potential and guides the particle's motion accordingly so that it has a deterministically well-defined position $r(t)$ and velocity $v(t)$ at each instant of time t. It is also postulated that the particle's equation of motion is

$$v(t) \equiv dr(t)/dt = v(r, t)_{r=r(t)} = \left.\frac{J(r, t)}{\rho(r, t)}\right|_{r=r(t)} . \tag{5.1}$$

For a Dirac electron (see Appendix)

$$\rho(r, t) \equiv \psi^\dagger(r, t)\,\psi(r, t)\,,\, J(r, t) \equiv c\psi^\dagger(r, t)\,\widehat{\alpha}\psi(r, t)\,, \tag{5.2}$$

where the four-component guiding field $\psi(r, t)$ is the appropriate solution of Dirac's relativistic wave equation and $c\widehat{\alpha}$ is the Dirac velocity operator.[2] We are using the minimal approach of Bell [2, 6] in which the dynamical properties of an electron usually associated with spin follow solely from the assumption that the spatial motion of the point-like particle is guided by a multicomponent wave function. In particular, there is no assumption of a mechanical spin somehow associated with the particle. We are also assuming that there is a very low energy regime for which Dirac's original interpretation of $\rho(r, t)$ and $J(r, t)$ as single-particle probability and probability current densities, respectively, provides an adequate approximation to reality. In any case, the relativistic velocity defined by (5.1) and (5.2), regardless of its physical meaning, cannot exceed the vacuum speed of light c in absolute value [3, 7].

---

[2] The local expectation value of a property represented by an operator $\widehat{O}$ is defined to be $\mathrm{Re}[\psi^\dagger(r, t)\widehat{O}\psi(r, t)]/\psi^\dagger(r, t)\psi(r, t)$ [7]. Although the particle velocity $v(r, t)$ in (5.1) is identical to the local expectation value of the velocity operator $c\widehat{\alpha}$, the relevant expression for the square of the particle velocity is $\{[\psi^\dagger(r, t)c\widehat{\alpha}\psi(r, t)]/\psi^\dagger(r, t)\psi(r, t)\}^2 \leq c^2$ rather than the local expectation value $\mathrm{Re}[\psi^\dagger(r, t)(c\widehat{\alpha})^2\,\psi(r, t)]/\psi^\dagger(r, t)\psi(r, t) = 3c^2$. (See Appendix.)

The equation $v(r,t) = J(r,t)/\rho(r,t)$ for the velocity field and (5.7)–(5.24) below have also been applied to Schrödinger electrons using the standard nonrelativistic expressions

$$\rho(r,t) = \psi^*(r,t)\psi(r,t) \ , \quad J(r,t) = (\hbar/m)\,\mathrm{Im}[\psi^*(r,t)\boldsymbol{\nabla}\psi(r,t)] \ , \qquad (5.3)$$

where the guiding field $\psi(r,t)$ is the appropriate solution of the t-dependent Schrödinger equation. The resulting equation of motion is the simplest that is Galilean and time-reversal invariant [9], but *for systems with more than one spatial dimension* the probability current and hence the particle equation of motion (5.1) are not uniquely defined within nonrelativistic quantum mechanics [21, 22]. However, Holland [23] has recently shown that the probability current density deduced from the continuity equation is uniquely defined for Dirac electrons when Lorentz covariance is imposed and indeed given by the standard result $c\psi^\dagger(r,t)\widehat{\alpha}\psi(r,t)$. In addition, he emphasized that its nonrelativistic limit for a spin eigenstate in zero magnetic field is not just the standard Schrödinger result but contains a spin-dependent term

$$J(r,t;\hat{s}) = (\hbar/m)\{\mathrm{Im}[\psi^*(r,t)\boldsymbol{\nabla}\psi(r,t)] + \mathrm{Re}[\psi^*(r,t)\boldsymbol{\nabla}\psi(r,t)] \times \hat{s}\}, \quad (5.4)$$

where

$$s \equiv (\hbar/2)\,\hat{s} \equiv (\hbar/2)\,\chi^\dagger\widehat{\sigma}\chi \qquad (5.5)$$

is the spin vector associated with the spin eigenstate χ, a two-component spinor normalized to unity $(\chi^\dagger\chi = 1)$. The nonrelativistic particle is, in effect, guided by the two-component wave function $\psi(r,t;\hat{s}) \equiv \psi(r,t)\chi$ and the original expression for the nonrelativistic velocity field should then be replaced by

$$v(r,t;\hat{s}) \equiv J(r,t;\hat{s})/|\psi(r,t)|^2 \ . \qquad (5.6)$$

Hence, (5.1) and (5.7)–(5.24) apply to Schrödinger electrons only when the spin-dependent term in the nonrelativistic probability current density is either zero or negligible. Section 5.5 considers some situations in which this term is important. Since the publication of the first edition of this book, Ali, Majumdar, Home, and Sengupta [24] have generalized Holland's uniqueness result so that it applies to the quantum particles described by any consistent relativistic wave equation.

Given the initial wave function $\psi(r,0)$ and particle position $r^{(0)} \equiv r(0)$, the subsequent motion is uniquely determined by integration of the time-evolution equation for $\psi(r,t)$ and the equation of motion (5.1) for $r(t)$ to obtain the particle trajectory $r(r^{(0)},t)$. A very useful property of these trajectories follows directly from the fact that the equation of motion is first order in time: trajectories with different $r^{(0)}$, but the same $\psi(r,0)$, do not intersect (or even touch) each other – if $r_1^{(0)} \neq r_2^{(0)}$ then $r(r_1^{(0)},t) \neq r(r_2^{(0)},t)$ for any t.

In Bohm's ontological interpretation $\rho(r,t)\,dr$ is, for an ensemble of identically prepared electrons, interpreted as the probability of the particle component of the electron *being* in the volume element dr at time t.

This is postulated only for the initial time $t = 0$, the continuity equation $\partial\rho/\partial t + \mathbf{\nabla} \cdot \mathbf{J} = 0$ and the equation of motion (5.1) together guaranteeing that it is then true for all subsequent times. This realistic, "*being*" rather than "*being found*," interpretation is not based on the concept of "measurement." Hence, the internal consistency of Bohm's theory – in contrast to conventional quantum theory – does not necessitate that it be possible (in principle at least) to *find* the single electron of interest at time t in some infinitesimal volume $\mathbf{dr}$ and thereby localize its wave function within $\mathbf{dr}$. This is problematic in the case of Dirac electrons where attempting such confinement can generate particle–antiparticle pairs. Within Bohmian mechanics there is, at least in principle, a simple way around this difficulty if both the initial state $\psi(\mathbf{r}, 0)$ and the potential $V(\mathbf{r}, t)$ are known to sufficient accuracy. Considering the 1D case for simplicity, suppose that one wants to know the position of the particle component of an electron at a specific time t' to an accuracy δx that is not very much larger than the Compton wavelength $\lambda_c \equiv \hbar/mc$ so that one needs to worry about the disruptive effect of particle–antiparticle creation and annihilation processes induced during the attempted measurement. A solution is to wait until a time t'' at which the wave packet has spread by a factor F that is sufficiently large so that a coarse position measurement with a spatial resolution of $F\delta x$ is not a problem in this regard. One measures the particle's position at time t'' and finds that it is within a particular interval of width $F\delta x$. Using the nonintersecting property of Bohm trajectories one calculates the two trajectories that go through the endpoints of that interval at time t'', backward in time to t', to retrodict the endpoints of the much smaller interval, of width $O(\delta x)$, in which, according to the theory, the particle was located at time t'.

Despite the radical departure of Bohm's theory from the conventional one, for the standard impulsive "measurement" of any system observable at an instant of time chosen by the experimentalist, it predicts precisely the same statistical distribution of apparatus pointer positions.[3] Moreover, for the important case of a strong, projective measurement it provides a resolution of the measurement problem: because of decoherence there is an effective but not actual collapse of the entangled system-apparatus wave function to the eigenstate component(s) selected by the actual positions of point-like particles – the other components, although still existing, play no further role in guiding the particles. The arrival time is not in the above class of quantities: typically, the experimentalist chooses the plane $x = X$, not the time T, of arrival [25]. Hence, it is not a foregone conclusion that the distribution of particle arrival times predicted by the two theories will be identical.

In Bohm's causal theory, uncertainty enters only through the probability distribution $\rho(\mathbf{r} = \mathbf{r}^{(0)}, t = 0)$ for the unknown initial position $\mathbf{r}^{(0)}$ of the particle. There is no explicit reference to initial momentum $\mathbf{p}^{(0)}$ because it is uniquely specified by $\mathbf{r}^{(0)}$ for a given initial wave function $\psi(\mathbf{r}, 0)$ and

---

[3] The interpretation of pointer positions in terms of what they reveal about the microscopic system – if anything – can be very different however.

potential $V(r,t)$. Other than this, calculation of the statistical properties of a particle proceeds in much the same way as in classical statistical mechanics. For example, the probability distribution for some particle property f, which is defined for all trajectories, is given by

$$\Pi(f) \equiv \int_{all\ space} dr^{(0)}\ \rho(r^{(0)},0)\ \delta[f - f(r^{(0)})]\ , \qquad (5.7)$$

where $f(r^{(0)})$ is the value of the property for a particle following the trajectory $r(r^{(0)},t)$. (For particle properties that are not defined for some trajectories it is necessary to restrict the range of integration in (5.7) to exclude those trajectories and to normalize the resulting distribution accordingly.) A good example is the distribution of first-exit times from a 3D region bounded by a surface S and the distribution of the corresponding exit positions on S. This is one of the problems considered by Daumer and collaborators [26] who applied Bohmian mechanics to the scattering of a quantum particle in 3D. They identified $|J(r,t) \cdot dS(r)|$ with the (unnormalized) distribution of particle crossing times through the surface element $dS(r)$ and also constructed a rigorous proof of the free flux-across-surfaces theorem, which is fundamental to a convincing derivation of the standard expression for the differential cross section [27]. In Sect. 5.3.2 we consider the simpler problem of deriving the arrival-time distribution for the 1D case.

5.3.2 Derivation of a General 1D Intrinsic Arrival-Time Distribution

In Bohmian mechanics wave/particle duality means that a single electron is at all times both a point-like particle *and* a guiding wave. In the Bohm trajectory approach to timing electrons it is only the particle component of the electron that is being clocked. Because the particle is point-like and has a well-defined continuous trajectory it is straightforward to define and derive an expression for $\Pi(T;X)$. If the arrival-time detector is not included in the Hamiltonian, as is the case here, then this is an expression for the intrinsic arrival-time distribution. In 1D the derivation [28] is greatly simplified when one takes into account the nonintersecting property of the trajectories. This means that only a single distinct Bohm trajectory contributes to the probability current density $J(x,t)$ and determines its sign at any particular space–time point (x,t). This is completely different from the situation for classical ensembles where the flux $J(x,t)$ can have contributions from a variety of distinct trajectories passing simultaneously through x at time t, from either the same or both directions.

Consider the complete set of starting points $x^{(0)}$ for each of which the associated trajectory $x(x^{(0)},t)$ reaches $x = X$ at least once at some time(s) subsequent to $t = 0$. Because the trajectories do not cross or touch each other, this set must consist of a single continuous interval, say $[x_a^{(0)}, x_b^{(0)}]$. Again because of the nonintersecting property, there is one and only one value of

$x^{(0)}$ in the range $[x_a^{(0)}, x_b^{(0)}]$ for which the trajectory $x(x^{(0)}, t)$ reaches X at a *particular value of* T within the support of $\Pi(T; X)$. Even if that trajectory reaches X more than once, only one of its arrival times, say $T(x^{(0)})$, is equal to the specified value of T. Hence, the (unnormalized) arrival-time distribution is

$$\int_{x_a^{(0)}}^{x_b^{(0)}} dx^{(0)} \, \rho(x^{(0)}, 0) \, \delta[T - T(x^{(0)})] \,. \tag{5.8}$$

Now,

$$\delta[x(x^{(0)}, t) - X]|_{t=T} = \left.\frac{\delta[t - T(x^{(0)})]}{|dx(x^{(0)}, t)/dt|}\right|_{t=T} = \left.\frac{\delta[t - T(x^{(0)})]}{|v[x(x^{(0)}, t), t]|}\right|_{t=T} \tag{5.9}$$

contains only a single term and (5.8) becomes

$$|v(X, T)| \int_{x_a^{(0)}}^{x_b^{(0)}} dx^{(0)} \, \rho(x^{(0)}, 0) \, \delta[x(x^{(0)}, T) - X] \,. \tag{5.10}$$

The integral is just the probability density $\rho(X, T)$ and (5.10) reduces to

$$|v(X, T)| \, \rho(X, T) = |J(X, T)| \,, \tag{5.11}$$

using (5.1). Normalization then gives

$$\Pi(T; X) = \frac{|J(X, T)|}{\int_0^\infty dt \, |J(X, t)|} \tag{5.12}$$

for the probability distribution of arrival times *for those particles that actually reach* X *subsequent to* $t = 0$. This is not defined if the denominator is zero, i.e. no particle in the ensemble ever reaches X subsequent to $t = 0$. Nor it is defined if the denominator is infinite as, for example, would be the case for periodic motion of a set of trajectories back and forth through $x = X$ forever [29]. In the latter case, the arrival-time density $|J(X, T)|$ is still of interest, even though it is not normalizable.

In general the distribution should be supplemented with a number giving the fraction of particles in the ensemble that reach X at least once.

It follows from (5.1) for the velocity field $v(x, t)$ that $J(X, T) > 0$ corresponds to a particle arriving at $x = X$ at $t = T$ from the left $(+)$ and $J(X, T) < 0$ corresponds to a particle arriving at X at time T from the right $(-)$. Hence, (5.12) can be rewritten as

$$\Pi(T; X) = \Pi_+(T; X) + \Pi_-(T; X) \tag{5.13}$$

with

$$\Pi_\pm(T; X) = \pm J_\pm(X, T) \left/ \int_0^\infty dt \, [J_+(X, t) - J_-(X, t)] \geq 0 \,, \right. \tag{5.14}$$

where

$$J_\pm(x,t) = J(x,t)\,\Theta[\pm J(x,t)]\,. \tag{5.15}$$

It should be noted that although (5.13) and (5.14) also hold for classical particles, the decomposition (5.15), as a general result valid for any ensemble of particles and for any potential, is peculiar to Bohmian mechanics. Wigner's results for $\Pi_\pm(T;X)$ are given by (5.14) with

$$J_\pm(x,t) = J_\pm^W(x,t) \equiv (1/2)[J(x,t) \pm c\rho(x,t)]\,. \tag{5.16}$$

Substituting (5.16) into (5.13) gives

$$\Pi(T;X) = \Pi^W(T;X) \equiv \rho(X,T)/\int_0^\infty dt\,\rho(X,t) \tag{5.17}$$

which is identical to the so-called "time-of-presence" distribution [30].

The arrival-time distribution (5.12) can also be readily decomposed into contributions from first, second, third, etc. arrivals using efficient numerical methods based on the nonintersecting property of the trajectories [31, 32]. For a few examples of calculated intrinsic arrival-time distributions, see [33, 28].

Grübl and Rheinberger [29] have applied Bohmian mechanics to the problem of calculating measured as opposed to intrinsic arrival-time distributions. They did, however, use the simplifying assumption that the presence of the detector does not significantly perturb the Bohm trajectories of the particles of interest. Regarding (5.12) for the intrinsic arrival-time probability distribution they said the following: "If one assumes that the detector clicks each time it intersects with the particle's Bohmian trajectory," (5.12) "indeed yields the probability density of clicks. This seems to be a reasonable idealization if the detector is active during a short time interval." This time interval is presumably so short that the probability of it spanning two arrivals for a single Bohm trajectory is negligible relative to that for a single arrival. This idealization also involves the active volume of the detector being sufficiently well localized about $x = X$. They then formulated a "detection probability for detectors sensitive to quite arbitrary space–time domains" paying special attention to the very different and more practical situation in which the detector, activated at $t = 0$, remains active until and only until the detection of the first arrival of the particle. Calculated detection probabilities for various case studies are presented in [29, 34].

Nogami, Toyama, and van Dijk [35] recently used Bohmian mechanics to provide a pictorial representation of various interesting features, such as multiple exits, of the escape of a particle from a potential well through a barrier. They considered a quantum particle prepared at $t = 0$ in the state $\psi(\mathbf{r},0) = \psi(r,0) = (1/r)(2/a)^{1/2}\sin(\pi r/a)\Theta(a-r)$ in the central potential $V(\mathbf{r}) = V(r) = (\lambda/a)\delta(r-a)$ with $\lambda > 0$ and $a > 0$. Their starting point was an exact analytic expression [36] for $\psi(r,t)$ for this model of a decaying system that can be evaluated accurately for any value of r and t.

So far we have said nothing about timing the arrival of the guiding wave. It is clear that clocking a specific feature, e.g., the centroid, of a wave packet undergoing unitary evolution results in a distribution of arrival times for a pure state ensemble that is $n^{-1}\Sigma_{i=1}^{n}\delta(T-T_i)$ where in *each* member of the ensemble the chosen feature reaches X at *the same* n times $\{T_i\,|\,i=1,...,n\,\}$ with $n>0$, if possible, for a sensible choice of arrival-time marker.[4] In general, such a distribution has little to do with experimental arrival-time distributions, even when $n=1$.

5.3.3 Intrinsic Transmission and Reflection Times

Consider an ensemble of a very large number of identically prepared single-particle 1D scattering experiments. In each, an electron with the same initial wave function $\psi(x, t=0)$ is incident from the left on the potential barrier $V(x,t)$ assumed to be zero outside the range $0 \le x \le d$ and to be nonnegative inside. It is assumed that the initial wave function is normalized to unity and is localized far enough to the left of the barrier so that the integrated probability density $\rho(x,0)$ from $x=0$ to ∞ is negligibly small compared to the transmission probability

$$P_T \equiv \int_d^\infty dx\, \rho(x, t_\infty)\,, \tag{5.18}$$

where the scattering process is essentially completed for $t \ge t_\infty$. A question of long-standing interest is "What is the average time $\tau_{T(R)}(x_1, x_2)$ spent in the region $x_1 \le x \le x_2$ subsequent to $t=0$ by those electrons that are ultimately transmitted (reflected) in such scattering experiments?".[5] The quantity $\tau_T(0, d)$ is often referred to as a "tunneling time." The author prefers not to use this name because it can imply that quantum mechanical tunneling is at the heart of the so-called "tunneling time problem." As should become clear in the next section, timing quantum particles is problematic even for the simplest case of free evolution.

We now determine the mean transmission and reflection times for the point-like particle component of an electron. For a particle that is at $x = x^{(0)}$ at $t = 0$ the time that it spends thereafter in the region $[x_1, x_2]$ is given by the classical stopwatch expression

$$t(x_1, x_2; x^{(0)}) = \int_0^\infty dt \int_{x_1}^{x_2} dx\, \delta[x - x(x^{(0)}, t)]\,. \tag{5.19}$$

---

[4] Obviously, the centroid of the probability density would not be a good choice for X on the far side of a potential barrier when the transmission probability P_T is less than $1/2$.

[5] For the special case of the stationary-state limit in which the dispersion Δk in wave number k is zero, the notation $\tau_{T(R)}(k; x_1, x_2)$ is used.

For the moment, we do not discriminate between transmitted and reflected particles and determine the so-called mean dwell time by averaging (5.19) over all $x^{(0)}$ to obtain (after changing the order of integration)

$$\tau_D(x_1, x_2) = \int_0^\infty dt \int_{x_1}^{x_2} dx \int_{-\infty}^\infty dx^{(0)} \, \rho(x^{(0)}, 0) \, \delta[x - x(x^{(0)}, t)] \,. \qquad (5.20)$$

The innermost integral is the distribution of particle positions at time t, i.e., $\rho(x, t)$. Hence

$$\tau_D(x_1, x_2) = \int_0^\infty dt \int_{x_1}^{x_2} dx \, \rho(x, t) = \int_0^\infty dt \, t \, [J(x_2, t) - J(x_1, t)] \,. \qquad (5.21)$$

The far right-hand side of (5.21) is obtained by multiplying the continuity equation by t and integrating over t from 0 to ∞ and over x from x_1 to x_2.

Now, because the Bohm trajectories associated with a given wave function do not cross each other, there is a special trajectory $x_c(t) \equiv x(x_c^{(0)}, t)$ that separates transmitted trajectories (those with $x^{(0)} > x_c^{(0)}$) from reflected ones (those with $x^{(0)} < x_c^{(0)}$). It is defined implicitly by

$$P_T = \int_{x_c(t)}^\infty dx \, \rho(x, t) \,. \qquad (5.22)$$

Using $x_c(t)$ to decompose $\rho(x, t)$ and $J(x, t)$ into components associated with transmission and reflection, e.g. $\rho_T(x, t) \equiv \rho(x, t)\Theta[x - x_c(t)]$ and $\rho_R(x, t) \equiv \rho(x, t)\Theta[x_c(t) - x]$, respectively, upon insertion of $\Theta[x - x_c(t)] + \Theta[x_c(t) - x] \equiv 1$ into the integrands of (5.21), leads immediately to

$$\tau_D(x_1, x_2) = P_T \, \tau_T(x_1, x_2) + P_R \, \tau_R(x_1, x_2) \,, \qquad (5.23)$$

where

$$P_T \, \tau_T(x_1, x_2) = \int_0^\infty dt \int_{x_1}^{x_2} dx \, \rho(x, t) \, \Theta[x - x_c(t)]$$
$$= \int_0^\infty dt \, t \, [J(x_2, t)\Theta[x_2 - x_c(t)] - J(x_1, t)\Theta[x_1 - x_c(t)]]$$

$$(5.24)$$

with similar expressions for $P_R \, \tau_R(x_1, x_2)$.

Within Bohm's theory τ_D, τ_T, and τ_R are *real-valued*, *non-negative*, and *additive*[6] quantities. However, there is a fundamental distinction between τ_D and its components $P_T\tau_T$ and $P_R\tau_R$: each of the integrands of (5.21) for τ_D is bilinear in $\psi(x, t)$ while those for the latter two quantities are not because they depend on the trajectory $x_c(t)$, which, from (5.22), itself is an implicit functional of $\psi(x, t)$. This has the important consequence that the mean

---

[6] $\tau_D(x_1, x_3) = \tau_D(x_1, x_2) + \tau_D(x_2, x_3)$ with $x_1 < x_2 < x_3$ and similarly for τ_T and τ_R.

transmission and reflection times $\tau_T(x_1, x_2)$ and $\tau_R(x_1, x_2)$ for wave packets are not linear functionals of their stationary-state counterparts $\tau_T(k; x_1, x_2)$ and $\tau_R(k; x_1, x_2)$, respectively. This loss of a useful relation appears to be the price one has to pay for the attractive general properties of τ_T and τ_R given above.

In the nonrelativistic case, for stationary-state scattering of "incident" electrons of precisely defined wave number $k > 0$ and kinetic energy $E \equiv \hbar^2 k^2/2m$, expression (5.21) for the mean dwell time becomes

$$\tau_D(k; x_1, x_2) = \frac{1}{J_{k,inc}} \int_{x_1}^{x_2} dx \, |\psi_k(x)|^2 , \qquad (5.25)$$

a result proposed by Büttiker [37] within standard quantum mechanics. Here the stationary-state wave function $\psi_k(x) \exp(-iEt/\hbar)$ is normalized so that the incident probability current density $J_{k,inc}$ associated with the $\exp(ikx)$ component of $\psi_k(x < 0)$ is $\hbar k/m$. Spiller, Clark, Prance, and Prance [38] postulated that for $|T(k)|^2 > 0$ the corresponding mean transmission time is given by

$$\tau_T(k; x_1, x_2) = \int_{x_1}^{x_2} dx/v_k(x) , \qquad (5.26)$$

where $v_k(x) \equiv J_k/|\psi_k(x)|^2$ with $J_k \equiv |T(k)|^2 J_{k,inc}$, the stationary-state flux. It follows from (5.25) and (5.26) that $\tau_T(k; x_1, x_2) = |T(k)|^{-2}\tau_D(k; x_1, x_2)$ and then from (5.23) that $|R(k)|^2 \tau_R(k; x_1, x_2) = 0$ for any value of $x_1 > -\infty$! Leavens and Aers [39] expressed doubts about this result. These reservations have been put to rest [40]. For the special case of an initial gaussian wave packet, approximate closed-form expressions were derived for $\psi(x = 0, t)$, $J(x = 0, t)$, and hence for $v(x = 0, t)$ that are accurate in the regime in which $\Delta k/\langle k \rangle$, the small parameter of the theory, is much less than unity. The explicit dependence of $J(x = 0, t)$ on $\Delta k/\langle k \rangle$ and on t graphically confirms the "freezing out" of the $P_R \tau_R(0, d)$ contribution to the mean dwell time $\tau_D(0, d)$ as $\Delta k/\langle k \rangle$ approaches zero. It also explains and removes the apparent inconsistency between the stationary-state results of Spiller et al. [38] and the numerical results for time-dependent wave packets of Leavens and Aers [39]: for an opaque rectangular barrier $P_R \tau_R(0, d)$ is almost constant at a value very close to $\tau_D(k = \langle k \rangle; 0, d)$ over a wide range of decreasing Δk before eventually plummeting toward zero at an extremely small value of $\Delta k/\langle k \rangle$, orders of magnitude smaller than the smallest value (0.01) considered in the numerical wave packet calculations. The author attributes the strangeness of the stationary-state result for $\tau_R(k)$ to the extreme idealization of the stationary-state limit, which requires a single-particle wave function to be coherent over all space–time from $t = -\infty$.

For examples of calculated Bohm trajectory transmission and reflection time distributions for a variety of systems, see [32, 39].

5.4 Conventional Approaches to Timing Quantum Particles from the Perspective of Bohmian Mechanics

5.4.1 Arrival Time in the Quantum Backflow Regime

The quantum backflow effect is important because it clearly reveals that the problem of timing quantum events can be problematic even in the absence of quantum tunneling. Quantum backflow refers to the following remarkable fact: for a *freely evolving* nonrelativistic quantum particle with a wave function $\psi(x,t)$ having nonzero Fourier components $\phi(k,t)$ only for $k > 0$ it is possible for $J(x = X,t)$ to be negative over a finite time interval, say $[t_1(X), t_2(X)]$. The existence of this effect has been shown explicitly by direct calculation of $J(x,t)$ for various simple wave functions [41, 10]. A more global way of showing the possibility of the effect is to calculate the Fourier transform $\tilde{J}(k,t)$ of $J(x,t)$ for a freely evolving wave function with Fourier transform $\phi(k,t) = \Theta(k)\phi(k)\exp(-i\hbar k^2 t/2m)$. It is readily shown that $|\tilde{J}(-k,t)| = |\tilde{J}(+k,t)|$. Bracken and Melloy [42] have shown that the time interval during which $J(X,t) < 0$ can be arbitrarily long but finite. They obtained an estimate of 0.04 for the least upper bound of the quantity $\int_{t_1(X)}^{t_2(X)} dt |J(X,t)| / \int_{-\infty}^{\infty} dt J(X,t)$. They also treated the relativistic case [43].

The backflow effect is the source of the following puzzle: it follows from the continuity equation that during the time interval $t_1(X) < t < t_2(X)$, when $J(X,t)$ is negative, the probability of finding the particle in the spatial region $[-\infty, X]$ is increasing with time even though many apparently regard it as a self-evident fact that a freely evolving quantum particle described by a wave function with only $k > 0$ wave number components can arrive at X only from the left.

Within Bohmian mechanics the above "self-evident fact" is not always true. In particular, according to (5.1), particles arrive at X only from the right during any time interval when $J(X,t)$ is negative. In place of the backflow paradox one has the *counterintuitive* property that free particles can evidently turn around.[7] There is, however, no inconsistency within Bohmian mechanics where the evolution of a particle is in general not truly free when $V(x,t) = 0$ because it is still under the influence of its guiding wave.

It is also illuminating to look at the above puzzle from the point of view of the "standard" arrival-time distribution [1]:

$$\Pi(T;X) = \Pi_+(T;X) + \Pi_-(T;X) \; ;$$

$$\Pi_\pm(T;X) = \frac{\hbar}{2\pi m} \left| \int_{-\infty}^{\infty} dk \, \Theta(\pm k) \, |k|^{1/2} \, \exp(ikX) \, \phi(k,T) \right|^2 . \quad (5.27)$$

---

[7] A much more dramatic example of a free particle turning around is provided by a freely evolving wave function that is either symmetric or antisymmetric about $x = X$ at all times t. A particle guided by such a wave function can never arrive at $x = X$ and if initially moving toward that point must reverse its direction of motion before reaching it.

The derivations of Allcock [41], Kijowski [10], Grot, Rovelli, and Tate [11], and of Delgado and Muga [12] are for the special case of free evolution in which case $\phi(k, t) = \phi_0(k, t) \equiv \phi(k) \exp(-i\hbar k^2 t/2m)$ where $\phi(k)$ is the Fourier transform of the initial wave function $\psi(x, 0)$. The derivation of Baute, Sala Mayato, Palao, Muga, and Egusquiza (BSPME) [13] is for arbitrary potentials $V(x)$ and for both stationary and nonstationary states.

It should be noted that according to (5.27) arrivals from the left (right) are associated with $k > 0$ ($k < 0$) and there is no interference between $k > 0$ and $k < 0$ contributions to $\Pi(T; X)$. A key issue is how to interpret the absence of such interference terms.[8] Several years ago – when (5.27) had been constructed only for the special case of free evolution – Muga, Leavens, and Palao [44] suggested the following two options[9]: "One possibility is that the interference terms do not in fact contribute to the intrinsic arrival-time distribution. The other is that the distribution (5.27) is appropriate only when the apparatus measures the sign of the momentum of each incident particle, thus collapsing the wave function of that particle either to ψ_+ or ψ_-, and then switches on the appropriate one-sided detecting screen." Here $\psi_\pm(x, t) = N_\pm \hat{P}_\pm \psi(x, t)$ where $\hat{P}_\pm$ is a projector onto positive (negative) wave numbers and $N_\pm$ is a normalization factor. The timing of the sign-of-k projection is of critical importance, especially in the presence of a nonzero potential $V(x, t)$, and is discussed in the concluding section.

Now, according to (5.27), when $\phi(k, t) = \phi_0(k, t)$ is nonzero only for $k > 0$, *the arrival-time density for arrivals from the right at $x = X$ at time $t = T$ is zero* for any (X, T), even within a backflow regime where *the probability of finding the particle to the left of $x = X$ is increasing with time*. The second of the above options reconciles these conflicting statistical statements essentially by aborting the first: because $\phi(k < 0, t) = 0$ the sign-of-k projection does not change the wave function but the active detector is sensitive only to arrivals from the left and hence it is not meaningful to talk about arrivals from the right.

There is another interesting difference between the Bohm trajectory and "standard" results for $\Pi_\pm(T; X)$. The former is local, depending only on the value of the wave function at the space–time point $(x, t) = (X, T)$ of interest in the relativistic case[10] and only on the value of the wave function and its first spatial derivative at (X, T) in the nonrelativistic case. The latter, which can be cast in the form [45]

$$\Pi_\pm(T; X) =$$
$$\frac{\hbar}{32\pi m} \left| \int_{-\infty}^{+\infty} dx \, \frac{[1 \pm i \, \text{sign}(x - X)]}{|x - X|^{3/2}} \, [\Psi(x, T) - \Psi(X, T)] \right|^2 , \quad (5.28)$$

---

[8] This issue does not arise for the Bohm trajectory and Wigner arrival-time distributions that do not associate arrivals from the left (right) with positive (negative) wave numbers.

[9] The notation and equation numbers have been changed to conform with that of this chapter. Also, by "momentum" is meant $p = \hbar k$.

[10] This is also the case for Wigner's directed arrival-time distributions.

is clearly temporally local but spatially nonlocal. This is the basis of various interesting properties of the density (5.27) discussed in [46, 47, 48].

5.4.2 Superluminal Phase Times

The well-known phase time result [49, 50] for $\tau_T(0, d)$, (2.94), is based on the assumption [51] that the peak (or centroid) of the incident wave packet evolves into the peak (or centroid) of the transmitted wave packet. It has been known for decades that this assumption can lead to superluminal and even negative values of the phase time [52]. In Bohmian mechanics, on the other hand, to-be-transmitted electrons are associated with the leading P_T part, $\rho_T(x, 0) \equiv \Theta[x - x_c(0)]\rho(x, 0)$, of the initial probability density, which for $P_T << 1$ can be far to the right of its peak (or centroid) and, hence, much closer to the barrier. This provides a concrete and successful example of the wave packet "reshaping" often invoked to avoid superluminal transmission times. An interesting question is whether there are other general prescriptions for reshaping wave packets that also prohibit Dirac electrons from having superluminal transmission times.

It should be noted that Landauer and Martin [53] are severely critical of both of the above ideas concerning the initial origin of transmitted electrons. They replace it with the "prevailing common sense view" that $\rho_T(x, 0) = P_T \rho(x, 0)$, where I have taken the liberty of expressing this view in a mathematical form. Since this democratic decomposition of the initial probability density apparently requires no wave packet reshaping it is at least as susceptible to superluminal transmission times as the phase time approach.

It should be noted that in the Bohm trajectory approach the question of superluminality is addressed directly via the expression in (5.1) for the relativistic velocity field $\boldsymbol{v}(\boldsymbol{r}, t)$. On the other hand, this question is usually addressed in a roundabout way involving a mean transmission time in approaches based on conventional, often nonrelativistic, quantum mechanics.

In Sects. 5.4.3 and 5.4.4, two general methods for deriving (nonrelativistic) expressions for mean transmission and reflection times are looked at from the point of view of Bohmian mechanics. Both of these can, in principle, generate an infinite number of expressions for τ_T and τ_R in contrast to the Bohm trajectory approach, which leads to a unique expression for each of them.

5.4.3 The Systematic Projector Approach of Brouard, Sala, and Muga

Brouard, Sala, and Muga (BSM) [54] introduced an approach capable of systematically generating an infinite hierarchy of possible expressions for τ_T and τ_R. A very attractive feature of this method is that it reproduces, as simple special cases, the expressions obtained with several well-known and, at first sight, quite different approaches. Whether or not the Bohm trajectory approach can also be included as a special case within this unifying formalism is an interesting question that is answered in this section [55].

The systematic projector approach is based on the projection operators $\widehat{D}(x_1, x_2)$, $\widehat{T}$, and $\widehat{R}$ defined by

$$\widehat{D}(x_1, x_2)\psi(x,t) \equiv \Theta(x - x_1)\Theta(x_2 - x)\psi(x,t) , \qquad (5.29)$$

$$\widehat{T}\psi(x,t) \equiv \psi_T(x,t) ; \ \widehat{R}\psi(x,t) \equiv \psi_R(x,t) . \qquad (5.30)$$

The operator $\widehat{D}(x_1, x_2)$ projects from the wave function $\psi(x,t)$, the part that is "located" in the region $[x_1, x_2]$ at time t. $\widehat{T}$ projects from $\psi(x,t)$, the part $\psi_T(x,t)$ that will have only positive wave number components at $t = t_\infty$ and is thus associated with transmission; $\widehat{R} \equiv \widehat{1} - \widehat{T}$ projects from $\psi(x,t)$, the part $\psi_R(x,t)$ associated with reflection. The T and R components of the initial wave function have interesting properties: even when $\psi(x,0)$ is well localized to the left of the barrier, for $P_T < 1$ both $\psi_T(x,0)$ and $\psi_R(x,0)$ are nonnegligible to the right of it. For an opaque barrier the part of $\psi_T(x,0)$ with $x > d$ completely dominates the part with $x < 0$ while the opposite is true for $\psi_R(x,0)$. Figures 10 and 11 of [56] show calculated snapshots of $\psi_T(x,t)$ and $\psi_R(x,t)$ for the case of an initial gaussian wave packet scattering from a rectangular barrier.

The starting point of the analysis is (5.21) written in the form

$$\tau_D(x_1, x_2) = \int_0^\infty dt \int_{-\infty}^\infty dx\, \psi^*(x,t)\, \widehat{D}(x_1, x_2)\, \psi(x,t). \qquad (5.31)$$

Since $\widehat{D}$ is a projector (i.e., $\widehat{D}^2 = \widehat{D}$) and $\widehat{T} + \widehat{R} = \widehat{1}$ with $\widehat{T}^2 = \widehat{T}$ and $\widehat{R}^2 = \widehat{R}$, one can replace $\widehat{D}$ in (5.31) with any one of the infinite number of equivalent expressions given by

$$\widehat{D} = \widehat{D}^n = (\widehat{T} + \widehat{R})^{m_0}\, \widehat{D}\, (\widehat{T} + \widehat{R})^{m_1}... (\widehat{T} + \widehat{R})^{m_{n-1}}\, \widehat{D}\, (\widehat{T} + \widehat{R})^{m_n} \qquad (5.32)$$

$$[\, n = 1, 2, ... \ ; \ m_i = 0, 1 \ (i = 0, 1, ..., n)\,]$$

or with any suitably weighted linear combination of these primitive expressions. After expansion and suitable rearrangement, the resulting decompositions of $\widehat{D}$ have the generic form

$$\widehat{D} = F(\widehat{T}, \widehat{D}) + F(\widehat{R}, \widehat{D}) + G(\widehat{T}, \widehat{R}, \widehat{D}) \qquad (5.33)$$

with $G(\widehat{T}, \widehat{R}, \widehat{D}) = G(\widehat{R}, \widehat{T}, \widehat{D})$. Substituting this into (5.31) gives

$$\tau_D(x_1, x_2) = P_T\, \tau_T^F(x_1, x_2) + P_R\, \tau_R^F(x_1, x_2) + \tau_{T-R}^G(x_1, x_2) , \qquad (5.34)$$

where

$$P_T\, \tau_T^F(x_1, x_2) = \int_0^\infty dt \int_{-\infty}^\infty dx\, \psi^*(x,t)\, F(\widehat{T}, \widehat{D})\, \psi(x,t) , \qquad (5.35)$$

$$P_R \, \tau_R^F(x_1, x_2) = \int_0^\infty dt \int_{-\infty}^\infty dx \, \psi^*(x,t) \, F(\widehat{R}, \widehat{D}) \, \psi(x,t) \,, \qquad (5.36)$$

$$\tau_{T-R}^G(x_1, x_2) = \int_0^\infty dt \int_{-\infty}^\infty dx \, \psi^*(x,t) \, G(\widehat{T}, \widehat{R}, \widehat{D}) \, \psi(x,t) \,. \qquad (5.37)$$

For $n = 1$ there are three nontrivial primitive decompositions of $\widehat{D}$: $\widehat{T}\widehat{D} + \widehat{R}\widehat{D}$, $\widehat{D}\widehat{T} + \widehat{D}\widehat{R}$, and $\widehat{T}\widehat{D}\widehat{T} + \widehat{R}\widehat{D}\widehat{R} + \widehat{T}\widehat{D}\widehat{R} + \widehat{R}\widehat{D}\widehat{T}$. Since $\widehat{T}$ and $\widehat{D}$ do not commute, the first two decompositions lead in general to inequivalent expressions for τ_T, and similarly for τ_R. BSM also considered the (Hermitian) symmetric linear combination $[(\widehat{T}\widehat{D} + \widehat{D}\widehat{T}) + (\widehat{R}\widehat{D} + \widehat{D}\widehat{R})]/2$ of the first two $n = 1$ primitive decompositions. Going from $n - 1$ to n generates four new primitive decompositions of $\widehat{D}$, namely those with $m_1, ..., m_{n-1} = 1$ and $(m_0, m_n) = (0,0), (0,1), (1,0), (1,1)$. Of the four new ones generated for $n = 2$ only $\widehat{D}\widehat{T}\widehat{D} + \widehat{D}\widehat{R}\widehat{D}$ contains no $T - R$ interference terms; for $n \geq 3$ all of the new primitive decompositions contain such interference terms. We are now in a position to prove that the unique mean transmission and reflection times of Bohmian mechanics are not included among the infinite number of possibilities generated by the systematic projector approach [55]. This is accomplished by showing that none of these possibilities has all four of the general properties of the Bohm trajectory times discussed in Sect. 5.3.3. The only decompositions of $\widehat{D}$ that lead to mean transmission and reflection times satisfying $\tau_D = P_T \tau_T + P_R \tau_R$ are the primitive ones $\widehat{T}\widehat{D} + \widehat{R}\widehat{D}$, $\widehat{D}\widehat{T} + \widehat{D}\widehat{R}$ and $\widehat{D}\widehat{T}\widehat{D} + \widehat{D}\widehat{R}\widehat{D}$, and their appropriately weighted linear combinations. Requiring that the resulting mean transmission and reflection times be spatially additive eliminates $\widehat{D}\widehat{T}\widehat{D} + \widehat{D}\widehat{R}\widehat{D}$ [54] and any linear combinations containing it. Requiring that the resulting times be real-valued eliminates both $\widehat{T}\widehat{D} + \widehat{R}\widehat{D}$ and its hermitian conjugate $\widehat{D}\widehat{T} + \widehat{D}\widehat{R}$ but not their symmetric linear combination. Finally, this last surviving decomposition is eliminated when one notes that the corresponding mean reflection time can be negative [57].

That none of the BSM expressions for τ_T can be identical to (5.24) also follows immediately from the fact that the double integral in (5.35) is a bilinear functional of $\psi(x,t)$ while that in (5.24) is not. It is interesting that the simple bilinear form is apparently not compatible simultaneously with all four of the simple classical properties just discussed. It should be noted, in this context, that Kijowski [10] derived his well-known arrival-time distribution by requiring that certain properties of the classical distribution, including a "surprising" one with no obvious intuitive basis, should also hold for the quantum distribution.

What is the origin of the "orthogonality" between the Bohm trajectory and systematic projector approaches to the determination of τ_T and τ_R? Both approaches have the same starting point, namely (5.21) for the mean dwell time. An important difference is that the former is based on the *particle-like* decomposition $\rho = \rho_T + \rho_R$ of the position probability density while the latter is based on the *wave-like* decomposition $\psi = \psi_T + \psi_R$ of the wave function.[11] There is no obvious inconsistency in the two approaches having the same starting point because $\rho(x,t)$ has the dual interpretation as the probability density for particle position and as the local intensity of the wave function.

In Bohmian mechanics what is being timed in the theory of Sect. 5.3.3 is the presence of an actual point-like particle in the region of interest in the absence of any external measuring device. Answering the important question of whether the Bohm trajectory and systematic projector approaches are incompatible with or complementary to each other hinges on what is being timed in the latter approach.

5.4.4 Approaches Based on the Probability Current

Olkhovsky and Recami [58] and Muga, Brouard, and Sala [54, 59] independently postulated that the mean transmission and reflection times for the barrier region $[0,d]$ can be expressed in terms of right-going and left-going components, $J_+(x,t)$ and $J_-(x,t)$, respectively, of the probability flux $J(x,t)$ as

$$\tau_T^J(0,d) = \frac{\int_0^\infty dt\, t\, J_+(d,t)}{\int_0^\infty dt\, J_+(d,t)} - \frac{\int_0^\infty dt\, t\, J_+(0,t)}{\int_0^\infty dt\, J_+(0,t)} , \tag{5.38}$$

$$\tau_R^J(0,d) = \frac{\int_0^\infty dt\, t\, J_-(0,t)}{\int_0^\infty dt\, J_-(0,t)} - \frac{\int_0^\infty dt\, t\, J_+(0,t)}{\int_0^\infty dt\, J_+(0,t)} . \tag{5.39}$$

(Olkhovsky and Recami use a convention in which the lower limit of integration is $-\infty$.) The first term on the right-hand side of (5.38) is interpreted as the mean exit time of particles from the barrier through its right edge at $x = d$ and the first term on the right-hand side of (5.39) as the mean exit time through its left edge at $x = 0$. The last term of (5.38) is equal to the last term of (5.39) and is interpreted as the mean time at which incident particles enter the barrier, democratically reflecting the fact that within conventional quantum mechanics, for x inside the barrier, one cannot separate $J(x,t)$, let alone $J_+(x,t)$, into "to-be-transmitted" and "to-be-reflected" components. For example, substitution of the wave function decomposition $\psi = \psi_T + \psi_R$ of the projector approach into the expression for J leads to $J = J_T + J_R + J_{T-R}$ where the $T - R$ interference term is in general nonzero.

An important feature of the present approach is that $J_+(x,t)$ and $J_-(x,t)$ are not uniquely defined within conventional quantum mechanics. Olkhovsky and Recami postulated that

$$J_\pm(x,t) = J(x,t)\,\Theta[\pm J(x,t)] , \tag{5.40}$$

---

[11] It should be noted that the latter decomposition plays no role in Bohmian mechanics in the sense that a to-be-transmitted particle is guided by ψ not just by ψ_T.

while Muga et al. assumed that

$$J_{\pm}(x, t) = \pm \int_0^{\pm\infty} dp \, (p/m) \, f_W(x, p, t) , \qquad (5.41)$$

where $p \equiv \hbar k/m$ is the free-particle momentum and $f_W(x, p, t)$ is the Wigner function. For both choices, $J(x, t) = J_+(x, t) + J_-(x, t)$ with no interference term. For $J(x, t) \neq 0$ the choice (5.40), by construction, always satisfies the requirement that $J_+(x, t) \geq 0$ and $J_-(x, t) \leq 0$. However, because the Wigner function $f_W(x, p, t)$ can be negative for some values of its arguments it is possible for (5.41) to give negative values of $J_+(x, t)$ or positive values of $J_-(x, t)$ for some values of x and t.

It is worthwhile to consider (5.38) and (5.39) from the point of view of Bohmian mechanics that leads to a unique decomposition that is precisely the one postulated by Olkhovsky and Recami. However, it is the decomposition $J = J_T + J_R$, not $J = J_+ + J_-$, that is involved in the derivation of the Bohm trajectory expressions (5.24) for τ_T and τ_R. Within Bohmian mechanics the mean time at which "to-be-transmitted" electrons enter the barrier is not equal in general to the mean time at which "to-be-reflected" electrons do. Consider the case in which at least some of the reflected electrons enter the barrier. The simplest such case involves two times t_T and t_R with $t_T < t_R$ such that only to-be-transmitted electrons enter the barrier for $0 \leq t \leq t_T$, only to-be-reflected ones enter for $t_T \leq t \leq t_R$, and none enter for $t > t_R$. Obviously, for this case the mean time at which to-be-transmitted electrons enter the barrier is less than the overall average entrance time. This suggests that the subtracted term can be too large in (5.38) and too small in (5.39) and, in the former case, can lead to anomalously small or even negative values for $\tau_T^J(0, d)$ [60]. Such negative values have indeed been calculated by Delgado, Brouard, and Muga [61].

5.4.5 The Quantile Approach

The trajectory $x(P; t)$ defined implicitly by

$$P \equiv \int_{x(P;t)}^{\infty} dx \, \rho(x, t) \quad (0 \leq P \leq 1) \qquad (5.42)$$

is mathematically identical to the Bohm trajectory $x(x^{(0)} = x(P; 0), t)$ [33, 62].

Brandt, Dahmen, Gjonaj, and Stroh [63] arrived at (5.42) independently by extending the concept of the quantile $x(P)$ associated in mathematical statistics with a time-independent probability density $\rho(x)$ to time-dependent probability densities $\rho(x, t)$ in physics and called $x(P; t)$ a quantile trajectory. For the nonrelativistic case they showed that the quantile position $x(P; t)$ for an initial wave packet (with only $k > 0$ Fourier components) incident on a potential barrier lags behind the corresponding quantile position in the absence of the barrier. They suggested that quantile trajectories could be used to time quantum particles within standard quantum mechanics without identifying them with the trajectories of point-like particles. This had, in fact, already been done for Dirac electrons by Challinor, Lasenby, Somaroo, Doran, and Gull [64] who calculated transmission time distributions without interpreting the streamlines of the probability flux (i.e. the quantile trajectories) as Bohm particle trajectories but rather by *assuming* that "the flow of the probability density reflects the temporal aspects of the tunneling process." From this it might be argued that the concept of Bohm particle trajectory is excess baggage in the calculation of distributions for the characteristic times of interest in this chapter because identical results can be obtained without the concept. However, such an argument would be based on the above rather vague assumption and is weakened by the existence of alternative results also based on standard quantum mechanics that do not agree with those obtained with the quantile approach. For example, in Sect. 5.3.3 it was suggested that it might be possible to rationalize the "orthogonality" of the systematic projector and Bohm trajectory approaches on the grounds that the former is timing waves and the latter point-like particles. Rationalizing the orthogonality of the projector and quantile methods will be a challenge because both involve only the wave function (none of theory involves at any stage its collapse to a point-like region). Moreover, the concept of particle trajectory provides a consistent picture of what might be happening in an individual member of an ensemble and thereby provides a necessary ingredient of Bohm's deterministic resolution of the measurement problem and the familiar paradoxes associated with two-slit and delayed-choice experiments [3, 7].

5.5 Spin-Dependent Arrival-Time Distributions for Nonrelativistic Electrons

For a nonrelativistic guiding wave of the form $\psi(\boldsymbol{r}, t; \hat{\boldsymbol{s}}) \equiv \psi(\boldsymbol{r}, t)\,\chi$, with $\psi(\boldsymbol{r}, t)$ a solution of the Schrödinger equation and χ a fixed spinor, the Bohm trajectory result for the probability distribution of particle arrival times T at the planar surface $x = X$ for the spot (y, z) on it is given by

$$\Pi(T; X, y, z; \hat{\boldsymbol{s}}) = |\,J_x(\,X, y, z, T\,;\,\hat{\boldsymbol{s}}\,)\,| \bigg/ \int_0^\infty dt\, |J_x(X, y, z, t; \hat{\boldsymbol{s}})|\,, \quad (5.43)$$

where J_x refers to the x-component of $\boldsymbol{J}$. Positive (negative) $J_x(X, y, z, T; \hat{\boldsymbol{s}})$ is associated with particles arriving at the plane $x = X$ from the half-space $x < X$ $(x > X)$, i.e. $J_{x,\pm} = J_x\,\Theta(\pm J_x)$.

Calculated results for 1D arrival-time distributions $\Pi(T; X)$ for nonrelativistic particles based on the original spin-independent equation of motion have appeared in the literature [33, 28]. Although spin does not exist in a strictly 1D system, by a 1D system one often means a 3D system in which there is no spatial variation of $\psi(\boldsymbol{r}, t)$ in the transverse (y and z) directions. This in turn is an idealization of some realistic system in which the spatial variation of $\psi(\boldsymbol{r}, t)$ in the transverse directions is relatively unimportant compared to that in the longitudinal (x) direction over the volume of interest, assumed large enough in the transverse directions to ignore edge effects.

It is important to investigate the effect of the spin-dependent term in (5.4) not only on the Bohm trajectory arrival-time distribution but also on other distributions in which the probability flux enters in an important way. (There is an entire class of "conventional" arrival-time theories based on the nonunique decomposition of the standard spin-independent probability flux into right-going and left-going components [15].) In order that the results of this section be of more general interest, the focus will be on the spin dependence of J_x.

For simplicity, consider a 3D system with a potential $V(\boldsymbol{r}) = V(x)$ with no variation in the $y - z$ plane and a wave function, which has the factorable form $\psi(\boldsymbol{r}, 0) = \psi_x(x, 0)\psi_y(y, 0)\psi_z(z, 0)$ at $t = 0$. Then for $t > 0$ the wave function is given by

$$\psi(\boldsymbol{r}, t) = \psi_x(x, t)\psi_y(y, t)\psi_z(z, t) , \tag{5.44}$$

with $\psi_x(x, t)$ satisfying $i\hbar\partial\psi_x(x, t)/\partial t = [-(\hbar^2/2m)\partial^2/\partial x^2 + V(x)]\psi_x(x, t)$ and $\psi_y(y, t)$ and $\psi_z(z, t)$ satisfying the corresponding free-particle Schrödinger equations. From (5.4) it immediately follows that

$$J_x(\boldsymbol{r}, t; \hat{\boldsymbol{s}}) = \frac{\hbar}{m}|\psi(\boldsymbol{r}, t)|^2\left[\mathrm{Im}\left(\frac{\partial \ln \psi_x(x, t)}{\partial x}\right) + \hat{s}_z\frac{\partial \ln |\psi_y(y, t)|}{\partial y}\right.$$
$$\left. - \hat{s}_y\frac{\partial \ln |\psi_z(z, t)|}{\partial z}\right] . \tag{5.45}$$

For the special case $\hat{\boldsymbol{s}} = \hat{\boldsymbol{x}}$ the spin-dependent terms in both (5.45) for J_x and the arrival-time distribution (5.43) are zero. For mixed ensembles in which $\psi(\boldsymbol{r}, t)$ is common to all members and χ varies from member to member in such a way that the average value of $\boldsymbol{s}$ is zero, the additional term makes no contribution to the probability current density for the entire ensemble. Similarly, if $\psi_y(y, t)$ and $\psi_z(z, t)$ are symmetric about, say, $y = 0$ and $z = 0$, respectively, then averaging $J_x(X, y, z, t; \hat{\boldsymbol{s}})$ over the planar region $[-L_y \leq y \leq +L_y , -L_z \leq z \leq +L_z]$ gives a quantity that is independent of $\hat{\boldsymbol{s}}$.

Consider the special case in which $V(x) = 0$ and the wave function of the freely evolving quantum particle is of the form (5.44) with $\psi_x(x, t)$ antisymmetric about $x = X$ so that both $|\psi(\boldsymbol{r}, t)|^2$ and $J_x(\boldsymbol{r}, t; \hat{\boldsymbol{s}})$ are zero at $x = X$ for all y, z, and t. According to Bohmian mechanics, such a freely evolving particle never arrives at the plane $x = X$ regardless of the direction of $\hat{\boldsymbol{s}}$.

For a more mundane, but important, example consider the free evolution $[V(x) = 0]$ of the initial 3D gaussian wave function

$$\psi(\boldsymbol{r},0) = \frac{1}{(2\pi)^{1/4}\Delta_x^{1/2}} \exp\left[-\left(\frac{x-x_0}{2\Delta_x}\right)^2 + ik_0 x\right]$$

$$\times \frac{1}{(2\pi)^{1/4}\Delta_y^{1/2}} \exp\left[-\left(\frac{y}{2\Delta_y}\right)^2\right] \frac{1}{(2\pi)^{1/4}\Delta_z^{1/2}} \exp\left[-\left(\frac{z}{2\Delta_z}\right)^2\right]$$

(5.46)

where it is assumed that the initial centroid x_0 is negative and the mean wave number k_0 is nonnegative. The x-component of the flux at time t is

$$J_x(\boldsymbol{r},t;\hat{\boldsymbol{s}}) =$$

$$\frac{\hbar|\psi(\boldsymbol{r},t)|^2}{m}\left[\frac{(2\Delta_x^2 k_0 + \hbar t(x-x_0)/2m\Delta_x^2)}{\gamma_x(t)} - \frac{\hat{s}_z y}{\gamma_y(t)} + \frac{\hat{s}_y z}{\gamma_z(t)}\right],\quad (5.47)$$

where

$$|\psi(\boldsymbol{r},t)|^2 = \frac{1}{[\pi^3\gamma_x(t)\gamma_y(t)\gamma_z(t)]^{1/2}}$$

$$\exp\left[-\frac{(x-x_0-\hbar k_0 t/m)^2}{\gamma_x(t)} - \frac{y^2}{\gamma_y(t)} - \frac{z^2}{\gamma_z(t)}\right] \quad (5.48)$$

and

$$\gamma_x(t) \equiv 2\Delta_x^2\left[1 + \left(\frac{\hbar t}{2m\Delta_x^2}\right)^2\right], \quad (5.49)$$

with $\gamma_y(t)$ and $\gamma_z(t)$ obtained from (5.49) by replacing x by y and z, respectively.

At first sight it appears that the size of the spin-dependent terms in (5.47) relative to the standard one can be made arbitrarily large by simply making $|y|$ and/or $|z|$ sufficiently large. However, once the corresponding particle velocity $\boldsymbol{J}(\boldsymbol{r},t;\hat{\boldsymbol{s}})/|\psi(\boldsymbol{r},t)|^2$ exceeds c in absolute value the nonrelativistic approximation for $\boldsymbol{J}$ should definitely not be used. Moreover, a large relative spin dependence is of little importance if the absolute size of the effect is negligible because of the factor $\exp[-y^2/\gamma_y(t) - z^2/\gamma_z(t)]$.

In order to cut down on the number of variables, cylindrical symmetry $(\Delta_y = \Delta_z)$ is assumed for $\psi(\boldsymbol{r},t)$ and s_y is fixed at 0 for the rest of this section. It should be noted that keeping y fixed and changing s_z to $-s_z$ then leads to the same change in J_x as keeping s_z fixed and changing y to $-y$. This might be important experimentally. With this in mind, it is of interest to consider the special case $\hat{\boldsymbol{s}} = \hat{\boldsymbol{z}}$ and introduce a global measure of the y-asymmetry in $J_x(x,y,z,t;\hat{\boldsymbol{z}})$:

$$A_x(x,t;\hat{\boldsymbol{z}}) \equiv \frac{J_x^{(-)}(x,t;\hat{\boldsymbol{z}}) - J_x^{(+)}(x,t;\hat{\boldsymbol{z}})}{J_x^{(-)}(x,t;\hat{\boldsymbol{z}}) + J_x^{(+)}(x,t;\hat{\boldsymbol{z}})}, \quad (5.50)$$

where $J_x^{(\pm)}(x,t;\hat{\boldsymbol{z}})$ is obtained by integrating $J_x(x,y,z,t;\hat{\boldsymbol{z}})$ over all z and over all positive or over all negative y, respectively, i.e.,

$$J_x^{(\pm)}(x,t;\hat{z}) \equiv \pm \int_0^{\pm\infty} dy \int_{-\infty}^{+\infty} dz \, J_x(x,y,z,t;\hat{z}) . \qquad (5.51)$$

The denominator in (5.50) is equal to $J_x(x,t)$ obtained by integrating the standard expression for $J_x(r,t)$ over all y and all z. For the wave function under consideration, the y-asymmetry arises solely from the spin dependence. Another nice feature of this wave function as regards a measurement of $A_x(x,t;\hat{z})$ is that the contribution is smallest for $y = 0$ and largest for $|y| = [\gamma_y(t)/2]^{1/2} \geq \Delta_y$, giving a little leeway in the accurate location of $y = 0$ on the detector plane, which we will take to be $x = X = 0$. For the initial wave function (5.46) one obtains

$$A_x(0,t;\hat{z}) = \frac{\gamma_x(t)}{\pi^{1/2}\gamma_y(t)^{1/2}(2\Delta_x^2 k_0 - \hbar t x_0/2m\Delta_x^2)} . \qquad (5.52)$$

For $t > 0$ the factor of $\gamma_y(t)^{-1/2}$ is zero in both $\Delta_y \to 0$ and $\Delta_y \to \infty$ limits and for a fixed positive value of t, say τ, takes on its maximum value of $(m/2\hbar\tau)^{1/2}$ when $\Delta_y = \Delta_y(\tau) \equiv (\hbar\tau/2m)^{1/2}$. To obtain a feeling for the importance of the asymmetry in the regime $k_0 >> \Delta k_x$ and $|x_0| >> \Delta_x$ of usual interest, let us maximize $A_x(0,t;\hat{z})$ as a function of Δ_y at the characteristic time $t_0 \equiv |x_0|/(\hbar k_0/m)$ for the motion of the centroid of $|\psi_x(x,t)|^2$ by choosing $\Delta_y = \Delta_y(t_0)$. It immediately follows that $A_x(0,t_0;\hat{z})^2 \leq 1/(2\pi|x_0|k_0) << \Delta_x/(\pi|x_0|) << 1/\pi$, where $\Delta_x \Delta k_x = 1/2$ has been used. Hence, in the regime under consideration the y-asymmetry is not expected to be large for t near t_0, where $J_x(x = 0,t)$ is expected to be relatively large. Let us now remove the restriction $k_0 >> \Delta k_x$ by choosing $k_0 = 0$ and maximize $A_x(0,t;\hat{z})$ as a function of Δ_y at the characteristic time $t_0' \equiv |x_0|/(\hbar\Delta k_x/m)$ for spreading of the packet to obtain $A_x(0,t_0';\hat{z}) \leq (\Delta_x/\pi|x_0|)^{1/2}$ (to leading order in $\Delta_x/|x_0|$). Hence, the importance of the spin-dependence of the arrival-time distribution (5.43) can be enhanced by relaxing the frequently imposed constraint $k_0 >> \Delta k_x$. This has been confirmed by explicit calculation for free electrons having the initial wavefunction (5.46) with $\Delta k_x = k_0$ and $\Delta_x = \Delta_y = \Delta_z$.

It is clear from the above analysis and from detailed calculations in the recent literature [24] that there are situations in which Bohm trajectory arrival-time distributions for nonrelativisitic electrons with a two-component wave function of the form $\psi(r,t)\chi$ can exhibit a significant dependence on spin direction. This is not necessarily the case for other approaches. For example, the "time-of-presence" result (5.17) for the arrival-time distribution is independent of $\hat{s}$. Hopefully, it will prove possible to investigate the spin-dependence of arrival-time distributions with experiments on spin-polarized electrons.

5.6 Protective Measurements and Bohm Trajectories

A recent paper by Aharonov, Englert, and Scully (AES) [65] with the above title "challenges any realistic interpretation of Bohm trajectories." These authors consider "a protective measurement of a particle in a box and find that the particle participates in a local interaction although its Bohm trajectory never comes near the interaction region." In particular, they claim the following: "[This] should have been quite disturbing to adherents of Bohmian mechanics because it implies that Bohm trajectories are forever hidden. If you cannot rely on local interactions to determine the 'actual position' of the particle, then you cannot determine it at all. The concept of position itself becomes shaky."

Now, that one cannot *directly* measure a Bohm trajectory is an inevitable consequence of the position–momentum uncertainty relation and has been known from the beginning – after all, the titles of Bohm's seminal papers were 'A suggested interpretation of the quantum theory in terms of "hidden variables": Part I; Part II.'[12] Moreover, the concept of particle position is already shaky in relativistic quantum mechanics and this is much more of a problem for conventional quantum mechanics than for a realistic theory such as Bohmian mechanics, which is founded on "beables" [6] rather than "observables" and hence does not mandate that a quantity cannot exist unless it is precisely measurable (at least in principle).

AES consider a thought experiment involving two quantum particles in which a protective measurement is made on the first with the second acting as a meter. The first has mass m_1 and is confined to the interior of a box extending from $x = -l$ to $x = +l$. It is prepared at $t = t_i$ in the lowest-energy eigenstate with initial wave function $\psi_1(x_1, t_i) = l^{-1/2} \cos(\pi x_1/2l) \Theta(x_1 + l) \Theta(l - x_1)$.[13] The second has mass m_2 and is not subject to any confining potential; its initial wave function is $\psi_2(x_2, t_i) = (2/\pi)^{1/4}(\Delta p_2/\hbar)^{1/2} \exp[-(x_2\Delta p_2/\hbar)^2]$, a Gaussian with mean position $\bar{x}(t_i) = 0$, mean momentum $\bar{p}(t_i) = 0$, spatial width $\Delta x_2(t_i)$, and momentum width $\Delta p_2 \equiv \Delta p_2(t_i) = 1/[2\Delta x_2(t_i)]$. The two quantum particles are subsequently coupled via the interaction

$$H_{int} = \epsilon \frac{\hbar}{T} f\left(\frac{t}{T}\right) \delta(\hat{x}_1) \hat{x}_2 , \qquad (5.53)$$

where $f(t/T)$ is zero for $|t| > T$ and positive for $|t| < T$ with $-T \geq t_i$, and the integral of $T^{-1}f(t/T)$ over all t is 1. For simplicity, $f(t/T)$ is also assumed to be symmetric about $t = 0$. AES consider a protective measurement

---

[12] If one knows the initial wave function $\psi(x,0)$ and the potential $V(x,t)$ then the result $X_a(t') \leq x(t') \leq X_b(t')$ of a measurement of the particle's position at a well-specified instant of time t' enables one to retrodict a band of nonintersecting Bohm trajectories for previous times t bounded by $X_a(t)$ and $X_b(t)$, one of which – according to the theory – is the actual trajectory. The t-dependent width of the band, of course, depends on the precision of the original (strong) position measurement at $t = t'$, which cannot be made arbitrarily small because of relativistic particle–antiparticle creation and annihilation effects.

[13] The Θ functions included here are essential. Otherwise one is led to the usual claim that the wave function is a superposition of two running waves, with wave numbers $k = \pm(\pi/2l)$, moving in opposite directions. In fact, the distribution of wave numbers for the correct wave function is $|\phi(k)|^2 = (\pi^2/2l^3)[1+\cos(2kl)]/[(\pi/2l)^2 - k^2]^2$ and takes on its maximum value at $k = 0$.

[66, 67, 68, 69, 70] in which (a) T is sufficiently large relative to other relevant timescales and $f(t/T)$ sufficiently smooth that the coupling is adiabatic and the first quantum particle remains in its time-dependent ground state and (b) the dimensionless parameter ϵ is sufficiently small that this ground state at $t \geq T$ is negligibly changed from the initial one. They showed that under these conditions the wave function for $t \geq T$ is given to a good approximation by

$$\Psi(x_1, x_2, t) =$$
$$\frac{1}{l^{1/2}} \cos\left(\frac{\pi x_1}{2l}\right) \Theta(x_1 + l) \Theta(l - x_1) \exp[i\phi(t)] \left[\frac{\alpha(t)\Delta p_2}{\hbar}\right]^{1/2}$$
$$\cdot \exp\left[-\frac{\alpha(t)([x_2 - \delta x_2(t)]\Delta p_2)^2}{\hbar^2} + i\frac{[x_2 - \delta x_2(t)]\delta p_2}{\hbar}\right] \qquad (5.54)$$

where $\alpha(t) \equiv 1/[1 + i2(\Delta p_2)^2(t - t_i)/\hbar m_2]$. The time-dependent spatial spread of the uncoupled meter is $\Delta x_2(t) = \Delta x_2(t_i)/|\alpha(t)|$; the final shift of the meter's mean momentum, $\delta p_2 \equiv \delta p_2(t \geq T)$, is given by

$$\delta p_2 = -\epsilon \hbar \int_{-\infty}^{\infty} dx_1 \, \psi_1^*(x_1, t_i)\delta(x_1)\psi_1(x_1, t_i)$$
$$= -\epsilon \hbar |\psi_1(0, t_i)|^2 = -\frac{\epsilon\hbar}{l} . \qquad (5.55)$$

The corresponding (time-dependent) shift of the meter's mean position is $\delta x_2(t > T) = (\delta p_2/m_2)t$. The time-dependent phase $\phi(t)$ is irrelevant.

To complete the measurement of $|\psi_1(0, t_i)|^2$, i.e., of $-\delta p_2/\epsilon\hbar$, a standard measurement of the momentum shift δp_2 is made for some instant of time $t > T$. If this shift is very much larger than the momentum width Δp_2 then one can, with high probability, obtain an accurate result for $|\psi_1(0, t_i)|^2$ by performing just one such measurement on a single member of the ensemble.

In the absence of any coupling between the two quantum particles, the Bohm velocity field $v_1(x_1, t)$ of the first is zero and it remains at rest at some unknown position within the box, i.e., its Bohm trajectory is $x_1(x_1^{(0)}, t) = x_1^{(0)}$. AES show that, unless the starting position $x_1^{(0)}$ at $t = t_i$ is very close to the origin $x_1 = 0$, the weak adiabatic coupling of the above protective measurement does not perturb the first particle sufficiently for its trajectory to reach the origin. They conclude as follows: "Accordingly, one has to concede either that the particle's Bohm trajectory and its position are unrelated, or that the particle's position is irrelevant for its participation in local interactions. The second concession cannot be considered seriously because it would put away with the phenomenological meaning of position altogether. Therefore we can hardly avoid the conclusion that the formally introduced Bohm trajectories are just mathematical constructs with no relation to the actual motion of the particle." I think that there is an alternative interpretation of their analysis that does not involve either theory making concessions.

I find the presentation of [65] misleading for two reasons. Firstly, it does not acquaint the reader with the ontology of Bohm's theory and is not phrased in a manner consistent with that ontology (e.g., "particle" is used when "wave" or "wave function" is the appropriate word). Secondly, there are crucial properties of protective measurements, including striking similarities regarding the status of the wave function in the two theories, which are not mentioned.

For a system of N quantum particles described by Bohm's theory the ontology is one of N real point-like particles that are always accompanied by a real wave – the wave function $\Psi(\boldsymbol{r}_1, ..., \boldsymbol{r}_N, t)$ – that probes the environment of each particle and choreographs [9] its motion accordingly so that it has a well-defined trajectory. For the nonrelativistic one-dimensional $N = 2$ system analyzed by AES the equations of motion for the particles are [2, 3, 4, 5, 6, 7, 8, 9]

$$
\begin{aligned}
v_j(t) &= v_j(x_1, x_2, t)_{x_1=x_1(t), x_2=x_2(t)} \\
&= \frac{J_j(x_1, x_2, t)}{\rho(x_1, x_2, t)}\bigg|_{x_1=x_1(t), x_2=x_2(t)} \quad (j = 1, 2)
\end{aligned} \tag{5.56}
$$

where

$$
J_j(x_1, x_2, t) \equiv \frac{\hbar}{m_j} \mathrm{Im}\left[\Psi^*(x_1, x_2, t)\, \frac{\partial \Psi(x_1, x_2, t)}{\partial x_j}\right] \quad (j = 1, 2)\,. \tag{5.57}
$$

The concern expressed by AES is that the first particle can participate in an interaction localized at the point $x_1 = 0$ by the delta function $\delta(\hat{x}_1)$ in the interaction Hamiltonian even though its trajectory never reaches that point. Let us first consider the one-particle case with external potential $V(x, t) = \epsilon'(1/T)f(t/T)V(x)$ where the subscript 1 has been dropped for simplicity and ϵ' is a fixed parameter with dimensions of time. The wave function $\psi(x, t)$ probes the environment described by $V(x, t)$ directly and locally, both spatially and temporally, as evidenced by the term $V(x, t)\psi(x, t)$ in the wave equation. The guidance of the point-like particle by the wave function is also direct and local as evidenced by the equation of motion $v(t) = [J(x, t)/\rho(x, t)]_{x=x(t)}$. (The instantaneous velocity of the particle component of a Dirac electron is determined solely by the (four components of the) wave function at the instantaneous space–time location of the particle.) The effect of the potential in the immediate neighborhood of a specific space–time point (x', t') on the particle when it is at $(x'', t'' \geq t')$ is mediated by the guiding wave. Since the functional derivative $\delta\psi(x'', t'')/\delta V(x', t')$ for times t'' subsequent to t' is not in general zero for $x'' \neq x'$, it is not necessary for the point-like particle of Bohm's theory to reach x' at some time subsequent to t' to be influenced by the potential at (x', t'). Hence, the interaction of the particle with the potential is in general indirect, retarded in time and, in the above sense, nonlocal in space. For the two-particle case the interaction of the first particle with the total potential $V(x_1, x_2, t)$ is mediated by the guiding field $\Psi(x_1, x_2, t)$ and there is no necessity for the first particle to reach x_1' subsequent to t' to be influenced by $V(x_1', x_2', t')$. This is particularly so in this case because of the well-known fact that unless the wave function is factorable, i.e., unless $\Psi(x_1, x_2, t)$ can be written in the form $\psi_1(x_1, t)\psi_2(x_2, t)$, the motion of each of the "entangled" particles depends on the instantaneous position of the other no matter how far apart they are at time t or how weak the interaction V has become. Bell [6] considered it a "merit" of Bohmian mechanics that this nonlocality is so manifest that it cannot be ignored. He did not ignore it and consequently discovered his famous inequalities [6]. It is rather ironic that if the nonlocal influence of the second particle on the first had been such as to cause the latter to pass through $x_1 = 0$ then the case made by AES for a Bohmian particle not participating properly in a local interaction might not have gone ahead.

It is clear that within Bohmian mechanics the first particle's position is relevant for its participation in interactions but the interactions in question, being mediated by the guiding wave, are not local in the sense of Aharonov, Englert, and Scully. Hence, their second choice of concession should be dropped simply because it is not relevant within Bohmian mechanics. As argued below, elimination of this concession does not necessarily imply that one should make the alternative one, namely that "the particle's Bohm trajectory and its position are unrelated."

The above postulated role for the wave function as a guiding wave makes little sense if the wave function of an individual system is not regarded as a real physical entity. Now, the originators and main proponents of the idea of a protective measurement have repeatedly claimed [66, 67, 68, 69, 70] that *the wave function of a single system is real because it can be observed with this new kind of measurement.* AES stress that the wave function possesses a physical significance not only for an ensemble of identically prepared systems but also for a single system. However, they do not mention that this dual role for the wave function is also an important feature of Bohmian mechanics, one that has been held against it. Moreover, Anandan [68] has extended the theory of protective measurements to systems of many quantum particles and concluded that this "clearly shows that it is the wave in configuration space, and not physical space, which acquires an ontological meaning." This echoes a statement made about Bohmian mechanics by Bell in Chap. 15 of [6]: "No one can understand this theory until he is willing to think of ψ as a real objective field rather than just a 'probability amplitude'. Even though it propagates not in $3-$ space but in $3N-$ space."

I see no reason why an extremely slow and weak protective measurement *on a single system or on a single member of an ensemble* especially designed to measure the local intensity of the wave function, that is, the guiding wave of Bohm's theory, should simultaneously provide a local measurement of the presence of the point-like particle of Bohm's ontology *for that single system.* After all, the two components are very different entities – it is only the former that is expected to have identical behavior in each member of a pure state ensemble. The behavior of the point-like particle, on the other hand, varies randomly from member to member. In fact, in [66] Aharonov, Anandan, and

Vaidman conclude " A [quantum] particle manifests itself through its entire wave function during a protective measurement instead of manifesting like a point particle as in the usual measurement." From the point of view of Bohmian mechanics this is a very attractive result because it means that the two components of an electron, namely the point-like particle and the guiding wave, can be experimented upon separately in the strong-fast and weak-slow coupling limits, respectively. Presumably, in intermediate regimes of coupling both components would contribute, making the extraction of useful information about either difficult.

Consider a single standard (i.e., impulsive, strong, and wave function collapsing) measurement of the position of a particle –what Mielnik [25] likens to a large number of simultaneous police raids– on one member of an ensemble. It would be illogical to claim that the failure to find the particle in a specific localized region meant that the (precollapse) wave function had zero intensity in that region. Given the ontology of Bohm's theory it seems equally illogical to demand for a single measurement on an individual ensemble member that the particle component must spend in some specified localized region a time proportional to the protectively measured wave function intensity for that region. This view finds support in papers published prior to [65] by Aharonov and collaborators. In particular, they discounted a (generic) hidden-particle trajectory explanation of the above protective measurement result by considering the first excited energy eigenstate, which has a node at the centre of the potential well. They argued that if the measured result for any small spatial region was in fact a direct reflection of the fraction of the time interval $-T \leq t \leq +T$ spent there by the single particle, then the particle must spend half of its time in each half of the box and consequently pass through the node with infinite velocity.[14] They also pointed [66] out that the particle of Bohm's theory, being at rest, " cannot move to and fro as required" in this (refuted) explanation.

The mathematical description of the protective measurement of [67, 69], which is essentially the same[15] as that of [65] except that the localized regions of interest are not restricted to the center of the box, reveals an important property that is directly relevant to what is actually being measured. For any very short period of time δt, within an extended temporal range in which the coupling with the measuring device is independent of time, the measuring device shifts by an amount proportional to $|\psi(x = 0)|^2 \delta t$ [67]. Similarly, one can carry out the measurement much more quickly by coupling a single meter simultaneously to a very large number of independent quantum particles

---

[14] A problem with this argument is that it does not apply to Dirac electrons, for which energy eigenstates do not have nodes in 1D, and one must show instead that the tentative explanation can lead to superluminal particle speeds. One would presumably consider a well with walls not so high that the single-particle interpretation of Dirac's equation ceased to be a good approximation.

[15] $\hat{x}_2$ is replaced by $\hat{p}_2$ and $\delta(x_1)$ by $(1/v_n)$ if $x_1 \in V_n$ and 0 otherwise, where v_n is the volume of the nth small region V_n.

each prepared with the same initial wave function ψ_1 and, according to the theory, contributing equally to the total shift of the meter [70]. Both of these clearly indicate that for a single quantum particle it is the local intensity of the wave function of a single system that is being measured in the thought experiment of [65] and not the integrated presence of a hypothetical point-like particle in the region of interest. For example, by coupling the single meter to a sufficiently large number of quantum particles one can make the required coupling time so short that the times of presence of the associated point-like particles, with initial positions distributed according to $|\psi_1|^2$, in any specified small region cannot possibly be identical unless they always move with infinite speed. In Bohm's theory, for the system considered by AES, they do not move at all before the coupling is turned on and very slowly afterward. Hence, there is no necessary inconsistency between Bohm's theory and the claim that it is the local intensity of the wave function in each system that is being probed in the protective measurement. One could also conclude that the mathematical description is consistent with the nonexistence of a point-like particle between those instants of time in which it is observed as such, for example, in a standard position measurement. However, this interpretation of the mathematics cannot be verified experimentally because in the first case considered above the spatial shift of the meter in the time δt is too small to be observed and in the second the individual contributions to the net shift of the meter are not identified and measured. That is, in both cases, what is needed to possibly refute Bohm's hidden variable theory is itself hidden.

It should be noted that Aharonov, Englert, and Scully's conclusion that Bohm trajectories are "just mathematical constructs with no relation to the actual motion of the particle" confirms earlier conclusions [71, 72] based on standard and weak measurements. The conclusion based on standard measurements has recently been criticized by Hiley, Callaghan, and Maroney [73]. Moreover, there is an inconsistency between the protective and weak measurement conclusions: an important general conclusion of protective measurement theory [66, 67, 68, 69, 70] is that the wave function is real while the negative conclusion of Aharonov and Vaidman [72] regarding the reality of Bohm trajectories based on their weak measurement analysis hinges on their "particular approach to the Bohm theory in which the wave is not considered to be a 'reality'." Their particular approach is a possibility but is not the approach adopted in this chapter. In my opinion, protective measurements on energy eigenstates support the latter approach.

5.7 Concluding Comments

It is a mathematical fact that the momentum distribution $\hbar^{-1} |\phi(k,t)|^2$ is independent of time for an ensemble of freely evolving quantum particles all prepared in the same state $\psi(x,0)$. Now, the question "Can the velocity of an individual freely evolving particle change with time?" is a meaningful one within

Bohmian mechanics and the answer is "yes." I suspect that, for a variety of good reasons, most quantum physicists would regard this as a meaningless question within conventional quantum mechanics. If that is the case, then for the same reasons they should also regard the question "Is the velocity of an individual freely-evolving particle constant in time?" as a meaningless one within conventional quantum mechanics. But, there are many statements in the literature that suggests that many do not share this view. For example, in his classic monograph "General Principles of Quantum Mechanics" Pauli [74] explicitly stated that the probability density in momentum space is constant in time for free particles since the momentum of the particle itself is constant. This claim that the momentum of a free particle is constant in time is implicit in the well-known gedanken experiment in which the instantaneous momentum $p(t_0)$ of a single particle localized in some confining potential $V(x)$ is determined by switching off the potential at the specified instant of time $t = t_0$ and then measuring the time of flight of the liberated particle over a flight path L very much longer than the spatial extent of the wave function at t_0 [75]. Then, $p(t_0) = mL/(T - t_0)$ with an accuracy that depends primarily on that of the measured arrival time T at the detector. It seems that this claim is also implicit in a recent derivation [11] of the arrival-time distribution (5.27) for free particles, which is based on the assumption – presented as a self-evident fact – that every free particle in an ensemble described by a wave function with $\phi(k = 0, t) = 0$ must arrive once and only once at any spatial point X at some time T between $-\infty$ and $+\infty$.[16] It is this assumption that leaves no room for $k > 0 / k < 0$ interference terms in $\Pi(T; X)$. The claim that the momentum of an individual freely evolving particle – not in a momentum eigenstate – is constant in time is, in my opinion, an unjustified assumption within conventional quantum mechanics. Doesn't this assumption imply that such a particle has a trajectory $x(t) = x(0) + p(0)t/m$ with two hidden variables $x(0)$ and $p(0)$?

To say that something arrives at $x = X$ at a certain instant of time $t = T$ implies that it existed for a finite interval of time before the moment of arrival. This leads naturally to the problem posed by Wigner [20] of deriving the probability that a particle arrives at a specified point $x = X$ at time T from the left and also the probability that it arrives at time T from the right. Neither the Bohm trajectory nor Wigner's answers to this question, i.e., the directed arrival-time densities $\Pi_\pm(T; X)$ of Sect. 5.3.2, associates arrivals from the left (right) with $k > 0$ ($k < 0$) wave number components of the wave function. On the other hand, the "standard" answer [13] *does* make such an association. As a result, an important issue is providing an operational meaning for the arrival-time distribution (5.27) consistent with it containing no

---

[16] The usual definition of the arrival-time distribution has been extended to include arrival times in the range $[-\infty, 0]$ by imagining that the particle was prepared at $t = -\infty$ in the state, which, in the assumed absence of any interaction, would evolve to the desired initial state $\psi(x, t = 0)$ at $t = 0$. It is claimed that for the extended problem, just as in the classical case (with $p(t) = p(0) \neq 0$), every particle must reach X once and only once.

158   C. R. Leavens

term associated with interference between $k > 0$ and $k < 0$ components. The following thought experiment to measure the distribution (5.27) for the case of free evolution has been proposed [15]. A strong measurement of the sign of the momentum of each incident quantum particle is made, collapsing the wave function to either ψ_+ or ψ_-. In each member of the ensemble, the result of this measurement triggers the switching on of the appropriate one-sided detector at $x = X$. In order not to miss any arrivals the sign-of-momentum measurement should be made as early as possible. The most efficient way of performing the experiment would be simply to prepare the desired mixture of the states ψ_+ and ψ_- at $t = 0$ with the appropriate one-sided detectors already in place. The experiment for the case in which the potential $V(x, t)$ is not zero would be much more difficult, perhaps impossible, to implement because for each ensemble member it is necessary to measure the sign of the momentum and turn on the corresponding one-sided arrival-time detector an instant before the particle will be found to have arrived at X, i.e., before the potential can introduce a nonnegligible contribution to the wave function with the "wrong" sign of the momentum. Since one cannot predict when an individual detection will occur it will be necessary to discard all those measured arrival times that occur outside some specified temporal window $[t_\pm, t_\pm + \Delta_t]$ where $t_\pm$ is the time at which the sign-of-momentum measurement is made, assuming that such a measurement can be made at an instant of time and the appropriate detector inserted immediately afterward. Hopefully, a suitable choice for Δ_t could be determined empirically by measuring $\Pi(T)$ for decreasing values of Δ_t until convergence is reached. In addition, there will be nondetections when the choice of $t_\pm$ is too large. Moreover, there are interesting conceptual and practical issues associated with the collapse of the wave function ψ at $t = t_\pm$ to either ψ_+ or ψ_- accompanied by an instantaneous change of integrated probability density for the two half-spaces $x > X$ and $x < X$.

Generalizing various approaches for deriving arrival-time distributions for Schrödinger particles to the case of Dirac particles could be an interesting avenue for future research, especially if there is the possibility that arrival-time experiments on spin-polarized electrons might discriminate between different theories. This will probably require that the theories be specialized to the case of first arrivals, which is relatively simple for the Bohm trajectory approach but has not been implemented for the standard approach.

Finally, in my opinion, there is the possibility that the concept of protective measurement on an energy eigenstate might enhance, rather than weaken, the version of Bohmian mechanics adopted here in which the reality of the wave function is an important element.

Acknowledgments

The author is indebted to his NRC collaborators, G.C. Aers, and W.R. McKinnon, to S. Goldstein, G. Iannaccone, and his CERION (Canadian European Research Initiative on Nanostructures) partners D. Alonso, J.G. Muga,

J.P. Palao, and R. Sala Mayato for their contributions to the work presented here. He is also thankful for the opportunity to participate in three of the workshops on time in quantum mechanics held at the University of La Laguna and organized by J.G. Muga, R. Sala Mayato, and S. Brouard.

Appendix

The Dirac equation describing an electron minimally coupled to the electromagnetic four potential $[U(\boldsymbol{r}, t), \boldsymbol{A}(\boldsymbol{r}, t)]$ is [76, 77, 78, 79]

$$
i\hbar \frac{\partial}{\partial t} \psi(\boldsymbol{r}, t) =
$$
$$
\left\{ c\hat{\boldsymbol{\alpha}} \cdot \left[(\hbar/i)\boldsymbol{\nabla} - (e/c)\boldsymbol{A}(\boldsymbol{r}, t) \right] + \hat{\beta} mc^2 + \hat{\alpha}_0 eU(\boldsymbol{r}, t) \right\} \psi(\boldsymbol{r}, t) , \quad (5.58)
$$

where the wave function $\psi(\boldsymbol{r}, t)$ is a four-component column matrix,

$$
\psi(\boldsymbol{r}, t) \equiv \begin{pmatrix} \psi_1(\boldsymbol{r}, t) \\ \psi_2(\boldsymbol{r}, t) \\ \psi_3(\boldsymbol{r}, t) \\ \psi_4(\boldsymbol{r}, t) \end{pmatrix} \equiv \Big(\psi_1(\boldsymbol{r}, t), \psi_2(\boldsymbol{r}, t), \psi_3(\boldsymbol{r}, t), \psi_4(\boldsymbol{r}, t) \Big)^T , \quad (5.59)
$$

and $\hat{\alpha}_x$, $\hat{\alpha}_y$, $\hat{\alpha}_z$, $\hat{\beta}$, and $\hat{\alpha}_0 \equiv \hat{1}$ are 4x4 matrices. In the standard representation used here

$$
\hat{\alpha}_x \equiv \begin{pmatrix} 0 & 0 & 0 & 1 \\ 0 & 0 & 1 & 0 \\ 0 & 1 & 0 & 0 \\ 1 & 0 & 0 & 0 \end{pmatrix} , \quad \hat{\alpha}_y \equiv \begin{pmatrix} 0 & 0 & 0 & -i \\ 0 & 0 & i & 0 \\ 0 & -i & 0 & 0 \\ i & 0 & 0 & 0 \end{pmatrix} ,
$$

$$
\hat{\alpha}_z \equiv \begin{pmatrix} 0 & 0 & 1 & 0 \\ 0 & 0 & 0 & -1 \\ 1 & 0 & 0 & 0 \\ 0 & -1 & 0 & 0 \end{pmatrix} , \quad \hat{\beta} \equiv \begin{pmatrix} 1 & 0 & 0 & 0 \\ 0 & 1 & 0 & 0 \\ 0 & 0 & -1 & 0 \\ 0 & 0 & 0 & -1 \end{pmatrix} . \quad (5.60)
$$

The continuity equation

$$
\frac{\partial}{\partial t} \rho(\boldsymbol{r}, t) + \boldsymbol{\nabla} \cdot \boldsymbol{J}(\boldsymbol{r}, t) = 0 \quad (5.61)
$$

is satisfied for

$$
\rho(\boldsymbol{r}, t) \equiv \psi^\dagger(\boldsymbol{r}, t) \psi(\boldsymbol{r}, t) , \quad (5.62)
$$

$$
\boldsymbol{J}(\boldsymbol{r}, t) = c\psi^\dagger(\boldsymbol{r}, t) \, \hat{\boldsymbol{\alpha}} \, \psi(\boldsymbol{r}, t) , \quad (5.63)
$$

where

$$
\psi^\dagger(\boldsymbol{r}, t) \equiv \Big(\psi_1^*(\boldsymbol{r}, t), \psi_2^*(\boldsymbol{r}, t), \psi_3^*(\boldsymbol{r}, t), \psi_4^*(\boldsymbol{r}, t) \Big) . \quad (5.64)
$$

The range of validity of the single-electron interpretation of the Dirac equation (5.58) is not well-defined [78]. It is a basic assumption of this chapter that there exists a very low energy regime in which it is justified to interpret $\rho(\boldsymbol{r}, t)$ and $\boldsymbol{J}(\boldsymbol{r}, t)$ as single-electron probability and probability current densities, respectively, to an adequate approximation. By "adequate approximation" it is meant that the results obtained with the Dirac equation are a closer approximation to reality than the corresponding ones obtained with the Schrödinger (or Pauli) equation and are not significantly altered by second quantization of the Dirac theory to take accurately into account the effects of real and virtual electron–positron creation and annihilation.

It is well-known [3, 7] that the velocity field $\boldsymbol{v}(\boldsymbol{r}, t) \equiv \boldsymbol{J}(\boldsymbol{r}, t)/\rho(\boldsymbol{r}, t)$ of Bohm's ontological interpretation of relativistic quantum mechanics cannot exceed c in absolute value. This is easily proven by straightforward calculation of the quantity $\rho^2(\boldsymbol{r}, t)[1 - v^2(\boldsymbol{r}, t)/c^2]$ for the Dirac wave function (5.59). With the jth component of the wave function expressed in polar form, i.e., $\psi_j(\boldsymbol{r}, t) \equiv |\psi_j(\boldsymbol{r}, t)| \exp[i\varphi_j(\boldsymbol{r}, t)]$, the result can be written as

$$\rho^2[1 - v^2/c^2] = (\ |\psi_1|^2 + |\psi_2|^2 - |\psi_3|^2 - |\psi_4|^2)^2 \ +$$
$$4\left[|\psi_1||\psi_3|\sin(\varphi_1 - \varphi_3) + |\psi_2||\psi_4|\sin(\varphi_2 - \varphi_4)\right]^2 \ (5.65)$$

with $\rho = |\psi_1|^2 + |\psi_2|^2 + |\psi_3|^2 + |\psi_4|^2$. Since the right-hand side of this equation is nonnegative, v^2 cannot exceed c^2 whenever $\rho \neq 0$. It should be noted that $\hat{\boldsymbol{\alpha}}^2 = 3\,\hat{1}$ and hence that the local expectation value of the square of the velocity operator, that is, of $(c\hat{\boldsymbol{\alpha}})^2$, is $3c^2$.

References

1. I.L. Egusquiza, J.G. Muga, A.D. Baute: ' "Standard" Quantum Mechanical Approach to Times of Arrival'. In: *Time in Quantum Mechanics*, ed. by J.G. Muga, R. Sala Mayato, I.L. Egusquiza, 1st edn. (Springer, Berlin 2002) pp. 279–304
2. D. Dürr, S. Goldstein, N. Zanghì: Journal of Statistical Physics **116**, 959–1055 (2004). quant-ph/0308038
3. D. Bohm, B. J. Hiley: *The Undivided Universe: An Ontological Interpretation of Quantum Mechanics* (Routledge, London 1993)
4. L. de Broglie: *Non-linear Wave Mechanics* (Elsevier, Amsterdam 1960)
5. D. Bohm: Phys. Rev. **85**, 166; 180 (1952)
6. J.S. Bell: *Speakable and Unspeakable in Quantum Mechanics* (Cambridge University Press, Cambridge 1987)
7. P.R. Holland: *The Quantum Theory of Motion* (Cambridge University Press, Cambridge 1993)
8. J.T. Cushing, A. Fine, S. Goldstein (eds.): *Bohmian Mechanics and Quantum Theory: An Appraisal* (Kluwer, Dordrecht 1996)
9. D. Dürr, S. Goldstein, N. Zanghí: J. Stat. Phys. **67**, 843 (1992)
10. J. Kijowski: Rep. Math. Phys. **6**, 361 (1974); Phys. Rev. A **59**, 897 (1999)
11. N. Grot, C. Rovelli, R.S. Tate: Phys. Rev. A **54**, 4676 (1996)
12. V. Delgado, J.G. Muga: Phys. Rev. A **56**, 3425 (1997)

13. A.D. Baute, R. Sala Mayato, J.P. Palao, J.G. Muga, I.L. Egusquiza: Phys. Rev. A **61**, 022118 (2000)
14. J.G. Muga, R. Sala, J.P. Palao: Superlattices Microstruct. **23**, 833 (1998)
15. J.G. Muga, C.R. Leavens: Phys. Rep. **338**, 353 (2000)
16. J. Ruseckas, B. Kaulakys: Phys. Rev. A **66**, 052106 (2002)
17. R.P. Feynman, A.R. Hibbs: *Quantum Mechanics and Path Integrals* (McGraw-Hill, New York 1965), pp. 34–36
18. C.R. Leavens: Phys. Lett. A **272**, 160 (2000)
19. T. Jacobson, L.S. Schulman: J. Phys. A **17**, 375 (1984)
20. E.P. Wigner: 'On the Time–Energy Uncertainty Relation'. In: *Aspects of Quantum Theory*, ed. by A. Salem, E.P. Wigner (Cambridge University Press, London 1972) pp. 237–247
21. E. Deotto, G.C. Ghirardi: Found. Phys. **28**, 1 (1998)
22. J. Finkelstein: Phys. Rev. A **59**, 3218 (1999)
23. P.R. Holland: Phys. Rev. A **60**, 4326 (1999)
24. Md.M. Ali, A.S. Majumdar, D. Home, S. Sengupta: Phys. Rev. A **68**, 042105 (2003)
25. B. Mielnik: Found. Phys. **24**, 1113 (1994)
26. M. Daumer: 'Scattering Theory from a Bohmian Perspective'. In: *Bohmian Mechanics and Quantum Theory: An Appraisal*, ed. by J.T. Cushing, A. Fine, S. Goldstein (Kluwer, Dordrecht 1996) pp. 87–98
27. D. Dürr, S. Goldstein, T. Moser, N. Zanghì: Commun. Math. Phys. **266**, no.3, 665–697 (2006). quant-ph/0509010
28. C.R. Leavens: Phys. Rev. A **58**, 840 (1998)
29. G. Grübl, K. Rheinberger: J. Phys. A **35**, 2907 (2002)
30. G. Iannaccone, B. Pellegrini: Phys. Rev. B **49**, 16548 (1994)
31. W.R. McKinnon, C.R. Leavens: Phys. Rev. A **51**, 2748 (1995)
32. X. Oriols, F. Martín, J. Suñé: Phys. Rev. A **54**, 2594 (1996)
33. C.R. Leavens: Phys. Lett. A **178**, 27 (1993)
34. S. Kreidl, G. Grübl, H.G. Embacher: J. Phys. A **36**. 8851 (2003)
35. Y. Nogami, F.M. Toyama, W. van Dijk: Phys. Lett. A **270**, 279 (2000)
36. W. van Dijk, Y. Nogami: Phys. Rev. Lett. **83**, 2867 (2000)
37. M. Büttiker: Phys. Rev. B **27**, 6178 (1983)
38. T.P. Spiller, T.D. Clark, R.J. Prance, H. Prance: Europhys. Lett. **12**, 1 (1990)
39. C.R. Leavens, G.C. Aers: 'Bohm Trajectories and the Tunneling Time Problem'. In: *Scanning Tunneling Microscopy III*, ed. by R. Wiesendanger, H-J. Güntherodt, 2nd edn. (Springer, Berlin 1993) pp. 105–140
40. C.R. Leavens, G. Iannaccone, W. R. McKinnon: Phys. Lett. A **208**, 17 (1995)
41. G.R. Allcock: Ann. Phys. **53**, 253; 286; 311 (1969)
42. A.J. Bracken, G.F. Melloy: J. Phys. A: Math. Gen. **27**, 2197 (1994)
43. G.F. Melloy, A.J. Bracken: Ann. Phys. (Leipzig) **7**, 726 (1998)
44. J.G. Muga, C.R. Leavens, J.P. Palao: Phys. Rev. A **58**, 4336 (1998)
45. C.R. Leavens: Phys. Lett. A **338**, 19 (2005)
46. C.R. Leavens: Phys. Lett. A **303**, 154 (2002)
47. I.L. Egusquiza, J.G. Muga, B. Navarro, A. Ruschhaupt: Phys. Lett. A **313**, 498 (2003)
48. C.R. Leavens: Phys. Lett. A **345**, 251 (2005)
49. D. Bohm: *Quantum Theory* (Prentice-Hall, New York 1951)
50. E.P. Wigner: Phys. Rev. **98**, 145 (1955)

51. E.H. Hauge, J.P. Falck, T.A. Fjeldly: Phys. Rev. B **36**, 4203 (1987)
52. T.E. Hartman: J. Appl. Phys. **33**, 3427 (1962)
53. R. Landauer, Th. Martin: Rev. Mod. Phys. **66**, 217 (1994)
54. S. Brouard, R. Sala, J.G. Muga: Phys. Rev. A **49**, 4312 (1994)
55. C.R. Leavens: 'Bohmian Mechanics and the Tunneling Time Problem for Electrons'. In: *Tunneling and Its Implications*, ed. by D. Mugnai, A. Ranfagni, L.S. Schulman (World Scientific, Singapore 1997) pp. 100–120
56. C.R. Leavens: Found. Phys. **25**, 229 (1995)
57. C.R. Leavens, W.R. McKinnon: Phys. Lett. A **194**, 12 (1994)
58. V.S. Olkhovsky, E. Recami: Phys. Rep. **214**, 339 (1992)
59. J.G. Muga, S. Brouard, R. Sala: Phys. Lett. A **167**, 24 (1992)
60. C.R. Leavens: Phys. Lett. A **197**, 88 (1995)
61. V. Delgado, S. Brouard, J.G. Muga: Solid State Commun. **94**, 979 (1995)
62. Y. Aharonov, N. Erez, M.O. Scully: Phys. Scr. **69**, 81 (2004); quant-ph/0412068
63. S. Brandt, H.D. Dahmen, E. Gjonaj, T. Stroh: Phys. Lett. A **249**, 265 (1998)
64. A. Challinor, A. Lasenby, S. Somaroo, C. Doran, S. Gull: Phys. Lett. A **227**, 143 (1997)
65. Y. Aharonov, B.G. Englert, M.O. Scully: Phys. Lett. A **263**, 137 (1999)
66. Y. Aharonov, J. Anandan, L. Vaidman: Phys. Rev. A **47**, 4616 (1993)
67. Y. Aharonov, L. Vaidman: Phys. Lett. A **178**, 38 (1993)
68. J. Anandan: Found. Phys. Lett. **6**, 503 (1993)
69. Y. Aharonov, L. Vaidman: 'Protective Measurements'. In: *Fundamental Problems in Quantum Theory*, ed. by D.M. Greenberger, A. Zeilinger (New York Academy of Sciences, 1995) pp. 361–373
70. Y. Aharonov, J. Anandan, L. Vaidman: Found. Phys. **26**, 117 (1996)
71. B-G. Englert, M.O. Scully, G. Sussman, H. Walther: Z. Naturforsch. **47a**, 1175 (1992)
72. Y. Aharonov, L. Vaidman: 'About Position Measurements Which Do Not Show the Bohmian Particle Position'. In: *Bohmian Mechanics and Quantum Theory: An Appraisal*, ed. by J.T. Cushing, A. Fine, S. Goldstein (Kluwer, Dordrecht 1996) pp. 141–154
73. B.J. Hiley, R.E. Callaghan, O. Maroney: quant-ph/0010020
74. W. Pauli: *General Principles of Quantum Mechanics* (Springer-Verlag, Berlin 1980)
75. W. Heisenberg: *Physical Principles of the Quantum Theory* (University of Chicago Press, Chicago 1930)
76. P.A.M. Dirac: *Principles of Quantum Mechanics*, 4th edn. (revised) (Clarendon, Oxford 1967)
77. J.D. Bjorken, S.D. Drell: *Relativistic Quantum Mechanics* (McGraw-Hill, New York 1964)
78. B. Thaller: *The Dirac Equation* (Springer-Verlag, Berlin 1992)
79. W. Greiner: *Relativistic Quantum Mechanics-Wave Equations* (Springer-Verlag, Berlin 1990)

6

Decoherent Histories for Space–Time Domains

Jonathan J. Halliwell

Blackett Laboratory, Imperial College, London SW7 2BZ, UK

The decoherent histories approach is a natural medium in which to address problems in quantum theory that involve time in a non-trivial way. This chapter reviews the various attempts and difficulties involved in using the decoherent histories approach to calculate the probability for crossing the surface $x = 0$ during a finite interval of time. The commonly encountered difficulties in assigning crossing times arise here as difficulties in satisfying the consistency (no-interference) condition. This can be overcome by introducing an environment to produce decoherence, and probabilities exhibiting the expected classical limit are obtained. The probabilities are, however, dependent to some degree on the decohering environment. The results are compared with a recently proposed irreversible detector model. A third method is introduced, involving continuous quantum measurement theory. Some closely related work on the interpretation of the wave function in quantum cosmology is described.

6.1 Introduction

Although opinions differ as to the value and achievements of attempts to quantize the gravity, it is undeniable that this endeavour has inspired a considerable amount of work in a variety of related fields. In particular, the quantization of gravity puts considerable pressure on both the mathematical and conceptual foundations of quantum theory, so it is perhaps not surprising that many researchers in quantum gravity have been drawn into working on the foundations of quantum mechanics.

One of the key issues that arises in the quantization of gravity is the "Problem of Time". In the quantization of cosmological models, the wave function of the universe satisfies not a Schrödinger equation but the Wheeler–DeWitt equation,

$$\mathcal{H}\Psi[h_{ij}, \phi] = 0 . \tag{6.1}$$

The wave function Ψ depends on the three-metric h_{ij} and the matter field configurations ϕ on a closed space-like three-surface [25, 37, 39]. There is no time

J.J. Halliwell: *Decoherent Histories for Space–Time Domains*, Lect. Notes Phys. **734**, 163–193 (2008)
DOI 10.1007/978-3-540-73473-4_6

label. Its absence is deeply entwined with the four-dimensional diffeomorphism invariance of general relativity. It is often conjectured that "time" is somehow already present amongst the dynamical variables h_{ij}, ϕ, although to date it has proved impossible to extract a unique, globally defined time variable.

Although a comprehensive scheme for interpreting the wave function is yet to be put forward, a prevalent view is that the interpretation will involve treating all the dynamical variables h_{ij}, ϕ on an equal footing, rather than trying to single out one particular combination of them to act as time. For this reason, it is of interest to see if one can carry out a similar exercise in non-relativistic quantum mechanics. That is, to see what predictions quantum mechanics makes about *space–time* regions, rather than regions of space at fixed moments of time.

Such predictions are not the ones that quantum mechanics usually makes. In standard non-relativistic quantum mechanics, the probability of finding a particle between points x and $x + dx$ at a fixed time t is given by

$$p(x, t)dx = |\Psi(x, t)|^2 \, dx \,, \qquad (6.2)$$

where $\Psi(x, t)$ is the wave function of the particle. More generally, the variety of questions one might ask about a particle at a fixed moment of time may be represented by a projection operator P_α, which is exhaustive,

$$\sum_\alpha P_\alpha = 1 \,, \qquad (6.3)$$

and mutually exclusive,

$$P_\alpha P_\beta = \delta_{\alpha\beta} P_\alpha \,. \qquad (6.4)$$

The projection operator appropriate to ask questions about position is $P = |x\rangle\langle x|$. The probability of a particular alternative is given by

$$p(\alpha) = \mathrm{Tr}\,(P_\alpha \varrho) \,, \qquad (6.5)$$

where ϱ is the density operator of the system at the time in question.

The key feature of the above standard formulae is that they do not treat space and time on an equal footing. Suppose one asks, for example, the same sort of question with space and time interchanged. That is, what is the probability of finding the particle at point x in the time interval t to $t + dt$? The point is that the answer is *not* given by $|\Psi(x, t)|^2 dt$. The reason for this is that, unlike the value of x at fixed t, the value of t at fixed x does not refer to an exclusive set of alternatives. The position of a particle at fixed time is a well-defined quantity in quantum mechanics, but the time at which a particle is found at a fixed position is much more difficult to define because of the possibility of multiple crossings.

This question is clearly a physically relevant one since time is measured by physical devices that are generally limited in their precision. It is therefore never possible to say that a physical event occurs at a precise value of time, that it occurs only in some range of times. Furthermore, there has been considerable recent experimental and theoretical interest in the question of tunnelling times [40, 48]. This is the question: given that a particle has tunnelled through a barrier region, how much time did it spend inside the barrier?

Space–time questions tend to be rather non-trivial. As stressed by Hartle, who has carried out a number of investigations in this area [35, 36, 38], time plays a "peculiar and central role" in non-relativistic quantum mechanics. It is not represented by a self-adjoint operator and there appears to be no obstruction to assuming that it may be measured with arbitrary precision. It enters the Schrödinger equation as an external parameter. As such, it is perhaps best thought of as a label referring to a classical, external measuring device, rather than as a fundamental quantum observable. Yet time is measured by physical systems, and all physical systems are believed to be subject to the laws of quantum theory.

Given these features, means more elaborate than those usually employed are required to define quantum-mechanical probabilities that do not refer to a specific moment of time, and the issue has a long history [2, 64]. One may find in the literature a variety of attempts to define questions of time in a quantum-mechanical way. These include attempts to define time operators [21, 41, 59], the use of internal physical clocks [36, 38] and path-integral approaches [14, 35, 47, 68]. The literature on tunnelling times is a particularly rich source of ideas on this topic [40]. Many of these attempts also tie in with the time-energy uncertainty relations [46, 50]. For a nice review of many of these issues, see [54]. All these subjects are treated in more detail in other chapters of this volume.

This chapter is concerned with the attempts to solve problems of a space–time nature using the decoherent histories approach to quantum theory [17, 18, 20, 57]. It is perhaps of interest to note that, in addition to inspiring work on the question involving time, considerations of quantum gravity were also partly responsible for the development of the decoherent histories approach. For our purposes, the particular attraction of this approach is that it directly addresses the notion of a "history" or a "trajectory" and in particular shows how to assign probabilities to them. It is therefore very suited to the question of space–time probabilities considered here. This is because the question of whether a particle did or did not enter a given region at *any* time in a given time interval clearly cannot be reduced to a question about the state of the particle at a fixed moment of time, but depends on the entire history of the system during that time interval.

The decoherent histories approach, for space–time questions, turns out to be most clearly formulated in terms of path integrals over paths in configuration space [68, 35, 39]. The desired space–time amplitudes are obtained by summing $\exp\left(\frac{i}{\hbar}S[x(t)]\right)$, where $S[x(t)]$ is the action, over paths $x(t)$ passing through the space-time region in question and consistent with the initial state. The probabilities are obtained by squaring the amplitudes in the usual way. (The decoherent histories approach is not inextricably tied to path integrals, however. Operator approaches to the same questions are also available, but are often more cumbersome.)

When computed according to the path integral scheme outlined above, the probability of entering a space–time region added to the probability of not entering that region is not equal to 1, in general. This is because of interference. The question of whether a particle enters a space–time region, when carefully broken down, is actually a quite complicated combination of questions about the positions of the particle at a sequence of times. It is therefore, in essence, a complicated combination of double-slit situations. Not surprisingly, there is therefore interference and probabilities cannot be assigned.

Therefore, from the point of view of the decoherent histories approach to quantum theory, the probability of entering a space–time region is quite simply *not defined* in general for a simple point particle system due to the presence of interference. It is here that the decoherent histories approach, like all the other approaches to defining time in quantum theory, runs up against its own particular brand of difficulties.

It is, however, a common feature of the decoherent histories approach that most of the histories of interest cannot be defined due to interference – histories defined by position at more than one time, for example. It is well known that the interference may be removed by coupling to environment, typically a bath of harmonic oscillators in a thermal state. We will therefore consider the above space–time problem in the presence of an environment.

The decoherent histories approach is reviewed in Sect. 6.2 and its application to simple space–time questions is discussed in Sect. 6.3. The inclusion of the environment to induce decoherence is described in Sect. 6.4.

The probabilities produced by the decoherent histories approach are in some sense somewhat abstract since they do not refer to a particular measuring device. In Sect. 6.5 we therefore introduce a model measuring device for the purposes of comparison. The decoherence model of Sect. 6.4 consists of quite a crude environment that has, however, been very successful in producing decoherence and emergent classicality. The measurements it effectively carries out are of a rather robust and crucially irreversible nature. Hence the most important sort of comparison is with an irreversible detector model. Interestingly, most of the arrival time models discussed in the literature are not of this type. It is therefore of interest to develop a model detector not dissimilar to the decoherence model but sufficiently modified to carry out a more precise measurement. The comparison between the decoherent histories approach and the detector model is then discussed out in Sect. 6.6. This also leads to the introduction of a third candidate for the crossing time probability, derived from continuous quantum measurement theory.

In Sect. 6.7 we briefly discuss another type of non-trivial time question, namely, given that a system is in an energy eigenstate, what is the probability that it will pass through a given region in configuration space at *any* time? The reason that is of interest is that it is, in essence, the question one needs to answer in order to interpret solutions to the Wheeler–DeWitt equation (6.1).

We summarize and conclude in Sect. 6.8. An update for the second edition is given in Sect. 6.9.

6.2 Decoherent Histories Approach to Quantum Theory

In this section we give a brief summary of the decoherent histories approach to quantum theory. It has been described in considerable depth in many other places [13, 17, 18, 20, 22, 23, 39, 42, 57].

In quantum mechanics, propositions about the attributes of a system at a fixed moment of time are represented by sets of projection operators. The projection operators P_α effect a partition of the possible alternatives α a system may exhibit at each moment of time. They are exhaustive and exclusive, as noted in (6.3) and (6.4). A projector is said to be *fine-grained* if it is of the form $|\alpha\rangle\langle\alpha|$, where $\{|\alpha\rangle\}$ are a complete set of states. Otherwise it is *coarse-grained*. A quantum-mechanical history (strictly, a *homogeneous* history [42]) is characterized by a string of time-dependent projections, $P^1_{\alpha_1}(t_1), \cdots P^n_{\alpha_n}(t_n)$, together with an initial state ϱ. The time-dependent projections are related to the time-independent ones by

$$P^k_{\alpha_k}(t_k) = e^{iH(t_k-t_0)} P^k_{\alpha_k} e^{-iH(t_k-t_0)} , \tag{6.6}$$

where H is the Hamiltonian. The candidate probability for these homogeneous histories is

$$p(\alpha_1, \alpha_2, \cdots \alpha_n) = \text{Tr}\left(P^n_{\alpha_n}(t_n) \cdots P^1_{\alpha_1}(t_1) \varrho P^1_{\alpha_1}(t_1) \cdots P^n_{\alpha_n}(t_n)\right) . \tag{6.7}$$

It is straightforward to show that (6.7) is both non-negative and normalized to unity when summed over $\alpha_1, \cdots, \alpha_n$. However, (6.7) does not satisfy all the axioms of probability theory, and for that reason it is referred to as a candidate probability. It does not satisfy the requirement of additivity on disjoint regions of sample space. More precisely, for each set of histories, one may construct coarser-grained histories by grouping the histories together. This may be achieved, for example, by summing over the projections at each moment of time,

$$\bar{P}_{\bar\alpha} = \sum_{\alpha \in \bar\alpha} P_\alpha \tag{6.8}$$

(although this is not the most general type of coarse graining – see below). The additivity requirement is then that the probabilities for each coarser-grained history should be the sum of the probabilities of the finer-grained histories of which it is comprised. Quantum-mechanical interference generally prevents this requirement from being satisfied. Histories of closed quantum systems cannot in general be assigned probabilities.

There are, however, certain types of histories for which interference is negligible, and the candidate probabilities for histories do satisfy the sum rules. These histories may be found using the decoherence functional:

$$D(\underline\alpha, \underline\alpha') = \text{Tr}\left(P^n_{\alpha_n}(t_n) \cdots P^1_{\alpha_1}(t_1) \varrho P^1_{\alpha'_1}(t_1) \cdots P^n_{\alpha'_n}(t_n)\right) . \tag{6.9}$$

Here $\underline\alpha$ denotes the string $\alpha_1, \alpha_2, \cdots, \alpha_n$. Intuitively, the decoherence functional measures the amount of interference between pairs of histories. It may be shown that the additivity requirement is satisfied for all coarse grainings if and only if

$$\mathrm{Re}D(\underline{\alpha}, \underline{\alpha}') = 0 \qquad (6.10)$$

for all distinct pairs of histories $\underline{\alpha}, \underline{\alpha}'$ [20]. Such sets of histories are said to be *consistent*, or *weakly decoherent*. The consistency condition (6.10) is typically satisfied only for coarse-grained histories, and this then often leads to satisfaction of the stronger condition of *decoherence*,

$$D(\underline{\alpha}, \underline{\alpha}') = 0 , \qquad (6.11)$$

for $\alpha \neq \alpha'$. The condition of decoherence is associated with the existence of so-called generalized records. This means that it is possible to add a projector R_β at the end of the chain such that decoherence is preserved and such that the label β is perfectly correlated with the history alternatives $\alpha_1, \cdots, \alpha_n$. There is therefore in principle some physical measurement that could be carried out at the end of the history from which complete information about the entire history can be recovered [18, 19, 26].

For histories characterized by projections onto ranges of position at different times, the decoherence functional may be represented by a path integral:

$$D(\alpha, \alpha') = \int_\alpha \mathcal{D}x \int_{\alpha'} \mathcal{D}y \, \exp\left(\frac{i}{\hbar}S[x] - \frac{i}{\hbar}S[y] \right) \varrho(x_0, y_0) . \qquad (6.12)$$

The integral is over paths $x(t)$, $y(t)$ starting at x_0, y_0, and both ending at the same final point x_f, where x_f, x_0 and y_0 are all integrated over and weighted by the initial state $\varrho(x_0, y_0)$. The paths are also constrained to pass through spatial gates at a sequence of times corresponding to the projection operators.

However, the path integral representation of the decoherence functional also points the way towards asking types of questions that are not represented by homogeneous histories [35]. In this article we are particularly interested in the following question. Suppose a particle starts at $t = 0$ in some quantum state. What is the probability that the particle will either cross or never cross $x = 0$ during the time interval $[0, \tau]$? In the path integral of the form (6.12) it is clear how to proceed. One sums over paths that either always cross or never cross $x = 0$, respectively, during the time interval.

How does this look in operator language? The operator form of the decoherence functional is

$$D(\alpha, \alpha') = \mathrm{Tr}\left(C_\alpha \varrho C_{\alpha'}^\dagger \right) , \qquad (6.13)$$

where

$$C_\alpha = P_{\alpha_n}(t_n) \cdots P_{\alpha_1}(t_1) . \qquad (6.14)$$

The histories that never cross $x = 0$ are represented by taking the projectors in C_α to be onto the positive x-axis and then taking the limit $n \to \infty$ and $t_k - t_{k-1} \to 0$. The histories that always cross $x = 0$ are then represented by the object

$$\bar{C}_\alpha = 1 - C_\alpha . \tag{6.15}$$

This is called an *inhomogeneous* history, because it cannot be represented as a single string of projectors. It can however, be represented as a *sum* of strings of projectors [35, 42].

The proper framework in which these operations, in particular (6.15), are understood is the so-called generalized quantum theory of Hartle [35] and Isham *et al.* [42]. It is called "generalized" because it admits inhomogeneous histories as viable objects, whilst standard quantum theory concerns itself entirely with homogeneous histories. We will make essential use of inhomogeneous histories in what follows.

In practice, for point particle systems, decoherence is readily achieved by coupling to an environment. Here, we will use the much studied case of the quantum Brownian motion model, in which the particle is linearly coupled through position to a bath of harmonic oscillators in a thermal state at temperature T and characterized by a dissipation coefficient γ. The details of this model may be found elsewhere [10, 15, 22, 23].

We consider histories characterized only by the position of the particle and the environmental coordinates are traced out. The path integral representation of the decoherence functional then has the form

$$D(\alpha, \alpha') = \int_\alpha \mathcal{D}x \int_{\alpha'} \mathcal{D}y \, \exp\left(\frac{i}{\hbar} S[x] - \frac{i}{\hbar} S[y] + \frac{i}{\hbar} W[x, y] \right) \varrho(x_0, y_0) , \tag{6.16}$$

where $W[x, y]$ is the Feynman–Vernon influence functional phase and is given by

$$W[x, y] = -m\gamma \int dt \, (x - y)(\dot{x} + \dot{y}) + i\frac{2m\gamma kT}{\hbar} \int dt \, (x - y)^2 . \tag{6.17}$$

The first term induces dissipation in the effective classical equations of motion. The second term is responsible for thermal fluctuations. It is also responsible for suppressing contributions from paths $x(t)$ and $y(t)$ that differ widely and produces decoherence of configuration space histories.

The corresponding classical theory is no longer the mechanics of a single point particle, but a point particle coupled to a heat bath. The classical correspondence is now to a stochastic process which may be described either by a Langevin equation or by a Fokker–Planck equation for a phase-space probability distribution $w(p, x, t)$:

$$\frac{\partial w}{\partial t} = -\frac{p}{m}\frac{\partial w}{\partial x} + 2\gamma\frac{\partial(pw)}{\partial p} + D\frac{\partial^2 w}{\partial p^2} , \tag{6.18}$$

where $w \geq 0$ and

$$\int dp \int dx \, w(p, x, t) = 1 . \tag{6.19}$$

When the mass is sufficiently large, this equation describes near-deterministic evolution with small thermal fluctuations about it.

6.3 Space–Time Coarse Grainings

We are generally interested in space–time coarse grainings that consist of asking for the probability that a particle does or does not enter a certain region of space during a certain time interval. However, the essentials of this question boil down to the following simpler question: What is the probability that the particle will either cross or not cross $x = 0$ at any time in the time interval $[0, \tau]$? We will concentrate on this question.

We briefly review the results of Yamada and Takagi [68], Hartle [35, 38, 39] and Micanek and Hartle [52]. We will compute the decoherence functional using the path integral expression (6.12), which may be written as

$$D(\alpha, \alpha') = \int dx_f \; \Psi_\tau^\alpha(x_f) \left(\Psi_\tau^{\alpha'}(x_f) \right)^* , \qquad (6.20)$$

where $\Psi_\tau^\alpha(x_f)$ denotes the amplitude obtained by summing over paths ending at x_f at time τ, consistent with the restriction α and consistent with the given initial state, so we have

$$\Psi_\tau^\alpha(x_f) = \int_\alpha \mathcal{D}x(t) \; \exp\left(\frac{i}{\hbar} S[x]\right) \Psi_0(x_0) . \qquad (6.21)$$

Suppose the system starts out in the initial state $\Psi_0(x)$ at $t = 0$. The amplitude for the particle to start in this initial state and end up at x at time τ, but without ever crossing $x = 0$, is

$$\Psi_\tau^r(x) = \int_{-\infty}^{\infty} dx_0 \; g_r(x, \tau | x_0, 0) \, \Psi_0(x_0) , \qquad (6.22)$$

where g_r is the restricted Green function, i.e., the sum over paths that never cross $x = 0$. For the free particle considered here (and also for any system with a potential symmetric about $x = 0$), g_r may be constructed by the method of images:

$$g_r(x, \tau | x_0, 0) = [\theta(x) \, \theta(x_0) + \theta(-x) \, \theta(-x_0)]$$
$$\times \; (g(x, \tau | x_0, 0) - g(x, \tau | - x_0, 0)) , \qquad (6.23)$$

where $g(x, \tau | x_0, 0)$ is the unrestricted propagator.

The amplitude to cross $x = 0$ is

$$\Psi_\tau^c(x) = \int_{-\infty}^{\infty} dx_0 \; g_c(x, \tau | x_0, 0) \, \Psi_0(x_0) , \qquad (6.24)$$

where $g_c(x, \tau | x_0, 0)$ is the crossing propagator, i.e., the sum over paths that always cross $x = 0$. This breaks up into two parts. If x and x_0 are on opposite sides of $x = 0$, it is clearly just the usual propagator $g(x, \tau | x_0, 0)$. If x and x_0 are on the same side of $x = 0$, it is given by $g(-x, \tau | x_0, 0)$. This may be seen by reflecting the segment of the path after last crossing about $x = 0$ [29]. (Alternatively, this is just the usual propagator minus the restricted one.) Hence,

$$g_c(x, \tau | x_0, 0) = [\theta(x)\theta(-x_0) + \theta(-x)\theta(x_0)] \ g(x, \tau | x_0, 0)$$
$$+ [\theta(x) \ \theta(x_0) + \theta(-x)\theta(-x_0)] \ g(-x, \tau | x_0, 0) \ . \quad (6.25)$$

The crossing propagator may also be expressed in terms of the so-called path decomposition expansion, a form that is sometimes useful [4, 5, 24, 29, 62].

Inserting these expressions in the decoherence function, Yamada and Takagi found that the consistency condition may be satisfied exactly by states that are antisymmetric about $x = 0$. The probability of crossing $x = 0$ is then 0 and the probability of not crossing is 1. What is happening in this case is that the probability flux across $x = 0$, which clearly has non-zero components going both to the left and the right, averages to zero.

Less-trivial probabilities are obtained in the case where one asks for the probability that the particle remains always in $x > 0$ or not, with an initial state with support along the entire x-axis [39]. The probabilities become trivial again, however, in the interesting case of an initial state with support only in $x > 0$.

Yamada and Takagi have also considered the case of the probability of finding the particle in a space–time region [68]. That is, the probability that the particle enters or does not enter the spatial interval Δ, at any time during the time interval $[0, t]$. Again the consistency condition is satisfied only for very special initial states and the probabilities are then rather trivial.

In an attempt to assign probabilities for arbitrary initial states, Micanek and Hartle considered the above results in the limit that the time interval $[0, \tau]$ becomes very small [52]. Such an assignment must clearly be possible in the limit $\tau \to 0$. They found that both the off-diagonal terms of the decoherence functional D and the crossing probability p are of order $\varepsilon = (\hbar t/m)^{\frac{1}{2}}$ for small t, and the probability $\bar{p}$ for not crossing is of order 1. Hence $p + \bar{p} \approx 1$. They therefore argued that probabilities can be assigned if t is sufficiently small. On the other hand, we have the exact relation,

$$p + \bar{p} + 2\mathrm{Re}D = 1 \ . \quad (6.26)$$

$\mathrm{Re}D$ represents the degree of fuzziness in the definition of the probabilities. Since it is of the same order as $\bar{p}$, one may wonder whether it is then valid to claim approximate consistency. Another condition that may be relevant is the condition

$$|D|^2 << p\bar{p} \ , \quad (6.27)$$

which was suggested in [13] as a measure of approximate decoherence, and is clearly satisfied in this case.

We conclude from these various studies that for a system consisting of a single point particle, crossing probabilities can be assigned to histories only in a limited class of circumstances.

There is one particularly important case in which this lack of probability assignment is perhaps unsettling. Consider a wave packet that starts at $x_0 > 0$ moving towards the origin. The amplitude for not crossing is given by the restricted amplitude (6.22) and the restricted propagator (6.23). However, in the case where the centre of the wave packet reaches the origin during the time interval, it is easily seen from the propagator (6.23) that after hitting the origin there is a piece of the wave packet that is reflected back into $x > 0$ (this is the image wave packet has come from $x < 0$). This means that we have the counterintuitive result that the probability for remaining in $x > 0$ is not in fact close to zero [35, 67] as one would expect. It is unsettling because one sometimes thinks of wave packets as being the closest thing quantum theory has to a classical path, yet the behaviour of the wave packet in this case is utterly different to the corresponding expected classical behaviour.

Although counterintuitive, it is not that disturbing, since with this initial state, the histories for crossing and not crossing do not satisfy the consistency condition, so we should not expect them to agree with our physical intuition. Still, it would be reassuring to see that the formalism set up so far yields the intuitively expected classical limit under appropriate circumstances. To obtain that, we need a decoherence mechanism, and this we now consider.

6.4 Decoherence of Space–Time Coarse-Grained Histories in the Quantum Brownian Motion Model

We have seen that crossing probabilities can be assigned in the decoherent histories approach only for very special initial states, and furthermore, we do not get an intuitively sensible classical limit for wave packet initial states. It is, however, well known that most sets of histories of interest do not in fact exhibit decoherence without the presence of some physical mechanism to produce it. In this section, we therefore discuss a modified situation consisting of a point particle coupled to a bath of harmonic oscillators in a thermal state. This model, the quantum Brownian motion model [1], produces decoherence of histories of positions in a variety of situations.

This explicit modification of the single particle system means that the corresponding classical problem (to which the quantum results should reduce under certain circumstances) is in fact a stochastic process described by either a Langevin equation or by a Fokker–Planck equation. It is therefore appropriate to first study the crossing problem in the corresponding classical stochastic process (see e.g. [63, 8, 9, 51, 69], and references therein).

6.4.1 The Crossing Time Problem in Classical Brownian Motion

Classical Brownian motion may be described by the Fokker–Planck equation (6.18) for the phase-space probability distribution $w(p, x, t)$. For simplicity we will work in the limit of negligible dissipation, hence the equation is

$$\frac{\partial w}{\partial t} = -\frac{p}{m}\frac{\partial w}{\partial x} + D\frac{\partial^2 w}{\partial p^2} \, , \tag{6.28}$$

where $D = 2m\gamma kT$. The Fokker–Planck equation is to be solved subject to the initial condition

$$w(p, x, 0) = w_0(p, x) \, . \tag{6.29}$$

Consider now the crossing time problem in classical Brownian motion. The question is this. Suppose the initial state is localized in the region $x > 0$. What is the probability that, under evolution according to the Fokker–Planck equation (6.28), the particle either crosses or does not cross $x = 0$ during the time interval $[0, \tau]$?

A useful way to formulate space–time questions of this type is in terms of the Fokker–Planck propagator, $K(p, x, \tau | p_0, x_0, 0)$. The solution to (6.28) with the initial condition (6.29) may be written in terms of K as

$$w(p, x, \tau) = \int_{-\infty}^{\infty} dp \int_{-\infty}^{\infty} dx \; K(p, x, \tau | p_0, x_0, 0) \, w_0(p, x) \, . \tag{6.30}$$

The Fokker–Planck propagator satisfies the Fokker–Planck equation (6.28) with respect to its final arguments and satisfies delta function initial conditions,

$$K(p, x, 0 | p_0, x_0, 0) = \delta(p - p_0) \, \delta(x - x_0) \, . \tag{6.31}$$

For the free particle without dissipation, it is given explicitly by

$$K(p, x, \tau | p_0, x_0, 0) = N \, \exp\left[-\alpha(p - p_0)^2 - \beta\left(x - x_0 - \frac{p_0\tau}{m}\right)^2 \right.$$
$$\left. + \, \varepsilon(p - p_0)(x - x_0 - \frac{p_0\tau}{m}) \right] \, , \tag{6.32}$$

where N, α, β and ε are given by

$$\alpha = \frac{1}{D\tau}, \quad \beta = \frac{3m^2}{D\tau^3}, \quad \varepsilon = \frac{3m}{D\tau^2}, \quad N = \left(\frac{3m^2}{4\pi D^2\tau^4}\right)^{\frac{1}{2}} \tag{6.33}$$

(with $D = 2m\gamma kT$). An important property it satisfies is the composition law

$$K(p, x, \tau | p_0, x_0, 0) = \int_{-\infty}^{\infty} dp_1 \int_{-\infty}^{\infty} dx_1 \; K(p, x, \tau | p_1, x_1, t_1)$$
$$K(p_1, x_1, t_1 | p_0, x_0, 0) \, , \tag{6.34}$$

where $\tau > t_1 > 0$.

For our purposes, the utility of the Fokker–Planck propagator is that it may be used to assign probabilities to individual paths in phase space. Divide the time interval $[0, \tau]$ into subintervals, $t_0 = 0, t_1, t_2, \cdots, t_{n-1}, t_n = \tau$. Then in the limit that the subintervals go to zero, and $n \to \infty$ but with τ held constant, the quantity

$$\prod_{k=1}^{n} K(p_k, x_k, t_k | p_{k-1}, x_{k-1}, t_{k-1}) \tag{6.35}$$

is proportional to the probability for a path in phase space. The probability for various types of coarse-grained paths (including space–time coarse grainings) can therefore be calculated by summing over this basic object.

We are interested in the probability $w_r(p_n, x_n, \tau)$ that the particle follows a path that always remains in the region $x > 0$ during the time interval $[0, \tau]$ and ends at the point $x_n > 0$ with momentum p_n. The desired total probabilities for crossing or not crossing can then be constructed from this object. w_r is clearly given by the continuum limit of the expression

$$w_r(p_n, x_n, \tau) = \int_0^\infty dx_{n-1} \cdots \int_0^\infty dx_1 \int_0^\infty dx_0 \int_{-\infty}^\infty dp_{n-1} \cdots \int_{-\infty}^\infty dp_1 \int_{-\infty}^\infty dp_0$$

$$\times \prod_{k=1}^{n} K(p_k, x_k, t_k | p_{k-1}, x_{k-1}, t_{k-1}) \, w_0(p_0, x_0) . \tag{6.36}$$

Now it is actually more useful to derive a differential equation and boundary conditions for $w_r(p, x, \tau)$, rather than attempt to evaluate the above multiple integral. First of all, it is clear from the properties of the propagator that $w_r(p, x, \tau)$ satisfies the Fokker–Planck equation (6.28) and the initial condition (6.29). However, we also expect some sort of condition at $x = 0$. From the explicit expression for the propagator (6.32), (6.33), we see that in the continuum limit, the propagator between p_{n-1}, x_{n-1} and the final point p_n, x_n becomes proportional to the delta function

$$\delta (x_n - x_{n-1} - p_n \tau / m) . \tag{6.37}$$

Since $x_{n-1} \geq 0$, when $x_n = 0$ this delta function will give zero when $p_n > 0$, but could be non-zero when $p_n < 0$. Hence we deduce that the boundary condition on $w_r(p, x, t)$ is

$$w_r(p, 0, t) = 0, \quad \text{if} \quad p > 0 . \tag{6.38}$$

This is the absorbing boundary condition usually given for the crossing time problem [51, 65] (although this argument for it does not seem to have appeared elsewhere).

It is now convenient to introduce a restricted propagator $K_r(p, x, \tau | p_0, x_0, 0)$, which propagates $w_r(p, x, \tau)$. That is, K_r satisfies the delta function initial conditions (6.38) and the same boundary conditions as w_r, (6.38). Since the original Fokker–Planck equation is not invariant under $x \rightarrow -x$, we cannot expect that a simple method of images (of the type used in Sect. 6.3) will readily yield the restricted propagator K_r. K_r has recently been found [9], using a modified method of image technique due to Carslaw [12], and we briefly summarize those results.

First consider the usual Fokker–Planck propagator (6.32). Introducing the coordinates

$$X = \frac{p}{m} - \frac{3x}{2\tau}, \quad Y = \frac{\sqrt{3}x}{2\tau}, \tag{6.39}$$

$$X_0 = -\frac{p_0}{2m} - \frac{3x_0}{2\tau}, \quad Y_0 = \frac{\sqrt{3}}{2}\left(\frac{p_0}{m} + \frac{x_0}{\tau}\right), \tag{6.40}$$

the propagator (6.32) becomes

$$K = \frac{\sqrt{3}}{2\pi\tilde{t}^2} \exp\left(-\frac{(X-X_0)^2}{\tilde{t}} - \frac{(Y-Y_0)^2}{\tilde{t}}\right). \tag{6.41}$$

Here, $\tilde{t} = D\tau/m^2$. Now go to polar coordinates,

$$X = r\cos\theta, \quad Y = r\sin\theta, \tag{6.42}$$
$$X_0 = r'\cos\theta', \quad Y_0 = r'\sin\theta'. \tag{6.43}$$

Then from (6.43), it is possible to construct a so-called multiform Green function [12],

$$g(r,\theta,r',\theta') = \frac{\sqrt{3}}{2\pi^{3/2}\tilde{t}^2} \exp\left(-\frac{r^2 + r'^2 - 2rr'\cos(\theta-\theta')}{\tilde{t}}\right) \int_{-\infty}^{a} d\lambda\, e^{-\lambda^2}, \tag{6.44}$$

where

$$a = 2\left(\frac{rr'}{\tilde{t}}\right)^{\frac{1}{2}} \cos\left(\frac{\theta-\theta'}{2}\right). \tag{6.45}$$

Like the original Fokker–Planck propagator, this object is a solution to the Fokker–Planck equation with delta function initial conditions, but differs in that it has the property that it is defined on a two-sheeted Riemann surface and has period 4π. The desired restricted propagator K_r is then given by

$$K_r(p,x,\tau|p_0,x_0,0) = g(r,\theta,r',\theta') - g(r,\theta,r',-\theta'). \tag{6.46}$$

The point $x = 0$ for $p > 0$ is $\theta = 0$ in the new coordinates, and the above object indeed vanishes at $\theta = 0$. Furthermore, the second term in the above goes to zero at $\tau = 0$, whilst the first one goes to a delta function as required.

The probability of not crossing the surface during the time interval $[0,t]$ is then given by

$$p_r = \int_{-\infty}^{\infty} dp \int_{0}^{\infty} dx \int_{-\infty}^{\infty} dp_0 \int_{0}^{\infty} dx_0\, K_r(p,x,\tau|p_0,x_0,0)\, w_0(p_0,x_0). \tag{6.47}$$

The probability of crossing must then be $p_c = 1 - p_r$, which can also be written as

$$p_c = \int_{-\infty}^{0} dp \int_{-\infty}^{\infty} dp_0 \int_{0}^{\infty} dx_0\, \frac{p}{m}\, K_r(p,x=0,\tau|p_0,x_0,0)\, w_0(p_0,x_0). \tag{6.48}$$

This completes the discussion of the classical stochastic problem.

6.4.2 The Crossing Time Problem in Quantum Brownian Motion

We now consider the analogous problem in the quantum case. We therefore attempt to repeat the analysis of Sect. 6.3, but using instead of (6.18) the decoherence functional appropriate to the quantum Brownian motion model. It may be written as

$$D(\alpha, \alpha') = \text{Tr}\,(\varrho_{\alpha\alpha'})\,, \tag{6.49}$$

where

$$\varrho_{\alpha\alpha'}(x_f, y_f) = \int_\alpha \mathcal{D}x \int_{\alpha'} \mathcal{D}y \, \exp\left(\frac{i}{\hbar}S[x] - \frac{i}{\hbar}S[y] + \frac{i}{\hbar}W[x, y]\right) \, \varrho_0(x_0, y_0)\,. \tag{6.50}$$

Here, $W[x, y]$ is the influence functional phase (6.5), but with the dissipation term neglected. The sum is over all paths x, y, which are consistent with the coarse graining α, α', and end at the final points x_f, y_f.

We will concentrate on the case in which the initial density operator has support only on the positive axis, and we ask for the probability that the particle either crosses or never crosses $x = 0$ during the time interval $[0, \tau]$. The history label α takes two values, which we denote $\alpha = c$ and $\alpha = r$ for crossing and not crossing respectively.

The objects $\varrho_{\alpha\alpha'}$ defined in (6.50) actually obey a master equation,

$$i\hbar\frac{\partial\varrho}{\partial t} = -\frac{\hbar^2}{2m}\left(\frac{\partial^2\varrho}{\partial x^2} - \frac{\partial^2\varrho}{\partial y^2}\right) - \frac{i}{\hbar}D(x - y)^2\varrho\,. \tag{6.51}$$

This is the usual master equation for the evolution of the density operator of quantum Brownian motion [10]. The objects $\varrho_{\alpha\alpha'}$ are then found by solving this equation subject to matching the initial state ϱ_0, and also to the following boundary conditions (which follow from the path integral representation):

$$\varrho_{rr}(x, y) = 0, \quad \text{for} \quad x \leq 0 \quad \text{and} \quad y \leq 0\,, \tag{6.52}$$

$$\varrho_{rc}(x, y) = 0, \quad \text{for} \quad x \leq 0\,, \tag{6.53}$$

$$\varrho_{cr}(x, y) = 0, \quad \text{for} \quad y \leq 0\,. \tag{6.54}$$

Given $\varrho_{rr}, \varrho_{rc}, \varrho_{cr}$, the quantity ϱ_{cc} may be calculated from the relation

$$\varrho_{rr} + \varrho_{rc} + \varrho_{cr} + \varrho_{cc} = \varrho\,. \tag{6.55}$$

In the unitary case, this problem was solved very easily using the method of images. The problem in the non-unitary case treated here, however, is that the master equation is *not* invariant under $x \to -x$ (or under $y \to -y$), hence $\varrho(-x, y)$ and $\varrho(x, -y)$ are *not* solutions to the master equation. The method of images is therefore not applicable in this case (contrary to the claim in [35]). As far as an analytic approach goes, this represents a very serious technical problem. Restricted propagation problems are very hard to solve analytically in the absence of the method of images. However, the presence of the decohering environment allows for an approximate solution of the problem. This is described in detail in [32]. The results are intuitively clear and we summarize them here.

First of all, decoherence of position histories in this model is extremely good, so $\varrho_{rc} \approx 0$, $\varrho_{cr} \approx 0$. We may therefore assign probabilities for not crossing and for crossing, and these are equal to $\mathrm{Tr}\varrho_{rr}$ and $\mathrm{Tr}\varrho_{cc}$ respectively. To see what these probabilities are, we make use of the Wigner representation of the density operator [6]:

$$W(p, x) = \frac{1}{2\pi\hbar} \int_{-\infty}^{\infty} d\xi \, e^{-\frac{i}{\hbar}p\xi} \, \varrho(x + \frac{\xi}{2}, x - \frac{\xi}{2}) \,. \tag{6.56}$$

The Wigner representation is very useful in studies of the master equation, since it is similar to a classical phase-space distribution function. Indeed, for quantum Brownian motion model with a free particle, the Wigner function obeys the same Fokker–Planck equation (6.28) as the analogous classical phase-space distribution function. What makes it fail to be a classical phase space distribution is that it can take negative values. However, it can be shown that the Wigner function becomes positive after a short time (typically the decoherence time), and numerous authors have discussed its use as an approximate classical phase-space distribution, under these conditions [31].

Given approximate decoherence, it was shown at some length in [32] using the path integral (6.50) that the Wigner transform of ϱ_{rr} is given by

$$W_{rr}(m\dot{X}_f, X_f) = \int_r \mathcal{D}X \, \exp\left(-\frac{m}{8\gamma kT} \int dt \ddot{X}^2\right) W_0(m\dot{X}_0, X_0) \,, \tag{6.57}$$

where the functional integral over $X(t)$ is over paths that lie in $X > 0$ and match X_f and $\dot{X}_f$ at the final time. If the paths $X(t)$ were not restricted, (6.57) would in fact be a path integral representation of the Fokker–Planck propagator (6.32) [44]. With the restriction $X > 0$, it may be shown that it is a representation of the restricted Fokker–Planck propagator (6.46) or (6.36).

It then follows that the probabilities for not crossing and for crossing $x = 0$ are given, to a good approximation, by the classical stochastic results (6.47), (6.48), with the classical phase-space distribution function w_0 replaced by the initial Wigner function W_0 in the quantum case. This result is the expected and intuitively obvious one, although as outlined in [32], it is a non-trivial matter to show that the boundary conditions on $\varrho_{\alpha\alpha'}$ in the quantum case reduce to the boundary conditions on W appropriate to the classical stochastic problem.

6.4.3 Properties of the Solution

Some simple properties of our results may be seen by examining the path integral form of the solution (6.57). The important case to consider is the motion of a wave packet, since this is the situation that gave problematic results in

Sect. 6.3. We take an initial state consisting of a wave packet concentrated at some $x > 0$, and moving towards the origin. We are interested in the probability of whether it will cross $x = 0$ or not during some time interval, under the evolution by the path integral (6.57).

The integrand in (6.57) is peaked about the unique path for which $\ddot{X} = 0$ with the prescribed values of X_0 and $\dot{X}_0$. This is of course the classical path with the prescribed initial data. From (6.57), the spatial width $(\Delta X)^2$ of the peak is of order $\gamma kT/(m\tau^3)$. If the classical path does not cross $x = 0$ and approaches $x = 0$ no closer than a distance ΔX during the time interval, then it will lie well within the integration range $X > 0$, and the propagation is essentially the same as unrestricted propagation, since the dominant contribution to the integral comes from the region $X > 0$. It is then easy to see, from the normalization of the Wigner function, that the probability of not crossing is approximately 1, the intuitively expected result.

If the classical path crosses $x = 0$ during the time interval, it will lie outside the integration range of X for time slices after the time at which it crossed. If it crosses sufficiently early that an entire wave packet of width ΔX may enter $x < 0$ before time τ, then the functional integration will sample only the exponentially small tail of the integrand, so W_{rr} will be very small. The probability of not crossing will therefore be close to zero, again the intuitively expected result.

The inclusion of the environment therefore restores the intuitively sensible classical limit to the quantum case of Sect. 6.3.

In the above simple examples, the crossing probabilities are independent of the details of the environment, to a leading order approximation. It is clear that in a more precise expression, the crossing probabilities will in fact depend on the features of the environment (e.g., its temperature). One might find this slightly unsettling, at least in comparison to quantum-mechanical probabilities at a fixed moment of time, which depend only on the state at that time and not on the details of how the property in question might be measured. This is in keeping with an opinion sometimes expressed on questions of time in quantum mechanics – that to specify times one has to specify the physical mechanism by which it is measured [48].

6.5 A Detector Model

Although the results of the previous sections produced mathematically viable candidates for the probabilities of crossing and not crossing $x = 0$, it is by no means clear how they correspond to a particular type of measurement. As noted in Sect. 6.2, general theorems exist showing that decoherence of histories implies the existence at the final end of the histories of a record storing the information about the decohered histories [18, 19]. This means that there is *some* quantity at a fixed moment of time, which is correlated with the property of crossing or not crossing $x = 0$ during the time interval $[0, \tau]$ and which could in principle be measured. Records associated with decoherence have, however, been explicitly found only in a few simple cases (see e.g., [26]). For these reasons, it is of interest to compare the approaches involving the decoherent histories approach with a completely different approach involving a specific model of a detector.

We therefore introduce, following [27], a detector model that is coupled to the particle in the region $x < 0$, and such that it undergoes a transition when the coupling is switched on. Such detectors have certainly been considered before (see, e.g., [3]). The particle could, for example, be coupled to a simple two-level system that flips from one level to the other when the particle is detected. One of the difficulties of many detector models, however, is that if they are modelled by unitary quantum mechanics, the possibility of the reverse transition exists. Because quantum mechanics is fundamentally reversible, the detector could return to the undetected state under its self-dynamics, even when the particle has interacted with it.

To get around this difficulty, we appeal to the fact that realistic detectors have a very large number of degrees of freedom, and are therefore effectively *irreversible*. They are designed so that there is an overwhelming large probability for them to make a transition in one direction rather than its reverse. We consider a simple model detector that has this property. This is achieved by coupling a two-level system detector to a large environment, which makes its evolution effectively irreversible.

The detector is a two-level system, with levels $|1\rangle$ and $|0\rangle$, representing the states of no detection and detection, respectively. Introduce the raising and lowering operators

$$\sigma_+ = |1\rangle\langle 0|, \quad \sigma_- = |0\rangle\langle 1|, \tag{6.58}$$

and let the Hamiltonian of the detector be $H_d = \frac{1}{2}\hbar\omega\sigma_z$, where

$$\sigma_z = |1\rangle\langle 1| - |0\rangle\langle 0|, \tag{6.59}$$

so $|0\rangle$ and $|1\rangle$ are eigenstates of H_d with eigenvalues $-\frac{1}{2}\hbar\omega$ and $\frac{1}{2}\hbar\omega$, respectively. We would like to couple the detector to a free particle in such a way that the detector makes an essentially irreversible transition from $|1\rangle$ to $|0\rangle$ if the particle enters $x < 0$, and remains in $|1\rangle$ otherwise. This can be arranged by coupling the detector to a large environment of oscillators in their ground state, with a coupling proportional to $\theta(-x)$. This means that if the particle enters the region $x < 0$, the detector becomes coupled to the large environment causing it to undergo a transition. Since the environment is in its ground state, if the detector initial state is the higher energy state $|1\rangle$ it will, with overwhelming probability, make a transition from $|1\rangle$ to the lower energy state $|0\rangle$. A possible Hamiltonian describing this process for the three-component system is

$$H = H_s + H_d + H_{\mathcal{E}} + V(x)H_{d\mathcal{E}}, \tag{6.60}$$

where the first three terms are the Hamiltonians of the particle, detector and environment, respectively, and $H_{d\mathcal{E}}$ is the interaction Hamiltonian of the detector and its environment. The simplest choice of environment is a collection of harmonic oscillators,

$$H_{\mathcal{E}} = \sum_n \hbar \omega_n a_n^\dagger a_n \ , \tag{6.61}$$

and we take the coupling to the detector to be via the interaction

$$H_{d\mathcal{E}} = \sum_n \hbar \left(\kappa_n^* \sigma_- a_n^\dagger + \kappa_n \sigma_+ a_n \right) \ . \tag{6.62}$$

An environment consisting of an electromagnetic field, for example, would give terms of this general form. $V(x)$ is a potential concentrated in $x < 0$ (and we will eventually make the simplest choice, $V(x) = \theta(-x)$, but for the moment we keep it more general). The important feature is that the interaction between the detector and its environment, causing the detector to undergo a transition, is switched on only when the particle is in $x < 0$.

A similar although more elaborate model particle detector has been previously studied by Schulman [60] (see also [61]). The advantage of the present model is that it is easier to solve explicitly.

We are interested in the reduced dynamics of the particle and the detector with the environment traced out. Hence we seek a master equation for the reduced density operator ϱ of the particle and the detector. With the above choices for $H_{\mathcal{E}}$ and $H_{d\mathcal{E}}$, the derivation of the master equation is standard [11, 16] and will not be repeated here. There is the small complication of the factor of $V(x)$ in the interaction term, but this is readily accommodated. We assume a factored initial state, and we assume that the environment starts out in the ground state. In a Markovian approximation (essentially the assumption that the environment dynamics is much faster than the detector or the particle dynamics) and in the approximation of weak detector–environment coupling, the master equation is

$$\dot{\varrho} = -\frac{i}{\hbar}[H_s + H_d, \varrho] - \frac{\gamma}{2} \left(V^2(x)\sigma_+\sigma_-\varrho + \varrho\sigma_+\sigma_- V^2(x) - 2V(x)\sigma_- \varrho\sigma_+ V(x) \right) \ . \tag{6.63}$$

Here, γ is a phenemonological constant determined by the distribution of oscillators in the environment and underlying coupling constants. The frequency ω in H_d is also renormalized to a new value ω'.

Equation (6.63) is the sought-after description of a particle coupled to an effectively irreversible detector in the region $x < 0$. In the dynamics of the detector plus environment only (i.e., with $V = 1$ and $H_s = 0$), it is readily shown that every initial state tends to the state $|0\rangle\langle 0|$ on a timescale γ^{-1}. With the particle coupled in, if the initial state of the detector is chosen to be $|1\rangle\langle 1|$, it undergoes an irreversible transition to the state $|0\rangle\langle 0|$ if the particle enters $x < 0$, and remains in its initial state otherwise.

Equation (6.63) is in fact of the Lindblad form (the most general Markovian master equation preserving density operator properties [49]). A similar detection scheme based on a postulated master equation similar to (6.63) was previously considered in [43].

The master equation (6.63) is easily solved by writing

$$\varrho = \varrho_{11} \otimes |1\rangle\langle 1| + \varrho_{01} \otimes |0\rangle\langle 1| + \varrho_{10} \otimes |1\rangle\langle 0| + \varrho_{00} \otimes |0\rangle\langle 0| . \tag{6.64}$$

We suppose that the particle starts out in an initial state $|\Psi_0\rangle$, hence the master equation is to be solved subject to the initial condition,

$$\varrho(0) = |\Psi_0\rangle\langle\Psi_0| \otimes |1\rangle\langle 1| . \tag{6.65}$$

The probability that the detector does not register during $[0, \tau]$ is

$$p_{nd} = \text{Tr}\varrho_{11} = \int_{-\infty}^{\infty} dx \, \varrho_{11}(x, x, \tau) , \tag{6.66}$$

and the probability that it registers is

$$p_d = \text{Tr}\varrho_{00} = \int_{-\infty}^{\infty} dx \, \varrho_{00}(x, x, \tau) \tag{6.67}$$

(where the trace is over the particle Hilbert space). Clearly $p_{nd} + p_d = 1$, since $\text{Tr}\varrho = 1$.

The explicit solution to the master equation is straightforward and was carried out in [27]. There, it was shown that, when $V(x) = \theta(-x)$, the solution for ϱ_{11} may be written as

$$\varrho_{11}(t) = \exp\left(-\frac{i}{\hbar}H_s t - \frac{\gamma}{2}Vt\right) \varrho_{11}(0) \, \exp\left(\frac{i}{\hbar}H_s t - \frac{\gamma}{2}Vt\right) . \tag{6.68}$$

What is particularly interesting about this expression is that it can be factored into a pure state. Let $\varrho_{11} = |\Psi\rangle\langle\Psi|$. Then, noting that $\varrho_{11}(0) = |\Psi_0\rangle\langle\Psi_0|$, (6.68) is equivalent to

$$|\Psi(t)\rangle = \exp\left(-\frac{i}{\hbar}H_s t - \frac{\gamma}{2}Vt\right) |\Psi_0\rangle . \tag{6.69}$$

The probability for no detection is then

$$p_{nd} = \int_{-\infty}^{\infty} dx \, |\Psi(x, \tau)|^2 . \tag{6.70}$$

The pure state (6.69) evolves according to a Schrödinger equation with an imaginary contribution to the potential, $-\frac{1}{2}i\hbar\gamma V$. Complex potentials of precisely this type have been used previously in studies of arrival times, as phenomenological devices, to imitate absorbing boundary conditions (see, e.g., [2, 55, 58]). Here, the appearance of a complex potential is *derived* from the master equation of a particle coupled to an irreversible detector, which in turn may be derived from the unitary dynamics of the combined particle–detector–environment system.

In summary, this detector model nicely reproduces earlier phenomenological results on arrival times. In [56] it is also shown that the expression (6.69), (6.70), is very closely related to the "ideal" arrival time distribution of Kijowski [45]. An improved, more physically realistic irreversible detector model (although more difficult to solve analytically) was recently put forward by Muga et al. [54], see Chap. 8.

6.6 A Comparison of the Decoherent Histories Result with the Detector Result

We may now compare the two candidate expressions for the crossing time probabilities, one from decoherent histories with an environment and the other from an irreversible detector model. We will quickly see that the two results are not in fact very close, but it is perhaps of interest to see exactly why and how they may be improved.

We first massage the decoherent histories result into a more suitable form. Consider the probability for remaining in $x > 0$. From (6.50) it is given by

$$
p_r = \int_r dx(t) \int_r dy(t) \exp\left(\frac{i}{\hbar}S[x(t)] - \frac{i}{\hbar}S[y(t)]\right)
$$
$$
\times \ \exp\left(-a\int dt\ (x-y)^2\right)\ \varrho_0(x_0, y_0)\,, \tag{6.71}
$$

where $a = D/\hbar^2$. Following [32], we make the observation that the last exponential may be deconvolved:

$$
\exp\left(-a\int dt\ (x-y)^2\right) = \int \mathcal{D}\bar{x}\ \exp\left(-2a\int dt\ (x-\bar{x})^2 - 2a\int dt\ (y-\bar{x})^2\right). \tag{6.72}
$$

Hence, assuming a pure initial state, the probability (6.71) may be written as

$$
p_r = \int d\bar{x}(t) \int_r dx(t)\ \exp\left(\frac{i}{\hbar}S[x(t)] - 2a\int dt\ (x-\bar{x})^2\right)\Psi_0(x_0)
$$
$$
\times \int_r dy(t)\ \exp\left(-\frac{i}{\hbar}S[y(t)] - 2a\int dt\ (y-\bar{x})^2\right)\Psi_0^*(y_0)\,. \tag{6.73}
$$

In these integrals, $\bar{x}(t)$ is integrated over an infinite range, but $x(t)$ and $y(t)$ are integrated only over the positive real line. This restriction is quite difficult to implement in practice [32]. However, because of the exponential factors, negative values of $\bar{x}(t)$ are strongly suppressed, so we may take its range to be over positive values only, with exponentially small error. Furthermore, having done this we may then (for technical simplicity) allow the range of $x(t)$ and $y(t)$ to be over the entire real line, again with exponentially small error. Therefore, we have that

$$p_r \approx \int_r d\bar{x}(t) \int dx(t) \ \exp\left(\frac{i}{\hbar}S[x(t)] - 2a \int dt \ (x - \bar{x})^2\right) \Psi_0(x_0)$$

$$\times \int dy(t) \ \exp\left(-\frac{i}{\hbar}S[y(t)] - 2a \int dt \ (y - \bar{x})^2\right) \Psi_0^*(y_0) \ . \qquad (6.74)$$

This may finally be written as

$$p_r \approx \int_r d\bar{x}(t) \ \langle \Psi_{\bar{x}}|\Psi_{\bar{x}}\rangle \ , \qquad (6.75)$$

where

$$\Psi_{\bar{x}}(x_f, \tau) = \int dx(t) \ \exp\left(\frac{i}{\hbar}S[x(t)] - 2a \int dt \ (x - \bar{x})^2\right) \Psi_0(x_0) \ . \qquad (6.76)$$

Written in this way the probability has a natural interpretation in terms of continuous quantum measurement. Equation (6.76) is the wave function for a system undergoing continuous measurement of its position along a trajectory $\bar{x}(t)$ to within a precision proportional to $a^{-\frac{1}{2}}$. The probability for any such trajectory is $\langle \Psi_{\bar{x}}|\Psi_{\bar{x}}\rangle$, hence the probability to remain in the region $x > 0$ is obtained by integrating over $\bar{x}(t) > 0$. The probability (6.71), derived from the decoherent histories approach, is therefore, to an excellent approximation, the same as the result naturally obtained from continuous quantum measurement theory.

Now we compare with the detector model. The probability for no detection is computed from the wave function (6.69). In a path integral representation, this may be written as

$$\Psi_{nd}(x, \tau) = \int \mathcal{D}x(t) \exp\left(\frac{i}{\hbar}S[x(t)] - \frac{\gamma}{2}\int_0^\tau dt \ V(x(t))\right) \Psi_0(x_0) \ . \qquad (6.77)$$

The sum is over all paths $x(t)$ connecting x_0 at $t = 0$ to x_f at $t = \tau$. The probability for no detection is then quite simply

$$p_{nd} = \langle \Psi_{nd}|\Psi_{nd}\rangle \ . \qquad (6.78)$$

Whilst the two different expressions, (6.75), (6.76), versus (6.77), (6.78), are similar in some ways, they are not obviously close and suffer from a rather key difference. Equation (6.75) is obtained by summing the probability for any path $\bar{x}(t)$ over positive values of $\bar{x}$. In (6.77) and (6.78), by contrast, the restriction to paths in $x > 0$ is already imposed in the amplitude. The difference between the probabilities provided by the detector and those provided by the decoherent histories approach is, therefore, the difference between summing amplitudes and squaring versus squaring and then summing.

In the decoherent histories approach, the coupling to the environment produces an effective measurement of the system that is much finer than is required for the crossing time problem. It effectively measures the entire trajectory, which is clearly much more information than is required to determine whether or not the particle enters $x < 0$. In this sense this particular decoherent histories model is much cruder than the detector model, since it destroys far more interference than it really needs to in order to define the crossing time. This is due to the form of the particle–environment coupling, which is linear in the particle's position. It would be of interest to explore a decoherent histories model with a more refined type of coupling, which is more specifically geared to the crossing time problem.

It is of interest to note that continuous quantum measurement theory in fact suggests another candidate expression for the probability of not crossing, which is closer to the detector model. Suppose that *before* squaring, we sum the amplitude (6.76) over positive $\bar{x}(t)$:

$$\Psi_+(x_f, \tau) = \int dx(t) \, \exp\left(\frac{i}{\hbar} S[x(t)]\right)$$
$$\times \int_r d\bar{x}(t) \left(-2a \int dt \, (x - \bar{x})^2\right) \Psi_0(x_0) \, . \tag{6.79}$$

The probability is then

$$p_+ = \langle \Psi_+ | \Psi_+ \rangle \, . \tag{6.80}$$

This expression for the probability of not entering $x < 0$ is completely natural from the point of view of continuous quantum measurement theory. It does not follow from either the detector model or the decoherent histories approach presented here, but one can regard it as yet another *proposal* with which to define the arrival time probability. The amplitude (6.79) is now more closely analogous to the detector result (6.77). To see this, introduce the effective potential $V_{eff}(x)$ defined by

$$\exp\left(-\int dt \, V_{eff}(x(t))\right) = \int_r d\bar{x}(t) \left(-2a \int dt \, (x - \bar{x})^2\right) \, . \tag{6.81}$$

The integral can be evaluated exactly, but it is clear that $V_{eff}(x) \sim 0$ for $x \gg 0$, and $V_{eff}(x) \sim 2ax^2$ for $x \ll 0$. Equation (6.79) therefore has the same general form as (6.77). The potential is not exactly the same, but has the same physical effect, which is to suppress paths in $x < 0$.

6.7 Timeless Questions in Quantum Theory

We now briefly consider a related question in quantum theory that involves time in a non-trivial way, which is in fact more closely related to the Wheeler–DeWitt equation of quantum cosmology, (6.1). This equation may be thought of as the statement that the wave function of the system is in an energy eigenstate. As stated in Sect. 6.1, the equation contains no notion of time, and indeed "time" and the notion of trajectories are thought to somehow emerge from the wave function. To test this idea, and hence to provide some sort of interpretation for the wave function, we need to find an answer to the question, "What is the probability associated with a given region Δ of configuration space when the system is in an energy eigenstate, without any reference to time?"

Classically, the question is well defined. A system with fixed energy consists of a set of classical trajectories, perhaps with some probability distribution on them. The classical trajectories are just curves in configuration space, and the question is then quite simply one of determining whether or not these curves intersect the given region Δ. But, like the arrival time problem in non-relativistic quantum mechanics, the problem is considerably harder to phrase in quantum theory.

To see the beginnings of the difficulties, we briefly consider the following simple question for a two-dimensional system with coordinates x_1, x_2: given that the system is in an energy eigenstate, what is the value of x_1 given the value of x_2? Slightly rephrased, what is the probability that the system intersects the surface $x_2 = constant$ between x_1 and $x_1 + dx_1$, at any time? An operator approach to the problem, for example, takes the following form. For a free particle, the classical trajectories are

$$x_1(t) = x_1 + \frac{p_1 t}{m}, \quad x_2(t) = x_2 + \frac{p_2 t}{m} \tag{6.82}$$

and we may eliminate t between them to write

$$x_1(t) = x_1 + \frac{p_1}{p_2}(x_2(t) - x_2) . \tag{6.83}$$

This is the classical answer to the question, what is the value of x_1 at a given value of x_2? One may attempt to raise this to the status of an operator in the quantum theory. It commutes with the free particle Hamiltonian,

$$H = \frac{1}{2}(p_1^2 + p_2^2) , \tag{6.84}$$

so is in this sense an observable of the theory – measuring it will not displace the system from an energy eigenstate of H. This approach encounters problems, however, in defining (6.83). It cannot be made into a self-adjoint operator, due to the presence of the $1/p_2$ factor. In this way it is very similar to the problem of defining a time operator.

We will not pursue this approach any further here. Instead we briefly report on two other approaches, which, exactly like the approaches described in this article, use decoherent histories, or a detector model.

The decoherent histories approach to the question involves summing over paths, in configuration space, which either enter or do not enter a given region Δ at *any* moment of time. In practice this is achieved by summing over paths, which either enter or do not enter during a fixed time interval $[0, \tau]$, and then summing τ over an infinite range. The detailed construction of this is described in [30]. As in the crossing time problem described in Sect. 6.4, a decohering environment is required to make the probabilities well-defined, and we then expect the final result to be a reasonably simple formula involving the Wigner function, closely analogous to the classical case. The full details of this have yet to be worked out, but is perhaps useful to give the classical result here (which, although well-defined, is not totally trivial).

We consider a $2n$-dimensional phase space with coordinates $\mathbf{p}, \mathbf{x}$. Denote the classical trajectories by $\mathbf{x}^{cl}(t)$, and suppose that they match the initial data $\mathbf{p}_0, \mathbf{x}_0$ at some fiducial initial point $t = t_0$ (which is arbitrary). For a free particle,

$$\mathbf{x}^{cl}(t) = \mathbf{x}_0 + \frac{\mathbf{p}_0}{m}(t - t_0) . \tag{6.85}$$

Let $f_\Delta(\mathbf{x})$ be a characteristic function for the region Δ so is 1 inside Δ and 0 outside. We suppose that the classical system is described by a phase-space distribution function $w(\mathbf{p}, \mathbf{x})$. To be a true analogue of an energy eigenstate in the quantum case, w has to be stationary, so

$$w(\mathbf{p}(t), \mathbf{x}(t)) = w(\mathbf{p}(t + t_1), \mathbf{x}(t + t_1)) \tag{6.86}$$

for any t_1.

We may now write down the probability for a classical trajectory entering the region Δ. It is

$$p_\Delta = \int d^n\mathbf{p}_0 d^n\mathbf{x}_0 \; w(\mathbf{p}, \mathbf{x}) \; \theta \left(\int_{-\infty}^{\infty} dt \; f_\Delta(\mathbf{x}^{cl}(t)) - \varepsilon \right) . \tag{6.87}$$

Here, ε is a small parameter, which is taken to zero through positive values, and is present to avoid ambiguities in the θ function at zero argument. The integral inside the θ function is the total time spent by the trajectory $\mathbf{x}^{cl}(t)$ inside the region Δ, but we are only interested in whether this time is positive or zero. The initial data $\mathbf{p}_0, \mathbf{x}_0$ are therefore effectively integrated only over values for which the trajectory spends a time in excess of ε in the region Δ. It is easy to see that the whole construction is invariant under shifting the fiducial point t_0. This is the analogue of reparametrization invariance (or more generally, diffeomorphism invariance) in the Wheeler–DeWitt equation (6.1). It is expected that a decoherent histories analysis will yield a result of the approximate form (6.87) (with w replaced by the Wigner function).

Another approach to the question posed at the beginning of this section is to use a detector model (this is described in detail in [28]). The detector model arises from Barbour's observation [7] that a substantial insight into the Wheeler–DeWitt equation may be found in Mott's 1929 analysis of alpha-particle tracks in a Wilson cloud chamber [53]. Mott's paper concerned the question of how the alpha-particle's outgoing spherical wave state, e^{ikR}/R, could lead to straight line tracks in a cloud chamber. His explanation was to model the cloud chamber as a collection of atoms that may be ionized by the passage of the alpha-particle. They therefore act as detectors that measure the alpha-particle's trajectory. The probability that certain atoms are ionized is indeed found to be strongly peaked when the atoms lie along a straight line through the point of origin of the alpha-particle.

Mott had in mind a time-evolving process, but he actually solved the time-independent equation

$$(H_0 + H_d + \lambda H_{int}) |\Psi\rangle = E|\Psi\rangle \,. \tag{6.88}$$

Here H_0 is the alpha-particle Hamiltonian, H_d is the Hamiltonian for the ionizing atoms and H_{int} describes the Coulomb interaction between the alpha-particle and the ionizing atoms (where λ is a small coupling constant). Now the interesting point, as Barbour notes, is that Mott derived all the physics from this equation with little reference to time. Mott's calculation is therefore an excellent model for many aspects of the Wheeler–DeWitt equation. In [28] a model of this type is considered with a series of detectors, and it is shown how to produce a plausible formula for the probability that the system enters a series of regions in configuration space without reference to time. A comparison of this approach with the anticipated decoherent histories result (6.87) is yet to be carried out.

6.8 Discussion

We have reviewed a number of approaches to the crossing time problem in non-relativistic quantum theory, primarily using the decoherent histories approach. We have also briefly reviewed some attempts to extend these ideas to models more closely related to the Wheeler–DeWitt equation. On the face of it, the decoherent histories approach appears to be particularly well adapted to this problem, since it naturally incorporates the notion of trajectory, and hence readily accommodates questions of a non-trivial temporal nature. Having said that, however, good expressions for the crossing time probability are not acquired very easily.

As described in Sect. 6.3, the decoherence or consistency conditions are satisfied only for very special classes of initial states. For a system consisting of a single point particle, therefore, the decoherent histories approach does not supply an answer to the crossing time problem for *arbitrary* initial states. This is rectified by the inclusion of a thermal environment, as described in Sect. 6.4, and probabilities for the crossing time can then be obtained for arbitrary initial system states. They do, however, depend to some extent on the environment producing the decoherence, and moreover, they are essentially the same as the classical stochastic results. One might therefore criticize this result on the grounds that it is "not very quantum". This is largely true, but the essential achievement of Sect. 6.4 is to show that the decoherent histories approach can be made to give the anticipated classical result. This was not true of the earlier approaches reviewed in Sect. 6.3.

In Sect. 6.5, a detector model was introduced to give an alternative expression for the crossing time probability, for the purposes of comparison with the decoherent histories result. The detector model gave a better result, in that it agreed and substantiated an earlier result of Allcock [2], which in turn is closely related to the ideal distribution of Kijowski [45].

On comparison with the decoherent histories result, in Sect. 6.6, it was easy to see that the environment in Sect. 6.4 produced far more decoherence than is necessary to define the arrival time, and in that sense, that particular environment is a very crude model for the measurement of time. The comparison did, however, inspire the proposal of a third candidate expression from which the arrival time probability could be calculated, namely (6.79), which is based on continuous quantum measurement theory. This expression does not seem to have been considered previously and will be explored in more detail elsewhere.

One might be led from these results to a somewhat negative assessment of the decoherent histories approach's ability to provide the crossing time probability. The somewhat crude nature of the results of Sect. 6.4, is however, due to the choice of a rather indiscriminate system–environment coupling, which effectively measures the entire trajectory. It seems likely that a much-improved result could be obtained through choice of a more refined coupling better suited to this particular problem.

Furthermore, there is another aspect to the decoherent histories approach in this context, which has not yet been explored. Many approaches to the arrival time problem are based on model measuring devices, i.e., physical systems in which one of the dynamical variables is correlated with time in some way. The detector model of Sect. 6.5 was of this type: one could think of the two-state system as being some kind of clock or detector attached to the particle, which switches on when the particle enters the region $x < 0$. By physically measuring the two-state system at the end of the time interval $[0, \tau]$ of interest, one expects to be able to deduce that the particle was in $x < 0$, or not, during the time interval. The outstanding question, however, is this: How do we really know that the detector state is correlated with whether or not the particle entered $x < 0$?

This is where the decoherent histories approach comes in. We consider a system consisting of the particle and a detector (and possibly also an environment, if necessary). We then look at histories in which both the final state of the detector and the particle alternatives (whether or not it entered $x < 0$ during $[0, \tau]$) are specified. If these histories are decoherent, we then obtain a joint probability distribution for the histories of the particle and the final state of the detector, and we can ask to what degree these two things are correlated. If they are perfectly correlated, then the detector probability is exactly the same as the probability of the detector and the particle alternatives.

In brief, therefore, the decoherent histories approach will be a useful tool in assessing the extent to which a proposed detector really does its job [37]. Many model detectors are proposed essentially on the basis of classical arguments, but the decoherent histories approach allows their effectiveness to be checked in a genuinely quantum way. This possibility does not appear to have been explored in the context of arrival times, but will be considered elsewhere.

6.9 An Update for the Second Edition

Since the original publication of this volume, there have been a number of developments of the ideas described in this chapter.

The first is an appreciate of the role of the Zeno effect in restricted propagation problems such as the calculation of the decoherence functional, (6.50). The main point is that summing over a set of paths remaining always in $x > 0$, $y > 0$ in (6.50) is equivalent to inserting an excluding potential in the complementary region. As a consequence, all paths in $x > 0$, $y > 0$ never leave that region and even if there is an environment present producing decoherence, the probability of remaining in the region is always 1. This is clearly an unphysical answer and, furthermore, appears to be inconsistent with the semiclassical result (6.57), where there is a non-zero probability of crossing the origin.

The key point here is that restricted propagation in $x > 0$ is equivalent to hitting the state continuously in time with a projection operator $\theta(\hat{x})$. The state then remains entirely in the region $x > 0$, which is essentially the Zeno effect for continuous variables. Differently put, the dynamics is unitary in the subspaces of state in $x > 0$, so no probability can be "lost" from this region [66]. It was first noted by Hartle that coarse grainings of this type can be too strong to give physically useful results [35], but this conclusion (now seen to be correct) was somewhat uncertain in the light of an erroneous calculation of the restricted propagator and as a result failed to be fully appreciated by the authors [32], whose work is described in Sect. 6.4.2.

To get physically sensible results, it is necessary to "soften" the coarse grainings used to describe the way in which paths remain in $x > 0$. One simple way is to use projection operators $\theta(\hat{x})$ operating at a large but finite set of times $t_1, t_2 \cdots t_n$, with unitary evolution in between. It is clearly the continuum limit that causes the state to remain entirely in $x > 0$ so by keeping the time intervals $(t_{k+1} - t_k)$ non-zero the state has a chance to "escape" between each projection. In retrospect this is essentially what was calculated in [32] and in Sect. 6.4.2, which is why the correct semiclassical result (6.57) is obtained, although the use of semiclassical path integral methods obscured the difference between projections at discrete times and the continuum limit. Hence, despite the difficulties linked to the Zeno effect, the results of these earlier works are correct in essence if the path integral constructions are understood in terms of projectors operating at discrete set of moments of time. It would, however, be of interest to carry out this calculation in more detail.

One can of course imagine many different "softer" coarse grainings characterizing the histories in the decoherent histories approach. Histories are normally characterized by exact projection operators, but these are window functions of operators so have derivatives which discontinuous at some points and it is this that can cause problems when they act continuously in time. It is therefore natural to explore the use of POVMs instead of projectors, which are generally smoother in their behaviour (although at the expense of losing the exact exclusivity enjoyed by projection operators). This will be explored in future publications, although it is worth noting that detector-inspired constructions such as (6.77) and (6.79) give some interesting hints as to how such POVMs could be constructed.

The other area in which there has been significant activity concerns the application of the decoherent histories approach to quantum cosmology. There, the key question is, given that the system is in an eigenstate of the Hamiltonian (the Wheeler-DeWitt operator), what is the probability that the system enters a region of configuration space at any stage in its entire history? (i.e., without regard to "time", which does not exist in cosmological models). The decoherent histories approach naturally adapts to this sort of question and a number of papers have addressed these issues [33, 34]. These questions are also affected by the Zeno effect difficulty described above which makes the resolution of this issue particularly pressing.

Acknowledgements

I am extremely grateful to Gonzalo Muga for inviting me to contribute to this volume. I would also like to thank Gonzalo for inviting me to take part in the meeting *Time in Quantum Mechanics* at La Laguna, Tenerife, in May 1998, which stimulated a substantial proportion of the work described here.

References

1. G.S. Agarwal: Phys. Rev. A**3**, 828 (1971); **4**, 739 (1971); H. Dekker: Phys. Rev. A**16**, 2116 (1977); Phys. Rep. **80**, 1 (1991); G.W. Ford, M. Kac, P. Mazur: J. Math. Phys. **6**, 504 (1965); H. Grabert, P. Schramm, G.-L. Ingold: Phys. Rep. **168**, 115 (1988); V. Hakim, V. Ambegaokar: Phys. Rev. A**32**, 423 (1985); J. Schwinger: J. Math. Phys. **2**, 407 (1961); I.R. Senitzky: Phys. Rev. **119**, 670 (1960)
2. G.R. Allcock: Ann. Phys. (N.Y.) **53**, 253 (1969); **53**, 286 (1969);**53**, 311 (1969)
3. Y. Aharanov, J. Oppenheim, S. Popescu, B. Reznik, W. Unruh: Phys. Rev. A **57**, 4130 (1998); Phys.Rev. A **57**, 4130 (1998)
4. A. Auerbach, S. Kivelson: Nucl. Phys. B **257**, 799 (1985)
5. P. van Baal: 'Tunneling and the path decomposition expansion'. In *Lectures on Path Integration: Trieste 1991*, ed. by H.A.Cerdeira et al. (World Scientific, Singapore, 1993)
6. N. Balazs, B.K. Jennings: Phys. Rep. **104**, 347 (1984); M.Hillery, R.F. O'Connell, M.O. Scully, E.P. Wigner: Phys. Rep. **106**, 121 (1984); V.I. Tatarskii: Sov. Phys. Usp **26**, 311 (1983)

7. J. Barbour: Class. Qu. Grav. **11**, 2853 (1994); **11**, 2875 (1994)
8. M.A. Burschka, U.M. Titulaer: J. Stat. Phys. **25**, 569 (1981); **26**, 59 (1981); Physica A **112**, 315 (1982)
9. A. Boutet de Monvel, P. Dita: J. Phys. A **23**, L895 (1990)
10. A.O. Caldeira, A.J. Leggett: Physica A **121**, 587 (1983)
11. H. Carmichael: *An Open Systems Approach to Quantum Optics* (Springer-Verlag Lecture Notes in Physics, m18, Berlin, 1993)
12. H.S. Carslaw: Proc. Lond. Math. Soc. **30**, 121 (1899)
13. H.F. Dowker, J.J. Halliwell: Phys. Rev. D **46**, 1580 (1992)
14. H. Fertig: Phys. Rev. Lett. **65**, 2321 (1990)
15. R.P. Feynman, F.L. Vernon: Ann. Phys. (N.Y.) **24**, 118 (1963)
16. C.W. Gardiner: *Quantum Noise* (Springer-Verlag, Berlin, 1991)
17. M. Gell-Mann, J.B. Hartle: *Complexity, Entropy and the Physics of Information, SFI Studies in the Sciences of Complexity*, Vol. VIII, ed. by W. Zurek (Addison-Wesley, Reading, 1990); in *Proceedings of the Third International Symposium on the Foundations of Quantum Mechanics in the Light of New Technology*, ed. by S. Kobayashi, H. Ezawa, Y. Murayama, S. Nomura (Physical Society of Japan, Tokyo, 1990)
18. M. Gell-Mann, J.B. Hartle: Phys. Rev. D **47**, 3345 (1993)
19. M. Gell-Mann, J.B. Hartle: *Proceedings of the 4th Drexel Symposium on Quantum Non-Integrability – The Quantum-Classical Correspondence*, ed. by D.H. Feng, B.L. Hu (International Press, Boston/Hong Kong, 1996)
20. R.B. Griffiths: J. Stat. Phys. **36**, 219 (1984); Phys. Rev. Lett. **70**, 2201 (1993); Am. J. Phys. **55**, 11 (1987)
21. N. Grot, C. Rovelli, R.S. Tate: Phys. Rev. A **54**, 4676 (1996); Phys. Rev. A **54**, 4679 (1996)
22. J.J. Halliwell: 'Aspects of the decoherent histories approach to quantum theory'. In: *Stochastic Evolution of Quantum States in Open Systems and Measurement Processes*, ed. by L. Diósi, B. Lukács (World Scientific, Singapore, 1994)
23. J.J. Halliwell: 'A review of the decoherent histories approach to quantum mechanics'. In *Fundamental Problems in Quantum Theory*, ed. by D.Greenberger, A. Zeilinger, Annals of the New York Academy of Sciences, Vol. 775, 726 (1994).
24. J.J. Halliwell: Phys. Lett. A **207**, 237 (1995)
25. J.J. Halliwell: *General Relativity and Gravitation 1992*, ed. by R.J. Gleiser, C.N. Kozameh, O.M. Moreschi (IOP Publishers, Bristol, 1993)
26. J.J. Halliwell: Phys. Rev. D **60**, 105031 (1999)
27. J.J. Halliwell: Prog. Th. Phys. **102**, 707 (1999)
28. J.J. Halliwell: Phys.Rev. D **64**, 044008 (2001)
29. J.J. Halliwell, M.E. Ortiz: Phys. Rev. D **48**, 748 (1993)
30. J.J. Halliwell, J. Thorwart: Phys. Rev. D **64**, 124018 (2001)
31. J.J. Halliwell, A. Zoupas: Phys. Rev. D **52**, 7294 (1995) Phys. Rev. D **55**, 4697 (1997)
32. J.J. Halliwell, E. Zafiris: Phys. Rev. D **57**, 3351 (1998)
33. J.J. Halliwell, J. Thorwart: Phys. Rev. D **65**, 104009 (2002)
34. J.J. Halliwell, P. Wallden: Phys. Rev. D **73**, 024011 (2006)
35. J.B. Hartle: Phys. Rev. D **44**, 3173 (1991)
36. J.B. Hartle: Phys. Rev. D **38**, 2985 (1988)
37. J.B. Hartle: *Quantum Cosmology and Baby Universes*, ed. by S. Coleman, J. Hartle, T. Piran, S. Weinberg (World Scientific, Singapore, 1991)

38. J.B. Hartle: Phys. Rev. D **37**, 2818 (1988)
39. J.B. Hartle: *Proceedings of the 1992 Les Houches Summer School, Gravitation et Quantifications*, ed. by B. Julia, J. Zinn-Justin (Elsevier Science B.V., Amsterdams 1995)
40. E.H. Hauge, J.A. Stovneng: Rev. Mod. Phys. **61**, 917 (1989)
41. A.S. Holevo: *Probabilistic and Statistical Aspects of Quantum Theory* (North-Holland Publishing Corporation, Amsterdam, 1982)
42. C. Isham: J. Math. Phys. **23**, 2157 (1994); C. Isham, N. Linden: J. Math. Phys. **35**, 5452 (1994); **36**, 5392 (1995); C. Isham, N. Linden, S. Schreckenberg: J. Math. Phys. **35**, 6360 (1994)
43. A. Jadczyk: Prog. Theor. Phys. **93**, 631 (1995); Ph. Blanchard, A. Jadczyk: Helv. Phys. Acta. **69**, 613 (1996)
44. That the propagator for the Fokker–Planck equation can be represented in terms of a configuration space path integral of this form does not appear to be widely known. See, however, H. Kleinert: *Path Integrals in Quantum Mechanics, Statistics and Polymer Physics* (World Scientific, Singapore, 1990)
45. J. Kijowski: Rep. Math. Phys. **6**, 362 (1974)
46. D.H. Kobe, V.C. Aguilera-Navarro: Phys. Rev. A**50**, 933 (1994)
47. N. Kumar: Pramana J. Phys. **25**, 363 (1985)
48. R. Landauer: Rev. Mod. Phys. **66**, 217 (1994); Ber. Bunsenges: Phys. Chem **95**, 404 (1991)
49. G. Lindblad: Commun. Math. Phys. **48**, 119 (1976)
50. L. Mandelstamm, I. Tamm: J. Phys. **9**, 249 (1945)
51. T.W. Marshall, E.J. Watson: J. Phys. A **18**, 3531 (1985); J. Phys. A **20**, 1345 (1987)
52. R.J. Micanek, J.B. Hartle: Phys. Rev. A **54**, 3795 (1996)
53. N.F. Mott: Proc. R. Soc A **124**, 375 (1929)
54. J.G. Muga, A.D. Baute, J.A. Damborenea, I.L. Egusquiza: quant-ph/0009111 (2000)
55. J.G. Muga, S. Brouard, D. Macías: Ann. Phys. (N.Y.) **240**, 351 (1995)
56. J.G. Muga, C.R. Leavens: Phys. Rep. **338**, 353 (2000)
57. R. Omnès: J. Stat. Phys. **53**, 893 (1988); **53**, 933 (1988); **53**, 957 (1988); **57**, 357 (1989); **62**, 841 (1991); Ann. Phys. **201**, 354 (1990); Rev. Mod. Phys. **64**, 339 (1992); *The Interpretation of Quantum Mechanics* (Princeton University Press, Princeton, 1994)
58. J.P. Palao, J.G. Muga, S. Brouard, A. Jadczyk: Phys. Lett. A**233**, 227 (1997)
59. A. Peres: *Quantum Theory: Concepts and Methods* (Kluwer Academic Publishers, Dordrecht, 1993)
60. L.S. Schulman: *Time's Arrows and Quantum Measurement* (Cambridge University Press, Cambridge, 1997)
61. L.S. Schulman: Ann. Phys. **212**, 315 (1991); B. Gaveau, L.S. Schulman: J. Stat. Phys. **58**, 1209 (1990)
62. L. Schulman, R.W. Ziolkowiski: in *Path integrals from meV to MeV*, ed. by V. Sa-yakanit, W. Sritrakool, J. Berananda, M.C. Gutzwiller, A. Inomata, S. Lundqvist, J.R. Klauder, L.S. Schulman (World Scientific, Singapore, 1989)
63. A.J.F. Siegert: Phys. Rev. **81**, 617 (1951)
64. See, for example, E.P. Wigner: Phys. Rev. **98**, 145 (1955); F.T. Smith: Phys. Rev. **118**, 349 (1960); E. Gurjoy, D. Coon: Superlattices Microsctruct. **5**, 305 (1989); C. Piron: in *Interpretation and Foundations of Quantum Theory*, ed. by H. Newmann (Bibliographisches Institute, Mannheim, 1979)

65. M.C. Wang, G.E. Uhlenbeck: Rev. Mod. Phys. **17**, 323 (1945); reprinted in, *Selected Papers on Noise and Stochastic Processes*, ed. by N. Wax (Dover Publications, New York, 1954)
66. P. Wallden: gr-qc/0607072.
67. N. Yamada: Phys. Rev. A**54**, 182 (1996)
68. N. Yamada, S. Takagi: Prog. Theor. Phys. **85**, 985 (1991);**86**, 599 (1991); **87**, 77 (1992); N. Yamada: Sci. Rep. Tôhoku Univ. Ser. 8 **12**, 177 (1992)
69. E. Zafiris: Imperial College preprint TP/96-97/05 (1997)

7

Quantum Traversal Time, Path Integrals and "Superluminal" Tunnelling

Dmitri Sokolovski

Theoretical and Computational Physics Research Division, Department of Applied Mathematics and Theoretical Physics, Queen's University of Belfast, Belfast BT7 1NN, UK
d.sokolovski@am.qub.ac.uk

7.1 Introduction

To find out how much time a classical particle spends in a given region of space one only has to use a stopwatch. The same question posed in the context of quantum mechanics has caused controversy for several decades and remains controversial to date. As early as in 1932, McColl [1] noted that tunnelling must be characterised not only by the transmission rate but also by the speed of transmission. The problem attracted renewed attention with the progress in nano-technology [3, 4] and photonic tunnelling experiments [2]. It has been hoped that a properly defined "traversal time" τ (i.e., the time a tunnelling particle spends in the barrier) would, among other things, describe the response of a tunnelling device to a time modulation of the barrier, provide an insight into the nature of the image forces affecting the tunnelling rate and explain why a transmitted wave packet appears to arrive at a detector ahead of the one that propagates freely. However, anyone interested in the subject soon discovers that standard texts on quantum theory offer neither a clear definition nor a unique recipe for determining the duration τ. Numerous attempts have been made to obtain a suitable quantum mechanical generalisation of the classical traversal time (for reviews see [5]) with approaches ranging from using specially designed "clocks" to invoking non-standard interpretations of the quantum mechanics, such as that of Bohm [6].

The main difference between the traversal time and a conventional quantum mechanical observable, such as coordinate or momentum, is that the former refers to a duration, rather than to a single instant in time. We let a particle travel for, say, 1 s and then ask how much of this 1 s it has spent in some region Ω. For this reason, the traversal time problem is most conveniently dealt with within the Feynman path integral formulation of quantum mechanics [7]. This chapter is a review of the path integral approach developed in [8, 9, 10, 11, 12, 13, 14, 15, 16, 17, 18, 19, 20, 21, 22]. It is not, however, our intention to review all the work on the subject and a more comprehensive list of references can

D. Sokolovski: *Quantum Traversal Time, Path Integrals and "Superluminal" Tunnelling*, Lect. Notes Phys. **734**, 195–233 (2008)
DOI 10.1007/978-3-540-73473-4_7   © Springer-Verlag Berlin Heidelberg 2008

be found in [2, 3, 4, 5] and [14]. The paper is organised as follows: in Sect. 7.2 we use the path integral to construct the traversal time amplitude distribution and derive the clocked Schrödinger equation which governs its evolution. Sections 7.3, 7.4 and 7.7 deal with various aspects of traversal time measurements. In Sects. 7.5 and 7.6 we give detailed analysis of the tunnelling time problem. Section 7.8 discusses the use of the traversal time in problems involving a particle coupled to external degrees of freedom. In Sect. 7.9 we analyse the phenomenon of "superluminal tunnelling" and in Sect. 7.10 we extend our analysis to relativistic theories. Section 7.12 contains our conclusions.

7.2 Path Decomposition. The Clocked Schrödinger Equation. Coarse Graining

We start by defining the time τ a quantum particle spends in a chosen region of space Ω. Conventionally, a measurement of a physical quantity F (e.g., coordinate or momentum) requires constructing a Hermitian operator $\hat{F}$ and expanding the particle's wave function $\Psi(x, t)$ it its eigenstates,

$$\hat{F}\phi_i = F_i\phi_i , \tag{7.2.1}$$

$$\Psi(x, t) = \sum_i c_i\phi_i(x) . \tag{7.2.2}$$

The probability that F has a value F_i is then given by $|c_i|^2$. As shown by von Neumann, the probability distribution $|c_i|^2$ can be measured, at least in principle, by coupling the particle to an external degree of freedom [23].

A short consideration convinces us that there is no obvious operator to represent the traversal time τ. Indeed, τ represents the part of a certain time interval, say $0 < t < T$, which the particle has spent inside some region Ω. In classical mechanics, evaluation of this quantity requires the knowledge of the its trajectory $x(t)$ between $t = 0$ and $t = T$. The corresponding classical functional is

$$t_\Omega^{cl}[x(t)] = \int_0^T \Theta_\Omega(x(t))dt , \tag{7.2.3}$$

where $\Theta_\Omega(x)$ equals unity for x inside Ω and zero otherwise. It is readily seen that $t_\Omega^{cl}[x(t)]$ yields the time measured by a stopwatch which runs only when the particle is inside Ω.

Quantum mechanics assigns probability amplitudes to physical events. Accordingly, we should be looking for an amplitude (distribution) $\Phi(x, t|\tau)$ that a particle in x at $t = T$ has spent in Ω a duration τ. This is most easily done with the help of Feynman path integral. The solution of the Schrödinger equation

$$i\hbar\frac{\partial\Psi(x, t)}{\partial t} = -\frac{\hbar^2}{2m}\frac{\partial^2\Psi(x, t)}{\partial x^2} + V(x)\Psi(x, t) , \tag{7.2.4}$$

for a particle in a potential $V(x)$ can be written in the path integral form [7]

$$\Psi(x,T) = \int dx' \int_{x'(0);x(T)} Dx(t) \exp\{iS[x(t)]/\hbar\}\Psi(x',0) \ , \qquad (7.2.5)$$

where

$$S[x(t)] = \int_0^T [m\dot{x}^2/2 - V(x)]dt \ , \qquad (7.2.6)$$

is the classical action. In (7.2.5) $Dx(t)$ denotes a sum over paths starting in x' at $t = 0$ and ending in x at $t = T$ (Fig. 7.1). Most of Feynman paths are continuous but highly irregular [7]. They form a complete set of possible histories for a spinless point particle whose final and initial positions are x and x', respectively. Each path contributes $\exp\{iS[x(t)]/\hbar\}$ to the Schrödinger wave function $\Psi(x,t)$ which is found by summing over all initial positions. Now we can identify $\Phi(x,t|\tau)$ with the contribution from the subset of paths which spend in Ω precisely duration τ. Thus, $\Phi(x,t|\tau)$ is given by a restricted path integral

$$\Phi(x,T|\tau) = \int dx' \int_{x'(0);x(T)} Dx(t)\delta(t_\Omega^{cl}[x(t)] - \tau)\exp\{iS[x(t)]/\hbar\}\Psi(x',0) \ ,$$

$$(7.2.7)$$

where $\delta(x)$ is the Dirac delta function. Equation (7.2.7) can be cast in a more convenient form by writing $\delta(t_\Omega^{cl}[x(t)] - \tau))$ as a Fourier integral,

$$\delta(z) = \frac{1}{2\pi\hbar} \int dW \exp\{-iWz/\hbar\} \ , \qquad (7.2.8)$$

and noting that adding the term $-iWt_\Omega^{cl}[x(t)]$ to the action $S[x(t)]$ is equivalent to adding a rectangular potential $W\Theta_\Omega(x)$ to $V(x)$. This gives

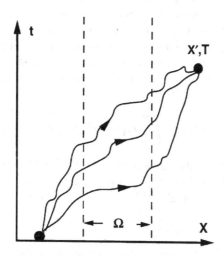

Fig. 7.1. Feynman path connecting $(x,0)$ with (x',T) spend different durations in Ω

$$\Phi(x,T|\tau) = \frac{1}{2\pi\hbar} \int dW \exp\{iW\tau/\hbar\}\Psi(x,T|W) , \qquad (7.2.9)$$

where $\Psi(x,T|W)$ denotes the result of evolving the initial state $\Psi(x,0)$ in a composite potential $V(x)+W\Theta_\Omega(x)$. It is easy to show that the traversal time distribution $\Phi(x,T|\tau)$ satisfies the "clocked" Schrödinger equation [16, 17, 18]

$$i\hbar\frac{\partial\Phi(x,t|\tau)}{\partial t} = -\frac{\hbar^2}{2m}\frac{\partial^2\Phi(x,t|\tau)}{\partial x^2} + V(x)\Psi(x,t) - i\hbar\Theta_\Omega(x)\frac{\partial\Phi(x,t|\tau)}{\partial\tau} ,$$
$$(7.2.10)$$

with the initial condition

$$\Phi(x,0|\tau) = \Psi(x,0)\delta(\tau) . \qquad (7.2.11)$$

Generalisation of (7.2.7)–(7.2.10) to three dimensions is straightforward. So far, we have sorted Feynman paths according to the duration τ each path spends in Ω and decomposed (unfolded) the Schrödinger amplitude $\Psi(x,T)$ into subamplitudes $\Phi(x,T|\tau)$ so that $\Psi(x,T)$ can be seen as a result of interference between different particle's histories,

$$\Psi(x,T) = \int_0^T \Phi(x,T|\tau)d\tau . \qquad (7.2.12)$$

In fact, our analysis is similar to that of a two-slit diffraction experiment [7], except that we have an infinite number of slits continually labelled by the value of τ. This analogy suggests that the Schrödinger wave function $\Psi(x,T)$ "knows" nothing about the value of the traversal time, as this information is lost through interference. We may also predict that a measurement of τ would destroy the value of $\Psi(x,T)$ just as an attempt to establish which of the two holes has been chosen by a particle destroys the interference pattern on the screen. Finally, we expect that a measured value of τ will occur with a probability $\approx |\Phi(x,T|\tau)|^2$. The last statement requires further attention. As τ is a continuous variable, we expect that a measurement of a finite accuracy would allow interference between certain values of τ. We therefore define the coarse grained amplitude $\Psi(x,t|\tau)$,

$$\Psi(x,T|\tau) = \int_{-\infty}^\infty G(\tau - \tau')\Phi(x,T|\tau')d\tau' , \qquad (7.2.13)$$

where $G(\tau)$ is sharply peaked around $\tau = 0$ with a width $\Delta\tau$ so that the (unnormalised) probability $\rho(x,T|\tau)$ for a particle found in x at $t = T$ to have τ in the range $[\tau - \Delta\tau/2, \tau + \Delta\tau/2]$ is

$$\rho(x,T|\tau) = |\Psi(x,T|\tau)|^2 . \qquad (7.2.14)$$

Equations (7.2.10)–(7.2.14) are the basis of our analysis.
As a first example consider a particle trapped in the ground state $\phi_0(x)$ of a potential box ($V = 0$ for $-L < x < L$ and 0 otherwise) [18]. The

last term in (7.2.10) tends to propagate the part of the distribution contained in Ω forward in τ, while the kinetic energy is responsible for diffusive spreading in the x variable. The latter effect has a characteristic time scale of $T_0 = 2mL^2/\hbar$, essentially the inverse of the particle's uncertainty in the kinetic energy. Contour plots of the probability density $\rho(x, T|\tau) = |\Psi(x, T|\tau)|^2$ calculated numerically with a Gaussian coarse graining function

$$G(\tau) = \Delta\tau/\pi^{1/2} \exp[-\tau^2/\Delta\tau^2] ,\qquad (7.2.15)$$

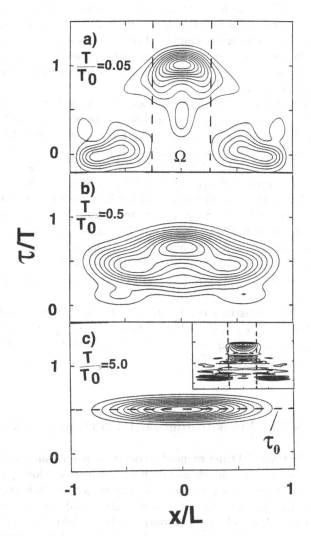

Fig. 7.2. Traversal time distributions for a particle in the ground state $\phi_0(x)$ of a potential box coarse grained with $\Delta\tau/T = 0.2$ in the short (**a**), medium (**b**) and long (**c**) time limits. The region Ω is chosen so that $\int_\Omega |\phi_0(x)|^2 dx = 0.5$. The plot in the inset is the same as in (**c**) but for $\Delta\tau/T = 0.02$

are shown in Fig. 7.2 for $\Delta\tau/T = 0.2, T/T_0 = 0.05(a), 0.5(b)$ and $5.0(c)$. In the short time limit (a), the diffusive term can be neglected and we find the particle close to its original position. For $T/T_0 = 0.5$ there is already a considerable mixing between Ω and the rest of the well, with still a tendency for larger τ's inside Ω. In the long term (Fig. 7.2c), the particle forgets its initial position and the distribution is centred around $\bar{\tau} = \int_\Omega |\phi_0(x)|^2 dx$. Overall, Fig. 7.2 suggests the existence of a "microscopic motion" responsible for the exchange of particles between Ω and the rest of the well, similar to the Brownian motion which carries a particle around a volume even though overall concentration remains unchanged [16]. However, contrary to the classical analogy, this exchange is hampered and a Zeno-type paradox arises if one reduces the amount of coarse graining, $\Delta\tau \to 0$ as shown in the inset in Fig. 7.2c. In the next section we will relate this behaviour to the influence of the clock required to measure τ.

We conclude with few general remarks. The Schrödinger wave function $\Psi(x, t)$ describes a quantum particle in terms of its position at a given time. This description can be extended, e.g., by adding an extra variable τ related to how the particle arrived in x, i.e., to the particle's past. The status of τ is different from that of x. Integrating $\Phi(x, t|\tau)$ over τ restores the wave function $\Psi(x, t)$ whereas $\int \Phi(x, t|\tau)dx$ yields no meaningful amplitude. In our approach, different traversal times may either interfere or be exclusive alternatives, depending on whether we attempt to measure τ or not. On the other hand, at any given time, the positions of different particles are always exclusive. The same analysis can be applied to other classical functionals, notably to the mean value of a dynamical variable [19, 20, 22] (p is the momentum) $F(p, x)$, $\langle F \rangle \equiv T^{-1} \int_0^T F(p, x)dt$. The von Neumann limit in (7.2.1) can then be obtained by letting $T \to 0$ [23].

Finally, we may construct conditional amplitude distributions for several functionals, $\Phi(x, t|y_1, y_2, ..., y_n)$ which satisfy Schrödinger-like equations in variables $x, y_1, y_2, ..., y_n$ and reduce to the Schrödinger wave function upon summation over $y_1, y_2, ..., y_n$, $\int \Phi(x, t|y_1, y_2, ..., y_n)dy_1, dy_2, ..., dy_n = \Psi(x, t)$.

7.3 Meters and Measurements. Uncertainty Relation

To measure the traversal time we need to devise a meter which would destroy interference between certain classes of Feynman paths. More precisely, we expect a meter of accuracy $\Delta\tau$ whose reading is τ_0 to include contributions from those paths with $\tau_0 - \Delta\tau/2 \leq \tau \leq \tau_0 + \Delta\tau/2$ and discard the rest. The recipe for constructing such meter is contained in the definition of the classical quantity to be measured or, more precisely, in the equation of motion (7.2.10) obeyed by $\Phi(x, t|\tau)$. First we note that the coarse grained amplitude $\Psi(x, t|\tau)$ also satisfies (7.2.10) (we write $\hat{H}(p, x)$ for the particle's Hamiltonian)

$$i\hbar\frac{\partial\Psi(x,t|\tau)}{\partial t} = [\hat{H}(p,x) - i\hbar\Theta_\Omega(x)\frac{\partial}{\partial\tau}]\Psi(x,t|\tau) , \qquad (7.3.1)$$

with the initial condition

$$\Psi(x,0|\tau) = \Psi(x,0)G(\tau) . \qquad (7.3.2)$$

This can be verified directly by substituting (7.2.13) into (7.2.10) and integrating by parts. Equation (7.2.10) describes a particle in terms of its position and the time spent in Ω. Alternatively, we may interpret it as a two-particle Schrödinger equation describing a particle coupled to a meter (pointer) which correlates its position τ with the value of the classical functional in (7.2.3). At $t = 0$ the meter is prepared in the state $G(\tau)$, and then runs until $t = T$. At $t = T$ the meter is read, i.e., its position is accurately determined. The (unnormalised) joint probability for finding the particle in x and a meter reading τ is given by $|\Psi(x,T|\tau)|^2$ in accordance with (7.2.14). Thus, the classification of Feynman paths based on the value of τ uniquely determines interaction with the meter required to destroy interference between the classes.

Now we may take a closer look at how a meter perturbs the particle's motion. The accuracy of a measurement (the amount of coarse graining in (7.2.13)) is determined by the initial state of the meter $G(\tau)$. The meter is inaccurate because we have no precise knowledge of its initial position. To improve the accuracy we may wish to choose $G(\tau)$ more narrowly peaked around $\tau = 0$. However, a decrease in the peaks width $\Delta\tau$ increases the spread of the meter's momenta. We note further that a meter prepared with initial momentum W, $\Psi(x,0|\tau) = \Psi(x,0)\exp(iW\tau/\hbar)$ retains it throughout interaction and adds $W\Theta_\Omega(x)$ to the potential $V(x)$. Therefore, a meter acts back on the particle by changing the potential in which it moves in such a way that a better accuracy leads to more perturbation. (This problem does not occur in classical mechanics where a pointer can have zero momentum as well as a well-defined initial position.) This can be summed up in the form of an uncertainty relation

$$\Delta\tau\Delta W \geq \hbar , \qquad (7.3.3)$$

which states that in order to measure the traversal time to accuracy $\Delta\tau$ one must create an uncertainty ΔW in the potential in the region of interest. In other words, W plays the role of the variable conjugate to τ.

The fact that destruction of coherence between Feynman paths is synonymous with the dynamical interaction between the particle and a meter can be demonstrated in a more direct manner. Choosing $G(\tau) = G_0(\alpha\tau)$ and scaling $\tau \to \alpha\tau$ in (7.2.10) yields

$$i\hbar\frac{\partial\Psi(x,t|\tau)}{\partial t} = [\hat{H}(p,x) - i\alpha\hbar\Theta_\Omega(x)\frac{\partial}{\partial\tau}]\Psi(x,t|\tau) , \qquad (7.3.4)$$

with $\Psi(x,0|\tau) = \Psi(x,0)G_0(\tau)$. Equation (7.3.4) shows that an increase in α and therefore in the meter's resolution can equally be described by an increase in the magnitude of the particle–meter coupling while the meter's initial state remains unchanged.

Finally, we relate our approach to the von Neumann projection postulate mentioned at the beginning of Sect. 7.2. For this purpose, it is convenient to consider the fractional traversal time

$$\xi = T^{-1} \int_0^T \Theta_\Omega(x(t)) dt .$$

Dimensionless ξ gives the fraction of the total time T spent inside Ω and $\Psi(x, 0|\xi)$ satisfies (7.3.4) with $\alpha = T^{-1}$. In the limit $T \to 0$ in (7.3.1) $H(p, x)$ can be neglected compared to $-i\alpha\hbar\Theta_\Omega(x)\partial\Psi(x, t|\tau)/\partial\tau$ and (7.2.10) reduces to that of von Neumann [23]. The fractional traversal time is seen to be represented by the projection operator $\Theta_\Omega(x)$. This operator has eigenvalues of 1 and 0 for the functions whose support lies inside and outside Ω and 0, respectively. Thus as $T \to 0$ the particle retains its position and a measurement of the fractional traversal time simply establishes whether the particle is inside Ω.

7.4 Averages. Complex Times. Weak Measurements

Probability distribution (7.2.14) can be used to construct various averages. For example, we may select the final position of a particle i.e., assume that the meter is read only if the particle is registered in x at $t = T$. This would require a large number of trials but can, in principle, be done. The expectation value in this case is

$$\langle \tau(x) \rangle = \int_{-\infty}^{\infty} \tau |\Psi(x, T|\tau)|^2 d\tau / \int_{-\infty}^{\infty} |\Psi(x, T|\tau)|^2 d\tau . \tag{7.4.1}$$

Alternatively, we may decide to read the meter regardless of where the particle arrives in the end. The expectation value becomes

$$\langle \tau \rangle = \int_{-\infty}^{\infty} \tau |\Psi(x, T|\tau)|^2 d\tau dx / \int_{-\infty}^{\infty} |\Psi(x, T|\tau)|^2 d\tau dx . \tag{7.4.2}$$

Note that the accuracy of the meter $\Delta\tau$ enters (7.4.1)–(7.4.2) through the coarse grained $\Psi(x, T|\tau)$. Equation (7.4.2) can be rewritten in a more appealing form if we use (7.2.10) to evaluate the time derivative of $\int \tau |\Psi(x, T|\tau)|^2 d\tau dx$. Integrating over time we have (we assume $\int \tau |G(\tau)|^2 d\tau = 0$)

$$\langle \tau \rangle = \int_0^T dt \int_{-\infty}^{\infty} dx \Theta_\Omega(x) \int_{-\infty}^{\infty} d\tau |\Psi(x, t|\tau)|^2 / \int_{-\infty}^{\infty} |\Psi(x, T|\tau)|^2 d\tau dx .$$
$$\tag{7.4.3}$$

Equation (7.4.3) is similar to the classical expression for the traversal time (7.2.3). The value of $\Theta_\Omega(x(t))$ on the particle's trajectory is, however, replaced by the conventional expectation value of the operator $\Theta_\Omega(x(t))$ evaluated with the particle–meter wave function $\Psi(x, T|\tau)$.

Next we ask whether it is possible to minimise the perturbation induced by a meter and still learn something about the quantum traversal time. This has been the approach of many early papers on the subject [5, 30, 31, 32]. Our two-slit analogy suggests that we are, in effect, trying to determine which hole was used by an electron without destroying the interference pattern on the screen and this cannot be done in a satisfactory manner [7]. It is, however, instructive to see how this approach fails. Consider $G(\tau)$ so broad that the meter all but decouples from the particle (cf. (7.3.4)). If so, expanding G in (7.2.12) to linear terms around $\tau' = 0$ yields

$$\Psi(x, t|\tau) \approx \Psi(x, t)[G(\tau) - G'(\tau)\bar{\tau}(x)] , \qquad (7.4.4)$$

where "the complex time" $\bar{\tau}(x)$ is the complex-valued first moment of the fine grained amplitude $\Phi(x, T|\tau)$,

$$\bar{\tau}(x) = \int_0^T \tau\Phi(x, T|\tau)d\tau / \Psi(x, t) , \qquad (7.4.5)$$

and we have used the folding property (7.2.12). The spread of the meter readings is extremely wide but we can still collect enough statistics to evaluate the expectation value. Selecting only particles in x and inserting (7.4.4) into (7.4.1) yields (again, $\int \tau|G(\tau)|^2 d\tau = 0$)

$$\langle \tau(x) \rangle \approx Re\bar{\tau}(x) . \qquad (7.4.6)$$

It is at this point that we experience a difficulty. The average $\bar{\tau}(x)$ is computed with a complex valued and, which is more important, oscillatory distribution $\Phi(x, T|\tau)$. The properties of such averages are very different from those obtained with non-negative probability distributions. For example, if a non-negative $\rho(x) \geq 0$ is contained within $[0, T]$, then $\bar{x} \equiv \int_0^T x\rho(x)dx / \int_0^T \rho(x)dx$ is guaranteed to lie between 0 and T. If, however, we allow $\rho(x)$ to take negative values, $\bar{x}$ may take any real value (e.g., $\int_0^T \rho(x)dx \to \pm\infty$ will send $\bar{x}$ to $\pm\infty$). As a result, $\bar{x}$ no longer represents the region which contains the support of $\rho(x)$. For this reason, although $\Phi(x, T|\tau)$ is contained inside $[0, T]$ (no Feynman path can spend in Ω duration which exceed T or is negative) $Re\bar{\tau}(x)$ may lie outside $[0, T]$. The ill-posed question provokes an unsatisfactory answer: on average, a particle appears to spend in Ω 10 s (or, possibly, 10 s) between $t = 0$ and $t = 1$ s. Clearly, this does not suggest that in quantum mechanics 1 s can be stretched into 10 s. Rather our approach fails, as we expected, to resolve the two-slit paradox. The situation just described corresponds to a weak measurement of the traversal time. The concept of a weak measurement was introduced by Aharonov et al. [24] and extended to the traversal time measurements in [26] and [27].

The just described problem with weak values seems to disappear if we choose not to control the final position of the particle. Replacing in (7.4.3) $\Psi(x,t|\tau)$ by $\Psi(x,t)G(\tau)$ gives (we assume $\int |\Psi(x,t)|^2 dx = 1$)

$$\langle \tau \rangle \approx \int_0^T dt \int_\Omega dx |\Psi(x,t)|^2 . \tag{7.4.7}$$

The "dwell time" [5] in (7.4.7) certainly has a value between 0 and T and looks like the classical expression (7.2.3) with $\Theta_\Omega(x(t))$ replaced by its expectation value with the unperturbed Schrödinger wave function $\Psi(x,t)$. A similar expression can be obtained in the Bohmian quantum mechanics [6]. There have been attempts to use (7.4.7) to define the quantum traversal time. However, as we have seen, traversal times with similar properties cannot be obtained if the averaging is done over subsets of the coordinate space, e.g., for the transmission channel in one-dimensional tunnelling [5, 32, 33, 35]. It should, therefore, be remembered that in the conventional quantum mechanics the mean traversal time is defined by (7.4.2) rather than by (7.4.7). The fact that $\langle \tau \rangle$ in (7.4.7) is real and positive can be attributed to a sum rule satisfied by $\bar{\tau}(x)$ which, in turn, is related to conservation of particles [8].

7.5 Examples: Free motion and Tunnelling

Duration of tunnelling has been of the most interest and next we will obtain the traversal time distribution $\Phi(x,t|\tau)$ for a particle with energy E tunnelling through a rectangular barrier ($\Omega \equiv [a,b]$)

$$V(x) = V\Theta_{ab}(x) . \tag{7.5.1}$$

At $t = 0$ the particle is described by a broad wave packet $\approx \exp(ikx), E = \hbar^2 k^2/2m$ to the left of the barrier. After tunnelling, at $t \to \infty$, the transmitted particle's state is essentially

$$\Psi(x,t) = T(k,V)\exp(ikx - iEt/\hbar) , \tag{7.5.2}$$

where $T(k,V)$ is the transmission amplitude. We recall from (7.2.9) that in order to construct $\Phi(x,t|\tau)$ we must consider transmission across all rectangular potentials $[V + W]\Theta_{ab}(x)]$ (barriers if $W > -V$ or wells if $W < -V$),

$$\Phi(x,t|\tau) = \exp(ikx - iEt/\hbar)\frac{1}{2\pi\hbar} \int dW \exp\{iW\tau/\hbar\}T(k, V + W)$$
$$\equiv \exp(ikx - iEt/\hbar)\eta_V(k,\tau) , \tag{7.5.3}$$

the problem reduces to evaluating the Fourier transform $\eta(k,\tau)$ of $T(k, V+W)$. For free motion, $V \equiv 0$,

$$\eta_0(k,\tau) = \frac{1}{2\pi\hbar} \int dW \exp\{iW\tau/\hbar\} T(k,W) \,, \qquad (7.5.4)$$

where the transmission amplitude $T(k,W)$ is given by

$$T(k,W) = \frac{-4i\beta z \exp(-i\beta)}{(z-i\beta)^2 \exp(z) - (z+i\beta)^2 \exp(-z)} \,, \qquad (7.5.5)$$

$$\beta \equiv k(b-a) \,,$$
$$z \equiv [2m(W-E)]^{1/2}(b-a)/\hbar \,.$$

The contour of integration in (7.5.4) can be closed in either upper (for $\tau > 0$) or lower (for $\tau < 0$) half-planes. The transmission amplitude $T(k,W)$ has no poles below the real W-axis and $\eta_0(k,\tau)$ vanishes, as it should, for negative traversal times. Above the real W-axis, $T(k,W)$ has an infinite number of poles $\{W_n, n = 1,2,3...\}$ whose locations are shown in Fig. 7.3. Summing the poles contribution, we obtain a series representation for $\eta_0(k,\tau), \tau > 0$.

$$\eta_0(k,\tau) = \sum_{n=1}^{\infty} (-1)^{n-1} a_n \exp(iW_n\tau/\hbar) \,. \qquad (7.5.6)$$

It can be demonstrated [14] that for large n, $a_n \to const$ and $W_n \to -(n-1)^2 \times const$. An interesting property of the series, whose fractal-like behaviour was studied by Berry and Goldberg [28], is that it is oscillatory so that $\Phi(x,t|\tau)$ must be interpreted as a distribution. Indeed, the r.h.s. of (7.5.6) contains arbitrary high frequencies and does not converge to a smooth

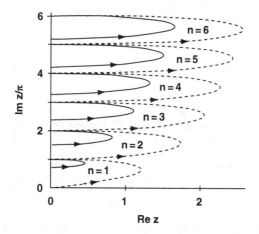

Fig. 7.3. Trajectories of the poles of the transmission amplitude $T(k,W)$ in the complex plane of $z \equiv [2m(W-E)]^{1/2}(b-a)/\hbar$ as $\beta \equiv k(b-a)$ increases from 0 to ∞ (dashed). Also shown by solid lines are the same trajectories for the double barrier structure discussed in Sect. 7.7

function of τ. The sum over poles does, however, converge after coarse graining with a suitably smooth $G(\tau)$.

More precisely, if the width of G is Δ, the number of terms required to converge the sum for the coarse grained $\Psi(x,t|\tau)$ is $\approx \Delta^{-1/2}$. Physically, this means that improving the accuracy of the measurement we will pick more and more terms in (7.5.6) and find more fine structure in the distribution of traversal times. The Gaussian coarse grained amplitude $\Phi_0(x,t|\tau)$ is shown in Fig. 7.4 for different values of β.

This remark equally applies to the case of tunnelling. The amplitude distribution $\Phi_V(x,t|\tau)$ for a rectangular barrier of height V differs from the that for a free particle only by the factor $\exp(-iV\tau/\hbar)$,

$$\Phi_V(x,t|\tau) = exp(-iV\tau/\hbar)\exp(ikx - iEt/\hbar)\eta_0(k,\tau) . \qquad (7.5.7)$$

This is because $\Phi_0(x,t|\tau)$ contains contributions only from those paths which spend τ inside $[a,b]$ (cf. (7.2.7)) and for any such path an additional

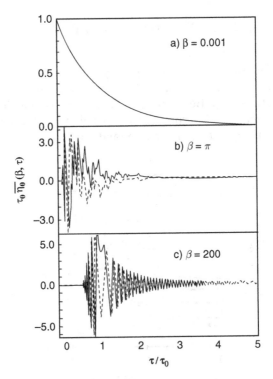

Fig. 7.4. Real (*solid*) and imaginary (*dashed*) parts of the free particle traversal time distribution after Gaussian coarse graining with $\Delta\tau/\tau_0 = 0.015$ for different values of $\beta \equiv k(b-a)$. In the semiclassical limit (**c**), note the stationary region near the classical value $\tau_0 = m(b-a)/\hbar k$

rectangular potential of height V modifies the Feynman amplitude $\exp(iS/\hbar)$ by $\exp(-iV\tau/\hbar)$.

Even though $|\Phi(x,t|\tau)|^2$ cannot be evaluated for any given τ we can estimate the probability $P(\Upsilon, \Upsilon + d\Upsilon)$ to find the readings of an infinitely accurate meter within the interval $[\Upsilon, \Upsilon + d\Upsilon]$. It can be demonstrated [14] that a high-resolution (ideal) measurement is distributed exponentially (the limits have to be taken in the given order)

$$P(\Upsilon, \Upsilon + d\Upsilon) \equiv lim_{d\Upsilon \to 0} lim_{\Delta \to 0} \int_{\Upsilon}^{\Upsilon + d\Upsilon} |\Psi_V(x,t|\tau)|^2 d\tau / \int_0^{\infty} |\Psi_V(x,t|\tau)|^2 d\tau$$

$$= 4\tau_0^{-1} \exp(-4\Upsilon/\tau_0) d\Upsilon \ , \tag{7.5.8}$$

where

$$\tau_0 \equiv m(b-a)/\hbar k \ , \tag{7.5.9}$$

is the time it takes a classical free particle with energy E to cross $[a, b]$. Notably, the result (7.5.8) does not depend on V and equally applies to both tunnelling and free motion. This is not surprising, since in the high-accuracy limit the original potential is overwhelmed by interaction with the clock (cf. (7.3.3)) and no longer matters.

7.6 Semiclassical Limit. How Long Does it Take for a Particle to Tunnel?

Quantally, the traversal time is a distributed quantity. A unique time is obtained in the classical limit when the size of the region $b - a$ exceeds the de Broglie wavelength of the particle

$$\beta \equiv k(b-a) \gg 1 \ . \tag{7.6.1}$$

To see how this happens, we write $T(k, W)$ in (7.5.5) as a geometric progression

$$T(k, W) = \sum_{n=0}^{\infty} T_n(k, W) \ , \tag{7.6.2}$$

$$T_n(k, W) \equiv \frac{-4iz\beta}{(z - i\beta)^2} \frac{(z + i\beta)^{2n}}{(z - i\beta)^{2n}} \exp[-(2n+1)z - i\beta/\hbar] \ . \tag{7.6.3}$$

In (7.6.2) the n-th term corresponds to a particle bouncing $(2n + 1)$ times between the discontinuities of the potential at $x = a$ and $x = b$ [29] and for free motion we only need $T_0(k, W)$. Inserting $T_0(k, W)$ into (7.5.4) and applying the stationary phase method, we find the phase of the integrand stationary at $W(\tau)$ defined by the classical relation

$$m(b - a)/\{2m[E - W(\tau)]\}^{1/2} = \tau \ . \qquad (7.6.4)$$

This means that the main contribution to (7.5.4) comes from a potential W such that it takes a classical trajectory exactly the time τ to cross $[a, b]$. For $\eta_0(k, \tau)$ we find

$$\eta_0(k, \tau) = -\frac{4\tau_0}{(\tau + \tau_0)^2} \left[\frac{\beta\tau_0}{2\pi\tau} \right]^{1/2} \exp\left[i\frac{\beta(\tau - \tau_0)^2}{2\tau\tau_0} - i\frac{\pi}{4} \right] \ . \qquad (7.6.5)$$

For large β the exponential in (7.6.5) oscillates very rapidly outside a narrow vicinity $\delta\tau \approx \beta^{1/2}\tau_0$ of the classical value τ_0 in (7.5). A coarse grained amplitude $\Psi_V(x, t|\tau)$ in (7.2.13) with $\Delta\tau \gg \delta\tau$ has the largest value when the maximum of G coincides with τ_0 and is cancelled elsewhere by the oscillations of $\Phi_0(x, t|\tau)$. As a result, the meter always finds the classical value and we conclude that a classical particle spends in $[a, b]$ duration τ_0.

The additional factor $\exp(-iV\tau/\hbar)$ which appears in the tunnelling time distribution (7.5.7) shifts the critical point of $\eta_0(k, \tau)$. In the case of semiclassical tunnelling, $V \gg E$, $\eta_V(k, \tau) = \exp(-iV\tau/\hbar)\eta_0(k, \tau)$ has two complex saddles

$$\tau_V = \pm i(b - a)m/[2m(V - E)]^{1/2} \ , \qquad (7.6.6)$$

and rapidly oscillates everywhere on the real τ-axis. There is, therefore, no unique real time associated with tunnelling. Rather, Feynman paths with all possible traversal times interfere destructively to produce exponentially small transmission amplitude [13]. The contour of integration in (7.2.12) can be deformed to run up the negatively imaginary τ-axis so that the lower saddle $-i|\tau_V|$ contributes to the transmitted part of the wave function. In this sense, the tunnelling particle "spends in the barrier imaginary time". (Note that the complex time of Sect. 7.3 is in this case purely imaginary, $\bar{\tau}(x) = -i|\tau_V|$.) All this does not however mean that semiclassical tunnelling takes approximately $|\tau_V|$ seconds. We may try to measure the tunnelling time in an experiment which for a free particle produces the result τ_0. Since τ_0 is real, it is possible that we measure either $Re\tau_0$ or $|\tau_0|$. In the tunnelling regime, however, the experiment of the first type would give us 0 and the experiment of the second type $|\tau_V|$. (For example, in Sect. 7.7 we will show that a Larmor clock can, in principle, determine $|\tau_V|$ while in Sect. 7.9 we will demonstrate that the complex saddle is responsible for what seems like infinitely fast transfer of a tunnelling wave packet across the barrier region.) Thus, the assumption that tunnelling has a well-defined duration similar to that of classical propagation would lead to a contradiction.

7.7 Larmor Clock as a Realistic Meter

Now we turn to the practical realisation of the traversal time meter described in Sect. 7.2. We recall that a clock would need to distinguish between Feynman paths with different values of τ, select those paths for which τ lies close to the measured value and discard the rest. Next we will show that such meter is, in fact, the Larmor clock often discussed in connection with the tunnelling time problem [5, 30, 31, 32, 33, 34, 35, 36, 37] and Chap. 8. The Larmor clock consist of a magnetic moment μ (proportional to the particle's spin or angular momentum) which travels with the particle and a constant magnetic field H created in the region of interest Ω. Inside the region the spin precesses around the direction of the filed with a constant angular velocity ω_L. Classically, the angle by which the spin rotates between, say, $t = 0$ and $t = T$ is proportional to the net duration the particle spends in Ω, $\Delta\phi = \omega_L t_\Omega^{cl}[x(t)]$. Quantally, this systems coarse grains the unfolded amplitude $\Phi(x, t|\tau)$ introduced in Sect. 2 [12, 14]. To demonstrate this, we note that the coupling between a spin an a magnetic field directed along the z-axis is given by

$$V_{int} = \omega_L \hat{j}_z \Theta_\Omega(x) , \qquad (7.7.1)$$

where $\hat{j}_z$ is the z-component of the spin. The wave function of the system $|\Upsilon\rangle$ is a $2j + 1$-component spinor, and the amplitude to find the particle in x and simultaneously observe the clock in $|\gamma^k\rangle$ at time t (we assume that $|\gamma^k\rangle, k = 0, 1, ..2j$ form a complete orthogonal set) is $A_k(x, t)\langle\gamma^k|\Upsilon\rangle$. If the magnetic field is switched on at $t = 0$ when the the particle state is $\Psi(x, 0)$ and the clock is prepared in $|\gamma^0\rangle$, i.e.,

$$A(x, 0) = \langle\gamma^k|\alpha\rangle\Psi(x, 0) , \qquad (7.7.2)$$

for $t > 0$ we find [12]

$$A(x, t) = \int_0^\infty d\tau F(\gamma^k|\gamma^0, j, \omega_L, \tau)\Phi(x, t|\tau) , \qquad (7.7.3)$$

where

$$F(\gamma^k|\gamma^0, j, \omega, \tau) \equiv \langle\gamma^k| \exp(-i\omega_L \hat{j}_z \tau/\hbar)|\gamma^0\rangle . \qquad (7.7.4)$$

Equation (7.7.3) looks similar to (7.2.12) and we only need to choose the states $|\gamma^k\rangle, k = 0, 1, ..2j$, the size of the spin j and the Larmor frequency ω_L to give the coarse graining function $F(\gamma^k|\gamma^0, j, \omega_L, \tau)$ a desired shape. It is convenient to choose $|\gamma^k\rangle, k = 0, 1, ..2j$ as follows [36, 37]. For $|\gamma^0\rangle$ we write $(|\gamma^k\rangle = \sum_{-j}^{j} \gamma_n^k|n\rangle, \hat{j}_z|n\rangle = n|n\rangle)$.

$$\gamma_n^0 = (2j + 1)^{-1/2} , \qquad (7.7.5)$$

and the remaining states are obtained by rotating $|\gamma^0\rangle$ by an angle $\phi_k \equiv 2\pi k/(2j + 1)$ around the direction of the field,

$$\gamma_n^k = (2j + 1)^{-1/2} \exp(-in\phi_k), k > 0 . \qquad (7.7.6)$$

(Note here the analogy with a stopwatch where the measured time is proportional to the angular displacement of the hand, Fig. 7.5.) With this choice we have

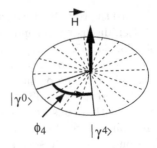

Fig. 7.5. The choice of the states $|\gamma^k\rangle$ for the Larmor clock

$$G(\tau_k - \tau) \equiv F(\gamma^k | \gamma^0, j, \omega_L, \tau) = (2j+1)^{-1} \frac{\sin[(j+1/2)\omega_L(\tau_k - \tau)]}{\sin[\omega(\tau_k - \tau)/2]} \ , \quad (7.7.7)$$

$$\tau_k \equiv \phi_k/\omega_L, k = 0, 1, ...2j \ .$$

As shown in Fig. 7.6, for large j and finite ω, the coarse graining function $G(\tau_k - \tau)$ is peaked around τ_k with the base width of $\Delta\tau \approx 4\pi/[\omega_L(2j + 1)]$. The Larmor clock is, therefore, similar to the meter of Sect. 7.2 except that its readings, τ_k like those of a digital watch [37], are discrete. It also provides a good illustration for the uncertainty principle (7.3.3) discussed in Sect. 7.2. Indeed, the potential energy (7.7.1) is such that if the projection of the spin on H is n, the particle experiences inside Ω an additional constant potential $\hbar n \omega_L \Theta_\Omega(x)$. The clock's initial state $|\gamma^0\rangle$ in (7.7.5) contains, with equal amplitudes, all $2j + 1$ values of n. The uncertainty in the potential is, therefore, $\Delta W = 2\hbar j \omega_L$ and the product of the two uncertainties, $\Delta W \Delta\tau$, is as it should, of the order of $\hbar$, $\Delta W \Delta\tau \approx 4\pi\hbar$.

Next we show in Figs. 7.7–7.10 the readings the clock would produce when applied to a free particle, transmission across a rectangular barrier and resonance tunnelling through a double barrier potential. In all cases (except

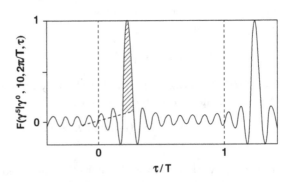

Fig. 7.6. Coarse graining function of a Larmor clock for $j = 10, k = 5, \omega_L = 2\pi/T$

in Fig. 7.10 where a better resolution is needed) we choose $j = 20$. We then plot the probability P_k to find the value τ_k for different values of the parameter $\beta \equiv (2mE)^{1/2}(b - a)$, $\Omega \equiv [a, b]$, essentially the ratio between the size of region and the de Broglie wavelength of the particle. For a free particle, in the classical limit $\beta \gg 1$ the clock measures the classical value $\tau_0 = (b - a)m^{1/2}/(2E)^{1/2}$. As the de Broglie wavelength becomes comparable to $b - a$, $\beta \approx 1$ the peak at τ_0 breaks into smaller multiple peaks and the readings spread. In the ultraquantum case $\beta \ll 1$, the readings are distributed exponentially around the origin. A similar plot for a rectangular barrier $V\Theta_{a,b}(x)$, $(2mV)^{1/2}(b - a)/\hbar = 150$ is shown in Fig. 7.7. For $\beta \gg 1$, we again find a single peak at the classical value on the trajectory passing above the barrier top. Near the barrier top, $E \approx V$, $\beta \approx 150$ the peak is almost destroyed as the transmission becomes dominated by the barrier top resonances [14]. Finally, in the tunnelling regime, $E \ll V$, $\beta \leq 150$ the clock measures a zero duration. We shall return to this result in the next section where we discuss the "superluminal" tunnelling.

Next we consider resonance tunnelling through a simple double barrier structure consisting of two δ-potentials,

$$V(x) = \hbar^2 \chi/m[\delta(x - a) + \delta(x - b)] . \tag{7.7.8}$$

For large value of χ the barriers are almost impenetrable and the potential (7.7.8) supports quasistationary states $E_i - i\Gamma_i$ which may trap the incident particle. The effect leads to a complete transparency of the structure when the particle's energy $E \approx E_i$. The resonance tunnelling is often described as accompanied by a long time delay as the particle bounces between the walls of the potential well. Both can be observed with the help of a Larmor clock. Figure 7.9 shows the readings of the clock with $\Delta\tau = 5\tau_0$ for E close

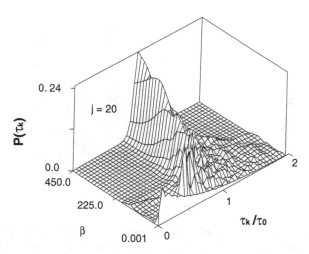

Fig. 7.7. Free particle traversal time distribution measured by a Larmor clock with accuracy $\Delta\tau/\tau_0 = 0.1$

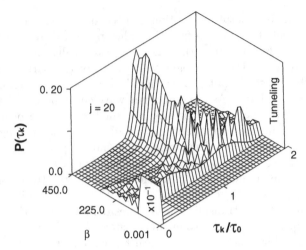

Fig. 7.8. Same as Fig. 7.7 except for a rectangular barrier such that $\beta < 150$ corresponds to tunnelling. In the tunnelling regime, the peak at $\tau_{k=0}$ is reduced by a factor of 10 to allow for better viewing

to the 32nd quasiationary level, $\beta = 32\pi$ whose position is indicated by a black triangle. One notices the long tails associated with resonance tunnelling (actually, the resonance tail is split into two tails on both sides of $\beta = 32\pi$). Further away from the resonance there are large peaks at $\tau \approx 0$, corresponding to "fast" off-resonance tunnelling. Improving resolution of the clock allows us to observe the multiple bounces mentioned above. In Fig. 7.9 a clock with

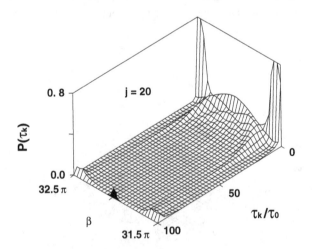

Fig. 7.9. Traversal time distribution for resonance tunnelling measured by a Larmor clock with accuracy $\Delta\tau/\tau_0 = 5$. Approximate position of the resonance is indicated by a black triangle

the resolution $\Delta\tau = 0.4\tau_0$ finds peaks near $\tau = \tau_0; 3\tau_0; 5\tau_0, ...$ corresponding to the particle crossing $[a, b]$ once, three times after two reflections, five times after four reflections, etc. Note that there is little change in the plot on passing through resonances. Thus the long tails at $E \approx E_i$ and the multiple bounces cannot be observed simultaneously – they require different resolutions.

Baz' was the first to propose the Larmor clock in order to measure the collision time [30, 31]. His approach, however, differed from the one just described and we conclude this section with its brief analysis. Baz' required the magnetic field to be small, $\omega_L \to 0$ so as not to perturb the particle. Initially, the spin is polarised along the x-axis. For a small field, precession of the spin in the xy-plane is defined as $\phi_\perp = \langle \hat{j}_y \rangle / \hbar j + O(\omega_L^2)$ and the traversal time is found as the ratio between ϕ and ω_L. As we have seen in Sect. 7.4 an attempt to minimise the effect of the meter results, in general, to a weak measurement. Thus we expect the Baz' clock to measure, in one way or another, the complex time $\bar{\tau}$ given by (7.4.5). To demonstrate this, we note first that if initially the spin is in the state $|\alpha\rangle$ polarised along the x-axis,

$$\alpha_n = [(2j)!/2^{2j}(j+n)!(j-n)!]^{1/2} , \qquad (7.7.9)$$

the final amplitude A_n to find the clock in the state $|n\rangle$ is given by (7.7.4) with

$$F^{Baz}(n|\alpha, j, \omega_L, \tau) = \alpha_n[1 - in\omega_L\tau + O(\omega_L^2)] . \qquad (7.7.10)$$

Using A_n to evaluate the mean value of $\hat{j}_y$ at some location x yields

$$\langle \hat{j}_y \rangle / \hbar j = \omega_L Re\bar{\tau}(x) + O(\omega_L^2) . \qquad (7.7.11)$$

What is more, the spin also acquires a non-zero component along the z-axis [8], proportional to $\bar{\tau}(x)$ (Fig. 7.11)

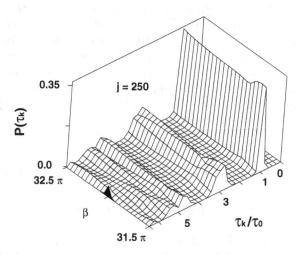

Fig. 7.10. Same as Fig. 7.9 except for $\Delta\tau/\tau_0 = 0.4$

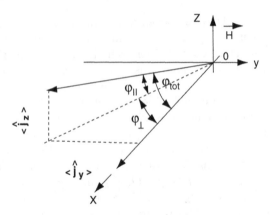

Fig. 7.11. After interacting with a weak magnetic field the spin, initially polarised along the x-axis appears to be rotated both in the xy- and xz-planes. The total rotation angle is proportional to the modulus of the complex traversal time $\bar{\tau}$

$$\langle \hat{j}_z \rangle / \hbar j = \omega_L Im \bar{\tau}(x) + O(\omega_L^2) \,. \tag{7.7.12}$$

Therefore, the Baz' clock operates in the weak measurement regime and allows one to determine $Re\bar{\tau}(x)$, $Im\bar{\tau}(x)$ or, if the total angle of rotation is considered, $|\bar{\tau}(x)|$.

7.8 Traversal Time Analysis

In classical mechanics, the amount of time a particle spends in the interaction region can be used to predict how the interaction affects its motion. Next, we ask whether the quantum traversal time, which is a distributed quantity, can be used in a similar way, e.g., to analyse tunnelling across time-dependent barriers or the effect of surface plasmons responsible for the image forces. The short answer to this question is no. The usefulness of the quantum traversal time is determined by its construction. To obtain the free particle fine grained amplitude $\Phi_0(x, t|\tau)$ we have sorted all Feynman paths according to the value of τ and evaluated the restricted path integral for each τ. Now, if an interaction adds to the free particle action $S_0[x(t)]$ a term which can be expressed via $\tau_\Omega^{cl}[x(t)]$, i.e., ($F(z)$ is an arbitrary function)

$$S_0[x(t)] \rightarrow S_0[x(t)] + F(\tau_\Omega^{cl}[x(t)]) \,, \tag{7.8.1}$$

we may write the Schrödinger wave function $\Psi(x, t)$ as a simple quadrature

$$\Psi(x, t) = \int \exp[iF(\tau)/\hbar]\Phi_0(x, t|\tau)d\tau \,. \tag{7.8.2}$$

(We have already used the property (7.8.2) (with $F(\tau_\Omega^{cl}[x(t)]) = \tau_\Omega^{cl}[x(t)]$ in (7.5.7) while discussing tunnelling across a rectangular barrier.) Unfortunately, neither interaction with a time-dependent barrier nor with the plasmons is of the required type and in each case a similar analysis would require a functional other than $\tau_\Omega^{cl}[x(t)]$.

We find the usefulness of the quantum traversal time greatly reduced. The fact that the integrand of (7.2.12) is oscillatory deserves further consideration. As an example, consider a particle with energy E which tunnels across a rectangular barrier $V\Theta_{ab}(x)$ and is coupled, while in the barrier, to a single harmonic oscillator with mass M, angular frequency ω and coordinate q [13, 14]. If the interaction has the form $\lambda\Theta_{ab}(x)q$, one finds the amplitude for a transition leaving the oscillator in its original ground state given by a path integral with the effective action [7]

$$S[x(t)] = S_0[x(t)] - V\tau_\Omega^{cl}[x(t)]$$
$$+2i\alpha^2 \int_0^t dt' \int_0^t dt'' \Theta_{ab}(x(t'))\Theta_{ab}(x(t'')) \exp[-i\omega(t' - t'')] ,$$

(7.8.3)

where $\alpha \equiv \lambda(4M\omega)^{-1/2}$. If the oscillator is very slow, $\omega \to 0$, the last term in (7.8.3) becomes $\alpha^2\tau_\Omega^{cl}[x(t)]^2$ and we can write the transmission amplitude as

$$T(k, V, \alpha) = \int_0^\infty \exp[-\alpha^2\tau^2/\hbar , -iV\tau/\hbar]\eta_0(k, \tau)d\tau \qquad (7.8.4)$$

where $\eta_0(k, \tau)$ is given by (7.5.4). Consider now how a slow oscillator affects the transmission probability $P(k, V, \alpha) \equiv |T(k, V, \alpha)|^2$. One might think that adding yet more interaction to a rectangular barrier would reduce already small tunnelling probability further. This is, however, not the case. The ratio $P(k, V, \alpha)/P(k, V, 0)$ plotted in Fig. 7.12 as a function of the coupling strength α shows initial rapid rise and then a more gradual decrease.

To understand this behaviour, we recall that for $\alpha = 0$ the tunnelling probability in (7.8.4) is exponentially small due to the cancellation between the negative and positive oscillations of the integrand, which itself is not small. Adding $\exp[-\alpha^2\tau^2]$ to the integrand at first destroys the cancellation and boosts transmission. A stronger coupling, however, restricts the integration range and, eventually, quenches tunnelling.

To conclude, we consider the time(s) associated with the transition in the semiclassical limit $\hbar \to 0$. These are given by the saddle points of the integrand in (7.8.4). With the help of (7.8.3) and (7.8.4) we find the saddles determined by a cubic equation

$$i\tilde{\alpha}\tau^3 - \tau^2 - |\tau_V|^2 = 0 , \qquad (7.8.5)$$

where $\tilde{\alpha} = 2\alpha^2/(V - E)^{1/2}$ and τ_V is the complex tunnelling time defined in (7.6.6). Without coupling, $\alpha = 0$, we have the two imaginary saddles $\pm i|\tau_V|$ discussed in Sect. 7.5. As the coupling increases, the upper saddle $\tau^{(1)}$ moves towards the origin. The lower saddle $\tau^{(2)}$ moves down the negative imaginary τ-axis until at some $\alpha = \alpha_{crit}$ it meets with the third root of (7.8.5), $\tau^{(3)}$ rising

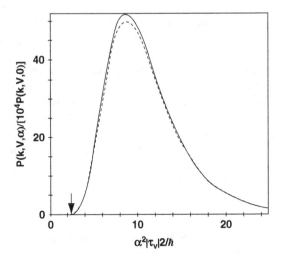

Fig. 7.12. The ratio of the transmission probabilities for a an opaque rectangular with and without coupling to a slow oscillatory mode as a function of the coupling strength α (*solid*). The *dashed line* is the saddle point result. An arrow indicates the value of α where two complex saddles coalesce

up the same axis from $-\infty$. After this, the two saddles leave the imaginary axis and describe loops in the third and the fourth quadrants of the complex τ-plane as shown in Fig. 7.13.

As $\alpha \to \infty$ all three saddles meet at the origin. A detailed analysis shows that for $\alpha \geq \alpha_{crit}$ the positive real axis in (7.8.4) can be deformed into a steepest descent contour involving only $\tau^{(3)}$ in the fourth quadrant. Thus we conclude that a tunnelling particle coupled to a low-frequency oscillator "spends in the barrier a complex duration $Re\tau^{(3)} + iIm\tau^{(3)}$" in the same sense that without coupling "tunnelling takes an imaginary time $-|\tau_V|$". The same analysis shows that for $\alpha \leq \alpha_{crit}$ both $\tau^{(2)}$ and $\tau^{(3)}$ contribute to the integral. The case $\alpha \approx \alpha_{crit}$ requires a uniform semiclassical treatment which we will not discuss here. The steepest descent approximation to $T(k, V, \alpha)$ is shown in Fig. 7.12 by a dashed line.

7.9 Traversal Time and the "Superluminal" Tunnelling

Can the traversal time be used, as is often suggested, to understand the speed at which a wave packet propagates across a classically forbidden region and, in particular, the phenomenon of "superluminal tunnelling"? The latter can be described as follows. Suppose we send a wave packet across a potential barrier of height V and width d. If the mean energy of the wave packet E_0 lies well above the barrier, the particle behaves classically. Therefore, the higher the barrier, the longer it takes the particle to cross it and, eventually, to arrive at a given location on the other side. Equivalently, a snapshot at a given time

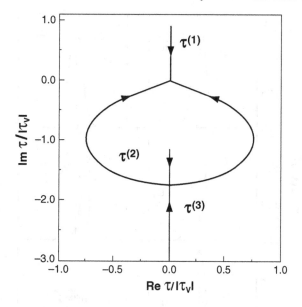

Fig. 7.13. Trajectories of the semiclassical complex times for an opaque rectangular barrier and a slow oscillatory mode as the coupling strength α increases from 0 to ∞. τ_V is the imaginary time for tunnelling across the barrier only

will always find the delayed particle behind the one that propagates freely (Fig. 7.14). If V is increased until $V > E$, a classical particle will bounce off the face of the barrier and reverse its motion. Quantally, a small part of the wave packet tunnels and a snapshot in Fig. 7.14 will find it approximately a distance d ahead of the free particle suggesting that a tunnelling particle spends almost no time in the barrier. It is tempting then to conclude that its velocity exceeds in the barrier the speed of light, hence the appearance of the term "superluminal" in the heading of this section.

All this is easily illustrated in the semiclassical limit, $\hbar \to 0$. Let the initial wave packet have a mean momentum p_0

$$\Psi(x,0) = \exp(ip_0 x/\hbar)G(x) = \int_{-\infty}^{\infty} A(p-p_0)\exp(ipx/\hbar)\,dp\,, \qquad (7.9.1)$$

where $G(x)$ is a suitable, e.g., Gaussian,

$$G([x+a]/\Delta x) = \exp[-(x+a)^2/\Delta x^2]\,, \quad a < 0\,, \qquad (7.9.2)$$

envelope, independent of $\hbar$. After scattering, its transmitted part becomes

$$\Psi(x,t|V) = \int_{-\infty}^{\infty} A(p-p_0)T(p,V)\exp[i(px-Et)/\hbar]dp\,, \qquad (7.9.3)$$

where $E(p) = p^2/2m$ and $T(p,V)$ is the transmission amplitude. Well above, $E \gg V$, and well below the barrier, $E \ll V$, we can replace $T(p,V)$ by the

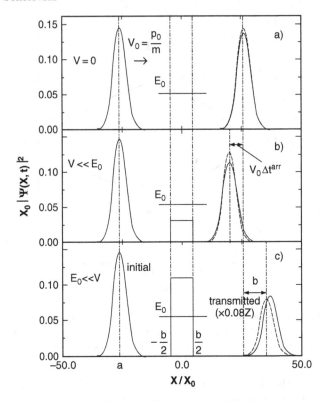

Fig. 7.14. Snapshot of a Gaussian wave packet with the mean momentum p_0, $p_0 d/\hbar = 90.25$ taken for $t/\tau_0 = 5.5$. The wave packet passing over the barrier (**b**) is delayed relative to the free one (**a**). The tunnelling wave packet (**c**) is very small (note the large factor of $Z = \exp[2(2mV - p_0)^{1/2}b/\hbar]$) but lies ahead of the free one. The dashed line is the semiclassical result. Other parameters are: $x_0 \equiv (\hbar b/p_0)^{1/2}$, $a/x_0 = 26$, $b/x_0 = 9.5$ and $\Delta x/x_0 = 5.5$

first term in the multiple reflection expansion (7.6.2)

$$T(p, V) \approx T_0(p, V) . \tag{7.9.4}$$

As $\hbar \to 0$, the width of the momentum distribution $A(p - p_0)$ decreases as $\hbar^{-1}$ and we expand the last two exponents in (7.9.3) around p_0 to first order in $p - p_0$ which yields

$$\Psi(x, t|V) = \exp[i(p_0 x - E(p_0)t)/\hbar]T_0(p_0, V)G(x - \delta x) , \tag{7.9.5}$$

where

$$\delta x(V) \equiv v_0 t - d[1 - p_0/(p_0^2 - 2mV)^{1/2}] . \tag{7.9.6}$$

By the time t a free wave packet travels the distance $v_0 t$. A wave packet passing above the barrier, $p_0^2 - 2mV \geq 0$ is delayed and lies a distance $d[1 -$

$p_0/(p_0^2 - 2mV)^{1/2}]$ behind. In the tunnelling regime, $p_0^2 - 2mV \ll 0$, the last term in (7.9.6) is imaginary, and the maximum of $|G(x - \delta x)|^2$ in (7.9.5) lies a distance d ahead of the freely propagating peak and the particle reaches a detector approximately d/v_0 earlier then a free one. Further discussion of superluminality can be found in [38].

We, however, are interested in the relation between the distribution of the times spent in the barrier and the final position (and shape) of the transmitted wave packet. In the classical limit there is a simple relation between τ and the distance travelled by the particle, δx

$$\delta x \equiv v_0(t - \tau) + d , \qquad (7.9.7)$$

so that finding the particle in x we also know the value of τ. A similar classical relation exists between τ and the angle by which a spin rotates in the magnetic field. By analogy, we may expect that quantally the amplitude to find the particle in x can be obtained by coarse graining the traversal time distribution. This turns out to be the case, but only in the semiclassical limit.

The transmitted wave packet can be constructed by summing the traversal time distribution over all τ. For a rectangular barrier, we have the identity

$$\Psi(x,t|V) = (2\pi\hbar)^{-1} \int_0^\infty d\tau \exp(-iV\tau/\hbar) \int_{-\infty}^\infty dW \exp(iW\tau/\hbar)\Psi(x,t|W) . \qquad (7.9.8)$$

where $\Psi(x,t|W)$ is the transmitted wave packet for the potential $W\Theta_{-d/2,d/2}(x)$ as given by (7.9.6). Again, for $T(p,W)$ it is convenient to use the multiple reflection expansion retaining only the first term. We then use (7.9.5) and evaluate the second integral in (7.9.8) by the stationary phase method. The stationary phase condition reads

$$md/(p_0^2 - 2mW(\tau))^{1/2} = \tau , \qquad (7.9.9)$$

and the result is

$$\Psi(x,t|V) = \int_0^\infty G(-[\bar{\tau}(x) - \tau]/\Delta\tau])\Phi_V(x,t,p_0|\tau)d\tau , \qquad (7.9.10)$$

where

$$\bar{\tau}(x) \equiv t - (\delta x - d)/v_0 , \qquad (7.9.11)$$

$$\Delta\tau \equiv \Delta x/v_0 , \qquad (7.9.12)$$

and

$$\Phi_V(x,t,p_0|\tau) = \exp(-iV\tau/\hbar)\Phi_0(x,t,p_0|\tau) , \qquad (7.9.13)$$

is the traversal time distribution for a particle with energy $E(p_0)$ in a potential $V\Theta_{-d/2,d/2}(x)$ defined earlier in (7.5.3) and (7.6.5). We note again that for any value of τ, $0 \le \tau \le \infty$ (7.9.9) finds a rectangular potential $\infty \le W(\tau) \le E$ (a barrier or a well) such that a classical trajectory with the initial momentum

p_0 spends inside $[-d/2, d/2]$ exactly τ. (No contradiction with the uncertainty principle of Sect. 7.3 as we consider $\hbar \to 0$.) Along this trajectory, the envelope G is translated as a whole and we obtain (7.9.10). The right-hand side of (7.9.10) has the form of a coarse grained amplitude (7.2.13) with the coarse graining function G determined by the envelope of the initial wave packet. Thus, constructing a broad initial wave packet and then finding a particle at a location x we perform a traversal time measurement on a particle with the energy $E(p_0)$. Note that in this case the system "measures itself" as we require no external degree of freedom to represent a meter.

Next, to find the position $\bar{x}$ of the peak of the transmitted probability $\Psi(x, t|V)$ we first look the most probable value of the traversal time measured with the coarse graining function $G([\bar{\tau}(x) - \tau]/\Delta\tau)$ $\bar{\tau}$ and then obtain $\bar{x}$ from the classical relation (7.9.7). For a free particle, $\Phi_0(x, t, p_0|\tau)$ has a stationary region near $\tau_0 = d/v_0$. As discussed at the end of Sect. 7.6, the integral (7.9.10) peaks when G is centred at τ_0 and the peak of the wave packet is at $\bar{x} = a + v_0 t$ (Fig. 7.14). For a finite barrier $E \gg V$, the stationary region of $\Phi_V(x, t, p_0|\tau)$ shifts to larger τ and the wave packet is delayed. For tunnelling, $V > E$, $\Phi_V(x, t, p_0|\tau)$ has no real stationary point, but (7.9.10) still applies and the integral (7.9.10) can be evaluated by the steepest descent method which yields the result (7.9.6) (Fig. 7.15).

The envelope G is evaluated at the complex saddle (7.6.6), $-idm/(2mV - p_0^2)^{1/2}$ and the right-hand side of (7.9.10) takes the largest value when G is positioned closest to the saddle, i.e., at $\bar{\tau} = 0$ (Fig. 7.14). Since $G([\bar{\tau}(x) - \tau]/\Delta\tau)$ restricts integration in (7.9.10) to small τ's, we may conclude that the trajectories which most contribute to the tunnelling experience in the barrier region a very deep potential well. As a result, the transmission is very fast, but since the paths interfere destructively, the transmitted pulse is typically exponentially small. We stress again that this analysis is valid only in the semiclassical limit $\hbar \to 0$ and, in general, the relation between the traversal time distribution and the shape of the Schrödinger wave function is limited to (7.2.12). One drawback of our approach is that it seems to rely on the classical trajectories which cross Ω faster than light. Such trajectories are forbidden by special relativity and appear only because we have employed the non-relativistic Schrödinger equation. The latter is sufficient to describe the propagation of a wave packet at the low energies we consider. Yet in the traversal time analysis we probe regimes which should be described by a relativistic wave equation. This will be done in the next section.

7.10 Relativistic Traversal Time

An obvious way to deal with the superluminal velocities we have encountered in the previous section is to replace in our analysis the Schrödinger equation by one of the relativistic wave equations [39]. Since we are only interested in the mechanism of tunnelling, we may choose the simplest one for the spinless particles, i.e., the Klein–Gordon equation [39, 40, 41]. In the relativistic notations the Klein–Gordon equation for a particle of mass m and charge e interacting with an electromagnetic field $A \equiv (A_0, \boldsymbol{A})$ reads $(\hbar = 1, c = 1)$

$$(i\partial/\partial x_\mu - eA_\mu)^2 \Psi = m^2 \Psi . \tag{7.10.1}$$

The propagator $G(x, x')$ of (7.10.1) can be written in the path integral form by considering $x_\mu(u)$ as function of the "fifth parameter" u [40, 41],

$$G(x, x') = \int_0^\infty du \exp(-im^2 u/2) \int Dx \exp[-i \int_0^u \mathcal{L}(x(u'))du'] , \tag{7.10.2}$$

with the Lagrangian

$$\mathcal{L}(x) \equiv (dx^\mu/du)^2/2 + eA_\mu dx^\mu/du . \tag{7.10.3}$$

With the help of an arbitrary field $B_\mu(x)$ we may now construct a (scalar) functional

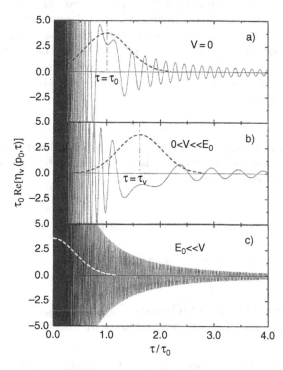

Fig. 7.15. Real part of the traversal time distribution for a particle with initial momentum p_0 for (**a**) free motion; (**b**) passage above the barrier and (**c**) tunnelling. Also shown is the Gaussian coarse grained function (*dashed*) determined by the wave packet's envelope and centred so as to maximise the integral (7.9.10). All parameters are as in Fig. 7.14

$$F[x(u)] = \int du B_\mu(x) dx^\mu / du , \tag{7.10.4}$$

which, for a particular choice

$$B_0 = \theta_\Omega(\boldsymbol{x}), B_1 = B_2 = B_3 = 0 , \tag{7.10.5}$$

coincides with the traversal time $\int \theta_\Omega(\boldsymbol{x}(t)) dt$. Next we use the delta function $\delta(F[x(u)] - \tau)$ to restrict integration in (7.10.2) only to those paths in $(t, \boldsymbol{x})$ which spend in Ω a duration τ. Proceeding exactly as in Sect. 7.2, we arrive at the "clocked" Klein–Gordon equation obtained from (7.10.1) by replacing $\partial/\partial t \to \partial/\partial t + \theta_\Omega(x)\partial/\partial \tau$. In particular, for a one-dimensional particle in a potential $V(x)$ we have

$$\left\{ c^{-2} \left[i\hbar \frac{\partial}{\partial t} + i\hbar\theta_\Omega(x)\frac{\partial}{\partial \tau} - V(x) \right]^2 + \hbar^2 \frac{\partial^2}{\partial x^2} \right\} \Phi_0(x, t|\tau) = m^2 c^4 \Phi_0(x, t|\tau) .$$
$$\tag{7.10.6}$$

We recall here that the paths in (7.10.2) are no longer required to proceed forward in time and may reverse themselves [40, 41]. The points at which these reversals occur are associated with the creation and annihilation of virtual particle–antiparticle pairs. Accordingly, we can no longer expect $\Phi_0(x, t|\tau)$ to vanish for $\tau < 0$ where it may describe antiparticles propagating back in time.

To find the traversal time distribution for a free wave packet we choose the initial condition (7.9.1). To exclude antiparticles we must require, in addition, that each plane wave component in (7.9.1) propagates with a positive energy (c is the speed of light)

$$\epsilon(p) = c(p^2 + m^2 c^2)^{1/2} . \tag{7.10.7}$$

We then continue as in previous section. The problem reduces to evaluation of the Fourier transform of the transmission coefficient $T_0(p, z)$ (7.6.2) with z given by the relativistic relation

$$z = [(\epsilon - V)^2 - m^2 c^4]^{1/2} d/\hbar . \tag{7.10.8}$$

In the limit $\hbar \to 0$, we obtain the analogue of (7.9.10)

$$\Psi(x, t|V) = \int_{-\infty}^{\infty} d\tau G(-[\bar{\tau}(x) - \tau]/\Delta\tau]) \Phi_V^{rel}(x, t, p_0|\tau) , \tag{7.10.9}$$

with

$$\bar{\tau}(x) \equiv t - (\delta x - d)/v_0, \quad v_0 = d\epsilon(p_0)/dp . \tag{7.10.10}$$

For a particle with initial momentum p_0 relativistic traversal time distribution is given by

$$\Phi_V^{rel}(x, t, p_0|\tau) = \exp[-i\epsilon(p_0)t/\hbar + ip_0 x/\hbar - ip_0 b/\hbar]$$
$$\times (2\pi\hbar)^{-1/2} \frac{4p_0 d\epsilon_0^{3/2} \tau_c^3}{(\tau - \tau_c)^{1/4}[p_0 d(\tau^2 - \tau_c^2)^{1/2} + \epsilon_0 \tau_c^2]^2}$$
$$\times \exp\{\frac{i}{\hbar}[\epsilon(p_0)\tau - V\tau - \epsilon_0(\tau^2 - \tau_c^2)^{1/2}] - i\pi/4\} , \tag{7.10.11}$$

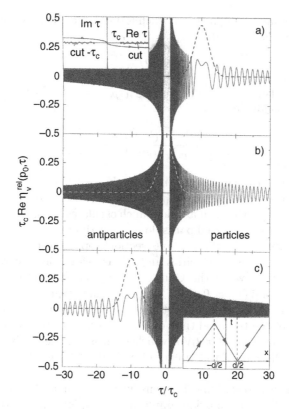

Fig. 7.16. Real part of the traversal time distribution for a relativistic particle with initial momentum p_0 for (**a**) free motion, (**b**) tunnelling and (**c**) transmission across the barrier with $V = 2\epsilon$. The Gaussian coarse grained function (dashed) is centred so as to maximise the integral (7.10.9). The contour of integration in (7.10.9) is shown in the inset in (**a**). Inset in (**c**) shows the time-reversed trajectory contributing to transmission.

where

$$\tau_c \equiv d/c, \epsilon_0 \equiv mc^2 . \tag{7.10.12}$$

The contour of integration in (7.10.11) is shown in the inset in Fig. 7.16, where the branch of the square root is chosen so that on the lower side of the right-hand cut $(\tau^2 - \tau_c^2)^{1/2}$ is real and positive.

A free particle distribution, $\eta_0^{rel}(p_0, \tau) \equiv \exp\{\frac{i}{\hbar}[\epsilon(p_0)t - p_0(x - b)]\}$ $\times(\Phi_0^{rel}(x, t, p_0|\tau)$ is shown in Fig. 7.16a. For $\tau \gg \tau_c$ $\Phi_0^{rel}(x, t, p_0|\tau)$ coincides with the non-relativistic distribution $\Phi_0(x, t, p_0|\tau)$ of the previous Section. No relativistic particle can traverse a distance d in less than d/c. Accordingly, for $-\tau_c \leq \tau \leq \tau_c$ $\Phi_0^{rel}(x, t, p_0|\tau)$ is exponentially small. For $\tau < \tau_c$, $\Phi_0^{rel}(x, t, p_0|\tau)$ contains an antiparticle branch associated with paths which cross Ω with time direction reversed. For the free motion, this branch is highly oscillatory and does not contribute to propagation which is dominated by τ's in the stationary region near the classical value

$$\tau_0 = d/v_0 \ .$$

We are interested in the tunnelling regime $(\epsilon(p_0) - V)^2 < \epsilon_0^2$ when $\Phi_V^{rel}(x, t, p_0 | \tau)$ has a complex saddle at

$$\tau_V = -i \frac{d(V - \epsilon(p_0))}{c[\epsilon_0^2 - (\epsilon(p_0) - V)^2]^{1/2}} \ .$$

For a non-relativistic particle $p_0^2/2m \ll \epsilon_0, V - p_0^2/2m \ll \epsilon_0$, τ_V reduces to $\approx -imd/[2m(V - E(p_0))]^{1/2}$ and $\Phi_V^{rel}(x, t, p_0 | \tau)$ is oscillatory both for $\tau > \tau_c$ and $\tau < \tau_c$. However, the antiparticle branch oscillates much more rapidly and, as expected, the time-reversed paths do not contribute to tunnelling. We note next that the analysis of the previous section applies even in the absence of the superluminal classical velocities. Indeed, as before, the integral in (7.9.10) has maximum value when the Gaussian G is placed opposite the complex saddle τ_V, i.e., for $\bar{\tau}(x) = 0$ (Fig. 7.16). The fact that $|\Psi(x, t|V)|^2$ has the largest value when the maximum of the coarse graining function coincides with the minimum of the traversal time distribution may at first seem surprising. We note, however, that $\Psi(x, t|V)$, which itself is exponentially small, builds from many oscillations of the integrand in (7.10.9) across the whole range of (positive) τ's.

It is interesting to see what happens when the barrier height becomes extremely large. The transmission amplitude (either full $T(p, V)$) or $T_0(p, V)$ depends on the potential only through $(\epsilon - V)^2$. This means, in particular, that $T(p, V)$ for $V = 2\epsilon$ is the same as for a free particle with energy ϵ, i.e., that a very high barrier is transparent. The τ-distribution for this case is shown in Fig. 7.16. Its particle branch is highly oscillatory and transmission is dominated by the vicinity of the stationary point at $-\tau_0$. Accordingly, the transmitted wave packet appears to spend in the barrier a negative amount of time and in the snapshot in Fig. 7.14 would be placed ahead of the tunnelled pulse. It is clear that the trajectory which contributes to transmission must experience in the barrier a time reversal (see inset in Fig. 7.16c). This effect is consistent with the following mechanism: as the particle approaches the barrier a particle–antiparticle pair is created at its right-hand side. The particle continues in the forward direction while the antiparticle, for which the potential energy has the opposite sign, travels to the left face of the barrier where it annihilates the incoming particle. The newly created particle has, therefore, a $2d/v_0$ head start on the original one. Note that no superluminal velocities are required to achieve this speedup. Unfortunately, this effect is an artefact of our one particle model and is unlikely to be observed. The reason for this is that the required barrier potential is strong enough to create real particle–antiparticle pairs. The antiparticles trapped in the inverted barrier potential will reduce its magnitude just is pair creation screens a large charge

[42]. In practice, it is impossible therefore to create a rectangular potential with a magnitude exceeding $2mc^2$. Together with the uncertainty principle (7.3.3), this suggests a limit of $\hbar/mc^2$ on the accuracy with which the quantum traversal time can be measured in a relativistic theory.

7.11 "Superluminal" Paradox and the Speed of Information Transfer

Next we follow [43] in order to further develop some of the conclusions of Sects. 7.9 and 7.10. Using a somewhat different definition of the time delay experienced by a particle in the scatterer and the analogy with Aharonov's weak measurements, we will demonstrate that "superluminality" occurs when the value of the duration spent in the barrier is uncertain, whereas when it is known accurately, no "superluminal" behaviour is observed. Using the same analogy, we will also show that the information encoded in non-analytic features (e.g., cut-offs) of an electromagnetic pulse cannot be transferred faster than light even across a fast light medium, where the group velocity of the pulse exceeds c.

As a prerequisite, we require a brief recourse to the Aharonov's weak measurements, already mentioned is Sect. 7.4 and fully described by the authors in Chap. 13 of this book. Consider a quantum system prepared at $t = 0$ in an initial state $|I >$ and then post-selected (observed) at $t = T$ in some final state $|F >$,

$$|I >= \sum_{\nu} a_\nu |\nu > \quad |F >= \sum_{\nu} b_\nu |\nu >, \quad \hat{A}|\nu >= A_\nu |\nu > . \qquad (7.11.13)$$

If at some t, $0 < t < T$, the system is subjected to a von Neumann-type measurement [23] of an operator $\hat{A}$, the state of the pointer with position τ after post-selection is given by [24, 25]

$$< \tau|M >= \sum_{\nu} G(\tau - A_\nu)\eta_\nu \qquad \eta_\nu \equiv b_\nu^* a_\nu \qquad (7.11.14)$$

where $G(\tau)$ is the initial, say, Gaussian state of the meter, whose width $\Delta\tau$ determines the accuracy of the measurement. The outcome of the measurement depends on the initial uncertainty in the pointer position, $\Delta\tau$. For $\Delta\tau$ sufficiently small

$$\Delta\tau << A_{\nu+1} - A_\nu,$$

the possible readings coincide with the eigenvalues of $\hat{A}$, which we will assume to be all non-negative and equally spaced, $A_\nu \geq 0$, For $\Delta\tau > A_{\nu+1} - A_\nu$, the situation is different. If η_ν were all positive, like the classical probabilities, $< \tau|M >$ would occupy the region $\tau \geq 0$ and rapidly decrease for $\tau < 0$.

However, η_ν are probability amplitudes, rather than probabilities, and may alternate. A careful choice of η_ν (i.e., of $|I>$ and $|F>$) can, therefore, produce a $< \tau|M >$ which would peak at some negative "anomalous" value $\tau_0 < 0$, while for $\tau > 0$ the pointer wave function will vanish due [24, 25] to the destructive interference between the Gaussians. Aharonov and co-workers have assesed that such anomalous outcome is possible although unlikely.

Consider next a one-dimensional wave packet with a central momentum k_0, transmitted across a scatterer (e.g., potential barrier, undersized waveguide, a slab of fast light material) whose width is b. For some large $t = T$ the transmitted pulse has the form

$$\Psi_F(x) \equiv < x|F >= \int_{-\infty}^{\infty} T(k)C(k - k_0)exp[ikx - iE(k)\mathrm{T}]dk, \quad (7.11.15)$$

where $T(k)$ is the transmission amplitude. Using the convolution property of the Fourier integral, (7.11.15) may be rewritted as

$$< x|F >= \int \Psi_\mathrm{T}(x - x')\xi(x')dx' \quad (7.11.16)$$

where

$$\Psi_\mathrm{T}(x) = \int_{-\infty}^{\infty} C(k - k_0)exp[ikx - iE(k)\mathrm{T}]dk, \quad (7.11.17)$$

is the pulse obtained from the initial state under the free evolution, and

$$\xi(x) \equiv (2\pi)^{-1} \int T(k) \exp[ikx]dk. \quad (7.11.18)$$

The transmitted pulse is now given by a weighted sum of the shapes $\Psi_\mathrm{T}(x-x')$ delayed, if $x' < 0$, or advanced if $x' > 0$, relative to the free propagation. Closing in (7.11.18), the contour of integration into the upper half-plane shows that

$$\xi(x) \equiv 0 \quad for \quad x > 0 \quad (7.11.19)$$

provided $T(k)$ does not have poles with positive imaginary parts. Such poles may be located only on the positive imaginary axis [42], where they correspond to bound states of the particle or field and without them the transmitted pulse builds up only from Ψ_T's *delayed* relative to the free pulse. This is not suprising, as in classical mechanics a particle only accelerates if passing over a potential well, which, in the quantum case, is the source of bound states.

The argument is readily rephrased in terms of scattering times by identifying

$$\tau(x') \equiv -x'/v_0 \quad (7.11.20)$$

with the time delay experienced by the particle in the scatterer. Replacing x' with τ yields

$$\Psi_F = \exp[ik_0x - iE(k_0)T] \int G(\tau(x) - \tau)\eta(\tau)d\tau \qquad (7.11.21)$$

where G is the envelope of Ψ_T,

$$G(\tau) \equiv \exp[iE(k_0)T + ik_0v_0\tau]\,\Psi_T(-v_0\tau) \qquad (7.11.22)$$

and

$$\eta(\tau) \equiv -v_0 \exp[ik_0v_0\tau]\xi(-v_0\tau). \qquad (7.11.23)$$

Comparing (7.11.21) with (7.11.14) shows that the relation between the time delay τ and the particle's position x is that between a measured quantity, whose amplitude distribution is $\eta(\tau)$ and the position of the pointer, whose initial state is determined by the envelope of the initial pulse $\Delta\tau$. This is not entirely unexpected, since we use the particle's position just like a pointer, by reading which we hope to gain information about the time delay τ. For a classical free particle, we recover the one-to-one correspondence betwen x' and τ by choosing a narrow initial wave packet, thus making the "measurement" more accurate, or "strong". Importantly, this cannot be done in the case of tunnelling. Indeed, making the initial pulse narrow in the position space will broaden its momentum distribution and eventually cause some momenta pass over the barrier rather than tunnel, thereby destroying the "superluminal" effect. For this reason, our "measurement" of the time delay of a tunnelling particle is inevitably "weak" and its uncertainty is inevitably large. Thus, the emergence of an apparently "superluminal" pulse in Fig. 7.14 with an estimated time delay of $-b/v_0$ when

$$\eta(\tau) \equiv 0 \quad \tau < 0 \qquad (7.11.24)$$

is completely analogous to obtaining an anomalous negative reading in an Aharonov's measurement of a variable with only non-negative eigenvalues, briefly discussed above. It is this large uncertainty in the value of the time delay which allows the negative result $-b/v_0$ to co-exist with the causality implied by (7.11.24), e.g., in the case of tunnelling across a potential barrier.

In Sect. 7.10 we have already demonstrated that the "superluminal" tunnelling does not contradict special relativity. Next, we will show that even under apparently "superluminal" conditions the information, encoded in a non-analytic feature, such as a sharp cut-off, of the initial pulse cannot travel with a velocity greater than c. Photonic wave packets, which do not spread in vacuum, are better suited for the purpose of information transfer and henceforth we will consider the optical propagation. The proof of the above statement is straightforward: as long as the transmission amplitude across the sample of, say, fast light material has no poles in the upper half of the complex k-plane, the most advanced term in (7.11.16) would be the one propagating freely with the vacuum light velocity c. This means that any feature in the envelope of the initial pulse may only advance at most by the distance cT, so that the information, encoded in such feature, may not travel with the speed greater then c. To achive a truly superluminal information transfer one would need a (one dimensional) material which supports both the scattering and the bound states of the photon, and would, therefore, speed the photon up in the same way a potential well accelerates an electron. We are not aware of the existence of such materials. Earlier, Aharonov and co-workers [25] have shown the weak measurements to be unsuitable for faster-than-light information transfer, and we have just extended their result to the case of "superluminal" propagation.

All the above can be illustrated with the help of a simple set-up exhibiting all the main features of apparently "superluminal" transmission. Consider the propagation of an electromagnetic pulse across two narrow semitransparent mirrors modelled by two δ-function of magnitude Ω located at $x = 0$ and $x = b$, respectively. The transmission amplitude can then be written as

$$T(k) = \sum_{m=0}^{\infty} T^{(m)}(k) \equiv (1 + R(k)) \sum_{m=0}^{\infty} R(k)^{2m} \exp(2imkb) \qquad (7.11.25)$$

where

$$R(k) = -i\Omega/(2k + i\Omega) \qquad (7.11.26)$$

is the reflection amplitude for a single δ-function placed at the origin $x = 0$. If the mirrors are opaque ($\Omega b \gg 1$), the amplitude distribution $\eta(\tau)$ is sharply peaked at $\tau_m = 2mb/c$ (Fig. 7.17a) and the incident pulse is split into a number of discrete path modes corresponding to $2m$, $m = 0, 1, 2...$ additional reflections experienced by the ray between $x = 0$ and $x = b$ [44] so that we deal with a discrete spectrum of non-negative delays, $\tau_m = 2mb/c$ (cf. (7.11.14)),

$$\Psi_F(x) \approx \sum_m G(\tau(x) - \tau_m)T^{(n)}(k_0). \qquad (7.11.27)$$

For a broad initial Gaussian pulse,

$$G(x) = \exp(-x^2/\Delta x^2), \quad \Delta x > b,$$

Fig. 7.17b shows that the terms in (7.11.27)), each delayed by $\tau_m > 0$ conspire to produce an advanced transmitted peak corresponding to a *negative* time delay of $\approx -b/c$. It is easy to see that choosing a narrow Gaussian pulse destroys tunnelling and with it the apparent "superluminality". The double barrier structure supports equally spaced resonances at $k = k_j \equiv \pi j/b$, between which the transmission amplitude is very small. For an incident pulse to tunnel, it must fit between two neighbouring resonances and its momentum spread Δk must not exceed π/b. Thus, if the "superluminal" behaviour is observed, the time delay is known with uncertainty of about $\Delta \tau \approx 1/c\Delta k \approx b/c$, which is just the "weakness" condition for the Aharonov's pointer, discussed in the first paragraph. And just as a "strong" von Neumann measurement reveals the spectrum of the variable $\hat{A}$, a narrow initial pulse is split into components exhibiting the positive time delays τ_m, $m = 0, 1, 2...$ contained in the amplitude distribution $\eta(\tau)$, as shown in Fig. 7.17e. Suppose now that we have

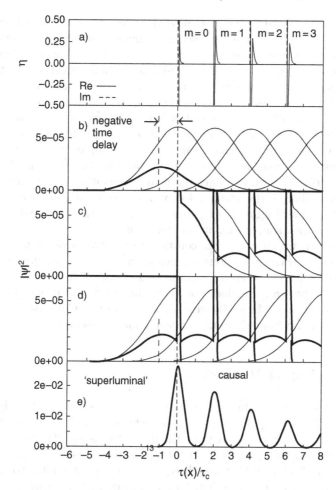

Fig. 7.17. (a) Discrete non-negative time delays $\tau_m \geq 0$ for two opaque mirrors. (b) Broad Gaussians representing Ψ_T, all delayed compared to the free pulse, intefere to produce a pulse advanced by b/c. (c) If the front part of initial pulse is removed, the field vanishes beyond the causal boundary x_B corresponding to $\tau(x) = 0$. (d) If the rear part of the pulse is removed, the field beyound x_B is identical to the uncut field, and contains no information as to whether the initial signal was cut or not. (e) A narrow initial pulse destroys "superluminality" and reveals the non-negative delays τ_m contributing to the transition

a detector which clicks when reached by a sharp front of the signal. With this in mind, we may cut the initial Gaussian pulse in two, discard the front half and send it through the mirrors, hoping that the front would emerge, like the peak of the full Gaussian, ahead of its freely propagating counterpart. Equation (7.11.27) and Fig. 7.17c show, however, that the field would always vanish beyond the causal boundary $x_B = cT + a$, where a denotes the initial position of the cut. Alternatively, we may remove the rear part of the initial pulse and ask whether using a "fast light" medium would allow us to distinguish between the cut and uncut pulses sooner, than we could in vacuum. Again, the answer is no, since for $x > x_B$ the transmitted field builds up from the front tails of Ψ_T's, which are the same for the cut and uncut Gaussians (Fig. 7.17d), so that the observer has to await the arrival of the information-carrying part of the signal, travelling at the speed of light. The experiment similar to one just described was actually made by Stenner and co-workers [44], who used potassium vapour, rather then mirrors, to achieve apparently "superluminal" propagation. They concluded that the information detection time for pulses propagating through the fast light medium is somewhat longer than that in vacuum, even though the group velocity in the medium is in the highly superluminal regime. Figure 1a of [44] is similar to Fig. 7.17d and the above analysis provides a simple explanation for their result.

In summary, we note that the contradiction between the impossibility of faster-than-light travel and observing an apparently "superluminal" transmitted pulse is resolved in a typically quantum mechanical fashion: when "superluminality" is present, one does not know the delay, and cannot claim that the duration spent in the scatterer is shorter than b/c. Conversely, when the delay is known, no "superluminal" transmission is observed.

7.12 Concluding Remarks

We had a twin task of defining the quantum traversal time and analysing its possible use for such applications as nanotechnology and wave packet propagation. The first part turns out to be the most interesting. The traversal time τ can be defined by equating it with the duration a Feynman path spends in the region of interest Ω. Since there are many paths leading to the same final particle's position, τ is a distributed quantity and the Schrödinger wave function $\Psi(x, t)$ is a sum of subamplitudes $\Phi(x, t|\tau)$ corresponding to all different durations. Conceptually, the situation is not much different from the two-slit diffraction experiment. (Feynman once wrote [45]: "any other situation in quantum mechanics, it turns, out can be explained by saying, 'You remember the case of the experiment with two holes? It's the same thing'".) However, since τ is a continuous variable, to construct observable probabilities one requires a coarse grained amplitude $\Psi(x, t|\tau)$ which allows for certain amount of interference between different classes of paths and can be used to construct measurable probabilities. These simple assumptions determine all properties of τ including a recipe for its measurement and the Heisenberg-like uncertainty relation (7.16). Mapping the equation of motion for $\Phi(x, t|\tau)$ onto a particle–meter Schrödinger equation determines the dynamical interaction required to destroy the coherence between different durations. The meter turns out to be just the Larmor clock consisting of a spin in a constant magnetic field $H\Theta_\Omega(x)$. The amount of coarse graining (and, therefore, the accuracy of the measurement) is related to the initial uncertainty of the meter's position and we obtain a finite time generalisation of the von Neumann measurement, which depending on the accuracy, can be either weak or strong. It is worth noting that such a measurement does not rely on expanding the Schrödinger state in eigenfunctions of any "traversal time operator". A similar scheme can, in principle, be applied in a relativistic quantum theory.

The second part of our conclusions is more negative. The quantum traversal time turns out to be a much less useful quantity than its classical counterpart. In particular for tunnelling we fail to provide a single timescale to be compared with the period with which the barrier goes up and down or with the frequency of surface plasmons. This leaves out most realistic interactions and limits us to a few cases where, as in scattering off rectangular potentials, the interaction can be expressed in terms of the classical functional (7.2.3). But even then we only succeed in replacing the sum over paths by a one-dimensional oscillatory integral which is difficult to evaluate beyond the semiclassical limit $\hbar \to 0$. For a classically allowed transition, a semiclassical analysis recoveres the unique classical duration, but for a classically forbidden transition, e.g., tunnelling, it finds complex saddle(s) of $\Phi(x, t|\tau)$ which cannot be identified as real duration(s).

In the end, "the actual time it takes to tunnel" is found to be as elusive as "the slit through which an electron actually went" and the tunnelling time problem remains a subject for someone more interested in the subtleties of the quantum measurement theory than in the practicalities of nanostructure engineering. Yet far from remaining a mathematical abstraction, the measurement theory is instrumental in explaining and classifying experiments related to to the tunnelling times in one way or another. The long-standing paradox of apparent "superluminality" can be resolved by analysing a particular type of interference, first associated with highly inaccurate von Neumann measurements. Such an analysis reveals the limits on the speed of information transfer in experiments on "superluminal" propagation, and once again demonstrates that it is impossible to describe quantum tunnelling by a single, albeit averaged, time parameter.

References

1. L.A. MacColl: Phys. Rev. **40**, 621 (1932)
2. R.Y. Chiao, A.M. Steinberg: Prog. Opt. **37**, 345 (1997)
3. A.P. Jauho. In: *Hot Carriers in Semiconductor Nanostructures: Physics and Applications*, ed. by J. Shah, (Academic Press, Boston, 1992), pp. 121–151
4. M. Jonson: In: *Quantum Transport in Semiconductors*, ed. by D.K. Ferry, C. Jacoboni (Plenum Press, New York, 1992) pp. 193–238
5. E.H. Hauge, J.A. Stoevneng: Rev.Mod.Phys. **61**, 917 (1989); C.R. Leavens, G.C. Aers: In: *Scanning Tunneling Microscopy and Related Methods*, Vol. 184 of *NATO Advanced Study Institute, Series E: Applied Sciences*, ed. by R.J. Behm, N. Garcia, H. Rohrer (Kluwer, Dordrecht, 1990), pp. 59–76; M.Büttiker:

In: *Electronic Properties of Multilayers and Low Dimensional Semiconductor Structures*, Vol. 231 of *NATO Advanced Study Institute, Series B: Physics*, ed. by J.M. Chamberlain, L. Eaves, J.-C. Portal (Plenum Press, New York, 1990), pp. 297–315; R. Landauer, Ber. Busenges: Phys. Chem. **95**, 404 (1991); V.S. Olkhovsky and E. Recami: Phys. Rep. **214**, 339 (1992); R. Landauer, Th. Martin: Rev. Mod. Phys. **66**, 217 (1994)

6. C.R. Leavens: Found.Phys. **25**, 229 (1995)
7. R.P. Feynman, A.R. Hibbs: *Quantum Mechanics and Path Integrals* (McGraw-Hill, New York 1965)
8. D. Sokolovski, L.M. Baskin: Phys. Rev. A **36**, 4604 (1987)
9. D. Sokolovski, P. Haenggi: Europhys. Lett. **7**, 7 (1988)
10. D. Sokolovski, J.N.L Connor: Phys. Rev. A **42**, 6512 (1990)
11. D. Sokolovski, J.N.L Connor: Phys. Rev. A **44**, 1500 (1991)
12. D. Sokolovski, J.N.L Connor: Phys. Rev. A **47**, 4677 (1993).
13. D. Sokolovski, J.N.L Connor: Solid State Comm. **89**, 475 (1994)
14. D. Sokolovski, S. Brouard, J.N.L Connor: Phys. Rev. A **50**, 1240 (1994)
15. D. Sokolovski: Phys. Rev. A **52**, R5 (1995)
16. D. Sokolovski: Phys. Rev. E **54**, 1457 (1996)
17. D. Sokolovski: 'Path Integrals and the Quantum Traversal Time Problem'. In: *Proceedings of the Adriatico Research Conference on Tunnelling and its Implications*, ed. by D. Mugnai, A. Ranfagni, L.S. Schulman (World Scientific, Singapore, New Jersey, London, New York, 1997), pp. 206–222
18. D. Sokolovski: Phys. Rev. Lett. **25**, 4946 (1997)
19. D. Sokolovski: Phys. Rev. A **57**, R1469 (1998)
20. D. Sokolovski: Phys. Rev. A **59**, 1003 (1999)
21. D. Sokolovski, Y. Liu: Phys. Rev. A **63**, 012109 (2000)
22. Y. Liu, D. Sokolovski: Phys. Rev. A **63**, 014102 (2001)
23. J. von Neumann: *Mathematical Foundations of Quantum Mechanics* (Princeton University Press, Princeton, NJ, 1955)
24. Y. Aharonov, D.Z. Albert, L. Vaidman: Phys. Rev. Lett. **60**, 1351 (1988)
25. Y. Aharonov, L. Vaidman: Phys. Rev. A **41**, 11 (1990)
26. A.M. Steinberg: Phys. Rev. Lett. **74**, 2405 (1995)
27. G. Iannacone: 'Weak Measurements and the Tunnelling Time Problem'. In: *Proceedings of the Adriatico Research Conference on Tunnelling and its Implications*, ed. by D. Mugnai, A. Ranfagni, L.S. Schulman (World Scientific, Singapore, New Jersey, London, New York, 1997), pp. 292–309
28. M.V. Berry, J. Goldberg: Nonlinearity **1**, 1 (1988)
29. B.R. Holstein, A.R. Swift: Am.J.Phys. **50**, 833 (1982)
30. A.I. Baz': Yad. Fiz. **4**, 252 (1966) [Sov. J. Nucl. Phys. **4**, 182 (1967)]; **5**, 229 (1967) [**5**, 161 (1967)]
31. V.F. Rybachenko: Yad. Fiz. **5**, 895 (1967) [Sov. J. Nucl. Phys. **5**, 635 (1967)]
32. M. Büttiker: Phys. Rev. B **27**, 6178 (1983)
33. C.R. Leavens, G.C. Aers: Phys. Rev. B **40**, 5387 (1989)
34. J.P. Falck, E.H. Hauge, Phys. Rev. B **38**, 3287 (1988)
35. J.G. Muga, S. Brouard, R. Sala: J. Phys.: Condens. Matter **4**, L579 (1992)
36. C. Foden, K.W.H. Stevens: IBM J. Res. Dev. **32**, 99 (1988)
37. A. Peres: Am. J. Phys. **48**, 552 (1980)
38. Special issue of Annalen der Physik **7** (1998). *Proceedings of the Workshop on Superluminal (?) Velocities: Tunneling Time, Barrier Penetration, Non-Trivial Vacua, Philosophy of Physics*, ed. by P. Mittelstaedt, G. Nimtz (Koeln, 1998)

39. W. Greiner: *Relativistic Quantum Mechanics. Wave Equations* (Springer, Berlin, 1977)
40. R.P. Feynman: Phys. Rev. **80**, 440 (1950)
41. L.S. Schulman: *Techniques and Applications of Path Integration* (Wiley, New York, 1981), pp. 225–236
42. V.B. Berestetsii, E.M. Lifshitz, L.P. Pitaevskii: *Quantum Electrodynamics* (Pergamon, Oxford, 1982)
43. D. Sokolovski, A.Z. Msezane, V.R. Shaginyan: Phys. Rev. A **71**, 064103 (2005) quant-ph/0401159
44. Wang Yun-ping, Zhang Dian-lin, Phys. Rev.A **52**, 2597 (1995)
45. M.D. Stenner, D.J. Gauthier, M.A. Neifeld: Nature **425**, 695 (2003)
46. R.P. Feynman: *The Character of Physical Laws* (Penguin, London, 1992)

8

Quantum Clocks and Stopwatches

Rafael Sala Mayato[1], Daniel Alonso[2], and Iñigo L. Egusquiza[3]

[1] Departamento de Física Fundamental II, Universidad de La Laguna, La Laguna, Tenerife, Spain
rsala@ull.es
[2] Departamento de Física Fundamental y Experimental, Universidad de La Laguna, La Laguna, Tenerife, Spain
dalonso@ull.es
[3] Fisika Teorikoaren Saila, Euskal Herriko Unibertsitatea, 644 P.K., Bilbao, Spain
wtpegegi@lg.ehu.es

Time is to clock as mind is to brain. The clock or watch somehow contains the time. And yet time refuses to be bottled up like a genie stuffed in a lamp. Whether it flows as sand or turns on wheels within wheels, time escapes irretrievably, while we watch. Even when the bulbs of the hourglass shatter, when darkness withholds the shadow from the sundial, when the mainspring winds down so far that the clock hands hold still as death, time itself keeps on. The most we can hope a watch to do is mark that progress. And since time sets its own tempo, like a heartbeat or an ebb tide, timepieces don't really keep time. They just keep up with it, if they are able.

Dava Sobel [1]

8.1 Introduction

In the formal development of quantum mechanics any observable is mathematically represented by an operator acting in an appropriate functional space. In this spirit the position, momentum, energy, and other quantities are associated with operators from which predictions can be made following the rules of the theory. In certain cases, for instance the decay of a metastable state, for which it is natural to ask about its lifetime, the question that we ask has as an answer a quantity with units of time. In this way, a time variable is observed and measured and, therefore, we should construct an operator to which the answer experimentally observed can be attributed. That is to say, we are required to produce some time operator within the context of the theory. As has been repeatedly emphasized elsewhere in this volume, there have been many attempts to build such an operator [2], despite an early theoretical objection by Pauli [3]. One of the difficulties lies on the fact that the Hamiltonian operator plays a dual role in the theory: on the one hand it represents the energy of the system and, on the other hand, it is the generator of the time

R. S. Mayato et al.: *Quantum Clocks and Stopwatches*, Lect. Notes Phys. **734**, 235–278 (2008)
DOI 10.1007/978-3-540-73473-4_8   © Springer-Verlag Berlin Heidelberg 2008

development for wave functions [4]. A more operational approach consists in constructing some physical system from which we can extract some timescale associated with a physical process.

Another aspect to consider is that the times we measure are uncertain and prone to error, as any kind of measurement is. Within quantum theory, there are two questions that we must pose concerning this point. In the first place, we should enquire whether there exists some intrinsically quantum reason for errors in the measurement of time. Secondly, if the measurement of time is indeed riddled with errors, be it for classical or quantum reasons, this could give rise to peculiar effects in the evolution of the quantum systems we study.

This second aspect we will be dealing with last in this chapter, and we shall first devote our attention to the study of quantum clocks and stopwatches, in order to understand better the intrinsically quantum aspects of the measurement of time, by including in our description of quantum evolution a physical system that plays the role of a clock. According to Peres, "A clock is a dynamical system which passes through a succession of states at constant time intervals" [5]. We shall review the construction of such devices using quantum systems, with different couplings that allow the clock to start or stop in such a way that it works as a stopwatch to measure durations of processes. The first historical landmark is a classic paper by Salecker and Wigner in 1958 [6], in which they proposed a clock to analyze the limitations which the quantized nature of microscopic systems imposes on the possibility of measuring distances between space–time events. In 1966, Baz' [7, 8] proposed the use of the Larmor precession of a spin in a weak magnetic field as a clock to measure the duration of quantum collision events. Rybachenko [9] applied this method to the case of particles in one dimension colliding with a rectangular barrier. Many other theoretical works have followed Baz' and Rybachenko, motivated by conceptual difficulties with the tunneling time reviewed in several chapters of this book, and because of the prospect of constructing high speed devices based on tunneling semiconductor structures. Later on, Peres [5] applied the Salecker and Wigner device for three different purposes: first, to measure the time of free flight between two spatial points for the nonrelativistic case (the relativistic case was subsequently treated by Davies [10]); second, for timing the decay of excited atoms; and third, to control the duration of a physical process.

In many of the proposals discussed above (proposed and further analyzed by a number of authors [11, 12, 13]) the system of interest was an electron. However, the experimental implementation of a quantum clock for electrons is difficult and the interpretation of the results may not be clear. In fact, a good part of the wide range of different viewpoints regarding this problem is due to the ultrashort experimental times involved (see Chap. 11). Motivated by this problem, some researchers have used existing thin-film technology [14, 15, 16] to realize and operate an optical Larmor clock based on the analogy between tunneling electrons and evanescent electromagnetic waves. Open questions and fundamental aspects such as clock precision, superluminal times, and nonclassical properties of evanescent photons, are being elucidated by these optical quantum clocks.

An important problem with many practical and scientific applications [17] that will not be treated here is clock synchronization. This problem is deeply important in modern technology and its formulation is very simple: "determine the time difference between two spatially separated clocks" [17, 18, 19].

8.2 What is a Clock?

As we stated in the introduction a good definition of a clock can be found in the article by Peres [5]: "A clock is a dynamical system which passes through a succession of states at constant time intervals." In this definition, we find no reference to the dynamical law governing the motion of the system. The operational definition given defines time according to the states of the clock dynamics. In fact we can think the other way around, that is to say, a given definition of time interval according to some system implies a dynamical law as discussed in [20].

In the definition given above it is important to take into account that the clock is by itself "a dynamical system": this means that the clock has a dynamical variable, the pointer, with a time dependence so simple that one can infer directly the value of t from the pointer position [21]. For example, in a classical clock the time dependence of the pointer is periodic and is given by the angle $\theta = \omega t (\mathrm{mod}\, 2\pi)$. In quantum mechanics the pointer, as a dynamical variable, must be represented by an operator, which implies that any quantum measurement of such an angle θ will have a spread. The relation, if any exists, between the spread of this pointer variable and the energy of the system being studied could be termed "time–energy" relation, which, however, will be of a very different kind from the one that quantum mechanics specifies for any pair of operators, $\widehat{A}$ and $\widehat{B}$, obeying the commutation relation $[\widehat{A}, \widehat{B}] = i\hbar$ (see Chap. 11). The problem of defining an uncertainty principle for energy and time, and the meaning we can give to it is then a fundamental question for these devices.

Let us center our attention on the measurement of a time quantity with a clock. The quantum clock will in the course of its time evolution exchange some energy with the system for which we want to extract some characteristic time, for instance the time of flight of a particle between two detectors. The exchange of energy between the clock and the system will eventually modify the system's state and its later evolution, so that we can ask whether the time measurement is related to the system or, rather, whether it is a feature of the clock–system complex. If we think of a sequence of time measurements in which the coupling between the clock and the system varies over a range from large to small values of the coupling constant, the sequence of time values so obtained will ideally reach an almost constant value for sufficiently small values of the coupling constant (perturbative regime). When this occurs the limiting time property is regarded as intrinsic to the system time property of the combined clock–particle system.

The conditions to be fulfilled in order to give some intrinsic character to a time measurement are discussed in [5] in the context of time-of-flight measurements. It was found there that the time interval between two spatial points is well defined although the time of passage through a spatial point is uncertain. The fact that the disturbance produced by the clock is inversely related to the time resolution of the clock imposes a limitation on the accuracy with which particle velocities can be measured over a given distance. Whence it follows that in this case the time measurement is more adequately understood as a particle–clock property, rather than as a particle property.

All of the clocks that we are going to discuss are constructed by coupling one extra degree of freedom to the system. Depending on the nature of the coupling Hamiltonian, we have different types of quantum clocks. Essentially, they are all described by a Hamiltonian that contains the system degrees of freedom and a coupling Hamiltonian that couples the degrees of freedom of the clock with those of the system. Let us denote the Hamiltonian of the system by $\widehat{H}_s(x, p_x)$, with x a spatial coordinate and p_x the conjugate momentum. The full Hamiltonian $\widehat{H}$ is given by $\widehat{H} = \widehat{H}_s + \widehat{H}_{c-s}$ with $\widehat{H}_{c-s}$ the coupling Hamiltonian of clock and system (notice that the clock normally will only evolve when some condition is fulfilled by the system). This Hamiltonian is a function of the clock coordinates, z and p_z, say, as well as the particle coordinates x and p_x.

8.3 The Salecker–Wigner Clock

As mentioned in the introduction of this chapter, the first historical landmark in the construction of a quantum clock was a theoretical paper written by Salecker and Wigner [6], where they proposed a microscopic clock to measure distances between space–time events (in fact, they examined several alternatives, with a view to the limitations on accuracy due to quantum behavior).

In this section we will address the mathematical description of a quantum clock and study a more realistic case studied by Peres [5, 22]. We will also discuss some of its applications to measuring characteristic and event times (for definitions of characteristic and event times see, e.g., [23]).

8.3.1 Mathematical Description of a Quantum Clock

As is well known, in the Hamiltonian formalism t is a parameter, not a dynamical variable. However, it is possible to find dynamical variables which depend linearly on t in such a way that we can infer the value of t by looking at them. For example, the position of the hand of a clock is not the time parameter but depends on it in a very simple manner: the angle θ swept out by the hand of the clock is a simple function of t, i.e., $\theta = \omega t \,(\mathrm{mod}\, 2\pi)$.

The canonical variable conjugate to this angle is the so-called "action," J, which always has dimensions of angular momentum.

In quantum mechanics both these variables will be associated to operators in a Hilbert space (for details, see the review of Carruthers and Nieto [24]). Let $\widehat{\Theta}$ and $\widehat{J}$ be the operators representing angle and action (or angular momentum), respectively. If one defines the range of θ from $-\infty$ to ∞ then both operators can be regarded as conjugate and they satisfy the canonical commutation relation

$$[\widehat{\Theta}, \widehat{J}] = i\hbar. \tag{8.1}$$

However, $\widehat{J}$ is self-adjoint only in the subspace of periodic functions with period 2π. Accordingly, we introduce a periodic coordinate θ on the interval $[0, 2\pi]$ in such a way that the Hilbert space in the angle representation consists of the square-integrable and differentiable functions f of θ satisfying $f(0) = f(2\pi)$ [24]. Of course, this raises serious problems for the status of the commutation relation (8.1), as is well known, but we shall not be concerned by those here, since they can be remedied by the use of positive operator valued measures (e.g., see [25], or the short description of POVMs in Chaps. 3 and 10).

The eigenvalue equation of the angle operator $\widehat{\Theta}$ is

$$\widehat{\Theta}|\theta\rangle = \theta|\theta\rangle \,,$$

where the eigenstates $\{|\theta\rangle\}$ form a continuous orthonormal basis satisfying $\langle\theta|\theta'\rangle = \delta(\theta - \theta')$. On the other hand, the eigenvalue equation for the angular momentum operator $\widehat{J}$ reads

$$\widehat{J}|m\rangle = \hbar m|m\rangle \,,$$

with $m = 0, \pm 1, \pm 2, \ldots$, and the basis $\{|m\rangle\}$ satisfies the corresponding orthonormalization relation, i.e. $\langle m|m'\rangle = \delta_{m,m'}$.

As pointed out by Hilgevoord [21, 26], the formal analogy with the position and momentum operators in the coordinate representation is very close. In the θ representation the states $|\theta\rangle$ and $|m\rangle$ are given by the Dirac's delta function

$$\langle\theta|\theta'\rangle = \delta(\theta - \theta')$$

and "plane wave" relations

$$u_m(\theta) = \langle\theta|m\rangle = (2\pi)^{-1/2}e^{im\theta} \,,$$

respectively. In particular, using the closure relation in $\{|m\rangle\}$, we can write

$$|\theta\rangle = (2\pi)^{-1/2}\sum_{m=-\infty}^{\infty} e^{-im\theta}|m\rangle \,. \tag{8.2}$$

The dynamics of the clock is governed by the Hamiltonian $\widehat{H}_c = \omega\widehat{J}$, where ω is a constant frequency. Clearly we have

$$\widehat{H}_c u_m(\theta) = -i\hbar\omega\frac{\partial}{\partial\theta}u_m(\theta) = m\hbar\omega u_m(\theta) \ . \tag{8.3}$$

In particular, $\widehat{H}_c$ is the generator of translations in time by means of the unitary operator $\widehat{U}(t) = e^{-i\widehat{H}_c t/\hbar}$. Taking into account the expression for $|\theta\rangle$ given above, it is easy to show that

$$\widehat{U}(t)|\theta\rangle = e^{-i\widehat{H}_c t/\hbar}|\theta\rangle$$

$$= (2\pi)^{-1/2}\sum_{m=-\infty}^{\infty}e^{-im\theta-im\omega t}|m\rangle$$

$$= |\theta + \omega t\rangle \ . \tag{8.4}$$

And this is just a clock: the evolution operator $e^{-i\widehat{H}_c t/\hbar}$ advances the "hand" of the clock through successive orthogonal states at constant angular velocity. Notice that $\langle\theta|\widehat{U}(t)|\psi\rangle = \psi(\theta, t) = \psi(\theta - \omega t, 0)$.

8.3.2 The Quantum Clock Analysis of Peres

Peres [5, 22] studied a more realistic case than the one considered in the subsection above. He restricted the sum in (8.2) to values of m satisfying $-j \leq m \leq j$, where j is a positive integer and considered the set of orthonormal states

$$|k\rangle = N^{-1/2}\sum_{m=-j}^{j}e^{-i2\pi km/N}|m\rangle$$

with $k = 0, 1, ..., N - 1$ and where $N = 2j + 1$ is the total number of states of our clock. The associated wave functions in the θ representation are

$$v_k(\theta) = \langle\theta|k\rangle = N^{-1/2}\sum_{m=-j}^{j}e^{-i2\pi km/N}u_m(\theta). \tag{8.5}$$

According to (8.4) the time evolution of these states satisfies

$$\widehat{U}(t = \tau = 2\pi/N\omega)v_k(\theta) = v_{k+1(\mathrm{mod}\ N)}(\theta) \ ,$$

or, in a slightly different notation, $v_k(\theta, \tau) = v_{k+1(\mathrm{mod}\ N)}(\theta, 0)$, and they are eigenfunctions of the clock time operator

$$\widehat{T}_c \equiv \tau\sum_{k'=0}^{N-1}k'\widehat{P}_{k'} \tag{8.6}$$

with eigenvalues $k\tau$. The projection operators $\widehat{P}_{k'}$ satisfy $\widehat{P}_{k'}v_k(\theta) = \delta_{k',k}v_k(\theta)$.

For large N, the basis function $v_k(\theta)$ has a sharp peak of width $2\pi/N$ at $\theta = k(2\pi/N)$. The states with wave functions $v_k(\theta, l\tau)$, with l an integer modulo N, are mutually orthogonal. Since $\tau = 2\pi/N\omega$ is also the time resolution of the clock, the functions defined in (8.5) can be understood as pointing to the kth hour with an uncertainty of $2\pi/N\omega$. Another important property is that the expectation value of the clock Hamiltonian $\widehat{H}_c$ in the clock state (8.5) is zero, from which it can be shown that for large values of j the energy uncertainty is given by $\Delta\widehat{H}_c = (\pi\hbar/\sqrt{3}\tau)$. Peres compares this quantity with the maximum value of the energy, $\pi\hbar/\tau$, and infers that this clock is basically a nonclassical object. As was only to be expected, the uncertainty product relation holds true, $\tau\Delta\widehat{H}_c \geq \hbar/2$. A more severe condition has been obtained by Hilgevoord and Uffink [21] (see also the contribution of P. Busch to this volume).

In the next subsections we will show some of the applications [5, 10, 12, 13, 27, 28], of the Salecker–Wigner clock to the measurement of characteristic times, such as the time of flight of a particle, the mean dwell time, or the lifetime of an excited state. We will also show how this clock has been used to measure event times, in particular the arrival time of a free particle in a one-dimensional experiment [13]. We shall consider other variants of the Salecker–Wigner clock in Sect. 8.5.2.

8.3.3 Time-of-Flight Measurements

Peres [5] applied the clock devised by Salecker and Wigner to measure the average time that particles spend in some specific region of space, for an ensemble consisting of a large number of identically prepared, single-particle, one-dimensional, stationary-state scattering experiments.

The classical picture associated with this measurement process is that of a stopwatch that runs only whenever the particle happens to be in the region of interest. Given this image, a convenient Hamiltonian for the particle (with incident kinetic energy $E = \hbar^2 K^2/2M$ in the absence of the clock) and the clock is

$$\widehat{H} = \widehat{H}_p + \widehat{H}_{c-p} = \widehat{H}_p + \widehat{H}_c\Theta(x)\Theta(d-x)$$
$$= -\frac{\hbar^2}{2M}\frac{\partial^2}{\partial x^2} - i\hbar\omega\Theta(x)\Theta(d-x)\frac{\partial}{\partial\theta} . \tag{8.7}$$

Here the region of interest is $0 \leq x \leq d$ and the Heaviside functions Θ indicate that the clock runs only when the particle "is" between 0 and d.

As initial state of the clock, Peres chose $v_0(\theta)$ (see (8.5)). However, for simplicity, he first solved the clock's equation of motion for an eigenstate u_m of $\widehat{J}$, and he then added the solutions for all m to get the solution associated with $v_0(\theta)$. As we stated in the preceding subsection, for large values of N this state is sharply peaked at $\theta = 0$. By taking (8.3) into account, for each m in the sum over m the second term in the Hamiltonian (8.7) can be replaced numerically by the constant $m\hbar\omega\Theta(x)\Theta(d-x)$, and hence the restriction of the Hamiltonian in (8.7) to this sector represents an otherwise free particle colliding with a square potential barrier of height $V_m = m\hbar\omega$ and length d.

The textbook solution of this stationary scattering problem, outside the region of interest, $0 \leq x \leq d$, is given by

$$\Psi_K^m(x) = \begin{cases} e^{iKx} + R_m(K)e^{-iKx}, & \text{if } x \leq 0 \\ T_m(K)e^{iKx}, & \text{if } x \geq d \end{cases}. \tag{8.8}$$

We assume that "initially" the particle and clock are not coupled so that the wave function of the entire system factorizes:

$$\Psi_i(x, \theta) = e^{iKx} v_0(\theta) = e^{iKx} \frac{1}{N^{1/2}} \sum_{m=-j}^{j} u_m(\theta) . \tag{8.9}$$

The term $\widehat{H}_{c-p}$ in (8.7) couples the particle and the clock coordinates in such a way that the wave function of the system is no longer factorizable in coordinates x and θ. Therefore, the final (stationary) wave function must be written as $\Psi(x, \theta) \exp(-iEt/\hbar)$ with

$$\Psi(x, \theta) = \frac{1}{N^{1/2}} \sum_{m=-j}^{j} \Psi_K^m(x) u_m(\theta) .$$

Clearly, $E = \hbar^2 K^2 / 2M$.

Peres focused on the case in which the maximum available energy of the clock is negligible compared with the energy of the particle, i.e., $E \gg V_j = j\hbar\omega$. This means that the measurement produces negligible disturbance on the particle. In this situation, we can approximate $R_m(K) \simeq 0$ and $T_m(K) \simeq |T_m(K)| \exp[i(K_m - K)d]$, under the condition that $|T_m(K)| \simeq 1$ and $K_m = \hbar^{-1}[2M(E - m\hbar\omega)]^{1/2}$. In this case, the phase shift caused by the barrier is $(K_m - K)d \simeq M\omega d/\hbar K$ and the final (stationary) wave function for the particle plus clock system is

$$\Psi_f(x, \theta) \simeq e^{iKx} \frac{1}{N^{1/2}} \sum_{m=-j}^{j} e^{-imM\omega d/\hbar K} u_m(\theta)$$

$$= e^{iKx} v_0(\theta - M\omega d/\hbar K) ,$$

for $x > d$ (transmitted wave), and approximately zero for $x \leq 0$ (there is no reflected wave). Initially (see (8.9)) this function was peaked at $\theta = 0$ and now it is peaked at $\theta = M\omega d/\hbar K$, indicating the time-of-flight $\tau_T(0, d) = Md/\hbar K$ through the region of interest. Note that this result gives the expected velocity $\hbar K/M$ for an unperturbed particle with energy $E = \hbar^2 K^2/2M$. Note also that combining the condition for negligible perturbation of the particle, $E \gg j\hbar\omega$, and for high resolution of the clock, $\tau_T(0, d) \gg \tau$, one obtains the criterion that the clocked flight path must be very large compared with the de Broglie wavelength, i.e., $d \gg 2\pi/K$, if one wants to measure accurately the velocity of a free particle.

Leavens [12] applied this scheme to calculate the mean dwell time (for details see Chap. 2) $\tau_D(0, d; E)$ for a particle scattered by a rectangular barrier $V(x) = V_0 \Theta(x)\Theta(d - x)$. The original expression postulated by Büttiker [11] for the mean dwell time, for the special case of stationary-state scattering, is

$$\tau_D(0, d; E) = \frac{1}{\hbar K/M} \int_0^d dx |\Psi_K(x)|^2 \, , \qquad (8.10)$$

where $E = \hbar^2 K^2/2M$ is the energy of the particle, $\hbar K/M$ is the incident probability current density and $\Psi_K(x)\exp(-iEt/\hbar)$ is the stationary-state wave function (see also 8.4.1 below).

Applying the quantum clock to this case means adding the barrier potential energy to the $\widehat{H}_p$ part of the Hamiltonian (8.7). The result is that the coefficients $T_m(K)$ and $R_m(K)$ in (8.8) become the transmission and reflection probability amplitudes for the rectangular barriers $V_m = V_0 + m\hbar\omega$ ($m = -j, ..., j$). It should be noted that the second term in the Hamiltonian (8.7) does not distinguish between transmitted and reflected electrons. Hence, repeating the previous analysis with the new potential barriers V_m should lead to the clocked mean dwell time $\tau_D^c(0, d; E)$ simply by calculating the expectation value of the clock time operator (8.6) in the final state $\Psi_f(x, \theta)$ minus that in the initial state $\Psi_i(x, \theta)$. As a preliminary to such a calculation consider the example of an opaque rectangular barrier ($\kappa_0 d \gg 1$ where $\kappa_0 \equiv \hbar^{-1}[2M(V_0 - E)]^{1/2}$). Büttiker's expression (8.10) gives

$$\tau_D(0, d; E) = \frac{\hbar}{V_0}\left(\frac{E}{V_0 - E}\right)^{1/2} . \qquad (8.11)$$

A necessary condition for negligible perturbation of the particle scattering process of interest by the clock during the measurement is $V_0 - E \gg j\hbar\omega$. Using $\tau = 2\pi/(2j+1)\omega$ and (8.11) this condition becomes $\tau_D(0, d; E) \ll \tau/4$ which is incompatible with the condition $\tau_T(0, d) \gg \tau$ for good time resolution. Hence, for a particle scattered by an opaque barrier the mean dwell time as measured with a Salecker–Wigner clock could be significantly different from the intrinsic quantity which, in principle, could be measured by an ideal apparatus.

To avoid this problem, Leavens [12, 27] introduced the concept of *calibration* of the clock. To do this, consider an ensemble of a large number of freely evolving Salecker–Wigner quantum clocks ($\widehat{H} = \widehat{H}_c$). Each clock is prepared in the same initial state $\chi(\theta, t_i) \equiv \exp(-i\widehat{H}_c t_i/\hbar)v_0(\theta)$. The averaged free quantum clock value for the elapsed time $\Delta t \equiv t - t_i$ at time t, namely $t_{free}^c(\Delta t)$, is obtained by subtracting the expectation value of the clock time operator $\widehat{T}_c$ in the initial state from its expectation value in the time-evolved state:

$$t_{free}^c(\Delta t) \equiv F(\Delta t) \equiv \langle\chi(\theta, t)|\widehat{T}_c|\chi(\theta, t)\rangle - \langle\chi(\theta, t_i)|\widehat{T}_c|\chi(\theta, t_i)\rangle .$$

The clock parameters are chosen such that F is a monotonic function of Δt in the temporal interval of interest. Then the inverse function F^{-1} exists and can be used to calibrate the clocked (c) dwell time

$$\tau_D^c(0, d; E) \equiv \langle \Psi_f(x, \theta) | \widehat{T}_c | \Psi_f(x, \theta) \rangle - \langle \Psi_i(x, \theta) | \widehat{T}_c | \Psi_i(x, \theta) \rangle \,,$$

which is not necessarily accurate, to obtain the calibrated-clock (cc) value

$$\tau_D^{cc}(0, d; E) \equiv F^{-1}[\tau_D^c(0, d; E)] \,.$$

With this calibration, τ has lost its meaning as the time resolution of the clock. There is no need to impose $\tau_T(0, d) \gg \tau$ and there is no advantage to choosing a very large value of $N\omega$ (there is an example in [27] with $N = 3$). For an arbitrary barrier, the calibrated clock result $\tau_D^{cc}(0, d; E)$, for the dwell time converges to the intrinsic one (8.10) when the limit $\hbar\omega \to 0$, i.e., when a negligibly small perturbation energy $\hbar\omega$ is taken. This is possible if the initial clock state is suitably chosen, i.e., $\chi(\theta, t_i)$ with t_i not too close to $k\tau$ ($k = 0, 1, ..., N - 1$) so that F is monotonically increasing and linear in Δt for $\Delta t / \tau \ll 1$.

The same calibration procedure [27] can be applied to obtain expressions for τ_T and τ_R (see Chap. 2), the mean times that particles that will finally be transmitted or reflected, respectively, spend in the region of interest. The complete ensemble of clocks can be distributed into two subensembles, one for transmitted and one for reflected particles. The expressions for these times are identical to those obtained by Rybachenko [9] using Larmor precession, by Iannacone and Pellegrini [29] using a path integral technique, by Steinberg [30] using the weak measurement theory of Aharonov and coworkers [31, 32], and by Brouard et al. (see [33] and the Introduction of this book) using a single formal framework based on projection operators (BSM approach).

The times τ_T and τ_R that apparently emerge naturally from the theoretical analysis can have unphysical properties. For example, in [27] Leavens and McKinnon show that the mean transmission speed at the center of the barrier for electrons of energy 1 eV incident on a rectangular barrier of height $V_0 = 10$ eV and width $d = 5$ Å, is about four times the speed of light. It can be argued that this does not matter because the calculation is based on the Schrödinger equation and that upon applying the Dirac equation such super-luminal effects would disappear. The first analysis for timing Dirac particles using a model quantum clock was made by Davies [10]. He showed, with an analysis analogous to that of Peres for the nonrelativistic case, that the velocity of a free relativistic particle can be measured using a Salecker–Wigner clock. Following [10, 27], Leavens and Sala Mayato [28] used the Bohm trajectory and Salecker–Wigner clock approaches to timing quantum particles to derive expressions for the mean transmission and reflection times for Schrödinger and Dirac electrons scattering in one dimension from an electrostatic potential barrier. They found realistic situations, involving barrier heights and incident electron kinetic energies both many orders of magnitude less than the rest energy mc^2, for which the calculated mean transmission times imply superluminal velocities for Schrödinger electrons. To the question of whether or not this is an artifact of using the nonrelativistic Schrödinger equation the answer is "yes" for the trajectory approach and "no" for the quantum clock approach.

8.3.4 Other Examples

Peres [5, 22] applied the Salecker–Wigner design for a quantum clock to other fundamental problems. Examples are the lifetime of an excited state and a prescription for controlling the duration of a physical process. We briefly review here the first one and for a detailed account of the second recommend the references cited above.

In the first example, the Hamiltonian associated with the problem is

$$\widehat{H} = \widehat{H}_a + \widehat{P}_0 \widehat{H}_c .$$

Here, $\widehat{H}_a$ is the Hamiltonian of the atom, $\widehat{P}_0$ is the projection operator for the initial undecayed state and $\widehat{H}_c$ is the clock Hamiltonian as usual. The projection operator $\widehat{P}_0$ permits the clock to run only when the state of the system is the initial one. It is stopped for any other state.

For the Hamiltonian of the atom, Peres used a very simple exponential decaying model given by

$$\widehat{H}_a = \widehat{H}_0 + \widehat{V} ,$$

where $\widehat{H}_0$ has just one discrete eigenstate ϕ_0, with an energy $E_0 > E_{min}$, which is included in the domain of the continuous spectrum

$$\widehat{H}_0 \phi(E) = E\phi(E), \qquad E > E_{min} .$$

The eigenstates are normalized according to $\langle \phi(E)|\phi(E')\rangle = \delta(E - E')$ (ϕ_0 is orthogonal to all the $\phi(E)$ states). Peres also assumed that there are no transitions between states belonging to the continuous spectrum, i.e., the only nonvanishing matrix elements of $\widehat{V}$ are

$$V(E) = \langle \phi_0|\widehat{V}|\phi(E)\rangle ,$$

and that these matrix elements have a slow dependence on E.

Subject to all these assumptions and coupling the atom to the clock (which essentially shifts the initial energy from E_0 to $E_0 + n\hbar\omega$) he showed first (for details see references [5, 22]) that there is a lower limit for the lifetime associated with the time resolution of the clock

$$\tau \gg \hbar \frac{d[\log |V(E_0)|^2]}{dE_0} .$$

Peres pointed out that finer resolution will eventually change the decay law to a Zeno effect in the limit $\tau \to 0$. Second, the probability of finding the clocks stopped at time $t_k = k\tau = 2\pi k/N\omega$ is in accordance with the exponential decay law.

Aharonov and coworkers [13] have applied the Salecker–Wigner clock ("toy model") to measure the time of arrival in one dimension of a free particle to a given point $x = x_A$ (for an extended treatment of this problem, see [34], Chap. 10 and references therein). In classical mechanics particles follow definite trajectories, so the meaning of the distribution of arrival times for an ensemble of particles at a given place is clear. In standard quantum mechanics we have no such concept of trajectory, so the meaning of arrival time is rather controversial. Actually, [13] provides a negative conclusion, claiming that the time of arrival cannot be precisely defined and measured in quantum mechanics. The toy model of Aharonov et al. is based on the classical Hamiltonian:

$$H = \frac{1}{2m} P^2 + \Theta(-x) P_c .$$

In this example the particle is moving in one dimension and the clock runs only when the particle is to the left of $x = 0$. The equations of motion for the particle's position x and for the pointer of the clock x_c are

$$\dot{x} = \frac{P}{m}, \qquad \dot{P} = -P_c \delta(x)$$
$$\dot{x}_c = \Theta(-x), \qquad \dot{P}_c = 0$$

As $t \to \infty$ the clock shows the time of arrival,

$$x_c(\infty) = x_c(t_0) + \int_{t_0}^{\infty} \Theta(-x(t)) dt .$$

In the classical case the back reaction on the particle can be made negligible by choosing $P_c \to 0$. Then the undisturbed solution for the particle is given by $x(t) = x(t_0) + (P/m)(t - t_0)$ and the clock finally reads

$$t_A = x_c(\infty) = -m \frac{x(t_0)}{P} ,$$

which is the classical result for the time of arrival measured from t_0. However, in quantum mechanics one expects an important back reaction because in the limit $\Delta x_c = \Delta t_A \to 0$, P_c must have a large uncertainty, i.e., the measurement affects strongly the particle.

Aharonov et al. computed a wave function for the quantum problem and showed that the probability to stop the clock takes the form

$$\left(\frac{E + E_c}{E} \right)^{1/2} \left[\frac{2\sqrt{E}}{\sqrt{E} + \sqrt{E + E_c}} \right]^2 ,$$

where E and E_c are the energies of the incident wave function and the clock, respectively[1] (for a more accurate treatment of the quantum system and comments on the way of arriving at this formula, see [38]). This probability is

---

[1] Allcock obtained essentially this result [35, 36, 37].

almost one only if $E/E_c > 1$. On the other hand, the minimum uncertainty principle for the clock implies that the possible values obtained by E_c are of order $\hbar/\Delta t_A$ if the average clock energy goes to zero (as in the classical case), so in order to trigger the clock the relation

$$\langle E \rangle \Delta t_A > \hbar \tag{8.12}$$

must be satisfied. In other words, this inequality implies large uncertainties for low-energy wave packets. Actually, Baute et al. [39] have showed that the limitations described by the inequality obtained above using this toy model are already present in the intrinsic time-of-arrival distribution of Kijowsky [40]. Notice however that the viewpoint of Aharonov et al. is markedly different: (8.12) is understood as a dynamical limit on the accuracy of the measurement of time, totally unrelated to the measurement of any other quantity, thus being conceptually distinct from a kinematical uncertainty limit, which relates the errors in the measurement distributions of two observables.

In the same reference [13], the authors consider a different model where the particle detector consists of a number of two-level spin degrees of freedom, and arrive at the same difficulty found for the previous model. The behavior of this spin model has been critically examined in [34].

Further analyses of the effect of the back reaction of the clock on the system can be found for instance in [41].

8.3.5 Simultaneous Measurement of Time and Energy

Since Arthurs and Kelly [42] first proposed a model for the simultaneous measurement of position and momentum of a quantum particle, their idea has been widely used and extended. A particularly fruitful avenue has been to consider the Arthurs–Kelly model as a generator of quasidistributions in phase space [43, 44, 45, 46, 47, 48]. These quasidistributions are then used to generate probability distributions for observables which are representable as functions on phase space [49, 50, 51, 52].

Kochánski and Wódkiewicz [53] suggested using these quasidistributions to define "operationally" a time of arrival distribution, thus providing a new kind of quantum clock. The reading of the clock is a pair of numbers, which more or less faithfully reflect the position and momentum of the underlying quantum particle, and from this dual reading a time of arrival is computed. On analyzing further this "indirect" quantum clock [54], it became apparent that the result is not covariant. That is, the measurements with this clock do not flow smoothly with parametric time, and are dependent on the choice of initial instant for parametric time. A number of other operational definitions were examined by Baute et al. [54], and they found that within a wide class it was not feasible to have both covariance and positivity for the distribution of times of arrival.

Therefore, a new kind of quantum clock was proposed [54], by coupling two additional degrees of freedom to the quantum particle whose time-of-arrival distribution is sought for. This coupling was performed in such a way that one of the additional pointer variables would directly record an estimate of the time of arrival. The resulting distributions were covariant and positive, and were found to be more or less complicated convolutions with Kijowski's distribution of times of arrival for the free particle case (for a discussion of Kijowski's distribution, see Chap. 10).

This latter result is of course more general: whenever we couple a clock to a physical system we should expect that the distribution of measured times should be a coarsening of the corresponding ideal distribution of times, were this to exist.

Notice that the quantum clocks presented in this subsection are not of the type of the Salecker–Wigner clock. Nonetheless, they share the common characteristic that the reading of the time evolution of the system is performed on the clock or ancillary variables. This common trait distinguishes them clearly from the set of clocks we shall next examine, namely, the family of Larmor clocks.

8.4 The Larmor Clock

One of the most popular clocks is the Larmor clock.[2] Consider first a classical particle with a nonzero magnetic moment. If the particle enters in a region where a uniform magnetic field is applied, the component of the magnetic moment perpendicular to the field will precess with the Larmor frequency, and the precession will stop as soon as the particle leaves the region. This suggests the use of the magnetic moment as a stopwatch dial to measure the duration of the particle stay in the selected region. Baz' translated directly this idea to a quantum system [7, 8]. To determine the time spent in a region $r < R$ by a spherically symmetric beam of monoenergetic spin 1/2 particles, polarized in the x-direction and incident on a central potential of finite range $r_0 < R$, he imagined an infinitesimal uniform magnetic field (to perturb minimally the particle by the presence of the clock) applied in the spherical region along the y-axis. The collision time was then defined as the ratio between the average precession angle and the Larmor frequency ω_L. Rybachenko [9] applied the same concept to obtain mean Larmor precession transmission and reflection times for particles incident on a one-dimensional barrier of support (a, b) with the magnetic field covering the interval $(x_1 < a, x_2 > b)$ as

$$\tau_{yA}(x_1, x_2) = -\frac{2}{\hbar} \lim_{\omega_L \to 0} \frac{\langle s_y \rangle_A}{\omega_L}. \tag{8.13}$$

Here and in the following the shorthand notation $A = T, R$ will denote "transmission or reflection" depending on the case. The averages of the spin component s_y are obtained from the transmitted and reflected waves, respectively.

---

[2] This section is written with J. G. Muga and S. Brouard.

Later Büttiker [11] revisited the Larmor clock, which has subsequently received much further attention, due to possible applications in the description of characteristic times for quantum processes in ultrafast electronic devices, and also because of its importance to fundamental questions concerning the measurement of time-like quantities in quantum mechanics.

Steinberg reviews several experiments related to the Larmor clock in Chap. 11. Experiments that realize quite closely the ideal conditions of the gedanken Larmor clock have been performed quite recently by Hino et al. [55]. They have succeeded in measuring precisely spin precession angles of neutrons tunneling through a Permalloy45 ($Fe_{55}Ni_{45}$) ferromagnetic film.

8.4.1 Büttiker's Analysis of the Larmor Clock

Büttiker considered a stationary wave for a particle of mass M with energy $E = \hat{p}^2/2M$ under the influence of a rectangular potential $\hat{V}(y) = V_0\chi(\hat{y})$, where y is the spatial coordinate and χ the characteristic function for the barrier region (a, b) of width $d = b-a$ [11]. The particle has spin $\hbar/2$, (modulus of the) magnetic moment μ and is fully polarized in the x-direction in the incident beam. In the barrier region there is a uniform magnetic field in the z-direction, $\mathbf{B} = Be_z$, with e_z the unit vector in the z-direction (Fig. 8.1). The (matrix) Hamiltonian of the system is given by

$$\hat{H} = \frac{\hat{p}^2}{2M}\hat{1} + \chi_d(\hat{y})(\hat{1}V_0 + \mu\mathbf{B}\cdot\boldsymbol{\sigma}) \,,$$

where $\hat{1} = diag(1,1)$, and $\boldsymbol{\sigma} = (\hat{\sigma}_x, \hat{\sigma}_y, \hat{\sigma}_z)$ are the Pauli matrices, given by

$$\hat{\sigma}_x = \begin{pmatrix} 0 & 1 \\ 1 & 0 \end{pmatrix}, \hat{\sigma}_y = \begin{pmatrix} 0 & -i \\ i & 0 \end{pmatrix} \text{ and } \hat{\sigma}_z = \begin{pmatrix} 1 & 0 \\ 0 & -1 \end{pmatrix} . \tag{8.14}$$

Therefore the Hamiltonian may be written as

$$\hat{H} = \frac{\hat{p}^2}{2M}\hat{1} + \chi_d(\hat{y})(\hat{1}V_0 + \mu B\hat{\sigma}_z) \tag{8.15}$$

and acts on a spinor

$$\Psi = \begin{bmatrix} \psi_+ \\ \psi_- \end{bmatrix} .$$

The Hamiltonian (8.15) is diagonal in the spinor basis and therefore the scattering problem can be solved for each of the spin components ($-\hbar/2$ and $+\hbar/2$) separately. Let us remark that the quantum particle will in principle be deflected by the magnetic field, so that consideration of the motion in only one dimension is an approximation. In the small field limit, however, the most important effect of the magnetic field is to change the effective barrier height by $\mp\hbar\omega_L/2$, and the deflection can be ignored.

For each spinor component, the problem is solved in the standard way, taking as incident wave e^{iky}, as reflected wave $R_\pm e^{-iky}$ and as transmitted

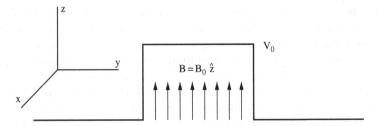

Fig. 8.1. Schematic graph representation of the Larmor clock

wave $T_\pm e^{iky}$. Then we impose the usual matching conditions with the wave inside the barrier, that has the form $B_\pm e^{\kappa_\pm y} + C_\pm e^{-\kappa_\pm y}$ (we are assuming $V_0 > E$). In these formulae, $k = \sqrt{2mE}/\hbar$, with E the energy of the incident particle, and $\kappa_\pm = (k_0^2 - k^2 \mp m\omega_L/\hbar)^{1/2}$, where $k_0 = \sqrt{2mV_0}/\hbar$.

The expectation values of the spinor components can be obtained analytically. The transmission amplitude and probability in the absence of magnetic field are given, respectively, by

$$T = |T|e^{i\delta\phi}e^{-ikd}, \; \mathcal{T} = |T|^2 = \frac{1}{1 + [(k^2 + \kappa^2)^2/4k^2\kappa^2]\sinh^2(\kappa d)} \; ,$$

where $\delta\phi$ is the "phase increase" across the barrier,[3]

$$\tan(\delta\phi) = \frac{k^2 - \kappa^2}{2\kappa k}\tanh(\kappa d) \; .$$

To obtain $T_\pm$, $\delta\phi_\pm$ and $\mathcal{T}_\pm$ we replace κ by $\kappa_\pm$ in the expressions for T, $\delta\phi$, and $\mathcal{T}$.

After some manipulations, the spin expectation values for the transmitted component of the particle yield

$$\langle s_z \rangle_T = \frac{\hbar}{2}\langle \Psi|\sigma_z|\Psi\rangle = \frac{\hbar}{2}\frac{\mathcal{T}_+ - \mathcal{T}_-}{\mathcal{T}_+ + \mathcal{T}_-} \; ,$$

$$\langle s_y \rangle_T = \frac{\hbar}{2}\langle \Psi|\sigma_y|\Psi\rangle = -\hbar\sin(\delta\phi_+ - \delta\phi_-)\frac{(\mathcal{T}_+\mathcal{T}_-)^{1/2}}{\mathcal{T}_+ + \mathcal{T}_-} \; , \qquad (8.16)$$

$$\langle s_x \rangle_T = \frac{\hbar}{2}\langle \Psi|\sigma_x|\Psi\rangle = \hbar\cos(\delta\phi_+ - \delta\phi_-)\frac{(\mathcal{T}_+\mathcal{T}_-)^{1/2}}{\mathcal{T}_+ + \mathcal{T}_-} \; .$$

In the small field limit, the precession time for transmission is thus

$$\tau_{yT} = -\frac{m}{\hbar\kappa}\partial_\kappa\delta\phi \; , \qquad (8.17)$$

---

[3] Comparing with (2.58), the total phase of the transmission amplitude is $\Phi_T = \delta\phi - kd$. Notice the difference between Φ_T and $\delta\phi$. In particular, $\hbar d\Phi_T/dE$ is the time delay (2.101) [or second summand in (2.94)], whereas $\hbar d\delta\phi/dE$ is the extrapolated phase time (2.94). These times differ from each other by the free motion term $Md/k\hbar$.

which is Rybachenko's result. However, Büttiker argued that the conventional Larmor precession does not apply in the presence of tunneling. Spin up and spin down components feel different barrier heights, which leads to different transmittances and to an additional spin rotation in the direction of the applied field. He associated a new time $\tau_{z,T}$ with this rotation,

$$\langle s_z \rangle_T = \frac{\hbar}{2} \omega_L \tau_{zT} , \qquad (8.18)$$

which is the most important effect for an opaque barrier. Taking the small field limit one obtains,

$$\tau_{zT} = -\frac{m}{2\hbar\kappa} \partial_\kappa \ln \mathcal{T} . \qquad (8.19)$$

The complete motion of the polarization during the barrier traversal is reflected, according to Büttiker, in the x-component of the polarization,

$$\langle s_{xT} \rangle_T = \frac{\hbar}{2} (1 - \omega_L^2 \tau_{xT}^2) . \qquad (8.20)$$

Since $\langle s_x \rangle_T^2 + \langle s_y \rangle_T^2 + \langle s_z \rangle_T^2 = \hbar^2/4$, it follows that

$$\tau_{xT} = (\tau_{yT}^2 + \tau_{zT}^2)^{1/2} . \qquad (8.21)$$

For the reflected part of the wave similar expressions are found,

$$\tau_{yR} = \tau_{yT} , \qquad (8.22)$$
$$\tau_{zR} = -\tau_{zT} \mathcal{T}/\mathcal{R} , \qquad (8.23)$$
$$\tau_{xR} = (\tau_{yR}^2 + \tau_{zR}^2)^{1/2} . \qquad (8.24)$$

For the square barrier the precession time equals the dwell time, $\tau_d = \tau_{yA}$ [11]. This is a particular feature of the rectangular or symmetric barriers that does not hold in general for unsymmetrical barriers, as shown, e.g., by Leavens and Aers [56]. The dwell time τ_d is defined as the ratio of $\int_a^b dy |\psi(y)|^2$ to the incident flux $J_I = \hbar k/m$ (see (8.10)). Büttiker stressed that the mean dwell time is not equal in general to the "extrapolated phase time", see (2.94), which in the present notation is given by

$$\tau_T^{Ph}(a,b) = \hbar \, d\delta\phi/dE , \qquad (8.25)$$

even though they may be comparable for energies above the barrier. More on their relation in 8.4.3 below. For small energy τ_T^{Ph} diverges whereas τ_{yT} tends to zero. Both quantities share the property of becoming independent of d for opaque barriers (this is the Hartman effect, see Sect. 2.4.1), in contrast with the linear dependence of τ_{xT}.

8.4.2 Generalizations and Relations to Other Approaches

The discussion carried out in the previous subsection is appropriate for rectangular barriers and for a magnetic field applied strictly in the barrier region. Leavens and Aers generalized Büttiker's treatment for arbitrary barriers and for a magnetic field confined to only part of the potential ("local Larmor clock") [56, 57, 58]. They also combined the precession and rotation times into a single complex time $\tau_A^{com} = \tau_A^{prec} - i\tau_A^{rot}$. To obtain a compact expression it is convenient to introduce the auxiliary barrier potential

$$\tilde{V}(y) \equiv V(y) + \Delta V \chi(y_1, y_2) \, , \tag{8.26}$$

so that

$$\tau_A^{com}(k; y_1, y_2) = i\hbar \frac{\partial \ln |A|^2}{\partial \Delta V}\bigg|_{\Delta V=0} . \tag{8.27}$$

In fact this expression is applicable for arbitrary y_1 and y_2, inside or outside the barrier.[4]

These complex times were also obtained using the path integral analysis of Sokolovski and Baskin [59] reviewed in Chap. 7. It is convenient for later reference to write out the explicit forms of the real and imaginary parts,

$$\tau_A^{prec}(k; y_1, y_2) = -\hbar \frac{\partial \Phi_A}{\partial \Delta V}\bigg|_{\Delta V=0} , \tag{8.28}$$

$$\tau_A^{rot}(k; y_1, y_2) = -\hbar \frac{\partial \ln |A|}{\partial \Delta V}\bigg|_{\Delta V=0} , \tag{8.29}$$

where $A = |A| \exp(i\Phi_A)$. We shall use the term "Büttiker–Landauer times" for the moduli of these complex quantities, $\tau_A^{BL} = |\tau_A^{com}|$. In the opaque barrier limit τ_T^{BL} coincides with the "traversal time" obtained by Büttiker and Landauer using semiclassical (WKB) arguments, see also [60], or as the time characterizing the transition from sudden to adiabatic regimes in an oscillating barrier, $\tau_T^{BL} = Md/\hbar\kappa$ for the opaque square potential.

It is interesting to note that for an electron colliding with a double symmetric barrier, the infinitesimal magnetic field being confined to an arbitrary region, the transmission time (8.27) for that region is real at resonance, that is to say when the incident energy of the electron agrees with the quasienergies of the well between barriers (see Fig. 8.2). In this case Leavens and Aers have also shown that the local transmission speed can be larger that the speed of light near the nodal points of the quasistationary wave functions in the well region. This phenomenon does not take place for Dirac's equation, however.

In [61] Golub, Felber, Gäler, and Gutsmiedl discussed the possibility of adding a small imaginary part ΔV_{Im} to the potential in a region of space so that a mean dwell time could be obtained from the exponential decay of the

---

[4] In this chapter the incident wave is always assumed to have positive momentum so we skip the superscript "l" (left incidence) used in Chap. 2 for the reflection amplitude R.

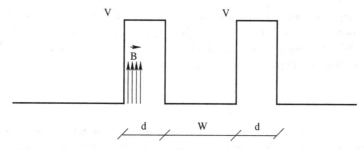

Fig. 8.2. Double well barrier treated by Leavens et al

particle norm due to the absorption. Later Huang and Wang extended this approach by defining absorption times for reflection and transmission [62]. As for the Larmor clock analysis, the (now imaginary) perturbation is assumed to be infinitesimal. The dwell time in the space region selected by the projector $\widehat{D} = \int_{y_1}^{y_2} |y\rangle dy\langle y|$ will determine the final absorption according to

$$|T|^2 + |R|^2 = \exp(-\tau_D/\tau_c) , \tag{8.30}$$

where the characteristic decay time is $\tau_c = -\hbar/2\Delta V_{Im}$. In the small ΔV_{Im} limit,

$$\tau_D = \frac{\hbar}{2} \frac{\partial(|T(p; \Delta V_{Im})|^2 + |R(p; \Delta V_{Im})|^2)}{\partial(\Delta V_{Im})}\bigg|_{\Delta V_{Im}=0} \tag{8.31}$$

whereas, from the relation

$$\frac{|A(\Delta V_{Im})|^2}{|A(\Delta V_{Im} = 0)|^2} = \exp(-\tau_A^a/\tau_c) \tag{8.32}$$

the absorption transmission and reflection times are defined as

$$\tau_A^a = \frac{\hbar}{2} \frac{\partial \ln |A(p; \Delta V_{Im})|^2}{\partial \Delta V_{Im}}\bigg|_{\Delta V_{Im}=0} . \tag{8.33}$$

These absorption times are equal to the precession times of the Larmor clock [62, 63], as may be seen by adding to the physical potential barrier the auxiliary potential $\Delta V = \Delta V_{Re} + i\Delta V_{Im}$, and using the Cauchy–Riemann conditions

$$\frac{\partial T_{Re}}{\partial \Delta V_{Re}}\bigg|_{\delta V=0} = \frac{\partial T_{Im}}{\partial \Delta V_{Im}}\bigg|_{\delta V=0} = \frac{2m\pi}{p}\mathrm{Im}\langle p^-|\widehat{D}|p^+\rangle|_{\Delta V=0} , \tag{8.34}$$

$$\frac{\partial T_{Im}}{\partial \Delta V_{Re}}\bigg|_{\delta V=0} = -\frac{\partial T_{Re}}{\partial \Delta V_{Im}}\bigg|_{\delta V=0} = -\frac{2m\pi}{p}\mathrm{Re}\langle p^-|\widehat{D}|p^+\rangle|_{\Delta V=0} , \tag{8.35}$$

$$\frac{\partial R_{Re}}{\partial \Delta V_{Re}}\bigg|_{\delta V=0} = \frac{\partial R_{Im}}{\partial \Delta V_{Im}}\bigg|_{\delta V=0} = \frac{2m\pi}{p}\mathrm{Im}\langle -p^-|\widehat{D}|p^+\rangle|_{\Delta V=0} , \tag{8.36}$$

$$\frac{\partial R_{Im}}{\partial \Delta V_{Re}}\bigg|_{\delta V=0} = -\frac{\partial R_{Re}}{\partial \Delta V_{Im}}\bigg|_{\delta V=0} = -\frac{2m\pi}{p}\mathrm{Re}\langle -p^-|\widehat{D}|p^+\rangle|_{\Delta V=0} . \tag{8.37}$$

The subscripts Re and Im denote real and imaginary parts of the amplitudes A or the potential ΔV. To obtain the last equalities in each of the equations (8.34)–(8.37) in terms of scattering states $|p^{\pm}\rangle$ [63] (see their definition, (2.22), and properties in Chap. 2) use has been made of

$$\frac{d\hat{V}}{d\Delta V} = \hat{D}(y_1, y_2) , \qquad (8.38)$$

and the relations between A and the on-shell matrix elements of the transition operators, Sect. 2.2.5. These equalities allow to express the Larmor times also in terms of the systematic projector approach of Brouard, Sala, and Muga [33], see Chap. 1. Using (1.17), (2.26), (2.32), (2.42), and (8.34)–(8.37) one finds that the complex times τ_A^{AD} for transmission and reflection,

$$\tau_A^{AD}(p) \equiv \frac{1}{J_I|A|^2} \langle p^+|\hat{A}\hat{D}|p^+\rangle , \qquad (8.39)$$

coincide with τ_A^{com} since their real and imaginary parts, corresponding to the anticommutator and commutator of $\hat{A}\hat{D}$, coincide [63],

$$\tau_A^{[A,D]_+/2} = \frac{1}{J_I|A|^2} \langle p^+|(\hat{A}\hat{D} + \hat{D}\hat{A})/2|p^+\rangle = Re\tau_A^{AD} = \tau_A^{prec} , \quad (8.40)$$

$$\tau_A^{[A,D]_-/2i} = \frac{1}{J_I|A|^2} \langle p^+|(\hat{A}\hat{D} - \hat{D}\hat{A})/2i|p^+\rangle = Im\tau_A^{AD} = -\tau_A^{rot} , \quad (8.41)$$

note that $\hat{A}\hat{D} = \frac{[\hat{A},\hat{D}]_+}{2} + i\frac{[\hat{A},\hat{D}]_-}{2i}$. The projector approach provides a simple way to derive some of the properties of the different Larmor times defined. For example, since the symmetrized operators $[\hat{A}, \hat{D}]_+/2 = (\hat{A}\hat{D} + \hat{D}\hat{A})/2$ are not positive, the precession times may indeed be negative; the same holds true for rotation times; from the relations $\hat{D} = \hat{T}\hat{D} + \hat{R}\hat{D} = [\hat{P}, \hat{D}]/2 + [\hat{R}, \hat{D}]/2$ it follows that the complex times and their real parts (precession times) satisfy the relation

$$\tau_D(x_1, x_2) = |T|^2\tau_T(x_1, x_2) + |R|^2\tau_R(x_1, x_2) , \qquad (8.42)$$

which is not satisfied in general by the Büttiker–Landauer times, and whose physical content or necessity has been much debated [56, 64, 65, 66], whereas the contribution of the rotation times (associated with the operators $[\hat{A}, \hat{D}]_-/2i$) adds to zero,

$$|T|^2\tau_T^{rot}(x_1, x_2) + |R|^2\tau_R^{rot}(x_1, x_2) = 0 . \qquad (8.43)$$

Also, since precession, rotation and complex times depend linearly on $\hat{D}$, all these times are additive, $\tau(y_1, y_3) = \tau(y_1, y_2) + \tau(y_2, y_3)$ for $y_3 > y_2 > y_1$, whereas the Büttiker–Landauer times (the moduli of the complex times) are not. $\tau_A^{BL}(y_1, y_2)$ is always positive but is not always a monotonically increasing function of y_2 [57].

A further connection between Larmor times and weak measurements [30] is discussed in Chap. 11 by Steinberg.

8.4.3 Relation to Phase Times

The relation between phase times and precession times is somewhat subtle. Explicit expressions for phase and precession times for the square barrier with the field confined between $x_1 < a$ and $x_2 > b$ show that $\tau_T^{Ph}(x_1, x_2)$ and $\tau_T^{prec}(x_1, x_2)$ do not coincide in general, see, e.g., [64]. The later is equal to the former plus oscillatory terms that vanish in the large energy limit. Moreover, by making the field region larger and larger both quantities tend to infinity.

It is sometimes stated, however, that phase and precession times are *equal* for a barrier confined between a and b when the field is applied to the whole space [62, 67]. Increasing the level of the potential function (from $-\infty$ to ∞) by a (real) amount ΔV for constant E is equivalent physically to decreasing E for a constant potential. The particle's momentum may in that case be written as $p = [2m(E - \Delta V)]^1/2$, so that [62]

$$\hbar \frac{\partial \delta \phi}{\partial E} = -\hbar \frac{\partial \delta \phi}{\partial \Delta V} \, . \tag{8.44}$$

The left-hand side is the extrapolated phase time between a and b, whereas the right-hand side *resembles* the real part of the complex time, but what is its interpretation in terms of a precession? It is definitely not the precession time between $x_1 = -\infty$ and $x_2 = \infty$, which should be infinite, nor it is the precession time when the field is applied in the barrier region only [63]. Huang et al. [67] could give it a meaning, computing the precession angle by comparing the spin polarizations of the incident, transmitted, and reflected waves right at the edges a and b of the barrier (contrary to the scattering problem discussed so far, the polarization does not acquire a constant value outside a selected region due to the omnipresent field). The procedure has been however criticized on the grounds that in a real time-dependent setting, the interference between incident and reflected components at the barrier edge cannot be ignored and would not allow a separation of the total polarization into "incident" and reflected components [58].

8.4.4 Wave Packets

All the previous Larmor times have been obtained from stationary states. Rigorously, it is necessary to examine physically realizable wave packets instead. This resembles the need to study wave packet collisions to justify the cavalier obtention of cross sections based on stationary scattering states. In fact, the analysis shows, as in the scattering theory case, that the stationary results are meaningful as components and/or limits of the wave packet results.

Assume that a wave packet with positive momentum impinges on the potential, composed as in the stationary case by a barrier and an infinitesimal field. Jaworski and Wardlaw used S-matrix perturbation theory to show that the precession and rotation angles may be expressed as the real and (minus) imaginary components of the complex quantities [68]

$$\tau_T^{com} = \frac{1}{P_T} \langle \phi_{in} | \widehat{S}^\dagger \widehat{F}_+ \widehat{S} \widehat{T}_D | \psi_{in} \rangle \tag{8.45}$$

$$\tau_R^{com} = \frac{1}{P_R} \langle \phi_{in} | \widehat{S}^\dagger \widehat{F}_- \widehat{S} \widehat{T}_D | \psi_{in} \rangle , \tag{8.46}$$

where $\widehat{S}$ is the scattering operator introduced in (2.10) or (2.26), $\widehat{T}_D$ the sojourn-time operator (2.69), ϕ_{in} the incoming asymptote, $\widehat{F}_\pm$ the projector onto positive or negative momenta, and $P_{R,T}$ the reflection or transmission wave-packet probability. Using (2.10) and (1.17), they may also be written in the language of the BSM projector theory, Sect. 1.6.3, as

$$\tau_A^{com} = \tau_A^{AD} = \frac{1}{P_A} \int_{-\infty}^{\infty} dt \langle \psi(t) | \widehat{A}\widehat{D} | \psi(t) \rangle . \tag{8.47}$$

In the limit of a wave packet sharply peaked around a given momentum these times reduce to the ones introduced in the stationary case (and similarly the precession and rotation terms). By inserting resolutions of the identity in terms of scattering states it can be shown that in the general case the relation between wave packet and stationary results is given by [33]

$$\tau_A^\alpha = P_A^{-1} \int_0^{\infty} dp \, \tau_A^\alpha(p) A(p) |\phi_{in}(p)|^2 , \tag{8.48}$$

where α is any of the combinations of A and D, such as AD, $[A, D]_+/2$, or $[A, D]_-/2i$.

8.5 Other Clocks

Following similar ideas to those presented above there are many different proposals of quantum clocks. We consider two of them: the Faraday clock and the micromaser clock and its variants.

8.5.1 The Faraday Clock

Several people have exploited the analogies between tunneling electrons and evanescent electromagnetic waves to construct an optical clock [14, 15, 16, 69]. The first proposal was presented by Gasparian et al. [14]. They used Faraday polarization rotation to measure the traversal and reflection times of an evanescent electromagnetic wave in a slab and in a magnetorefractive layered structure.

The Faraday effect is understood as follows: a linearly polarized electromagnetic plane wave (the results are also valid for a wave packet) enters the slab (of length L, Faraday constant g and refractive index n_0) along the x-axis from the left at normal incidence. The electric field **E** in this incident wave is orientated along the z-axis and a weak magnetic field **B** is applied along the x-direction across the slab. This magnetic field rotates the plane of polarization of the light with respect to the original direction of polarization and, additionally, generates ellipticity. Gasparian et al. [14] quantified both effects through the complex angle θ given by

$$\theta = \theta_1 - i\theta_2 \, ,$$

where, in terms of the complex amplitude of transmission $T_\pm = |T_\pm| \exp i\psi_\pm$ (here the $+$ and $-$ signs correspond to the outgoing right and left polarized light, respectively), the real part of the angle is

$$\theta_1 = \frac{\psi_+ - \psi_-}{2} \, ,$$

corresponding to Faraday rotation, and the imaginary part of θ is

$$\theta_2 = \frac{1}{4} \ln \frac{T_+}{T_-} \, ,$$

which corresponds to the ellipticity ratio.

If one neglects the influence of the boundaries of the slab, the standard Faraday rotation is

$$\theta_0 = \frac{\omega g B L}{2 c n_0} = \Omega \tau_0 \, , \tag{8.49}$$

where $\tau_0 = L n_0 / c$ is the time that light with velocity c/n_0 would take to cross the slab. Then, following Büttiker's analysis for electrons and by analogy with (8.49), Gasparian et al. [14] defined the characteristic time

$$\tau = \tau_1 - i\tau_2 \, ,$$

where

$$\tau_1 = \frac{\theta_1}{\Omega} = \frac{n_0}{\omega} \frac{\delta\psi}{\delta n_0}$$

and

$$\tau_2 = \frac{\theta_2}{\Omega} = \frac{n_0}{2\omega} \frac{\delta \ln T}{\delta n_0} \, .$$

These results and those obtained for a quantum particle are very similar: the real component of this time, τ_1, is analogous to the time associated with the Larmor precession in the electronic case, and the imaginary part, τ_2, is analogous to the rotation time associated with Zeeman splitting in the electronic case.

The Faraday clock presents some advantages with respect to the Larmor one: advances on photonic tunneling (see Chap. 11) make experimental measurements of the Faraday rotation and ellipticity much easier than the observation of the spin rotation, and the interactions between particles (photons) are completely negligible. Also, Deutsch and Golub [16] argue that the times provided by the Larmor clock are, at most, a definition and do not emerge as naturally as the times related to the Faraday clock do.

8.5.2 The Micromaser (Rabi) Clock

Let us suppose that we have a two-level atom and a radiation field and that the effective coupling between the atomic degrees of freedom and those of the radiation is only different from zero in a certain region, a cavity. When the atom crosses the cavity, which can be tuned at the transition energy between the two atomic levels of the atom, it will undergo Rabi oscillations and the corresponding Bloch vector will suffer a rotation. The micromaser clock is based on the relation between the above-mentioned rotation angle and the interaction time [70].

In this clock a beam of two-level atoms coming from the left to a micromaser cavity is considered. The two levels are denoted by $|+\rangle, |-\rangle$. The cavity is tuned to the transition energy $\hbar\omega$. If the other modes of the cavity can be neglected, the micromaser wave function can be given in terms of the eigenstates of the photon number operator $|0\rangle, |1\rangle, \cdots, |n\rangle, \cdots$. The corresponding Hamiltonian of the system is given by

$$\widehat{H} = \frac{\widehat{p}^2}{2M} + V(x) + \hbar\omega(\widehat{a}^+\widehat{a} + \widehat{\pi}^+\widehat{\pi}) + \hbar\nu(x)(\widehat{a}^+\widehat{\pi} + \widehat{\pi}^+\widehat{a}),$$

under the dipole and rotating wave approximations [71]. The first two terms correspond to the kinetic energy of the atom and some potential $V(x)$, which pertains to the center of mass motion of the atom. Next come the terms involving $\{\widehat{\pi}, \widehat{\pi}^+\}$ and $\{\widehat{a}, \widehat{a}^+\}$, which are the photonic and atomic annihilation and creation operators, whose action on states is defined by $\widehat{\pi}^+|n\rangle = \sqrt{n+1}|n+1\rangle$ and $\widehat{a}^+|-\rangle = |+\rangle$. The coupling constant $\nu(x)$ depends on the spatial coordinate x so that ν is only different from zero within the cavity. The last term is responsible for the change in the internal degrees of freedom of the atoms (two-level systems) accompanied by the emission or absorption of a single photon. In this way Rabi oscillations between the excited and the ground state of the atom take place, and from these the characteristic times of the motion are extracted [70].

This procedure can be taken together with the phase time approach to provide a way of measuring traversal times of a free cavity [72]. These traversal times, however, can be negative, and they are measured by "following the peak," thus not being included in the context of quantum clocks as such.

Another alternative way of using Rabi oscillations as a quantum clock is the setup discussed by Bužek et al. [73, 74], in which the motion of the system is disregarded (Raman–Nath approximation). Consider first the very simplified situation of one two-state atom or ion initially in state $|\psi(0)\rangle = (|0\rangle + |1\rangle)/\sqrt{2}$, which evolves under the Hamiltonian $\widehat{H}_0 = \hbar\omega|1\rangle\langle 1|$. After a time t has elapsed, the state of the system is $|\psi(t)\rangle = (|0\rangle + \exp(-i\omega t)|1\rangle)/\sqrt{2}$, and the probability of finding the system in the initial state at this moment is $P(t) = (1 + \cos(\omega t))/2$. By repeating the experiment several times, if we know for sure that the same interval of time has happened, we can measure its length through the function $P(t)$. This does not look like a very efficient way of measuring time, since we would need some way of making sure that the same time interval has passed in all instances of the experiment, and this is precisely what we would like to measure. However, this system is equivalent to that presented in Sect. 8.3.2, under the restriction $N = 2$, as can be readily seen. The relevant projectors are $\hat{P}_0 = |\psi(0)\rangle\langle\psi(0)|$ and $\hat{P}_1 = |\psi(\tau)\rangle\langle\psi(\tau)|$, where $\tau = 2\pi/N\omega$. In this manner, the system works as a clock as follows: let the system begin in the state $|\psi(0)\rangle$, and after the time interval we wish to measure has elapsed, we check whether the state is in the same state. If it is, we *estimate* the time elapsed as $0 \mod(2\tau)$. If it is not in the state $|\psi(0)\rangle$, we estimate the time elapsed as $\tau \mod(2\tau)$.

Is there any way of improving these estimates? Consider a set of n two-state ions or atoms. Bužek et al. pose the following question: Is there an initial state and a projector or positive operator valued measure that optimizes the estimates of time? The underlying hypothesis is that on measuring that the system is in the eigenspace of a given projector $\hat{P}_r$ we would estimate the elapsed time to be t_r (modulus a period). The answer is, as was only to be expected, affirmative. This affirmative result is an elaboration of the theorems by Holevo ([75], Chap. 10) concerning optimality, extended to finite POVMs [76], and applied to the present situation [73].

The Hamiltonian of the system is $\hat{H} = \hat{H}_0 \otimes 1 \otimes 1 \cdots \otimes 1 + 1 \otimes \hat{H}_0 \otimes 1 \cdots \otimes 1 \cdots$. There are $\binom{n}{k}$ states with energy $\hbar k\omega$, given by the presence of k atoms in state $|1\rangle$ and $n - k$ in the ground state $|0\rangle$. Define the states $|k\rangle$ as the symmetric superposition of all $\binom{n}{k}$ states with energy $\hbar k\omega$. Let the reference state $|\psi_0\rangle$ be

$$|\psi_0\rangle = \frac{1}{n+1} \sum_{k=0}^{n} |k\rangle \,,$$

and define the states

$$|\psi_r\rangle = e^{i2\pi r\hat{H}/(n+1)\hbar\omega} |\psi_0\rangle \,.$$

We thus have a set of orthogonal projectors $P_r = |\psi_r\rangle\langle\psi_r|$. We prepare the set of n atoms in an initial reference state $\hat{\rho}(0)$. After the time interval we desire to measure has elapsed, we measure the projection of the evolved state over the projectors. If we find that the system is in state $\hat{P}_r$, then we *estimate* the elapsed time as $t_r = 2\pi r/((n+1)\omega)$ (modulo the period $2\pi/((n+1)\omega)$). The operator being measured is thus equivalent to the time operator (8.6), i.e., equivalent to the phase operator of Pegg and Barnett [77].

The problem of optimality is reduced to the adequate determination of the initial state $\hat{\rho}(0)$, which is dependent on the cost function used. The cost function f gives the weight $f(t - t_r)$ assigned to the error we make if we assign the estimate t_r when the true elapsed (parametric) time is t. If this cost function is the delta function, then the best initial state is $\hat{P}_0$. However, if great disparities between the estimation and the actual time are more strongly penalized than small ones, then other initial states provide better clocks. Bužek et al. [73] give as optimum state $\widehat{P}_{opt} = |\psi_{opt}\rangle\langle\psi_{opt}|$, in the case of large n, where

$$|\psi_{opt}\rangle = \sqrt{\frac{2}{n+1}} \sum_{k=0}^{n} \sin\left(\frac{\pi(k+1/2)}{n+1}\right) |k\rangle .$$

A further development concerns successive measurements of time. Bužek et al. [74] proved that for large atom ensembles the system is robust with respect to subsequent measurements.

Quite a different use of this kind of system was put forward by Bollinger et al. [78]: a number of entangled ions is used to improve the measurement of a frequency, going beyond the shot noise limit. Huelga et al. [79] criticized this proposal on physical grounds, asserting that taking into account decoherence properly entails that the improvement in the measurement of frequency does not take place for the maximally entangled state.

8.5.3 Irreversible Stopwatch Models

In the example of the Rabi clock and derived models, irreversibility appears through the measurement process. This is in fact general in all quantum clocks, starting from Salecker–Wigner's on. There is however a different kind of irreversibility, which has to do with decoherence in the form of coherences being lost because of large number of degrees of freedom. In the protracted history of the study of measurement in relation to quantum mechanics, one approach has been that of modeling detectors by means of quantum systems with such a large number of degrees of freedom, so the models of the kind proposed for determining time quantities follow naturally.

One such proposal was put forward by Schulman [80] in the context of his "two time theory" (see his book [80] and his contribution to this volume in Chap. 4). A further development, with the added advantage that the new model was exactly solvable, was carried out by Halliwell [81]. This model is explained in Halliwell's contribution to this volume in Chap. 6.

In Halliwell's model, the detector is a two-level system, which is coupled to the particle whose arrival at point $x = 0$ we want to detect, and to a (fast) environment, through a term of the form $\lambda\theta(\widehat{x})\widehat{H}_{d\varepsilon}$, where $\widehat{x}$ is the position operator of the system particle, and $\widehat{H}_{d\varepsilon}$ is an interaction term between the environment and the detector. This term, whenever activated, leads to a fast decay of the latter from the excited state $|1\rangle$ to the ground state $|0\rangle$. If we then start with the particle in $x < 0$ and the detector in its excited state, we will know that the particle has crossed $x = 0$ if the detector is found in its ground state. Notice that in this model the measuring of time is as follows: we prepare the system, let go, and at some time t measure whether the detector is or is not in the ground state. If it is, then we assert that the system particle crossed $x = 0$ at some instant previous to t. Therefore, we do not directly obtain the density of arrivals, but a derived quantity: the probability that at time t the detector remains in the excited state. The derivative of this quantity with respect to time is minus the probability density that the system particle crosses $x = 0$ at instant t.

A further point to remark concerning Halliwell's model is that the measurement of time itself is performed with an external clock, and the measurement of a time quantity pertaining to the system is carried out through the measurement of a time quantity pertaining to the detector. Another model, somewhat more involved, which has this property as well, was proposed by Muga et al. [82]. Even so, this model has the advantage that, treated with quantum jump techniques, it provides us with a way of computing the statistics concerning correlations between other time quantities, as we shall see. The model consists of an atom or ion moving in one dimension, which finds in its way either the vacuum electromagnetic state, or a laser tuned to a particular transition of the atom or ion. In this case, and using the rotating wave and dipole approximations [83], we can restrict ourselves to the internal Hamiltonian as

$$\widehat{H}_{\text{internal}} = \frac{\hbar\omega}{2}\left(|1\rangle\langle 1| - |0\rangle\langle 0|\right),$$

so that, by passing to the interaction picture, and neglecting recoil effects, the master equation takes the form

$$\frac{d\widehat{\rho}}{dt} = -\frac{i}{\hbar}\left[\widehat{H},\widehat{\rho}\right] + \frac{\gamma}{2}\left(2\widehat{\sigma}^-\widehat{\rho}\widehat{\sigma}^+ - \widehat{\sigma}^+\widehat{\sigma}^-\widehat{\rho} - \widehat{\rho}\widehat{\sigma}^+\widehat{\sigma}^-\right),$$

where

$$\widehat{H} = \frac{\widehat{p}^2}{2m} + \frac{\hbar}{2}\Omega(\widehat{x})\left(\widehat{\sigma}^+ + \widehat{\sigma}^-\right),$$
$$\widehat{\sigma}^- = |0\rangle\langle 1|,$$
$$\widehat{\sigma}^+ = |1\rangle\langle 0|,$$
$$\Omega(\widehat{x}) = \Omega_0\theta(\widehat{x}).$$

$\Omega(x)$ is the position-dependent Rabi frequency, and γ the decaying constant of the excited level $|1\rangle$ due to the coupling to the background vacuum state.

This master equation can be interpreted as follows: the atom moves in one dimension, in the internal ground state $|0\rangle$, until it perceives the laser field at $x > 0$. This laser field can take the atom to the excited state. The speed with which this is performed is related to the intensity of the laser and the coupling, in the combination given by Rabi's frequency. The atom, nonetheless, is still coupled to the background vacuum state, and thus can and will decay, emitting photons that will be detected (in the process of deducing the master equation one runs into subtleties regarding emitted versus detected photons [84, 85, 86] that we will not be considering further).

In this model, then, the time at which the *first* photon is detected is a good proxy for the time of first arrival of the atom to the region $x > 0$, i.e., a good measurement of that quantity. If only the instant of the first photon detection is of importance, the solution of the master equation may be simplified by considering the (two component) amplitude ψ_c conditioned to no-photon detection. It satisfies an effective Schrödinger equation with effective Hamiltonian

$$\widehat{H}_{eff} = \widehat{H} - i\hbar\frac{\gamma}{2}|2\rangle\langle 2| . \tag{8.50}$$

The distribution of times of detection of the first photons can be computed exactly as $-dN/dt$, where $N = \langle\psi_c|\psi_c\rangle$ is the probability of not having a detection up to time t.

Numerical comparison has shown [82] that there may be very good agreement between Kijowski's distribution and the distribution of first detected photons. Perfect agreement cannot be obtained, however. An intuitive explanation is that inevitably "charging" and "decaying" delays will be introduced: the system needs some time in presence of the laser field to change into the excited state, and some time to decay to the ground state. Additional to this fact, it is clear that particles will be reflected by the laser field if it is too intense, thus producing a depletion in the density of arrivals. Conditions have been found that establish a regime where delay in detecting the atom and reflection can be neglected [87]. In this respect, one should notice that the idea that first comes to mind of tuning the decay by taking γ to infinity (or, more realistically, choosing a system for which the decay is extremely fast) in order to avoid delay in the detection of the atom is not adequate: the driving of the system by the laser becomes so inefficient that actually no transition ever takes place. If, on the other hand, a weak driving laser is chosen, there is a big delay due to the "charging" of the system. In [87] the idea was put forward that, at least in some regimes, one might be able to assume that the observable we measure is the sum of two independent quantities: the arrival time proper, and the time it takes for the atom to emit from the arrival time on. Furthermore, it is likely that in some regimes the latter distribution be the same that an atom would present when stationary in a laser field. Under these joint assumptions, the probability density our gedanken experiment would be measuring would be the convolution of the density for time of emission of the first photon by a stationary atom in a laser field and the (ideal) probability density of time of arrivals. This last interesting quantity can then be recovered by deconvolution with a known distribution, and is amenable to analytical exploration. An exploration which leads to the interesting result that in the limit of no reflection the ideal density is actually nothing but the current! This obviously means that the joint hypotheses of independence, on the one hand, and delay distribution being the one of the stationary case, on the other, are actually not true, since the function one recovers need not be positive semidefinite, and is therefore not a probability density.

One of the richness of this model is that the quantum jump technique [84, 85, 86, 88, 89] allows us to consider other quantities, such as the time a particle leaves the laser region, or dwell times, or other techniques for the same arrival time density. One such idea is to use a laser in a finite region [90]: this is no different from a laser in an infinite region in those regimes in which the penetration before activation and emission is much smaller than the width of the region. Heuristically, this penetration length is given by

$$l = 5v \left(\frac{2}{\gamma} + \frac{\gamma}{\Omega^2} \right).$$

For finite regions of laser presence such that the width is smaller than l, the effectiveness of the measurement (as denoted by the fraction of atoms that actually emit) is dependent on the speed of the atom v and the intensity of the laser, in a fan-like structure, with some straight lines in the $v - \Omega$ space, with Ω the Rabi frequency, leading to total efficiency. The idea of separating the time of arrival of the first photon in two independent quantities, as before, requires good control of the hypothesized distribution for first emission. In the regime of small laser region, with size leading to resonant excitation of the atom, one could assume that decay only takes place outside of this region, so the distribution is that of first emission of photons by an excited stationary atom. The current density is again recovered as the ideal distribution in a number of regimes under those two hypotheses of sum of independent quantities and delay being given by emission density for an excited stationary atom.

These ideas can be refined to search for optimized detection for a class of kinematical states of the atom [91]. One surprising aspect is that the object which is recovered as "ideal" after the deconvolution is the current density and not Kijowski or some other probability density. However, using a different technique Kijowski's distribution is reobtained in some limit [92, 93]: this is operator normalization [94]. The idea is to use a filtering after and before the measurement device so as to enhance the signal obtained. The filtering is adapted to the signal one wants to detect. It is difficult however to propose simple models for the filtering procedure, although the mathematical idea is fully developed.

As one can see, the simple model of a two-level atom impinging on a laser-filled region, with the detection times of first photons being used as a proxy for arrival of the atom to the region, has revealed quite a wealth of phenomena. Furthermore, the one-dimensional analyses mentioned here have also been expanded to three dimensions [95], and one could hope that they might be of use for the experimentalist.

8.6 Simple "Time-dependent" Clocks: the Kick Clock

So far we have considered "time-independent clocks," in the sense that their Hamiltonians are time independent. Let us suppose now that the clock Hamiltonian $\widehat{H}_{s-c}$ is activated periodically during intervals of time τ, but remains zero otherwise. To achieve this behavior, we can imagine a periodic time signal $F(t)$ of period T with the following form:

$$F(t) = 1 + \frac{2\mathsf{T}}{\tau\pi} \sum_{n=1}^{\infty} \frac{1}{n} \sin\left(\frac{n\pi\tau}{\mathsf{T}}\right) \cos\left(\frac{2\pi nt}{\mathsf{T}}\right) . \tag{8.51}$$

This corresponds to a periodic series of stepwise functions activated during intervals of duration τ with intensity T/τ. Thus the time-dependent clock Hamiltonian is given by

$$\widehat{H} = \widehat{H}_s + \widehat{H}_{s-c}(t) = \widehat{H}_s + F(t)\widehat{H}_{s-c} . \tag{8.52}$$

To simplify the model we can further consider the limit $\tau \to 0$,

$$\widehat{H} = \widehat{H}_s + \widehat{H}_{s-c}(t) = \widehat{H}_s + \mathsf{T}\delta_\mathsf{T}(t)\widehat{H}_{s-c} , \tag{8.53}$$

with $\delta_\mathsf{T}(t) = \sum_{n=-\infty}^{\infty} \delta(n\mathsf{T} - t)$. "Kicked" systems like these are useful in different fields, e.g. in the study of chaotic behavior, because of their simplicity [96, 97, 98, 99]. In particular, the unitary evolution operator for the Hamiltonian (8.53) over a time T has the manageable form

$$\widehat{U}^\mathsf{T} = \exp\left(-\frac{i}{\hbar}\mathsf{T}\widehat{H}_s\right) \exp\left(-\frac{i}{\hbar}\mathsf{T}\widehat{H}_{s-c}\right) , \tag{8.54}$$

which defines a quantum map,

$$|\psi((n+1)\mathsf{T})\rangle = U^\mathsf{T}|\psi(n\mathsf{T})\rangle . \tag{8.55}$$

Thus the information of the evolution is obtained at integer multiples of T as a sequence of applications of the evolution operator of the system over a time T followed by an application of the evolution operator of the clock over a time T. Clearly, if $\mathsf{T} \to 0$ the infinitesimal evolution operator $U^{\mathsf{T}\to 0}$ of the kicked system coincides with the infinitesimal evolution operator of the time-independent clock with Hamiltonian $\widehat{H} = \widehat{H}_s + \widehat{H}_{s-c}$ up to $O(\mathsf{T}^2[\widehat{H}_s, \widehat{H}_{s-c}])$. This is a direct consequence of the definition of $F(t)$. We may expect a proper functioning of the kicked clock when the period of the the signal $F(t)$, T, is a fraction of the dwell time of the particle in the interaction region. The surprising fact is that very few kicks are enough to reproduce the non-kicked clock behavior.

The most common quantum clocks found in the literature are the Salecker–Wigner clock and the Larmor clock. The kicked versions of these clocks are described next.

• Kicked Salecker–Wigner clock
  Let us next consider a particle of mass m moving freely in one dimension with Hamiltonian $\widehat{H}_s = \widehat{p}_x^2/2m$. In a certain region of length d the particle is coupled to the clock as follows [5, 12, 27, 28]:

$$\widehat{H} = \widehat{p}_x^2/2m + \chi_d(\widehat{x})\omega\widehat{J} = \widehat{p}_x^2/2m - i\omega\hbar\chi_d(\widehat{x})\partial/\partial\theta, \tag{8.56}$$

where

$$\chi_d(x) = \begin{cases} 1, & \text{if } x \in [0, d] \\ 0, & \text{elsewhere} \end{cases} \qquad (8.57)$$

is the characteristic function indicating that the clock runs only when the particle "is" within the interval $[0, d]$.

Noting that $\widehat{H}$ commutes with $\widehat{J}$, the time-dependent wave function with initial state $\psi(x)v(\theta, t = 0)$ is given by

$$\frac{1}{N^{1/2}} \sum_n \psi_n(x, t) e^{-in\omega t/\hbar} u_n(\theta), \qquad (8.58)$$

where $\psi_n(x, t)$ is the (partial) wave that evolves from $\psi(x)$ with the Hamiltonian

$$\widehat{H}_s + \widehat{V}_n = \widehat{p}_x^2/2m + n\hbar\omega\chi_d(\widehat{x}), \qquad (8.59)$$

which represents a particle that collides with a rectangular barrier (well) of height (depth) $V_n = n\hbar\omega$ and width d. For an incident particle with energy $E = p_x^2/2m$ and wave number $k = \sqrt{2mE}/\hbar$, the wave number inside the barrier is $k' = \sqrt{2m(E - V_n)}/\hbar$ so that the phase shift caused by the barrier is approximately given by

$$(k' - k)d \sim -n\omega t_f, \qquad (8.60)$$

where $t_f = d/(2E/m)^{1/2}$ is the classical time of flight. The right-hand side in (8.60) is a good approximation if $E >> |V_n|$, which means that the disturbance caused by the measurement is negligible. If the incident wave packet is very much peaked (in wave number) around

$$\psi_i(x, \theta) = e^{ikx} v_0(\theta) = e^{ikx} \sum_{n=-j}^{j} u_n(\theta)/\sqrt{2j + 1}, \qquad (8.61)$$

the outgoing one after the barrier for the particle plus the clock system will be very much peaked around

$$\psi_f(x, \theta) \sim e^{ikx} v(\theta - \omega t_f(k)), \qquad (8.62)$$

so that the hand points at the expected classical time of flight through the region of interest. In the clock, $|V_n|$ can be as large as $j\hbar\omega \sim \pi\hbar/\tau$ so that the condition of negligible disturbance is given by [5, 22]

$$\tau >> \frac{\pi\hbar}{E} \quad \text{or} \quad E >> \frac{\pi\hbar}{\tau}. \qquad (8.63)$$

The same result may be obtained by imposing that the transmission probability should be close to one [13].

Equation (8.63) imposes a lower limit on the time resolution of the clock. Equivalently, it imposes a lower bound on the incident energy of the particle such that a clock with resolution τ can be considered as a small disturbance to the incident particle during the measuring process. It is worth noticing that this limitation affects equally the measurement of dwell times in the region of length d [5], or arrival times at $x = d$ [13], which are obtained, respectively, by locating the initial wave packet outside or inside the selected interval $[0, d]$.

Let us now work out a pulsed version of the particle–clock system to avoid the excessive disturbance of the continuous clock. The simplest realization of a pulsed interaction is a succession of instantaneous kicks separated by a time T. (For examples of kicked systems see [96, 97, 98, 99]. An experimental realization of a *kicked rotor* can be found in [100]. We refer the interested reader to [96] for further details.) The Hamiltonian for the kicked Peres–Salecker–Wigner clock is [101]

$$\widehat{H}(t) = \frac{\widehat{p}_x^2}{2m} + \mathsf{T}\delta_\mathsf{T}(t)\omega\chi_d(\widehat{x})\widehat{J}, \qquad (8.64)$$

where we have defined $\delta_\mathsf{T}(t) = \sum_{n=-\infty}^{\infty} \delta(n\mathsf{T} - t)$, and the evolution operator of the kicked clock over each kick period is

$$\widehat{U}^\mathsf{T} = e^{-\frac{i}{\hbar}\mathsf{T}\widehat{p}_x^2/2m}e^{-\mathsf{T}\omega\chi_d(x)\partial/\partial\theta} = \widehat{U}_s^\mathsf{T}\widehat{U}_{c-s}^\mathsf{T}. \qquad (8.65)$$

If $\mathsf{T} \to 0$, the infinitesimal evolution operator, $\widehat{U}^{\mathsf{T}\to 0}$, of the kicked system coincides with the infinitesimal evolution operator of the continuous-coupling clock up to $\mathcal{O}(\mathsf{T}^2[\widehat{H}_s, \widehat{H}_{c-s}])$, which goes to zero with T. The difference between the pulsed and continuous-coupling clocks will therefore be seen for larger values of T. These larger values may also avoid an excessive perturbation, but the time interval between kicks cannot be arbitrarily large. If we want to extract a characteristic timescale of the particle, T must be smaller than the timescale we want to measure.

We have discussed elsewhere [101] that the kicked clock can be successfully used for energies that are smaller that $\pi\hbar/\tau$. The conditions are such that

$$E \gg \frac{2\pi\hbar}{\mathsf{T}}. \qquad (8.66)$$

However, for sufficiently small T the kick clock behavior resembles the one of the continuous-coupling clock and (8.63) holds instead of (8.66). Since we must also have $t_f > \mathsf{T}$, the proper working regime of the pulsed apparatus is defined by the conditions

$$t_f > \mathsf{T} > \frac{2j+1}{j}\tau. \qquad (8.67)$$

We may in addition set $2\pi/\omega$ as the maximum time to be measured by the apparatus, to avoid the possibility of multiple times corresponding to a single θ. A measurement of θ at an asymptotically large time well after the particle–clock interaction will not tell us the number of 2π cycles that have occurred, so that the time read is only known modulo $2\pi/\omega$. This ambiguity may be avoided by substituting the periodic hand motion by a linear one as in [13].

From our previous considerations, it is clear that the energy of the particle may violate the inequality in (8.63) and still lead to a succesfull time-of-flight measurement for $T > (2j + 1)\tau/j$, as we shall illustrate below with numerical examples in which a minimun uncertainty product Gaussian wave packet with negligible negative momenta is prepared at $t = 0$ outside the region where the clock (continuous or kicked) is activated. Of course, because of the momentum width we should not expect a single time but a distribution. Well after the packet collision with the interaction region the probability to find the value θ is calculated and the corresponding (operational) time of flight distribution is obtained from the scaling $t = \theta/\omega$.

Figure 8.3A shows the ideal distribution of flight times obtained for the system in isolation, $\mathcal{P}_d$, and the operational distributions obtained from a kicked clock and from the continuous clock. (Incidentally, note that the continuous-coupling clock distribution may also be obtained using D. Sokolovski's Feynman path-based theory [102]). The parameters are chosen so that the classical time of flight for the average momentum of the wave packet is 10 a.u., and in such a way that the inequality in (8.63) is not obeyed, i.e., the continuous-coupling clock does not work correctly: note the large early peak denoting an important reflection in its distribution, and the displacement of the second peak with respect to $\mathcal{P}_d$ to shorter times because of the filtering effect of the more energetic barriers and wells that hinder the passage of slower components and allows a dominant contribution of faster components [13].

The reference (ideal) curve $\mathcal{P}_d(t)$ is the distribution of the free particle probability distribution of dwell times. For positive momentum states the dwell time probability distribution is given by

$$\mathcal{P}_d(t) = \int_0^\infty dp\, \delta(t - md/p) P(p), \qquad (8.68)$$

where $P(p)$ is the momentum distribution. $\mathcal{P}_d$ is both the dwell time distribution for an ensemble of classical particles with momentum distribution $P(p)$, and the quantum dwell time distribution $\langle \delta(t - \widehat{\tau}_d) \rangle$, where

$$\widehat{\tau}_d = \int_{-\infty}^\infty dt\, e^{i\widehat{H}_s t/\hbar} \left(\int_0^d |x\rangle dx \langle x| \right) e^{-i\widehat{H}_s t/\hbar} \qquad (8.69)$$

is the dwell-time operator [103].

Figures 8.3B and 8.3C show the cumulative distributions for several values of T and the cumulative distributions for $\mathcal{P}_d$ and for the continuous-coupling clock. As T is increased there is a passage from the continuous-like

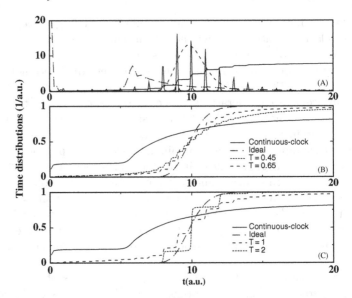

Fig. 8.3. (A): In *solid line* a typical $\mathcal{P}_d(t)$ and a cumulative integration for $\mathsf{T} = 1$. In *dashed line* the ideal time distribution, (8.68), and in *dotted–dashed* line the time distribution obtained from the continuous-coupling clock. **(B,C)**: Cumulative distributions for different values of T (*dotted and dashed lines*); cumulative distribution for the continuous-coupling clock (*solid line*) and for the ideal dwell time distribution (*long dashed line*). The simulations were performed for a particle of mass $m = 1$ a.u. represented initially by a minimum–uncertainty–product wave packet with width $\sigma = 1$ a.u., center at $x_0 = -30$ a.u., and average momentum $p_0 = 5.0$ a.u. The collision region is the interval $x \in (-25, 25)$ a.u. The method used was a split operator method and we took 2^{13} plane waves for the spatial coordinate and 2^{10} plane waves for the angular coordinate

regime to the truly kicked regime, where the cumulative distributions reproduce in a step-like fashion the behavior of the reference ideal curve. The perturbation of the kicks may be seen in the broader wings, which grow with decreasing T.

In summary, it is possible to extend the energy domain where a clock coupled to the particle's motion provides its (free motion) time of flight by using a pulsed particle–clock coupling rather than a continuous one. Due to recent developments in ultrafast spectroscopy and pulsed lasers this contibution may be relevant. Pulsed couplings are indeed possible and there is a need to understand the consequences of such couplings in the measurement of time quantities. In this respect our result is quite promising.

- Kicked Larmor clock

  As before we consider the motion of a quantum particle (an electron), and again in certain region of space we confine a magnetic field in the z-direction, in this case the Hamiltonian of the full system is given by

$$\widehat{H} = \widehat{H}_s + \widehat{H}_{s-c}(t)$$
$$= \frac{\widehat{p}^2}{2M}\widehat{1} + \widehat{1}\widehat{V}(\widehat{y}) + \chi_d(\widehat{y})\mathsf{T}\delta_\mathsf{T}(t)\mu\mathbf{B}\cdot\boldsymbol{\sigma}\,, \tag{8.70}$$

where $\widehat{1} = diag(1,1)$ and $\boldsymbol{\sigma} = (\widehat{\sigma}_x, \widehat{\sigma}_y, \widehat{\sigma}_z)$. The Pauli matrices are given by (8.14). We notice that in this case the system remains one dimensional, while the Salecker–Wigner clock is two dimensional. From now on we shall center our attention on the kicked Larmor clock.

To illustrate the functioning of a Larmor kick clock we have integrated numerically (8.54) for a rectangular barrier $\widehat{V}(\widehat{y}) = \chi_d(\widehat{y})V_0$. As we can see from (8.54) the integration involves two steps: first we have to propagate over a time T the spinor Ψ under the operator $\exp\left(-i\mathsf{T}(\frac{\widehat{p}^2}{2M}\widehat{1} + \widehat{1}\chi_d(\widehat{y})V_0)/\hbar\right)$, and then we multiply the spinor by the operator $\exp\left(-i\mathsf{T}\mu\chi_d(\widehat{y})\mathbf{B}\cdot\boldsymbol{\sigma}/\hbar\right)$. To carry out the first propagation we have used a split operator method (SOM) [104]. The initial state is an x-polarized spinor formed by two identical Gaussians centered at x_0 with momentum k and width σ,

$$\Psi(t=0) = \Psi = \begin{bmatrix} \psi_0 \\ \psi_0 \end{bmatrix}; \quad \psi_0(y) = \frac{1}{\sqrt{2}}\frac{e^{iky - \frac{(y-y_0)^2}{4\sigma^2}}}{(2\pi\sigma^2)^{1/4}}\,. \tag{8.71}$$

The barrier parameter are $V_0 = 0.16$ eV, and $d = 80$ Å, and the incident energy is $E_0 = 0.159$ eV. For the effective mass we chose $0.067m_e$, with m_e the mass of the electron. In the unit system we have used, space distances are measured in nanometers, energy in electron-volts, time in femtoseconds, and magnetic field in teslas.

During the time evolution, the electron state collides with the potential and at certain times it is affected by the action of the external magnetic field confined within the interaction region. The transmission part of the two spinor components starts being zero and increases until it reaches its final stationary value. An interesting feature is that the values of the transmission probability in time remain the same for a large set of values of the period T of the external signal $F(t)$ (Fig. 8.4). In this sense, the number of kicks can be reduced with no appreciable change to the transmission coefficient.

On looking at the behavior of the spin components with time, we see that they also reach a stationary value when the transmission part of the wave is out of the interaction region. Notice that the asymptotic value of the spin components remains stable for a large interval of values for the period T. In Table 8.1 we can see this phenomenon indeed taking place in the interval from $\mathsf{T} = 5$ fs to $\mathsf{T} = 40$ fs. There again we conclude that a few kicks are enough to extract the characteristic times related to those spin components being examined.

In Table 8.2 we give some numerical data of the typical times obtained from the kick clock. For the periods used these times are in good agreement with those expected from a standard (nonkick) Larmor precession clock.

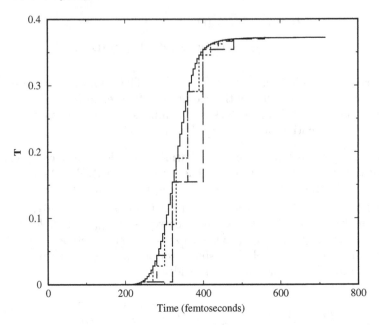

Fig. 8.4. Different transmission probabilities for the spinor component ψ_- as a function of time for different values of T. In full line T = 5 fs, in dotted line T = 30 fs, in dashed line T = 40 fs and in long dashed line T = 80 fs. $\sigma = 2^4 \Delta y$

Table 8.1. Different values of the spin components $\langle s_y \rangle = \hbar/2 \int dy\, \Psi^\dagger \hat{\sigma}_y \Psi$ and $\langle s_z \rangle = \hbar/2 \int dy\, \Psi^\dagger \hat{\sigma}_z \Psi$ for different periods, together with the characteristic times derived from the spin components

T(fs)	$\langle s_y \rangle$	τ_y(fs)	$\langle s_z \rangle$	τ_z(fs)
5	−0.005827	13.486	0.001828	4.231
30	−0.005826	13.486	0.001830	4.236
40	−0.005813	13.456	0.001831	4.238
80	−0.006198	14.346	0.001776	4.110

8.7 Decoherence in Time

There is a further aspect of time in quantum mechanics that we have not addressed so far. On first sight, one might think that the fact that all real clocks must naturally be prone to errors and fluctuations need not be related to quantum mechanics. However, these errors lead to decoherence, and can be modeled, in the case of good clocks, by thermal baths with a special coupling to the system [105]. In another section we shall also investigate other modelings of clocks in terms of [106] or related to [82] thermal baths.

Table 8.2. Typical characteristic times obtained for the kick clock, in this case for $\sigma = 2^7 \Delta y$. Let us notice that the initial wave packet is very well localized in momentum space, so the times given are very close to the ones obtained by the stationary-state analysis done by Büttiker [11]

$T(fs)$	$\tau_y(fs)$	$\tau_z(fs)$
1	10.301	12.319
2	11.193	12.498
3	11.193	12.497
4	11.191	12.497
5	11.189	12.496
6	11.188	12.495
7	11.187	12.494
8	11.187	12.493
9	11.186	12.492

The basic issue is that quantum and classical systems evolve in terms of parametric time, whereas our measurements are always located in time according to real clocks, which are of necessity faulty. This has led to an ever more refined search for better clocks, up to the ultimate time standard currently provided by atomic clocks [107]. Even so, the errors in measurement of time add up as parametric time goes by, until a point is reached when the typical error in time is comparable to the characteristic evolution timescale for the system under study, thus eliminating any predictability beyond this instant.

In the previous paragraph, four relevant timescales have appeared:

- Characteristic evolution timescale (ζ): This pertains to the system we desire to investigate, and it is the longest timescale among those present in the evolution of the system. For a quantum system, for instance, it is the inverse of the smallest energy difference between stationary states, multiplied by $\hbar$.
- Initial error width: As soon as the clock starts going, it is going to acquire an error, which will have a statistical width.
- (Relative) error correlation time (ϑ): It is clear that the absolute error made in measuring time has a growing spread as parametric time goes by. The rate of growth of this spread, roughly speaking, is related to (is the inverse of) the correlation time for relative errors, as we shall see below.
- Period of applicability: This is a derived scale of time, which corresponds to the instant when the absolute error width, due to the initial error width and the rate of growth of this error, is comparable to the characteristic evolution timescale.

From the presence of these four scales, it is clear that the question of what is a good clock has a complete answer only in terms of "a good clock for a given system." Thus, a clock will be good for the study of a given system if $\zeta \gg \vartheta$ and the period of applicability is much larger than ζ. Additional to this, some intrinsic properties are needed for a clock to be good: first, causality must be preserved, i.e., if two events take place in consecutive instants in parametric time, they also must take place in consecutive instants in clock time; second, on average the clock must be neither slow nor fast; third, it should always behave (statistically) in the same way.

The absolute errors in the clock are the difference between the clock readout, t, and the parametric time when the clock readout is t, $s(t)$, namely, $\Delta(t) = s(t) - t$ (we should point out that there is an alternative description which relies more on parametric time than on clock readout time; however, in order to tie in with experimental procedure, our presentation is much more accessible). The object we must consider is therefore a continuous stochastic sequence $\Delta(t)$, and the clock is characterized by the corresponding probability functional $\mathcal{P}\left[\Delta(t)\right]$. In fact, the continuous stochastic sequence of *relative* errors, $\alpha(t)$, and the corresponding functional $\mathcal{P}\left[\alpha(t)\right]$, are much more practical. Consider for instance the condition that causality be respected by the clock's readout. This implies that if $t_2 > t_1$, then $s(t_2) > s(t_1)$. For an infinitesimal dt, therefore, $s(t + dt) > s(t)$, from which it is clear that $\alpha(t)$ must be bigger than -1, when we define $\Delta(t + dt) - \Delta(t) = \alpha(t)dt$ (the possible subtleties associated with nondifferentiability of the corresponding stochastic processes can be obviated by first studying a clock with discrete positions, asserting that α is more useful than Δ in that context, and obtaining this bound that is kept in the continuum limit).

For the clock to be neither slow nor fast (nor both slow *and* fast at different instants), it is necessary that for each t the average value of $\alpha(t)$ be zero, i.e., $\langle \alpha(t) \rangle = 0$, which translates into $\langle \Delta(t) \rangle = 0$. The requirement that the clock behave in a statistically similar way at all instants is transposed into the requirement of stationarity of the stochastic process of relative errors. The reader should be aware that it is not possible for the stochastic process of absolute errors to be stationary, since these must accumulate in some way, in particular by increasing their statistical spread.

Consider now any given realization of the stochastic process, that is to say, a specific function $\alpha(t)$. Using the chain rule, we can write the von Neumann equation as

$$\partial_t \widehat{\varrho}_S(t + \Delta(t)) = -(\mathrm{i}/\hbar)(1 + \alpha(t))\left[\widehat{H}, \widehat{\varrho}_S(t + \Delta(t))\right] . \tag{8.72}$$

In this equation, $\widehat{\varrho}_S(s)$ is the density matrix for the system at parametric time s, in the Schrödinger picture. Since we are dealing with a particular realization, at clock time t the corresponding parametric time is $s(t) = t + \Delta(t)$, and the Schrödinger density matrix becomes a function of clock time t. The change of this function $\widehat{\widetilde{\varrho}}_S(t) = \widehat{\varrho}_S(s(t))$ as clock time flows is then computed as above.

Define now $\widehat{\varrho}(t)$ as the average over the stochastic process $\alpha(t)$ of $\widehat{\varrho}_S(t + \Delta(t))$, that is, $\widehat{\varrho}(t) = \int \mathcal{D}\alpha \, \mathcal{P}[\alpha(t)]\widehat{\varrho}_S(t + \Delta(t))$. The density matrix that is actually measured using the clock defined by the stochastic process $\alpha(t)$ is thus $\widehat{\varrho}(t)$. Adapting the standard procedures for treating quantum noise [108] to the case at hand, one obtains the following master equation [105]:

$$\dot{\varrho}(t) = -\frac{i}{\hbar}[\widehat{H}, \widehat{\varrho}(t)] - \frac{\kappa^2}{\hbar^2\vartheta}\left[\widehat{H}, \left[\widehat{H}, \widehat{\varrho}(t)\right]\right] , \tag{8.73}$$

where the small noise and Markovian approximations have been performed, ϑ is the correlation time for the stochastic process $\alpha(t)$, $\vartheta = \int_{-\infty}^{+\infty} d\tau \langle\alpha(t)\alpha(t-\tau)\rangle/2\langle\alpha(t)^2\rangle$, and $\kappa^2 = \vartheta^2\langle\alpha(t)^2\rangle$ (since the stochastic process $\alpha(t)$ is stationary, $\langle\alpha(t)^2\rangle$ is actually time independent).

The physical meaning of (8.73) is clear: there is a process of decoherence, whereby our knowledge of the system gradually decreases and becomes more probabilistic. This increase in our ignorance about the system is a direct consequence of our growing ignorance about parametric time. Moreover, the physically preferred description of the system is in terms of the stationary states, since the off-diagonal components in the basis of energy eigenstates are the coherences that progressively vanish. More explicitly, any given pure state $|\psi\rangle = \sum_j c_j|E_j\rangle$ evolves in time as

$$\widehat{\varrho}(t) = \sum_{j,k} c_j c_k^* e^{-i(E_j - E_k)t/\hbar - (E_j - E_k)^2\kappa^2 t/(\hbar^2\vartheta)}|E_j\rangle\langle E_k| \rightarrow \sum_j |c_j|^2|E_j\rangle\langle E_j| ,$$

under the assumption that $|E_j\rangle$ are the nondegenerate eigenstates of the Hamiltonian. The hypotheses underlying the derivation of (8.73) do not hold for longer than the time of applicability; nonetheless, it is clear that the limit written above is a stationary solution of (8.72) for any particular realization of the clock noise $\alpha(t)$, and that for long times the spread of $\Delta(t)$ will be so big that, effectively, we would be forced to consider only stationary states.

In fact, we can be more specific. The solution of (8.72) is clearly

$$\widehat{\varrho}_s(t + \Delta(t)) = e^{-i\int_0^t ds(1+\alpha(s))\widehat{H}/\hbar}\widehat{\varrho}_S(0)e^{i\int_0^t ds(1+\alpha(s))\widehat{H}/\hbar} ,$$

whence, again using the energy eigenstates, and assuming that $\widehat{\varrho}_S(0) = \widehat{\varrho}(0) = \sum_{j,k} \widehat{\varrho}_{jk}|E_j\rangle\langle E_k|$, we obtain

$$\widehat{\varrho}(t) = \sum_{j,k} \widehat{\varrho}_{jk} e^{-i(E_j - E_k)t/\hbar} \int \mathcal{D}\alpha\, \mathcal{P}[\alpha] e^{-i\int_0^t ds\alpha(s)(E_j - E_k)/\hbar}|E_j\rangle\langle E_k| .$$

The requirement of the clock being good entails that $\mathcal{P}[\alpha]$ is a functional very close to a well normalized, zero-mean Gaussian (even though, strictly speaking, it cannot be a Gaussian, since the constraint $\alpha(t) > -1$ would not be fulfilled in such a case), for which the characteristic functional (the functional Fourier transform) is again a zero-mean Gaussian. Thus, the restriction of the characteristic functional to constant functions is a Gaussian function, namely $\exp(-\kappa^2(E_j - E_k)^2 t/\hbar^2\vartheta)$! The departures from this form will be due to the non (strict) gaussianity of the probability functional, but, at any rate, they will be functions of $(E_j - E_k)$, and for large clock times (even past the time of validity of the master equation), the limit state will indeed be $\sum_j |c_j|^2 |E_j\rangle\langle E_j|$ for the initial pure state $|\psi\rangle = \sum_j c_j |E_j\rangle$.

Put much more simply, our lack of certainty of the instant of time forces our knowledge of the system to be restricted to stationary density matrices, i.e., those that do not evolve in time.

There is an important alternative way of understanding (8.73), which was proposed by Milburn [109], and later in a slightly different form by Bonifacio [110, 111]: imagine that the evolution of the system takes place randomly. That is, a system initially in a state $|\psi\rangle$ would remain in that state for an stochastic interval δt, after which it would change to the state $\widehat{U}(\delta t)$, with $\widehat{U}(t) = \exp(-i\widehat{H}t/\hbar)$ the unitary evolution operator. Under a set of suitable approximations, one is immediately led again to the master equation (8.73). This approach was termed "intrinsic decoherence." A unified description of several presentations of dephasing as stochasticity in time has been presented in [112].

Both formally and conceptually it is now not too difficult to establish the equivalence of the process due to a faulty clock, with an evolution for which the course of time flows smoothly according to parametric time, but with the system coupled to a noisy bath, which forces our knowledge about the system to be reduced. In fact, this bath can be assumed to be a thermal bath of harmonic oscillators at temperature T, coupled to the system in a rather unorthodox manner:

$$\widehat{H}_{\text{int}} = \hbar\widehat{H}_{\text{sys}} \int_0^\infty d\omega \left[\xi(\omega)a^\dagger(\omega) + \xi^*(\omega)a(\omega)\right] ,$$

where $\widehat{H}_{\text{sys}}$ is the Hamiltonian of the system, $a(\omega)$ and $a^\dagger(\omega)$ are the annihilation and creation operators, respectively, of the bath oscillator of frequency ω, and $\xi(\omega)$ is a function such that

$$|\xi(\omega)|^2 = \frac{1}{\pi} \tanh\left(\frac{\hbar\omega}{2k_BT}\right) \int_0^\infty d\tau \,\langle\alpha(t)\alpha(t - \tau)\rangle \cos(\omega\tau),$$

k_B being Boltzmann's constant.

It is by now well established [113, 114, 115, 116] that the introduction of a thermal bath of oscillators to account for the decohering fluctuations (in our case, for $\langle\alpha(t)\alpha(t - \tau)\rangle$), forces us to consider a further term, usually denominated "dissipative." This is the content of the fluctuation–dissipation theorem, and is related to the structure of the thermal bath, not to the specific coupling to the system nor to the system itself. However, the interpretation of the additional term as dissipative does indeed depend on the system being analyzed and the coupling between the system and the bath. In the particular case at hand, no such interpretation is feasible. The "dissipative" term has no influence on the classical equations of motion! That is to say, the extrema of the action of the system, when the effect of the bath is taken into account through the influence functional formalism [113, 114, 115, 116], are determined by the equations of motion of the system alone, plus a noise term associated to the fluctuations, but the "dissipative" term plays no role. Thus we can say that the fluctuation–dissipation theorem is indeed fulfilled, but that the system does not acquire dissipation of energy into the bath.

Acknowledgments

The authors wish to thank the fruitful discussions held with C. R. Leavens. Support by Gobierno Autónomo de Canarias (PI2004/025), Ministerio de Ciencia y Tecnología of Spain (FIS2004-05687), Ministerio de Ciencia y Tecnología (AEN99-0315), The University of the Basque Country (UPV 063.310-EB187/98 and 9/UPV 00039.310-13507/2001), and the Basque Government (PI-1999-28) (I. L. Egusquiza) is gratefully acknowledged.

References

1. D. Sobel: *Longitude* (Penguin, New York, 1995)
2. Y. Aharonov, D. Bohm: Phys. Rev. **122**, 1649 (1961)
3. W. Pauli: in *Encyclopedia of Physics*, edited by S. Flugge (Springer, Berlin, 1958), Vol. V/1, p. 60
4. B. Misra, I. Prigogine, M. Courbage: Proc. Natl. Acad. Sci. USA **76**, 4768 (1979)
5. A. Peres: Am. J. Phys. **48**, 552 (1980)
6. H. Salecker, E.P. Wigner: Phys. Rev. **109**, 571 (1958)
7. A.I. Baz': Sov. J. Nucl. Phys. **4**, 182 (1967)
8. A.I. Baz': Sov. J. Nucl. Phys. **5**, 161 (1967)
9. V.F. Rybachenko: Soc. J. Nucl. Phys. **5**, 635 (1967)
10. P.C.W. Davies: J. Phys. A: Math. Gen. **19**, 2115 (1986)
11. M. Büttiker: Phys. Rev. B **27**, 6178 (1983)
12. C.R. Leavens: Solid State Commun. **86**, 781 (1993)
13. Y. Aharonov, J. Oppenheim, S. Popescu, B. Reznik, W.G. Unruh: Phys. Rev. A **57**, 4130 (1998). quant-ph/9709031
14. V. Gasparian, M. Ortuño, J. Ruiz, E. Cuevas: Phys. Rev. Lett. **75**, 2312 (1995)
15. Y. Japha, G. Kurizki: Phys. Rev. A **60**, 1811 (1999)
16. M. Deutsch, J.E. Golub: Phys. Rev. A **53**, 434 (1996)
17. I.L. Chuang: Phys. Rev. Lett. **85**, 2006 (2000)
18. E.A. Burt, C.R. Ekstrom, T.B. Swanson: A reply to "quantum clock synchronization" (2000). quant-ph/0007030
19. M. Genovese, C. Novero: Quantum clock synchronization based on entangled photon pairs transmission (IEN September 2000). quant-ph/0009119
20. A. Cook: *Observational Foundations of Physics* (Cambridge University Press, Cambridge, 1994)
21. J. Hilgevoord: Am. J. Phys. **66**, 396 (1998)
22. A. Peres: *Quantum Theory: Concepts and Methods* (Kluwer, Dordrecht, 1993)

23. A.D. Baute, S. Brouard, I.L. Egusquiza, J.G. Muga, J.P. Palao, R. Sala Mayato: in *Time Arrows, Quantum Measurements and Superluminal Behaviour* (CNR - Italian National Research Council, 2000)

24. P. Carruthers, M.M. Nieto: Rev. Mod. Phys. **40**, 441 (1968)

25. P. Busch, M. Grabowski, P.J. Lahti: Phys. Lett. A **191**, 357 (1994)

26. J. Hilgevoord: Am. J. Phys. **64**, 1451 (1996)

27. C.R. Leavens, W.R. McKinnon: Phys. Lett. **A194**, 12 (1994)

28. C.R. Leavens, R. Sala Mayato: Ann. Phys. (Leipzig) **7**, 662 (1998)

29. G. Iannaccone, B. Pellegrini: Phys. Rev. B **50**, 14659 (1994)

30. A.M. Steinberg: Phys. Rev. Lett. **74**, 2405 (1995)

31. Y. Aharonov, D.Z. Albert, L. Vaidman: Phys. Rev. Lett. **60**, 1351 (1988)

32. Y. Aharonov, L. Vaidman: Phys. Rev. A **41**, 11 (1990)

33. S. Brouard, R. Sala, J. G. Muga: Phys. Rev. A **49**, 4312 (1994)

34. J.G. Muga, C.R. Leavens: Phys. Rep. **338**, 353 (2000)

35. G.R. Allcock: Ann. Phys. (N.Y.) **53**, 253 (1969)

36. G.R. Allcock: Ann. Phys. (N.Y.) **53**, 286 (1969)

37. G.R. Allcock: Ann. Phys. (N.Y.) **53**, 311 (1969)

38. A.D. Baute, I.L. Egusquiza, J.G. Muga: Phys. Rev. A **64**, 014101 (2001). quant-ph/0012051

39. A.D. Baute, R. Sala Mayato, J.P. Palao, J.G. Muga, I.L. Egusquiza: Phys. Rev. A **61**, 022118 (2000). quant-ph/9904055

40. J. Kijowski: Rept. Math. Phys. **6**, 361 (1974)

41. A. Casher, B. Reznik: Phys. Rev. A **62**, 042104 (2000). quant-ph/9909010

42. E. Arthurs, J. Kelly: Bell Syst. Tech. J. **44**, 725 (1965)

43. E.P. Wigner: Phys. Rev. **40**, 749 (1932)

44. L. Cohen: J. Math. Phys. **7**, 781 (1966)

45. M. Hillery, R.F. O'Connell, M. O. Scully, E. P. Wigner: Phys. Rep. **106** (1984)

46. N.L. Balazs, B.K. Jennings: Phys. Rep. **104**, 347 (1984)

47. H.W. Lee: Phys. Rep. **259**, 147 (1995)

48. R. Sala, J.P. Palao, J.G. Muga: Phys. Lett. **A231**, 304 (1997). quant-ph/9901042

49. E.B. Davies: *Quantum Theory of Open Systems* (Academic Press, London, 1976)

50. S. Stenholm: Ann. Phys. (N.Y.) **218**, 233 (1992)

51. K. Wódkiewicz: Phys. Rev. Lett. **52**, 1064 (1984)

52. B.G. Englert, K. Wódkiewicz: Intrinsic and operational observables in quantum mechanics (1995). quant-ph/9502013

53. P. Kochański, K. Wódkiewicz: Phys. Rev. A **60**, 2689 (1999). quant-ph/9902044

54. A.D. Baute, I.L. Egusquiza, J.G. Muga, R. Sala Mayato: Phys. Rev. A **61**, 052111 (2000). quant-ph/9911088

55. M. Hino et al.: Phys. Rev. A **59**, 2261 (1999)

56. C.R. Leavens, G.C. Aers: Solid State Commun. **63**, 1101 (1987)

57. C.R. Leavens, G.C. Aers: Solid State Commun. **67**, 1135 (1988)

58. C.R. Leavens, G.C. Aers: 'Tunneling time for one-dimensional barriers'. In: *S*canning Tunneling Microscopy and Related Techniques, edited by R. J. Behm, N. García, H. Rohrer (Kluwer, Dordrecht) pp. 59–76

59. D. Sokolovski, L.M. Baskin: Phys. Rev. A **36**, 4604 (1987)

60. B. Gottlieb, M. Kleber: Ann. Physik **1**, 369 (1992)

61. R. Golub, S. Felber, R. Gähler, E. Gutsmiedl: Phys. Lett. A **148**, 27 (1990)
62. Y.Z. Huang, C.M. Wang: J. Phys. Condens. Matter **3**, 5915 (1991)
63. J.G. Muga, S. Brouard, R. Sala: J. Phys. Condens. Matter **4**, L579 (1992)
64. E. Hauge, J.A. Stovneng: Rev. Mod. Phys. **61**, 917 (1989)
65. R. Landauer: Ber. Bunsenges. Phys. Chem. **95**, 404 (1991)
66. S. Brouard, R. Sala, J.G. Muga: Eur. Phys. Lett. **22**, 159 (1993)
67. Z. Huang et al.: J. Phys. Colloq **C6**, 17 (1988)
68. W. Jaworski, D.M. Wardlaw: Phys. Rev. A **43**, 5137 (1991)
69. J.Y. Lee, H.W. Lee, J.W. Hahn: J. Opt. Soc. Am. B **17**, 401 (2000)
70. C. Bracher: J. Phys. B: At. Mol. Opt. Phys. **30**, 2717 (1997)
71. M.O. Scully, M.S. Zubairi: *Quantum Optics* (Cambridge University Press, 1997)
72. R. Arun, G.S. Agarwal: Tunneling and traversal of ultra-cold atoms through vacuum induced potentials. Phys. Rev. A **64**, 065802 (2001). quant-ph/0103015
73. V. Bužek, R. Derka, S. Massar: Phys. Rev. Lett. **82**, 2207 (1999). quant-ph/9808042
74. V. Bužek, P.L. Knight, N. Imoto: Phys. Rev. A **62**, 062309 (2000). quant-ph/0006048
75. A.S. Holevo: *Probabilistic and Statistical aspects of Quantum Theory* (North-Holland, Amsterdam, 1982)
76. R. Derka, V. Bužek, A.K. Ekert: Phys. Rev. Lett. **80**, 1571 (1998). quant-ph/9707028
77. D.T. Pegg, S.M. Barnett: Europhys. Lett. **6**, 483 (1988)
78. J.I. Bollinger, W.M. Itano, D.J. Wineland, D.J. Heinzen: Phys. Rev. A **54**, R4649 (1996)
79. S.F. Huelga, C. Macchiavello, T. Pellizzari, A.K. Ekert, M.B. Plenio, J.I. Cirac: Phys. Rev. Lett. **79**, 3865 (1997). quant-ph/9707014
80. L.S. Schulman: *Time's Arrows and Quantum Measurement* (Cambridge University Press, Cambridge, 1997)
81. J.J. Halliwell: Prog. Theor. Phys. **102**, 707 (1999). quant-ph/9805057
82. J.G. Muga, A.D. Baute, J.A. Damborenea, I.L. Egusquiza: Model for the arrival-time distribution in fluorescence time-of-flight experiments (2000). quant-ph/0009111. Superseded by [87]
83. W.H. Louisell: *Quantum Statistical Properties of Radiation* (John Wiley & Sons, New York, 1973)
84. G.C. Hegerfeldt: Phys. Rev. A **47**, 449 (1993)
85. G.C. Hegerfeldt, M.B. Plenio: Phys. Rev. A **53**, 1164 (1996)
86. M.B. Plenio, P.L. Knight: Rev. Mod. Phys. **70**, 101 (1998). quant-ph/9702007
87. J.A. Damborenea, I.L. Egusquiza, G. Hegerfeldt, J. G. Muga: Phys. Rev. A **66**, 052104 (2002). quant-ph/0209027
88. H. Carmichael: *An Open Systems Approach to Quantum Optics* (Springer-Verlag, Berlin, 1993)
89. J. Dalibard, Y. Castin, K. Mölmer: Phys. Rev. Lett. **68**, 580 (1992)
90. J.A. Damborenea, I.L. Egusquiza, G.C. Hegerfeldt, J. G. Muga: J. Phys. B **36**, 2657-2669 (2003). quant-ph/0302201
91. B. Navarro, I.L. Egusquiza, J. G. Muga, G.C. Hegerfeldt: J. Phys. B **36**, 3899-3907 (2003). quant-ph/030575
92. G.C. Hegerfeldt, D. Seidel, J.G. Muga: Phys. Rev. A **68**, 022111 (2003). quant-ph/0308087

93. G.C. Hegerfeldt, D. Seidel, J.G. Muga, B. Navarro: Phys. Rev. A **70**, 012110 (2004) `quant-ph/0311107`
94. R. Brunetti, K. Fredenhagen: Phys. Rev. A **66**, 044101 (2002).
95. V. Hannstein, G.C. Hegerfeldt, J.G. Muga: J. Phys. B: At. Mol. Opt. Phys. **38** (2005) 409-420. `quant-ph/0412054`
96. G. Casati, B. Chirikov: *Quantum Chaos* (Cambridge University Press, Cambridge, 1995). And references therein
97. V.V. Sokolov, O. Zhirov, D. Alonso, G. Casati: Phys. Rev. Lett. **84**, 3566 (2000). `chao-dyn/9907033`
98. V.V. Sokolov, O. Zhirov, D. Alonso, G. Casati: Physica E **9**, 554 (2001)
99. V.V. Sokolov, O. Zhirov, D. Alonso, G. Casati: Phys. Rev. E **61**, 5057 (2000). `chao-dyn/9912009`
100. F.L. Moore, J.C. Robinson, C.F. Bharucha, B. Sundaram and M. G. Raizen, Phys. Rev. Lett. **75**, 4598 (1995).
101. Daniel Alonso, R. Sala Mayato and J.G. Muga. Phys. Rev. A **67**, 032105 (2003).
102. D. Sokolovski, in *Time in Quantum Mechanics*, 1st edn., edited by J.G. Muga, R. Sala Mayato and I.L. Egusquiza (Springer-Verlag, Berlin, 2002), Ch. 7.
103. J. G. Muga, in *Time in Quantum Mechanics*, 1st edn., edited by J.G. Muga, R. Sala Mayato and I.L. Egusquiza (Springer-Verlag, Berlin, 2002), Ch. 2.
104. M.D. Feit, J.A.F. Jr., A. Steige: J. Comp. Phys. **47**, 212 (1982)
105. I.L. Egusquiza, L.J. Garay, J.M. Raya: Phys. Rev. A **59**, 3236 (1999). `quant-ph/9811009`
106. J.J. Halliwell: Prog. Theor. Phys. **102**, 707 (1999). `quant-ph/9805057`
107. F.G. Major: *The Quantum Beat: the Physical Principles of Atomic Clocks* (Springer-Verlag New York, 1998)
108. C.W. Gardiner, P. Zoller: *Quantum Noise* (Springer-Verlag, Berlin, 2000), 2nd edn.
109. G.J. Milburn: Phys. Rev. A **44**, 5401 (1991)
110. R. Bonifacio: Il Nuovo Cimento **114 B**, 473 (1999)
111. S. Mancini, R. Bonifacio: Intrinsic decoherence in atomic diffraction by standing wave J. Phys. B **34**, 1909 (2001). `quant-ph/0104015`
112. I.L. Egusquiza and L.J. Garay: Phys. Rev. A **68**, 022104 (2003). `quant-ph/0301168`
113. R.P. Feynman, J.F.L. Vernon: Ann. Phys. **24**, 118 (1963)
114. R.P. Feynman, J.F.L. Vernon: Ann. Phys. **281**, 547 (2000). Reprinted version
115. R. Feynman, A. Hibbs: *Quantum Mechanics and Path Integral* (McGraw-Hill, New York, 1965)
116. A.O. Caldeira, A.J. Leggett: Physica A **121**, 587 (1983)

9

The Local Larmor Clock, Partial Densities of States, and Mesoscopic Physics

Markus Büttiker

Département de physique théorique, Université de Genève, 24 Quai Ernest-Ansermet, CH-1211 Genève, Switzerland
buttiker@karystos.unige.ch

9.1 Introduction

The Larmor clock is one of the most widely discussed approaches to determine the timescales of tunneling processes. The essential idea [1, 2, 3] of the Larmor clock is that the motion of the spin polarization in a narrow region of magnetic field can be exploited to provide information on the time carriers spend in this region. It is assumed that incident carriers are spin polarized and that they impinge on a region to which a small magnetic field is applied perpendicular to the spin polarization of the incident carriers (see Fig. 9.1). The spin polarization of the transmitted and reflected carriers can then be compared with the polarization of the incident carriers. Dividing the angle between the polarization of the exiting carriers and that of the incident carrier by the Larmor frequency ω_L gives a time. Originally, only spin precession (the movement of the polarization in the plane perpendicular to the magnetic field) was considered. However, [Büttiker] [3] pointed out, that especially if we deal with regions in which only evanescent waves exist (tunneling problems) the polarization executes not only a precession but also a rotation into the direction of the magnetic field. In fact, in the presence of a tunneling barrier, the spin *rotation*, is the dominant effect. Reference [3] considered a rectangular barrier and considered a magnetic field of the same spatial extend as the barrier. In the local version of the Larmor clock, introduced by Leavens and Aers [4], we consider an arbitrary region in which the magnetic field is nonvanishing and investigate again the direction of the spin polarization and rotation of the transmitted and reflected carriers. The magnetic field might be nonvanishing in a small region localized inside the barrier or in a small region outside the barrier on the side on which carriers are incident or on the far side of the barrier. We mention already here, that the response of the carriers is highly nonlocal: even carriers which are reflected are affected by a magnetic field that is nonvanishing only on the far side of the barrier where naively we would expect only transmitted carriers [5]. In this work, we use the local Larmor clock to derive a set of local density of states [6, 7, 8, 9, 10],

M. Büttiker: *The Local Larmor Clock, Partial Densities of States, and Mesoscopic Physics*, Lect. Notes Phys. **734**, 279–303 (2008)
DOI 10.1007/978-3-540-73473-4_9

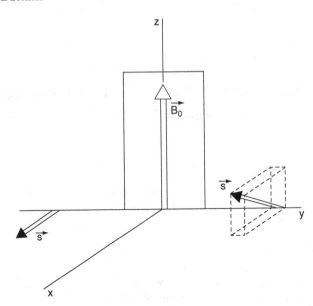

Fig. 9.1. Spin polarized carriers incident on a barrier subject to a weak magnetic field B_0. The transmitted carriers exhibit both a spin precession and and a spin rotation. After [3]

which we call *partial densities of states*, related to *spin precession*, and in terms of *sensitivities* which are related to *spin rotation* . The partial densities of states, below abbreviated as PDOS, are useful to understand a number of transport problems: the transmission probability from a tunneling microscope tip into a multiterminal mesoscopic conductor [10] can be expressed in terms of PDOS, the absorption of carriers by an optical potential (a potential with a small imaginary component), inelastic scattering and dephasing caused by a weak coupling voltage probe, and the low-frequency transport in mesoscopic conductors.

The partial densities of states are determined by functional derivatives of the scattering matrix [6, 7, 9, 10]. Only in certain limited situations can the density of states be expressed in terms of energy derivatives. Expressions for the density of states in terms of energy derivatives of the scattering matrix are familiar [11, 12]. In the discussion of characteristic times the distinction between time-scales found from energy derivatives (like the Wigner–Smith phase delay) and time scales found from derivatives with respect to the local potential (the dwell time) has found some recognition. In contrast, densities of states are almost invariably discussed in terms of energy derivatives. Here we emphasize that a more precise discussion of density of states also uses derivatives with respect to the (local) potential and not energy derivatives. It is the dwell time (or sums of relevant dwell times) which is related to the density of states [13, 14]. The use of energy derivatives always signals that approximations are involved.

The interpretation of the Larmor clock remains a subject of discussion. Reference [3] considered the total rotation angle divided by the Larmor precession frequency to be the relevant time. This interpretation brings the Larmor clock into agreement with the timescales obtained by considering tunneling through a barrier with an oscillating potential [15]. Subsequent works have argued that the precession angle and the rotation angle divided by the Larmor precession frequency separately should be viewed as timescales [16, 17]. The difficulty with such an interpretation is not only that one has two scales characterizing the same process, but the times defined in such a way are also not necessarily positive. Since we aim at characterizing a duration, that is a definite drawback. The two timescales can be combined into a complex time, with the real part referring to the precession time and the imaginary part to the time obtained from rotation. Like negative times, complex times are not part of the commonly accepted notions of time. Steinberg argues (cf. Chap. 11) that the clock presents only a "weak measurement" and that therefore complex times are permitted [18]. In quantum mechanics the questions "how much time has the transmitted particle spent in a given interval" is problematic since being in an "interval" and "to be transmitted" correspond to noncommuting operators [19]. Reasonably, we can only speak of a time duration if it is real and positive.

A comparison of the Larmor clock with the closely related linear ac response of an electrical conductor shows immediately the ambiguity of the clock: in the ac response of a conductor which is predominantly capacitive (tunneling limit) the voltage leads the current whereas for a highly transmissive conductor the response is inductive and the current leads the voltage.

The language used here implies similarly an extension of the usual notion of density of states. At the bottom of the hierarchy of density of states which we discuss are the *partial densities of states* (PDOS) which represents the contribution to the local density of states if we prescribe both the incident and the outgoing channel. It turns out that certain partial densities of states are not positive. (These are of course just the PDOS that correspond to negative precession times.) Thus the discussion presented below does not solve the interpretational questions related to the Larmor clock. Nevertheless, as we will show, even such negative PDOS are physically relevant. Using the partial densities of states, either by summing over the outgoing channels (or by summing over the incident channels) we obtain the *injectivity* of a contact into a point within the sample or the *emissivity* of a point within the sample into a contact. Both injectivities and emissivities are positive and in the language of tunneling times correspond to local dwell times for which either the incident channel (or the outgoing channel but not both) are prescribed. Finally, if we take the sum of all the injectivities or the sum of all the emissivities we obtain the local density of states.

9.2 The Scattering Matrix and the Local Larmor Clock

We start by considering a one-dimensional scattering problem [3, 4, 5]. We consider particles moving along the y-axis in a potential $V(y)$. The potential is arbitrary, except that asymptotically, for large negative and large positive values of y, it is assumed to be flat. We adopt here the language from mesoscopic transport discussions and call the region of large negative y the contact 1 (the left contact) and the region of large positive y contact 2 (the right contact). We assume that the quantum mechanical evolution is described by the Schrödinger equation. A particle with energy E has for large negative y a wave vector $k_1(E)$ and a velocity $v_1(E)$ and for large positive values of y a wave vector $k_2(E)$ and a velocity $v_2(E)$. We are interested in scattering states. A particle incident from the left is for large negative values of y described by a scattering state

$$\psi_1(E, y) = e^{ik_1 y} + S_{11} e^{-ik_1 y} , \tag{9.1}$$

and for large positive values of y is described by a transmitted wave

$$\psi_1(E, y) = \left(\frac{v_1}{v_2}\right)^{1/2} S_{21} e^{ik_2 y} . \tag{9.2}$$

Similarly, a particle incident from the right is for large positive values of y described by a scattering state

$$\psi_2(E, y) = e^{-ik_2 y} + S_{22} e^{-ik_2 y} , \tag{9.3}$$

and for large negative values of y is described by a transmitted wave

$$\psi_2(E, y) = \left(\frac{v_2}{v_1}\right)^{1/2} S_{12} e^{ik_2 y} . \tag{9.4}$$

Here the amplitudes $S_{\alpha\beta}$ determine the elements of the 2×2 scattering matrix of the problem.[1] Each scattering matrix element is a function of the energy E of the incident carrier and is a functional of the potential $V(y)$. To express this dependence we write $S_{\alpha\beta}(E, V(y))$. Conservation of the probability current requires that this matrix is unitary and in the absence of a magnetic field time-reversal invariance implies that it is also symmetric.

Next we now consider a weak magnetic field applied to a small region. The magnetic field shall point into the z-direction. For simplicity, we consider the case where the magnetic field is constant in a small interval $[y, y + dy]$

---

[1] N. of E.: There are two common but different conventions for the ordering of the S-matrix elements. M. Büttiker follows the ordering which is usually found in mesoscopic physics, whereas J.G. Muga in Chap. 2 follows the convention used in a scattering theoretical context, see (2.45). The relation between the matrix introduced here and the one defined in Chap. 2 is: $S_{11} = S_{-+}$; $S_{12} = S_{--}$; $S_{21} = S_{++}$; $S_{22} = S_{+-}$.

and takes there the value B. We consider only the effect of the Zeeman energy. Thus the motion of the particle remains one dimensional and is as in the absence of a magnetic field confined to the y-axis. In the setup for the Larmor clock, we consider particles with a spin. The spin of the incident particles (in the asymptotic regions) is polarized along the x-axis. For spin $1/2$ particles the wave functions are now spinors with two components $\psi_+(y, E)$ and $\psi_-(y, E)$. Carriers incident from the left, have a spinor with components given by $\psi_+(y, E) = \psi_-(y, E) = (1/\sqrt{2})exp(ik_1 y)$. The Zeeman energy which is generated by the local magnetic field is diagonal in the spin up and spin down components. Consequently, in the interval $[y, y + dy]$ for a particle with spin up, the energy is reduced by $\hbar\omega_L/2$ with $\omega_L/2 = g\mu B/\hbar$ and for a particle with spin down the potential is increased by $\hbar\omega_L/2$. Thus with the magnetic field switched on particles with spin up travel in a potential $V(y) - dV(y)$ and particles with spin down travel in potential $V(y) + dV(y)$. Here $V(y)$ is the potential in the absence of the magnetic field and $dV(y)$ is the potential generated by the magnetic field. Thus $dV(y)$ vanishes everywhere, except in the interval $[y, y + dy]$ where it takes the value $dV(y) = \hbar\omega_L/2 = g\mu B$.

We can evaluate the polarization of the transmitted particles and the reflected particles if we can determine the scattering matrix for spin up and spin down particles in the potential generated by the magnetic field. Thus we need the scattering matrices $S_{\alpha\beta}^\pm(E, V(y) \mp dV(y))$ where S^+ is the scattering matrix for spin up carriers and S^- is the scattering matrix for spin down carriers. Since the potential variation generated by the magnetic field is small we can expand these matrices away from the scattering matrix for the unperturbed potential. Thus we find to first order in the magnetic field for the scattering matrices

$$S_{\alpha\beta}^\pm(E, V(y)) \mp dV(y)) = S_{\alpha\beta}(E, V(y))$$
$$\mp [dS_{\alpha\beta}(E, V(y))/dV(y)]dV(y)dy + ... \quad (9.5)$$

The variation of the scattering matrix due to the magnetic field is proportional to the derivative of the scattering matrix with respect to the local potential at the location where the magnetic field is nonvanishing. More generally, we can consider a magnetic field which varies along the y-axis (but always points along the z-axis). This leads to a potential $\delta V(y)$ determined by the local magnetic field. The variation of the scattering matrix is then determined by a functional derivative $[\delta s_{\alpha\beta}(E, V(y))/\delta V(y)]$ of the scattering matrix with regard to the local potential,

$$S_{\alpha\beta}^\pm(E, V(y)) \mp \delta V(y)) = S_{\alpha\beta}(E, V(y))$$
$$\mp \int dy' [\delta s_{\alpha\beta}(E, V(y'))/\delta V(y')]\delta V(y') + ... \quad (9.6)$$

We emphasize that even though this equation looks quite simple, the evaluation of a functional derivative of a scattering matrix, while not difficult for a one-dimensional problem, can still be a laborious calculation.

Let us now find the precession and rotation angles of the polarization of the transmitted and reflected carriers. The normalized spinor of the transmitted particles, which determines the spin orientation of the transmitted a particles has the components

$$\psi_{1+}(E, y) = \frac{S_{21}^+}{(|S_{21}^+|^2 + |S_{21}^-|^2)^{1/2}} ,\tag{9.7}$$

$$\psi_{1-}(E, y) = \frac{S_{21}^-}{(|S_{21}^+|^2 + |S_{21}^-|^2)^{1/2}} .\tag{9.8}$$

First consider the polarization in the y-direction. It is found by evaluating the expectation value of the Pauli spin matrix σ_y,

$$\langle s_y \rangle_{21} = \frac{\hbar}{2} \langle \psi_1 | \sigma_y | \psi_1 \rangle = -i \frac{\hbar}{2} \frac{S_{21}^{+\dagger} S_{21}^- - S_{21}^+ S_{21}^{-\dagger}}{(|S_{21}^+|^2 + |S_{21}^-|^2)} .\tag{9.9}$$

Here the indices 21 indicate that we consider transmission from left (1) to right (2) and evaluate the the spin in the transmitted beam. We need the spin polarization only to first order in the applied magnetic field. Using (9.6) we find

$$\langle s_y \rangle_{21} = \frac{h}{T} \nu(2, y, 1) \omega_L dy ,\tag{9.10}$$

where $T = |S_{21}|^2$ is the transmission probability in the absence of the magnetic field and

$$\nu(2, y, 1) = -\frac{1}{4\pi i} \left(S_{21}^\dagger \frac{\delta S_{21}}{\partial V(y)} - \frac{\delta S_{21}^\dagger}{\delta V(y)} S_{21} \right)\tag{9.11}$$

is the partial density of states at y of carriers which emanate from contact 1 (the asymptotic region for large negative y) and eventually in the future reach contact 2. Since initially the spin polarization was in the x-direction $\langle s_y \rangle_{21}$ directly determines the angle of precession of the carriers in the $x - y$ plane. Thus by dividing $\langle s_y \rangle_{21}$ by the Larmor precession frequency we can formally introduce a quantity with the dimension of time, which we call $\tau_y(2, y, 1)$ and which is given by $\tau_y(2, y, 1) = (h/T)\nu(2, y, 1)dy$. Here the index y indicates that we deal with a time-scale obtained from the y-component of the spin polarization. We can now proceed to evaluate also the y-component of the spin polarization of the carriers which are reflected and can proceed to evaluate the y-component of the spin polarization of the carriers that are in the past incident from contact 2 (large positive y) and in the future will be transmitted into contact 1 (large negative 1) or will be reflected back into contact 1. We can summarize the results in the following manner: there are a total of four spin polarizations to be considered, each of them determined by a partial density of states

$$\nu(\alpha, y, \beta) = -\frac{1}{4\pi i}\left(S_{\alpha\beta}^\dagger\frac{\delta S_{\alpha\beta}}{\delta V(y)} - \frac{\delta S_{\alpha\beta}^\dagger}{\delta V(y)}S_{\alpha\beta}\right) \qquad (9.12)$$

of carriers that are incident at contact $\beta = 1, 2$ and eventually in the future are transmitted or reflected into contact $\alpha = 1, 2$. Formally, the timescales related to precession in the local magnetic field at y can be introduced which are related to the partial densities of states via, $\tau_y(\alpha, y, \beta) = (h/|S_{\alpha\beta}|^2)\nu(\alpha, y, \beta)dy$ where $|S_{\alpha\beta}|^2$ is the transmission probability T if α and β are not equal and is the reflection probability $\mathcal{R}$ if α and β are equal. Thus with *each* element of the scattering matrix we can associate a partial density of states. Later we discuss the properties of the partial densities of states in more detail.

Next we consider the spin polarization in the z-direction. For the carriers incident in contact 1 and transmitted into contact 2 we find that the z-component of the transmitted carriers is determined by

$$\langle s_z\rangle_{21} = \frac{\hbar}{2}\langle\psi_1|\sigma_z|\psi_1\rangle = \frac{\hbar}{2}\frac{|S_{21}^+|^2 - |S_{21}^-|^2}{(|S_{21}^+|^2 + |S_{21}^-|^2)}. \qquad (9.13)$$

Using (9.6) we find

$$\langle s_z\rangle_1 = \frac{h}{T}\eta(2, y, 1)\omega_L dy, \qquad (9.14)$$

where we call

$$\eta(2, y, 1) = -\frac{1}{4\pi}\left(S_{21}^\dagger\frac{\delta S_{21}}{\delta U(y)} + \frac{\delta S_{21}^\dagger}{\delta U(y)}S_{21}\right) \qquad (9.15)$$

the *sensitivity* of the scattering problem. Since the spin polarization of the incident particles was originally along the x-direction only a small z-component of the incident particle determines a spin *rotation* angle. We can formally introduce a timescale $\tau_z(2, y, 1)$ associated with spin rotation which is given by $\tau_z(2, y, 1) = (h/T)\eta(2, y, 1)dy$. Again we can ask about the z-polarization of reflected particles and can ask about the z-polarization of particles incident from the right. The results are summarized by attributing each scattering matrix element $S_{\alpha\beta}$ a sensitivity

$$\eta(\alpha, y, \beta) = -\frac{1}{4\pi}\left(S_{\alpha\beta}^\dagger\frac{\delta S_{\alpha\beta}}{\delta U(y)} + \frac{\delta S_{\alpha\beta}^\dagger}{\delta U(y)}S_{\alpha\beta}\right), \qquad (9.16)$$

which determine the timescales $\tau_z(\alpha, y, \beta) = (h/|S_{\alpha\beta}|^2)\eta(\alpha, y, \beta)dy$. Finally, we can determine the spin polarization in the x-direction. This component is reduced from its initial value both because of spin precession in the $x-y$ plane and because of the rotation of spins into the z-direction. Since we have at every space point

$$\langle s_x\rangle^2 + \langle s_y\rangle^2 + \langle s_z\rangle^2 = \hbar^2/4, \qquad (9.17)$$

it follows immediately that the time scale τ_x is related to the two timescales introduced above by

$$\tau_x = (\tau_y^2 + \tau_z^2)^{1/2} \, . \tag{9.18}$$

Using the expressions for τ_y and τ_z given above, we find for the timescales τ_x the following expressions:

$$\tau_x(\alpha, y, \beta) = \frac{h}{|S_{\alpha\beta}|^2} \left(\frac{\delta S_{\alpha\beta}}{\delta V(y)} \frac{\delta S_{\alpha\beta}^\dagger}{\delta V(y)} \right)^{1/2} . \tag{9.19}$$

We reemphasize that neither the partial densities of states nor the sensitivities are in general positive. In contrast, $\tau_x(\alpha, y, \beta)$ is positive for all elements of the scattering matrix.

9.3 Absorption and Emission of Particles: Injectivities and Emissivities

Before discussing the partial densities of states in more detail it is of interest to investigate the absorption of particles in a small scattering region [20]. We assume that in a narrow interval $[y, y+dy]$ there exists a nonvanishing absorption rate Γ. Thus the potential $V(y)$ is equal to $V_0(y) - i\hbar\Gamma$ in the interval $[y, y + dy]$ and is equal to $V_0(y)$ outside this interval. To solve the scattering problem, we need to find the scattering matrix $S_{\alpha\beta}^\Gamma(E, V(y))$ in the presence of this complex potential $V(y)$. Of interest here is, as in [20], the limit of small absorption. The case of strong absorption is also of interest but thus far has been used only to discuss global properties and not the local quantities of interest here [21, 22]. For a small absorption rate, we can expand the scattering matrix $S_{\alpha\beta}^\Gamma(E, V(y))$ in powers of the absorption rate away from the scattering problem in the original real potential V_0. We obtain

$$
\begin{aligned}
S_{\alpha\beta}^\Gamma(E, V(y)) = {} & S_{\alpha\beta}(E, V_0(y)) \\
& + i\hbar[\delta S_{\alpha\beta}(E, V(y))/\delta V(y)]|_{V(y)=V_0(y)}\Gamma dy + \dots..
\end{aligned}
\tag{9.20}
$$

We note that the adjoint scattering matrix has to be evaluated in the potential $V^*(y)$ and hence

$$
\begin{aligned}
S_{\alpha\beta}^{\Gamma\dagger}(E, V(y)) = {} & S_{\alpha\beta}^\dagger(E, V_0(y)) \\
& - i\hbar[\delta S_{\alpha\beta}^\dagger(E, V(y))/\delta V(y)]|_{V(y)=V_0(y)}\Gamma dy + \dots.
\end{aligned}
\tag{9.21}
$$

With these results it easy to show that the transmission and reflection probabilities in the presence of a small absorption in the interval $[y, y + dy]$ are

$$|S_{\alpha\beta}^\Gamma(E, V(y))|^2 = |S_{\alpha\beta}(E, V_0(y))|^2(1 - \Gamma\nu(\alpha, y, \beta)dy) \, , \tag{9.22}$$

where $|S_{\alpha\beta}(E, V_0(y))|^2$ is the transmission probability $\mathcal{T}$ of the scattering problem without absorption if α and β are different and is the reflection probability $\mathcal{R}$ of the scattering problem without absorption if $\alpha = \beta$. The incident current j_{in}, must be equal to the sum of the transmitted current j_T, the reflected current j_R and the absorbed current j_Γ,

$$j_{in} = j_T + j_R + j_\Gamma . \qquad (9.23)$$

Using (9.20) and taking into account that the incident flux is normalized to 1, we find for carriers incident from the left or right $\beta = 1, 2$ an absorbed flux given by

$$j_\Gamma(y, \beta) = \Gamma \nu(y, \beta) dy , \qquad (9.24)$$

where $\nu(y, \beta)$ is called the *injectivity* of contact β into point y. The injectivity of the contact is related to the partial densities of states via

$$\nu(y, \beta) = \sum_\alpha \nu(\alpha, y, \beta) . \qquad (9.25)$$

In our problem with two contacts the injectivity is just the sum of two partial densities of states.

Another way of determining the absorbed flux proceeds as follows. The absorbed flux is proportional to the integrated density of particles in the region of absorption (in the interval $[y, y + dy]$). The density of particles can be found from the scattering state $\psi_\beta(y)$ given by (9.1)–(9.4). For carriers incident from contact β, the absorbed flux is thus

$$j_\Gamma(y, \beta) = \Gamma \frac{1}{h v_\beta} |\psi_\beta(y)|^2 dy . \qquad (9.26)$$

Note that here the density of states $1/h v_\beta$ of the asymptotic scattering region appears. It normalizes the incident current to 1. Thus we have found a wave function representation for the injectivity. Comparing (9.23) and (9.22) gives

$$\nu(y, \beta) = \frac{1}{h v_\beta} |\psi_\beta(y)|^2 . \qquad (9.27)$$

The total local density of states $\nu(y)$ at point y is obtained by considering carriers incident from both contacts. In terms of wave functions, $\nu(y)$ is for our one-dimensional problem given by

$$\nu(y) = \sum_\beta \frac{1}{h v_\beta} |\psi_\beta(y)|^2 . \qquad (9.28)$$

Thus the total density of states is also the sum of the injectivities from the left and right contacts

$$\nu(y) = \sum_\beta \nu(y, \beta) . \qquad (9.29)$$

There is now an interesting additional problem to be addressed. Instead of a potential which acts as a carrier sink (as an absorber) we can ask about a potential which acts as a carrier source. Obviously, all we have to do to turn our potential into a carrier source is to change the sign of the imaginary part of the potential. With a a carrier source in the interval $[y, y + dy]$ we should observe a particle current toward contact 1 and a particle current toward contact 2. We suppose that carriers are incident both from the left and the right and evaluate the currents in the contact regions. The total current injected into the sample at y is

$$j_{in}(y) = \Gamma \nu(y) dy . \tag{9.30}$$

Taking into account that the incident current is normalized to 1 the current $j_{out}(\beta, y)$ in contact β due to a carrier source at y is given by

$$j_{out}(\alpha, y) = 1 - \sum_{\beta} |s^{\Gamma}_{\alpha\beta}(E, V(y))|^2 \tag{9.31}$$

due to the modification of both the transmission and reflection coefficients. Using (9.20) (with Γ replaced by $-\Gamma$) gives

$$j_{out}(\alpha, y) = -\Gamma \sum_{\beta} \nu(\alpha, y, \beta) \, dy = -\Gamma \nu(\alpha, y) \, dy . \tag{9.32}$$

The current in contact α is determined by the *emissivity* $\nu(\alpha, y)$ of the point y into contact α. Note the reversal of the sequence of arguments in the emissivity as compared to the injectivity. Thus the emissivity is like the injectivity a sum of two partial densities of states,

$$\nu(\alpha, y) = \sum_{\beta} \nu(\alpha, y, \beta) . \tag{9.33}$$

For a scattering problem in the absence of a magnetic field, the injectivity and emissivity are identical. If there is a homogeneous magnetic field present they are related by reciprocity: the injectivity from contact α into point y is equal to the emissivity of point y into the contact α in a magnetic field that has been reversed, $\nu_{+B}(y, \alpha) = \nu_{-B}(\alpha, y)$.

We have thus obtained a hierarchy of density of states: at the bottom are the partial densities of states $\nu(\alpha, y, \beta)$ for which we describe both the contact from which the carriers are incident and the contact through which the carriers have to exit. On the next higher level are the injectivities $\nu(y, \alpha)$ and the emissivities $\nu(\alpha, y)$. For the injectivity, we prescribe the contact through which the carrier enters but the final contact is not prescribed. In the emissivity, we prescribe the contact through which the carrier leaves but the incident contact is not prescribed. Finally, on the highest level is the local density of states $\nu(y)$ for which we prescribe neither the incident contact nor the contact through which carriers leaves.

For simple scattering problems (delta functions, barriers) the interested reader can find a derivation and discussion of partial densities of states in [9, 10, 23].

Returning to timescales: we have shown that the partial densities of states are associated with spin precession. It is tempting, therefore, to associate them with a time duration. However, as can be shown, the partial densities of states are not necessarily positive. (The simple example of a resonant double barrier shows that one of the diagonal elements $\nu(\alpha, y, \alpha)$ has a range of energies where it is negative [20]). The injectivities and emissivities are, however, always positive. The proof is given by (9.25). We can associate a local dwell time $\tau_D(y, \beta) = \hbar \nu_D(y, \beta)$ with the injectivity which gives the time a carrier incident from contact β spends in the interval $[y, y + dy]$ irrespective of whether it is finally reflected or transmitted. Similarly, we can associate a dwell time with the emissivity which is the time carriers spend in the interval $[y, y + dy]$ irrespective from which contact they entered the scattering region. There is little question that the dwell times have the properties which we associate with the duration of a process: they are real and positive. However, as explained they do not characterize transmission or reflection processes.

9.4 Potential Perturbations

Thus far our discussion has focused much on the partial densities of states. The sensitivity introduced as a measure of the spin rotation in the Larmor clock has, however, also an immediate direct interpretation. We have seen that the partial densities of states are obtained in response to a complex perturbation of the original potential $V(y)$. The sensitivity comes into play if we consider a *real* perturbation δV of the original potential. Thus if we consider a potential which is equal to $V(y) + \delta V$ in the interval $[y, y + dy]$ and equal to $V(y)$ elsewhere the transmission probability $\mathcal{T}^V$ in the presence of the perturbation is $\mathcal{T}^V = \mathcal{T} + 4\pi \eta(\alpha, y, \beta) \delta V dy$, with $\alpha \neq \beta$. The reflection probability is $\mathcal{R}^V = \mathcal{R} + 4\pi \eta(\alpha, y, \alpha) \delta V dy$. Since also $\mathcal{T}^V + \mathcal{R}^V = 1$ we must have $\eta(y) \equiv \eta(\alpha, y, \beta) = -\eta(\alpha, y, \alpha)$. For our scattering problem, described by a 2×2 scattering matrix there exists only one independent sensitivity $\eta(y)$. In the Larmor clock the sensitivity corresponds to spin rotation and the fact that there is only one sensitivity follows from the conservation of angular momentum: the weak magnetic field cannot produce a net angular momentum. If carriers in the transmitted beam acquire a polarization in the direction of the magnetic field then carriers in the reflected beam must have a corresponding polarization opposite to the direction of the magnetic field. In mesoscopic physics, in electrical transport problems, the sensitivity plays a role in the discussion of nonlinear current–voltage characteristics and plays a role if we ask about the change of the conductance in response to the variation of a gate voltage [24]. Below, we will not further discuss the sensitivity, but we will present a number of examples in which the partial densities of states play a role.

9.5 Generalized Bardeen Formulae

It is well known that with a scanning tunneling microscope (STM) we can measure the local density of states [25]. STM measurements are typically performed in a two terminal geometry, in which the tip of the microscope represents one contact and the sample provides another contact [25]. Here we consider a different geometry. We are interested in the transmission probability from an STM tip into the contact of a sample with two or more contacts as shown in Fig. 9.2. Thus we deal with a multiterminal transmission problem [10]. If we denote the contacts of the sample by a Greek letter $\alpha = 1, 2, ..$ and use tip to label the contact of the STM tip, we are interested in the tunneling probabilities $\mathcal{T}_{\alpha tip}$ from the tip into contact α of the sample. In this case, the STM tip acts as carrier source. Similarly we ask about the transmission probability $\mathcal{T}_{tip\alpha}$ from a sample contact to the tip. In this case, the STM tip acts as a carrier sink. Earlier work has addressed this problem either with the help of scattering matrices, electron wave dividers, or by applying the Fermi Golden Rule. Recently, Gramespacher and the author [10] have returned to this problem and have derived expressions for these transmission probabilities from the scattering matrix of the full problem (sample plus tip). For a tunneling contact with a density of states ν_{tip} which couples locally at the point x with a coupling energy $|t|$ these authors found

$$\mathcal{T}_{tip\alpha} = 4\pi^2 \nu_{tip} |t|^2 \nu(x, \alpha) , \qquad (9.34)$$

$$\mathcal{T}_{\alpha tip} = 4\pi^2 \nu(\alpha, x) |t|^2 \nu_{tip} . \qquad (9.35)$$

In a multiterminal sample the transmission probability from a contact α to the STM tip is given by the injectivity of contact α into the point x and

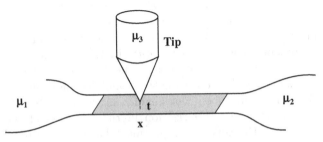

Fig. 9.2. Partial density of states measurement with a scanning tunneling microscope (STM). The tip of an STM couples at a point x with a coupling strength t to the surface of a multiterminal conductor. The contacts of the conductor are held at potentials $\mu_\alpha = eV_\alpha$ and the tip at potential $\mu_3 = eV_{tip}$. After [10]

the transmission probability from the tip to the contact α is given by the emissivity of the point x into contact α. Equations (9.34) and (9.35) when multiplied by the unit of conductance e^2/h are generalized Bardeen conductances for tunneling into multiprobe conductors. Since the local density of states of the tip is an even function of magnetic field and since the injectivity and emissivity are related by reciprocity we also have the reciprocity relation $\mathcal{T}_{tip\alpha}(B) = \mathcal{T}_{\alpha tip}(-B)$.

The presence of the tip also affects transmission and reflection at the massive contacts of the sample. To first order in the coupling energy $|t|^2$ these probabilities are given by

$$|S_{\alpha\beta}^{tip}|^2 = |S_{\alpha\beta}|^2 - 4\pi^2\, \nu(\alpha, x, \beta)\, |t|^2\, \nu_{tip} \; . \tag{9.36}$$

The correction to the transmission probabilities $\alpha \neq \beta$ and reflection probabilities $\alpha = \beta$ is determined by the partial densities of states, the coupling energy and the density of states in the tip. Note that if these probabilities are placed in a matrix then each row and each column of this matrix adds up to the number of quantum channels in the contacts.

9.6 Voltage Probe and Inelastic Scattering

Consider a two-probe conductor much smaller than any inelastic or phase breaking length. Carrier transport through such a structure can then be said to be coherent and its conductance is at zero temperature given by $G = (e^2/h)\mathcal{T}$, where $\mathcal{T}$ is the probability for transmission from one contact to the other. How is this result affected by events which break the phase or by events which cause inelastic scattering ? To investigate this question, [26] proposes to use an additional (third) contact to the sample. The third probe acts as a voltage probe, which has its potential adjusted in such a way that there is no net current flowing into this additional probe, $I_3 = 0$. The current at the third probe is set to zero by floating the voltage $\mu_3 = eV_3$ at this contact to a value for which I_3 vanishes. The third probe acts, therefore, like a *voltage probe*. Even though the total current at the voltage probe vanishes individual carriers can enter this probe if they are at the same time replaced by carriers emanating from the probe [26]. Entering and leaving a contact are *irreversible* processes, since there is no definite phase relationship between a carrier that enters the contact and a carrier that leaves a contact. In a three-probe conductor, the relationship between currents and voltages is given by $I_\alpha = \sum_\beta G_{\alpha\beta}V_\beta$ where the $G_{\alpha\beta}$ are the conductance coefficients. Using the condition $I_3 = 0$ to find the potential V_3 and eliminating this potential in the equation for I_2 or I_1 gives for the two-probe conductance in the presence of the voltage probe

$$G = -G_{21} - \frac{G_{23}\, G_{31}}{G_{31} + G_{32}} \; . \tag{9.37}$$

For a very weakly coupled voltage probe we can use (9.34)–(9.36). Taking into account that $G_{\alpha\beta} = -(e^2/h)|S_{\alpha\beta}|^2$ for $\alpha \neq \beta$ we find

$$G = \frac{e^2}{h}\left(T - 4\pi^2\,|t|^2\,[\nu(2,x,1) - \frac{\nu(2,x)\,\nu(x,1)}{\nu(x)}]\right). \qquad (9.38)$$

Here $\nu(x)$ is the local density of states at the location of the point at which the voltage probe couples to the conductor. Equation (9.38) has a simple interpretation [26]. The first term T is the transmission probability of the conductor in the absence of the voltage probe. The first term inside the brackets proportional to the local partial density of states gives the reduction of coherent transmission due to the presence of the voltage probe. The second term in the brackets is the incoherent contribution to transport due to inelastic scattering induced by the voltage probe. It is proportional to the injectivity of contact 1 at point x. A fraction $\nu(2,x)/\nu(x)$ of the carriers which reach this point, proportional to its emissivity, are scattered forward and, therefore, contribute to transport. Notice the different signs of these two contributions. The effect of inelastic scattering (or dephasing) can either enhance transport or diminish transport, depending on whether the reduction of coherent transmission (first term) or the increase due to incoherent transmission (second term) dominates.

Instead of a voltage probe, we can also use an optical potential to simulate inelastic scattering or dephasing. However, in order to preserve current, we must use both an absorbing optical potential (to take carriers out) and an emitting optical potential (to reinsert carriers). The absorbed and reemitted current must again exactly balance each other. From (9.23) it is seen that the coherent current is again diminished by $\Gamma\nu(2,x,1)$, i.e., by the partial density of states at point x. The total absorbed current is proportional to $\Gamma\nu(x,1)$, the injectance of contact 1 into this point. As shown in Sect. 9.3 a carrier emitting optical potential at x generates a current $-\Gamma\nu(1,x)$ in contact 1 and generates a current $-\Gamma\nu(2,x)$ in contact 2. It produces thus a total current $-\Gamma\nu(x)$. In order that the generated and the absorbed current are equal we have to normalize the emitting optical potential such that it generates a total current proportional to $\Gamma\nu(x,1)$ (equal to the absorbed current). The current at contact 2 generated by an optical potential normalized in such a way is thus $-\Gamma\nu(2,x)\nu(x,1)/\nu(x)$. The sum of the two contributions, the absorbed current and the reemitted current gives an overall transmission (or conductance) which is given by (9.38) with $4\pi|t|^2$ replaced by Γ.

Thus the weakly coupled voltage probe (which has current conservation built in) and a discussion based on optical potentials coupled with a current conserving reinsertion of carriers are equivalent [27]. There are discussions in the literature which invoke optical potentials but do not reinsert carriers. Obviously, such discussions violate current conservation. A recent discussion [28], which compares the voltage probe model and the approach via optical potentials, does reinsert carriers but does this in an ad hoc manner. In fact, [28] claims that the Onsager symmetry relations are violated in the optical potential approach. This is an incorrect conclusion arising from the arbitrary manner in which carriers are reinserted.

We conclude this section with a cautionary remark: we have found here that the weakly coupled probe voltage probe model and the optical potential model are equivalent. But this equivalence rests on a particular description of the voltage probe. There are many different models and even in the weak coupling limit our description of the voltage probe given here is not unique. The claim can only be that for sufficiently weak optical absorption and reinsertion of carriers there exits *one* voltage probe model which gives the same answer. Differing weak coupling voltage probes are discussed in [29].

9.7 AC Conductance of Mesoscopic Conductors

In this section we discuss as an additional application of partial densities of states briefly the ac conductance of mesoscopic systems. We consider a conductor with an arbitrary number of contacts labeled by a Greek index $\alpha = 1, 2, 3....$ The problem is to find the relationship between the currents $I_\alpha(\omega)$ at frequency ω measured at the contacts of the sample in response to a sinusoidal voltage with amplitude $V_\beta(\omega)$ applied to contact β. The relationship between currents and voltages is given by a dynamical conductance matrix [7] $G_{\alpha\beta}(\omega)$ such that $I_\alpha(\omega) = \sum_\beta G_{\alpha\beta}(\omega)V_\beta(\omega)$. All electric fields are localized in space. The overall charge on the conductor is conserved. Consequently, current is also conserved and the currents depend only on voltage differences. Current conservation implies $\sum_\alpha G_{\alpha\beta} = 0$ for each β. In order that only voltage differences matter, the dynamical conductance matrix has to obey $\sum_\beta G_{\alpha\beta} = 0$ for each α. We are interested here in the low-frequency behavior of the conductance and therefore we can expand the conductance in powers of the frequency [6],

$$G_{\alpha\beta}(\omega) = G^0_{\alpha\beta} - i\omega E_{\alpha\beta} + K_{\alpha\beta}\omega^2 + O(\omega^3) . \qquad (9.39)$$

Here $G^0_{\alpha\beta}$ is the dc conductance matrix. $E_{\alpha\beta}$ is called the *emittance* matrix and governs the displacement currents. $K_{\alpha\beta}$ gives the response to second order in the frequency. All matrices $G^0_{\alpha\beta}, E_{\alpha\beta}$ and $K_{\alpha\beta}$ are real.

We focus here on the emittance matrix $E_{\alpha\beta}$. The conservation of the total charge can only be achieved by considering the long-range Coulomb interaction. Here we describe the long-range Coulomb interaction in a random phase approach (RPA) in terms of an effective interaction. The effective interaction potential $g(x', x)$ has to be found by solving a Poisson equation with a non-local screening term. The effective interaction gives the potential variation at point x' in response to a variation of the charge at point x. With the help of the effective interaction we find for the emittance matrix [6]

$$E_{\alpha\beta} = e^2 \left[\int dx\nu(\alpha, x, \beta) - \int dx'dx\nu(\alpha, x')\, g(x', x)\, \nu(x, \beta) \right] . \qquad (9.40)$$

Here the first term, proportional to the integrated partial density of states, is the ac response at low frequencies, which we would have in the absence of interactions. The second term has the following simple interpretation: an ac voltage applied to contact β would (in the absence of interactions) lead to a charge built up at point x given by the injectivity of contact β. Due to interaction, this charge generates at point x' a variation in the local potential which then induces a current in contact α proportional to the emissivity of this point into contact α. The effective interaction has the property that at an additional charge with a distribution proportional to the local density of states gives rise to a spatially uniform potential, $\int dx' \nu(x') g(x', x) = 1$ for every x. This property ensures that the elements of each row and each column of the emittance matrix add up to zero. In particular, if screening is local (over a length scale of a Thomas Fermi wavelength) we have $g(x', x) = \delta(x' - x)\nu^{-1}(x)$. In this limit, the close connection between (9.40) and (9.35) is then obvious.

9.8 Transition from Capacitive to Inductive Response

The following example [30] provides an instructive application of the ac conductance formula (9.40). Consider the transmission through a narrow opening shown in Fig. 9.3. Carrier motion is in two dimension through a potential $U(x, y)$ which has the form of a saddle with a height U_0. The conductance (transmission) through such a narrow opening (a quantum point contact) has been widely studied and is found to rise steplike [31] as a function of the potential U_0 with plateaus at values $G = (2e^2/h)N$ corresponding to perfect transmission of N spin degenerate channels. Here we are interested in the emittance E as a function of U_0. We consider the case that U_0 is so large that transmission is completely blocked and then lower U_0 such that the probability of transmission probability $\mathcal{T}$ gradually increases from 0 to 1.

We introduce two regions Ω_1 and Ω_2 to the left and the right of the barrier, respectively. Instead of the local partial density of states we consider the partial density of states integrated over the respective volumes Ω_1 and Ω_2. Thus we introduce $D_{\alpha k \beta} = \int_{\Omega_k} dx dy \nu(\alpha, x, y, \beta)$. We furthermore introduce the total density of states D of the two regions. We assume that the potential has left-right symmetry and consequently the density of states in the regions Ω_1 and Ω_2 are $D_1 = D_2 = D/2$. Reference [30] evaluates the partial densities of states semiclassically. We find that carriers incident from contact 1 and transmitted into contact 2 give rise to a partial density of states in region 1 given by $D_{211} = \mathcal{T} D_1/2$. To determine D_{212}, we note that in the semiclassical limit considered here, there are no states in Ω_1 associated with scattering from contact 2 back to contact 2, hence it holds $D_{212} = 0$. With similar arguments one finds for the semiclassical PDOS

$$D_{\alpha k \beta} = D_k \left(\mathcal{T}/2 + \delta_{\alpha\beta} (\mathcal{R}\, \delta_{\alpha k} - \mathcal{T}/2) \right) \ , \quad \text{if } \alpha, \beta \neq 3 \ . \tag{9.41}$$

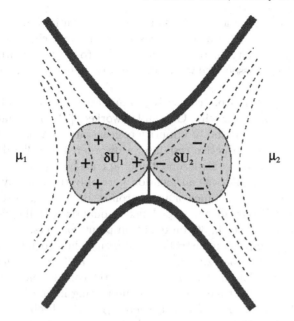

Fig. 9.3. Charge dipole across a saddle point constriction. $\mu_1 = eV_1$ and $\mu_2 = eV_2$ are the potentials of the contacts, δU_1 and δU_2 are local potentials. The dashed lines are the equipotential lines of the equilibrium potential. After [30]

From (9.41) we obtain for the emissivity into contact 1 from region 1 and injectivity from contact 1 into region 1, $D_{11}^e = D_{11}^i = (1/4)(1 + \mathcal{R})D$ and obtain for the emissivity into contact 1 from region 2 and the injectivity into region 1 from contact 2, $D_{12}^e = D_{12}^i = (1/4)\mathcal{T}D$. Instead of the full Poisson equation the effective interaction [30, 6, 32] is determined with the help of a geometrical capacitance C. For a detailed discussion we refer the reader to [30, 32]. Due to charge conservation we have $E \equiv E_{11} = E_{22} = -E_{12} = -E_{21}$ with

$$E = (RC_\mu - DT^2/4) \, . \tag{9.42}$$

Here $C_\mu^{-1} = \mathcal{R}^{-1}(C^{-1} + (e^2 D/4)^{-1})$ is the effective capacitance of the contact. It is proportional to the reflection probability and proportional to the series capacitance of the geometrical capacitance C and the "quantum capacitance" $e^2 D$. If the contact is completely closed we have $\mathcal{R} = 1$, $\mathcal{T} = 0$ and the emittance is completely determined by the capacitance C_μ. If the channel is completely open we have $\mathcal{R} = 0$, $\mathcal{T} = 1$ and the emittance is $E = -D/4$. It is negative indicating that for a completely open channel the ac response is now not capacitive but inductive . Thus there is a voltage U_0 for which the emittance vanishes. For a simple saddle point potential, the behavior of the capacitance and emittance is illustrated in Fig. 9.4. The dotted line shows the conductance, the dashed line is the capacitance C_μ, and the full line is the emittance as a function of the saddle point potential U_0. The emittance is capacitive (positive) for a nearly closed contact and changes sign as the transmission probability increases from near 0 to 1. The emittance shows additional structure associated with the successive opening of further quantum channels.

A similar transition from capacitive to inductive behavior is found in the emittance of a mesoscopic wire: Guo and coworkers [33] investigate the emittance of a wire as a function of impurity concentration. A ballistic wire with no impurities has an inductive response, a disordered metallic diffusive wire has a capacitive response. Experiments on ac transport in mesoscopic structures are challenging and we can mention here only the work by Pieper and Price [34] on the ac conductance of an Aharonov–Bohm ring, work by Melcher et al. [35] on the low-frequency impedance of quantized Hall conductors and recent work by Regul et al. [36] on quantum point contacts.

Our simple example demonstrates the difficulty in associating a time with a result obtained by analyzing a stationary (or as here) a quasistationary scattering problem. The emittance divided by the conductance quantum e^2/h has the dimension of a time. (In the noninteracting limit $e^2/C << 4/D$ we have [37] $E/(e^2/h) = (\mathcal{R} - \mathcal{T})\tau_D/4$ where τ_D is the dwell time in the two regions Ω_1 and Ω_2.) But since this "time" changes sign such an interpretation is not appropriate. Furthermore, in electrical problems the natural "times" are RC-times if the low-frequency dynamics is capacitive or an R/L-time if it is inductive. The fact that we describe here a crossover from capacitive to an inductive-like behavior demonstrates that neither of these two timescales can adequately describe the dynamics.

9.9 Partial Density of States Matrix

Thus far the main aim of our discussion has been to illustrate the concept of partial densities of states with a number of simple examples. We now would like to point to some important extensions of this concept.

Quantum mechanics is a theory of probability amplitudes. We can thus suppose that our initial state is a scattering experiment which is described by a superposition $\Psi(y) = a_\beta \Psi_\beta(y) + a_\gamma \Psi_\gamma(y)$. Here Ψ_β and Ψ_γ are two scattering states describing particles incident in channel β *and* in channel γ with amplitudes a_β and a_γ. Often such superpositions are eliminated since we can assume that each incident amplitude carries its own phase ϕ_β and ϕ_γ. We suppose that $a_\beta = |a_\beta|^{1/2} \exp(i\phi_\beta)$ and $a_\gamma = |a_\gamma|^{1/2} \exp(i\phi_\gamma)$ and that these phases are random and uncorrelated. Thus, if an average over many scattering experiments is taken, we have $\langle e^{i\phi_\beta} e^{-i\phi_\gamma} \rangle = \delta_{\beta\gamma}$ and consequently we find that $\langle |\Psi(y)|^2 \rangle = |a_\beta|^2 |\Psi_\beta|^2 + |a_\gamma|^2 |\Psi_\gamma|^2$ is the sum of scattering events in the different quantum channels. However, as soon as we are interested in quantities which depend to fourth (or higher) order on the amplitudes, then even after averaging over the random phases of the incident waves we find that

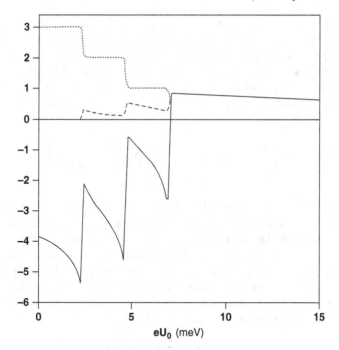

Fig. 9.4. Conductance of the saddle point constriction (dotted line) in units of $2e^2/h$, capacitance and emittance (dashed and full curve) as a function of the height of the saddle point potential U_0. In the range of voltages shown, three quantum channels are opened. After [30]

scattering processes which involve two (or more) incident waves matter. For instance consider $\langle |\Psi(y)|^2 |\Psi(y')|^2 \rangle - \langle |\Psi(y)|^2 \rangle \langle |\Psi(y')|^2 \rangle$ which is determined by $|a_\beta|^2 |a_\gamma|^2 \Psi_\beta^*(y) \Psi_\gamma(y) \Psi_\beta(y') \Psi_\gamma^*(y') + h.c..$ This expression describes a correlation of the particle density at the points y and y'. Obviously, for such higher-order correlations superpositions are very important even if we associate a random phase with each incident scattering states and take an average over these phases.

We can relate such density fluctuations to functional derivatives of scattering matrix expressions. To describe density fluctuations of scattering processes with an incident carrier stream from channel β and γ, we consider

$$\nu(\alpha, y, \beta, \gamma) = -\frac{1}{4\pi i} \left(S_{\alpha\beta}^\dagger \frac{\delta S_{\alpha\gamma}}{\delta V(y)} - \frac{\delta S_{\alpha\beta}^\dagger}{\delta V(y)} S_{\alpha\gamma} \right). \tag{9.43}$$

Note that two different scattering matrices enter into this expression. The connection to the scattering states is determined by the relation

$$\sum_\alpha \nu(\alpha, y, \beta, \gamma) = (1/h)(v_\beta v_\gamma)^{-1/2} \Psi_\beta^*(y) \Psi_\gamma(y). \tag{9.44}$$

Here v_β and v_γ are the (asymptotic) velocities of the carriers in the scattering channels β and γ far away from the scattering region. Equation (9.44) was given in [38] and a detailed derivation of this relation is presented in [39].

The expressions $\nu(\alpha, y, \beta, \gamma)$ can be viewed as the off-diagonal elements of a partial density of states matrix. In (9.44) we take the sum over outgoing channels. The resulting matrix is the *local* density of states matrix. Using (9.43) in (9.44) we find,

$$\nu(y, \beta, \gamma) \equiv \sum_\alpha \nu(\alpha, y, \beta, \gamma) = -\frac{1}{2\pi i} \sum_\alpha \left(S_{\alpha\beta}^\dagger \frac{\delta S_{\alpha\gamma}}{\delta V(y)} \right), \qquad (9.45)$$

where we have taken into account that the scattering matrix is unitary.

Let us now consider the total density of states matrix. The fluctuations of interest are then the total particle number fluctuations in the scattering region Ω, $\mathcal{D}(\beta, \gamma) = \int_\Omega dy \nu(y, \beta, \gamma)$. Furthermore, if the volume of integration is sufficiently large, we can, in WKB (Wentzel-Kramers-Brillouin) approximation, replace the functional derivative with respect to V with a derivative with respect to energy. The matrix which governs the fluctuations in the particle number in the scattering region then becomes,

$$\mathcal{D}(\beta, \gamma) = \frac{1}{2\pi i} \sum_\alpha \left(S_{\alpha\beta}^\dagger \frac{dS_{\alpha\gamma}}{dE} \right). \qquad (9.46)$$

Equation (9.46) is the Wigner–Smith delay time matrix (apart from a factor h) [40]. We have earlier emphasized that the appearance of the energy derivative is a consequence of approximations (here the fact that we consider the WKB limit). We also mention that strictly speaking, here we do not consider a "delay." We do not compare with a reference scattering problem (a free motion) as is typically done in nuclear scattering problems. Equation (9.46) determines a total time or absolute time and we should more appropriately call it the absolute time matrix instead of the delay time matrix.

We conclude by briefly discussing the Wigner–Smith matrix for a tunnel barrier. For a symmetric barrier with transmission and reflection probability $\mathcal{T}$ and $\mathcal{R}$ the scattering matrix has elements $S_{11} = S_{22} = -i\sqrt{\mathcal{R}}\exp(i\phi)$, and $S_{21} = S_{12} = \sqrt{\mathcal{T}}\exp(i\phi)$ where ϕ is the phase accumulated during a reflection or transmission process. Thus the elements of the Wigner–Smith delay time matrix (9.46) are

$$\mathcal{D}_{11} = \mathcal{D}_{22} = \frac{1}{2\pi}\frac{d\phi}{dE}, \quad \mathcal{D}_{12} = \mathcal{D}_{21} = \frac{1}{4\pi}\frac{1}{\sqrt{\mathcal{R}\mathcal{T}}}\frac{d\mathcal{T}}{dE}. \qquad (9.47)$$

To be specific consider now the case of a tunnel barrier. In the WKB limit we have $\mathcal{R} \approx 1$ and $\mathcal{T} = exp(-2S/\hbar)$ with $S = \int dy\sqrt{2m}\sqrt{V(y) - E}$ where the integral extends from one turning point to the other. We have $d\phi/dE = 0$ and using the expression $\tau_T = m \int dy(\sqrt{2m}\sqrt{V(y) - E})^{-1}$ for the traversal time of tunneling gives $dT/dE = (2\tau_T/\hbar)T$. Consequently, the diagonal elements of the Wigner–Smith matrix vanish and the nondiagonal elements are

$$\mathcal{D}_{12} = \mathcal{D}_{21} = \frac{\tau_T}{2\pi\hbar}\sqrt{T} \, . \tag{9.48}$$

Thus while the the average density inside the barrier vanishes (the trace of the Wigner–Smith matrix is zero) the off-diagonal elements are nonzero and indicate that the fluctuations of the charge will in general be nonvanishing even deep inside the classically forbidden region.

We mention a number of transport problems in mesoscopic physics in which the Wigner–Smith matrix, (9.46) (or more generally the partial density of states matrix, (9.45)) plays a prominent role. Random matrix theory has been extended to permit the calculation of the entire distribution function of delay times [41] for structures whose dynamic is in the classical limit chaotic (chaotic cavities). This work treats structures in which carrier propagation is in an allowed energy region.

The partial density of states matrix has been used in [38] to obtain the second order in frequency term of the ac conductance (see (9.39)). Reference [42] investigated the current induced into a nearby gate due to charge fluctuations in quantum point contacts and chaotic cavities. More recently, the charge fluctuations in two nearby mesoscopic conductors were treated and the effect of dephasing due to charge fluctuations was evaluated within this approach [43, 39]. The results can be compared with other theoretical works [44] and with experiments [45]. The quantum measurement problem, a two-state system coupled to a mesoscopic conductor described by a general scattering matrix is treated by Pilgram and Büttiker [46] who derive a general expression for the Heisenberg efficiency of the detector in terms of the Wigner–Smith matrix. The role of screening (interactions) is discussed in [47]. Clerk, Girvin, and Stone [48] discuss the information theoretical aspects of this problem and Clerk and Stone [49] investigate the effect of phase breaking and inelastic scattering on the efficiency of the detector.

Of considerable interest is a theory of quantum pumping in small systems. In quantum pumping one is interested in the current generated as two parameters (gate voltages, magnetic fluxes) which modulate the system are varied sinusoidally but out of phase. Brouwer [50], Avron et al. [51], Shutenko et al. [52], and Polianski and Brouwer [53] develop a theory which is based on the modulation of the partial densities of states discussed here. Expressions for the pump current, the heat flow, and the noise in terms of the diagonal and off-diagonal elements of the partial density of states matrix, (9.45) of an adiabatic quantum pump are given by Moskalets and Büttiker [54]. Particularly revealing is the investigation of a quantum pump in the presence of oscillating contact potentials. In this case the ac currents generated in the contacts of the sample interfere with the quantum pump [55]. Again the partial density of states play a prominent role.

9.10 Local Friedel Sum Rule

We have discussed the connection of partial density of states with the Wigner–Smith delay matrix, or more precisely, its *local* version. In solid state physics the connection between scattering data and the (charge)-density is known as Friedel sum rule [56]. We can now similarly derive a connection between partial densities and the *local* density matrix [38, 9, 57, 58]. Namely, if we combine (9.44) and (9.45) and integrate over an (arbitrary) volume Ω of interest we have

$$\frac{1}{h(v_\beta v_\gamma)^{1/2}} \int_\Omega d^3\mathbf{r} \Psi_\beta^*(\mathbf{r})\Psi_\gamma(\mathbf{r}) = -\frac{1}{2\pi i} \sum_\alpha \int_\Omega d^3\mathbf{r} \left(S_{\alpha\beta}^\dagger \frac{\delta S_{\alpha\gamma}}{\delta V(\mathbf{r})} \right). \qquad (9.49)$$

For $\gamma = \beta$ the left-hand side is just the contribution of the scattering state with carriers incident from contact β to local density in Ω. Equation (9.49) shows that similar relations apply even for products of two different wave functions, i.e., to the off-diagonal matrix elements of the charge density operator. A simplification occurs if we take the volume Ω to be large. In fact if Ω is large it will intersect our sample far in the asymptotic regions. Suppose then that we define the scattering matrix such that it describes exactly the phases and amplitudes for scattering from one of these intersections to another. We can then replace the functional derivative with respect to the local potential on the right-hand side with an energy derivative to obtain [9, 58],

$$\frac{1}{h(v_\beta v_\gamma)^{1/2}} \int_\Omega d^3\mathbf{r} \Psi_\beta^*(\mathbf{r})\Psi_\gamma(\mathbf{r}) = s^\dagger ds/dE + (s - s^\dagger)/4E. \qquad (9.50)$$

The first term on the right-hand side is just the Wigner–Smith matrix, (9.46). The second term represents an interference term which arises since we ask about the density in a finite, well-defined volume. Such interference terms were discussed for a reflection problem by Leavens and Aers [5], by Gasparian et al. [9], and more recently for scattering problems on networks by Texier and the author [58]. We remark that here it is the absolute densities and phases which are important, in contrast, to the typical discussion in scattering theory in which differences of densities and phases of a problem with and without scattering are evaluated. The integrated version of the Friedel sum rule, (9.50) is especially useful since it relates integrated densities directly to energy derivatives of the scattering matrix. On the other hand, many problems require knowledge of local densities and it is then necessary to calculate either wave functions or functional derivatives of the scattering matrix making use of (9.49). The scattering matrix contains knowledge only of states which connect to the contacts of the sample (asymptotic scattering regions). We cannot extract information from the scattering matrix on states which are entirely localized within the sample. Therefore, there are limits on the validity of (9.49) and (9.50). We refer the interested reader to [58].

9.11 Discussion

The Larmor clock and its close relatives have become one of the most widely investigated approaches mainly in order to understand the question: "How long does a particle traveling through a classically forbidden region (a tunnel barrier) interact with this region?". We have already in the previous sections pointed out that there is no consensus in the interpretation even of this simple clock. Regardless of these difficulties the investigation of the Larmor clock has been helpful in understanding a number of transport problems: In particular, we have discussed a hierarchy of density of states as they occur in open multiprobe mesoscopic conductors. These density of states are directly related to local Larmor times. We have shown that a small absorption or a small emission of particles can be described with these densities (or in terms of the Larmor times). We have shown that the transmission probabilities through weakly coupled contacts like the STM is related to these densities. We have shown that a weakly coupled voltage probe, describing inelastic scattering or a dephasing process can be treated in terms of these densities. We have also pointed out that the ac conductance of a mesoscopic conductor at small frequencies can be expressed with the help of these densities. Furthermore, we have indicated that it is useful to consider also the off-diagonal elements of a partial density of states matrix since this permits a description of fluctuation processes. Thus there is no question that the investigation of the Larmor clock has been a very fruitful and important enterprise.

References

1. A.I. Baz': Sov. J. Nuc. Phys. **4**, 182 (1967); **5**, 161 (1967)
2. V.F. Rybachenko: Sov. J. Nucl. Phys. **5**, 635 (1967)
3. M. Büttiker: Phys. Rev. B **27**, 6178 (1983)
4. C.R. Leavens, G.C. Aers: Solid State Commun. **63**, 1107 (1989)
5. C.R. Leavens, G.C. Aers: Phys. Rev. B **40**, 5387 (1989)
6. M. Büttiker: J. Phys.: Condens. Matter **5**, 9361 (1993)
7. M. Büttiker, H. Thomas, A. Prêtre: Z. Phys. B **94**, 133 (1994)
8. M. Büttiker, A. Prêtre, H. Thomas: Phys. Rev. Lett. **70**, 4114 (1993); M. Büttiker, H. Thomas, A. Prêtre: Phys. Lett. A **180**, 364 (1993)
9. V. Gasparian, T. Christen, M. Büttiker: Phys. Rev. A **54**, 4022 (1996)
10. T. Gramespacher, M. Buttiker: Phys. Rev. B **56**, 13026 (1997); Phys. Rev. B **60**, 2375 (1999); Phys. Rev. B **61**, 8125 (2000)
11. R. Dashen, S. Ma, H.J. Bernstein: Phys. Rev. **187**, 345 (1969)
12. Y. Avishai, Y.B. Band: Phys. Rev. B **32**, 2674 (1985)
13. G. Iannaccone: Phys. Rev. B **51**, 4727 (1995)
14. G. Iannaccone, B. Pellegrini: Phys. Rev. B **53**, 2020 (1996)
15. M. Büttiker, R. Landauer: Phys. Rev. Lett. **49**, 1739 (1982); Phys. Scr. **32**, 429 (1985)
16. D. Sokolovski, L.M. Baskin: Phys. Rev. A **36**, 4604 (1987); H.A. Fertig, Phys. Rev. B **47**, 1346 (1993)

17. V. Gasparian, M. Ortuno, J. Ruiz, E. Cuevas: Phys. Rev. Lett. **75**, 2312 (1995); Y. Japha, G. Kurizki: Phys. Rev. A **60**, 1811 (1999)
18. A.M. Steinberg: Phys. Rev. Lett. **74**, 2405 (1995)
19. S. Brouard, R. Sala, J.G. Muga: Phys. Rev. A **49** 4312 (1994)
20. M. Büttiker: in *Electronic Properties of Multilayers and Low Dimensional Semiconductors*, ed. by J.M. Chamberlain, L. Eaves, J.C. Portal (Plenum, New York, 1990), pp. 297–315
21. S.A. Ramakrishna, N. Kumar: Phys. Rev. B **61**, 3163 (2000)
22. C.W.J. Beenakker, in "Photonic Crystals and Light Localization in the 21st Century", ed. by C.M. Soukoulis, *NATO Science Series C563* (Kluwer, Dordrecht, 2001), pp. 489–508
23. X. Zhao: J. Phys.: Condens. Matter **12**, 4053 (2000)
24. P.W. Brouwer, S.A. van Langen, K.M. Frahm, M. Büttiker, C.W.J. Beenakker: Phys. Rev. Lett. **79**, 914 (1997)
25. G. Binnig, H. Rohrer: Helv. Phys. Acta **55**, 726 (1982); J. Tersoff, D.R. Hamann, Phys. Rev. B **31**, 805 (1985)
26. M. Büttiker: IBM J. Res. Develop. **32**, 63 (1988)
27. P.W. Brouwer, C.W.J. Beenakker: Phys. Rev. B **55**, 4695 (1997)
28. T.P. Pareek, S.K. Joshi, A.M. Jayannavar: Phys. Rev. B **57**, 8809 (1998)
29. M. Büttiker: in *Analogies in Optics and Micro-Electronics*, ed. by W. van Haeringen, D. Lenstra (Kluwer Academic Publishers, Dordrecht, Boston, London, 1990), pp. 185–202
30. T. Christen, M. Büttiker: Phys. Rev. Lett. **77**, 143 (1996)
31. B.J. van Wees et al.: Phys. Rev. Lett. **60**, 848 (1988); D.A. Wharam et al.: J. Phys. C: Solid State Phys. **21**, L209 (1988)
32. M. Büttiker, T. Christen: in *Mesoscopic Electron Transport, NATO Advanced Study Institute, Series E: Applied Science*, ed. by L. L. Sohn, L. P. Kouwenhoven, G. Schoen (Kluwer Academic Publishers, Dordrecht, 1997), Vol. 345, p. 259; cond-mat/9610025
33. Tiago De Jesus, Hong Guo, Jian Wang: Phys. Rev. B **62**, 10774 (2000)
34. J.P. Pieper, J.C. Price: Phys. Rev. Lett. **72**, 3586 (1994)
35. J. Melcher, J. Schurr, F. Delahye and A. Hartland: Phys. Rev. B **64**, 127301 (2001)
36. J. Regul, F. Hohls, D. Reuter, A.D. Wieck, R. J. Haug: Physica E **22**, 272 (2004)
37. S.A. Mikhailov, V.A. Volkov: JETP Lett. **61**, 524 (1995)
38. M. Büttiker: J. Math. Phys. **37**, 4793 (1996)
39. M. Büttiker: in *Quantum Mesoscopic Phenomena and Mesoscopic Devices*, ed. by I.O. Kulik, R. Ellialtioglu (Kluwer, Academic Publishers, Dordrecht, 2000). Vol. 559, p. 211; cond-mat/9911188
40. F.T. Smith: Phys. Rev. **118** 349 (1960)
41. Y. V. Fyodorov, H. J. Sommers: Phys. Rev. Lett. **76**, 4709 (1996); V. A. Gopar, P. A. Mello, M. Büttiker: Phys. Rev. Lett. **77**, 3005 (1996); P. W. Brouwer, K. M. Frahm, C. W. J. Beenakker, Phys. Rev. Lett. **78**, 4737 (1997); C. Texier and A. Comtet, Phys. Rev. Lett. **82**, 4220 (1999)
42. M.H. Pedersen, S.A. van Langen, M. Büttiker: Phys. Rev. B **57**, 1838 (1998)
43. M. Büttiker, A.M. Martin: Phys. Rev. B **61**, 2737 (2000)
44. Y.B. Levinson: Europhys. Lett. **39**, 299 (1997); L. Stodolsky, Phys. Lett. B **459**, 193 (1999)

45. E. Buks, R. Schuster, M. Heiblum, D. Mahalu, V. Umansky: Nature **391**, 871 (1998); D. Sprinzak, E. Buks, M. Heiblum, H. Shtrikman: Phys. Rev. Lett. **84**, 5820 (2000)
46. S. Pilgram, M. Büttiker: Phys. Rev. Lett. **89**, 200401 (2002)
47. M. Büttiker, S. Pilgram: Surface Science, **532**, 617 (2003)
48. A.A. Clerk, S.M. Girvin, A.D. Stone: Phys. Rev. B **67**, 165324 (2003)
49. A.A. Clerk, A.D. Stone: Phys. Rev. B **69**, 245303 (2004)
50. P.W. Brouwer: Phys. Rev. B **58**, R10 135 (1998)
51. J.E. Avron, A. Elgart, G.M. Graf, L. Sadun: Phys. Rev. B **62**, R10618 (2000): J. E. Avron, A. Elgart, G. M. Graf, L. Sadun, K.Schnee: Commun. Pure Appli. Math. **57** (4): 528 (2004)
52. T.A. Shutenko, I.L. Aleiner, B.L. Altshuler: Phys. Rev. B **61**, 10366 (2000)
53. M.L. Polianski, P.W. Brouwer: Phys. Rev. B **64**, 075304 (2001)
54. M. Moskalets, M. Büttiker: Phys. Rev. B **66**, 035306 (2002)
55. M. Moskalets, M. Büttiker: Phys. Rev. B **69**, 205316 (2004)
56. J. Friedel: Philos. Mag. **43**, 153 (1952)
57. M. Büttiker: Pramana-J. Phys. **58** 241-257 (2002)
58. C. Texier, M. Büttiker: Phys. Rev. B **67**, 245410 (2003)

"Standard" Quantum–Mechanical Approach to Times of Arrival

Iñigo L. Egusquiza[1], J. Gonzalo Muga[2], and Andrés D. Baute[1,2]

[1] Fisika Teorikoaren Saila, Euskal Herriko Unibertsitatea, 644 Posta Kutxa, 48080 Bilbao, Spain
inigo.egusquiza@ehu.es
[2] Dpto. de Química Física, Universidad del País Vasco, Apdo. 644, 48080 Bilbao, Spain
jg.muga@ehu.es

10.1 Introduction

As we already pointed out in the introduction, the understanding of time observables in quantum mechanics has been hindered by the early appreciation that there could be no such "observable" in the standard, von Neumann formalism. A statement of such a view is that provided by Pauli [1], whereby a self-adjoint positive operator cannot have a self-adjoint conjugate operator.

This did not prevent some adventurous souls from trying to propose operators with the required characteristics of time. Among these, a very special mention of Aharonov and Bohm is required [2]. They considered the symmetric quantization of the phase space function which, in the free particle case, provides us with the time of arrival at a point y of a particle that at the instant $t = 0$ has position x and momentum p, namely, $m(y - x)/p$. Their main concern was to estimate uncertainties and to investigate the status and meaning of time–energy uncertainty relations. Later on, several groups of researchers looked harder at the mathematical fine details of the operator [3, 4, 5, 6, 7, 8, 9]. As a result, the status of the Aharonov–Bohm time-of-arrival operator as a maximally symmetric operator, with an associated positive operator valued measure (POVM), has now been firmly established.

On a different tack, Kijowski [10] introduced a probability density for times of arrival of free particles from an axiomatic perspective. He thus concluded that he had obtained a time-of-arrival operator, and gave an explicit expression in the energy representation, pointing out that the correct commutation relation is indeed fulfilled. He also signaled the fact that no self-adjoint extension could exist (see also [11]), and pursued other possible routes (later retaken in [7]). He also studied the relativistic case.

I. L. Egusquiza et al.: *"Standard" Quantum–Mechanical Approach to Times of Arrival*, Lect. Notes Phys. **734**, 305–332 (2008)
DOI 10.1007/978-3-540-73473-4_10   © Springer-Verlag Berlin Heidelberg 2008

After introducing Kijowski's distribution for the free particle, we shall present a short introduction to the concept and mathematics of POVMs, including the case of continuous POVMs, which will be illustrated with the example of the momentum on the half-line. Without delving too deeply in the mathematics of Naimark's dilation theorem, we shall use it in the same physical context. Armed with all these concepts and examples, we shall then show that the Aharonov–Bohm time-of-arrival operator is a maximally symmetric operator to which those constructions are applicable. In order to show the generality of the concept of POVM for understanding time, we shall also discuss several other examples of time operators.

Once the time-of-arrival POVM for the free case has been well established, we will proceed to identify the associated probability distribution with that postulated by Kijowski on axiomatic grounds [10]. In this manner, Kijowski's operator is identified with Aharonov and Bohm's for the free particle case. An alternative construction using arrival states will be introduced that generalizes the time-of-arrival distributions for the interacting case. We shall see that in general it is not possible to associate POVMs to these distributions. However, the "operator normalization" of Brunetti and Fredenhagen [12] will allow us to recover a setting for them within standard quantum mechanics. We shall see also that this concept combined with a model for time-of-arrival detection using fluorescence photons provides an operational procedure to measure Kijowski's distribution.

The purpose of this chapter is to include at least some embodiment of time observables, in particular time-of-arrival observables, within a very slight mathematical extension of "standard" quantum mechanics, understood as defined by von Neumann [13] and others. We shall limit ourselves to this particular approach to times of arrival, and we direct the reader to a review of time of arrival in quantum mechanics with a wider scope than this work [14].[1]

10.2 Kijowski's Time-of-Arrival Distribution

Kijowski [10], in an attempt to formulate a realistic interpretation of the time–energy relation, proposed a probability density for times of arrival of a free particle, that we shall henceforth be calling Kijowski's distribution. His approach consisted in identifying some minimum properties that the distribution of times of arrival fulfills in the classical free case, and then demand them of a quantum–mechanical probability density, again for the free particle. In a more detailed manner, Kijowski proved the following theorem (in a three-dimensional version that we will obviate): consider the set of continuous positive bilinear functionals F of wave functions ψ with support restricted to positive momentum, bilinears which are invariant under space translations,

---

[1] For other recent, complementary views and discussions on the quantum arrival time see [15, 16] and further references below.

such that for any normalized wave function ψ it is the case that $\int dt\, F[\psi_t] = 1$, where ψ_t is the evolved state from the initial state $\psi_0 = \psi$. Additionally demand that $F[\bar{\psi}] = F[\psi]$ and that the dispersion, defined as

$$\int_{-\infty}^{+\infty} dt\, t^2 F[\psi_t] - \left(\int_{-\infty}^{+\infty} dt\, t F[\psi_t] \right)^2$$

be finite. Then there is a specific functional F_0 for which this variance is minimum; the average value $\int dt\, t F[\psi_t]$ is constant over this class of functionals; and the functional F_0 has the following expression:

$$F_0[\psi] = \int \frac{dp_1 dp_2}{2\pi m\hbar}\, \bar{\psi}(p_1)\sqrt{p_1 p_2}\psi(p_2)$$

(notice a slight deviation from the notation of Kijowski; we denote by $\psi(p)$ the wave function in the momentum representation, normalized to unity). The integration variables run from 0 to infinity, since this functional is defined only on functions with support on positive momenta. The probability density of arrivals for these states, with support on positive momenta only, is thus defined as

$$\Pi_+^K(t) = \Pi_+^K(t, \psi) = F_0[\psi_t] = \left| \int_0^\infty dp\sqrt{\frac{p}{2\pi m\hbar}}\, e^{-ip^2 t/2m\hbar}\psi(p) \right|^2 .$$

The average of t with Π_+^K, $\int dt\, t \Pi_+^K(t)$, coincides with the "average" computed with the current density $J(t)$, even though $J(t)$ is not necessarily positive.[2] Notice also that we are dealing with the case of free motion, which means that the p component of the evolved state ψ_t is related to the p component of the initial $(t = 0)$ state ψ through

$$\psi_t(p) = e^{-ip^2 t/2m\hbar}\psi(p) .$$

Kijowski pointed out that $\Pi_+^K(t)$ regards only arrivals from the left, but that arrivals from the right lead by symmetry to an analogous expression, $\Pi_-^K(t)$. In this way, he obtained a total probability density for arrivals at time t at position $x = 0$ of a freely moving particle in one dimension, as follows:

$$\Pi_\psi^K(t) = \left| \int_0^\infty dp\sqrt{\frac{p}{2\pi m\hbar}}\, e^{-ip^2 t/2m\hbar}\psi(p) \right|^2 + \left| \int_{-\infty}^0 dp\sqrt{\frac{-p}{2\pi m\hbar}}\, e^{-ip^2 t/2m\hbar}\psi(p) \right|^2 .$$

Since a (good) bilinear functional has been defined over the space of states of the free particle, Kijowski concludes that an operator has been obtained that fulfills the characteristics we desire of a time-of-arrival operator. However, this operator cannot have any self-adjoint extension, as he proves by going to the energy representation, which we will be reexamining later. How does this fact tally with the existence of a perfectly well-defined procedure for assigning probability densities of times of arrivals to each state? Is this not an observable? The answer to these questions lies in the concept of POVM.

---

[2] This property is used in Chap. 2.

10.3 POVMs

When first presenting quantum mechanics to undergraduates, it is rather usual to state that the correspondence between observables and self-adjoint operators is due to the fact that in experiments we measure real numbers. When the level of mathematical sophistication increases, the association between observability and self-adjointness is referred to the probabilistic definition of quantum mechanics (see [17] for an enlightening axiomatic presentation of quantum mechanics as a probabilistic theory—for a Bayesian perspective see, for instance, [18]). In actual fact, however, the real reason underlying the usual presentation is the spectral theorem that, in broad terms, states that the moments of the probability distribution for the measurement of an observable do coincide with the operator moments, if the associated operator is self-adjoint.

What is indeed demanded by the probabilistic character of quantum mechanics? The probability p_σ for the outcome of an experiment to lie in a subset σ of the range of possible values for an observable is linear with respect to the density matrix $\widehat{\rho}$, which implies that for each subset σ there exists an operator $\widehat{A}_\sigma$ in such a way that the probability is given by

$$p_\sigma = \text{Tr}(\widehat{A}_\sigma \widehat{\rho}) \ . \tag{10.1}$$

Now, for disjoint subsets of possible values of the observable, the probabilities must be additive for all states, whence it follows that $\widehat{A}_{\sigma \cup \sigma'} = \widehat{A}_\sigma + \widehat{A}_{\sigma'}$ (if $\sigma \cap \sigma' = \emptyset$). Even more, the additivity of the operators must be inherited from that of the subsets of possible values, so, for instance, $\widehat{A}_{\sigma \cup \sigma'} = \widehat{A}_\sigma$ if $\sigma' \subset \sigma$. Next, since probabilities sum to one for every state $\widehat{\rho}$, i.e.,

$$1 = \sum_\sigma p_\sigma = \text{Tr}\left(\sum_\sigma \widehat{A}_\sigma \widehat{\rho}\right)$$

for all $\widehat{\rho}$, it must be that

$$\sum_\sigma \widehat{A}_\sigma = \widehat{1} \ .$$

Furthermore, since probabilities are positive, $\widehat{A}_\sigma$ must be a positive operator for every σ which is a subset of the set of possible values of the observable.

It thus follows that an observable is characterized by its range of possible values and by a map from subsets of the set of possible values to the space of positive operators, in such a way that the additivity of the sets is respected by the operators and the positive operator associated with the whole set of possible values is the unit operator.[3]

---

[3] It is possible to extend even further this general result by considering the possibility that none of the possible values is measured, because of the unavailability of the system or for dynamical reasons. Such an extension is indeed considered by Hardy [17] and others [19], and, implicitly and in a different context, particularly

This result is encoded in the sentence "each observable is associated to a positive operator valued measure (POVM)," which is precisely that mapping from intervals of the real line (or a more general Borel space, if an abstract treatment is necessary) to positive operators that has been presented above.

As a simple example of a POVM, let us consider the following set of two by two matrices:

$$\widehat{M_1} = \frac{1}{2}\begin{pmatrix} 1 & 0 \\ 0 & 0 \end{pmatrix}, \quad \widehat{M_2} = \frac{1}{2}\begin{pmatrix} 0 & 0 \\ 0 & 1 \end{pmatrix}, \quad \widehat{M_3} = \frac{1}{4}\begin{pmatrix} 1 & 1 \\ 1 & 1 \end{pmatrix}, \quad \widehat{M_4} = \frac{1}{4}\begin{pmatrix} 1 & -1 \\ -1 & 1 \end{pmatrix}.$$
(10.2)

They are positive (since all four matrices are in fact a positive constant times a projector), and they add up to the unit matrix. In this case, the probability p_i of obtaining the outcome a_i associated with the i-th operator is $\text{Tr}(\widehat{M_i}\widehat{\rho})$ for the two by two density matrix $\widehat{\rho}$.

There is an operator associated with this example, namely $\widehat{A} = \sum_{i=1}^{4} a_i \widehat{M_i}$. Under the assumption that the possible outcomes are real numbers, it is indeed a self-adjoint operator. Wherein lies the difference with the older standard approach? The average value of the observable is of course

$$\sum_{i=1}^{4} p_i a_i = \sum_{i=1}^{4} a_i \text{Tr}(\widehat{M_i}\widehat{\rho}) = \text{Tr}\left(\left(\sum_{i=1}^{4} a_i \widehat{M_i}\right)\widehat{\rho}\right) = \text{Tr}\left(\widehat{A}\widehat{\rho}\right).$$

However, the second moment is

$$\sum_{i=1}^{4} p_i a_i^2 = \text{Tr}\left(\left(\sum_{i=1}^{4} a_i^2 \widehat{M_i}\right)\widehat{\rho}\right),$$

which is most definitely not equal to $\text{Tr}\left(\widehat{A}^2\widehat{\rho}\right)$ for all $\widehat{\rho}$. This is due to the fact that the positive operators above are not orthonormal projectors.

The common objection against understanding each of the four matrices as associated to distinct possible outcomes of the observable is that the lack of orthogonality would not allow such a discrimination between the different outcomes. This objection can however be surmounted by introducing either an ancillary system or an extended set of states for the measuring apparatus. Let us now consider the second possibility. Imagine that we have a quantum measuring apparatus and a two-level system coupled to this apparatus. Consider further that the measuring apparatus has four possible distinct orthogonal pointer states, and, associated with these, four projectors $\widehat{E_i}$. Denote with $\widehat{U}$ the unitary evolution of both system and measuring apparatus, and assume that the initial state is $\widehat{\rho}_s \otimes \widehat{\rho}_a$, that is to say, the product state of the apparatus state $\widehat{\rho}_a$ and the system state $\widehat{\rho}_s$. The probability that the apparatus provides us with the i-th outcome is

---

relevant for the topic of times of arrival, by [20]. Note that in this presentation, we have implicitly assumed the conditions required for Gleason's theorem and its extensions for POVMs, such as that of Busch [19].

$$p_i = \mathrm{Tr}_{s,a} \left(\widehat{U} \widehat{\rho}_s \otimes \widehat{\rho}_a \widehat{U}^\dagger \left(1 \otimes \widehat{E}_i \right) \right) =$$

$$= \mathrm{Tr}_s \left(\widehat{\rho}_s \left(\mathrm{Tr}_a \left(\widehat{\rho}_a \widehat{U}^\dagger \widehat{E}_i \widehat{U} \right) \right) \right) = \mathrm{Tr}_s \left(\widehat{\rho}_s \widehat{M}_i \right) ,$$

where

$$\widehat{M}_i = \mathrm{Tr}_a \left(\widehat{\rho}_a \widehat{U}^\dagger \widehat{E}_i \widehat{U} \right) .$$

Given this example, one could argue that POVMs are simply the consequence of having traced over some additional ancillary system, or that they appear because we are interposing some quantum measurement between the system and the observer. This is not the correct place to address the quantum theory of measurement. Let us simply note that indeed POVMs can be understood as associated to "unsharp" measurements [21] or "nonideal" measurements [22], and this has been a very active area of research since the fundamental insights of Davies, Ludwig, Holevo, and others [23, 24, 25]. A somewhat recent result in this context is for instance Paul Busch' extension of Gleason's theorem to "effect valuations" (i.e., functionals over POVMs) [19].

We should also mention a very nice example of an experimental setup which makes full use of the concept of POVMs (even though the authors assert their dislike for this name!), presented in [26].

There is a further use for the concept of POVM, in a context where the connection to the quantum theory of measurement is not at all clear: there are measurements which are intrinsically associated with POVMs, and for which there is no underlying sharp observable with which to compare the incompleteness of the resolution of the identity. For instance, as mentioned in Paul Busch' chapter, phase observables. Mathematically, we will be considering the case of maximally symmetric operators in an infinite dimensional Hilbert space.

10.3.1 Maximally Symmetric Operators: Momentum on the Half-Line

Consider the momentum operator $\widehat{p} = -i\hbar\partial_x$ defined on a dense domain of the Hilbert space of square integrable functions on the half-line, $\mathcal{H}_> = L^2\left(\mathbf{R}^+, \mathrm{d}x\right)$. More precisely, it is defined on $D(\widehat{p})$, the subspace of square integrable absolutely continuous functions ψ whose derivative is also square integrable, and such that $\psi(0) = 0$. Clearly, this operator is symmetric on its domain. Let us write down a set of generalized (weak) eigenfunctions that provide us with a resolution of the identity. The adequate set, which does not belong to the Hilbert space and is parametrized by real p, is given by $\psi_p(x) = \exp(ipx/\hbar)/\sqrt{2\pi\hbar}$, as is only to be expected. They form a complete basis, since $\int_{-\infty}^\infty \mathrm{d}p\, \overline{\psi_p(x')}\psi_p(x) = \delta(x'-x)$. However, they are not orthogonal (in the generalized sense):

$$\int_0^\infty \mathrm{d}x\, \overline{\psi_{p'}(x)}\psi_p(x) = \frac{1}{2}\delta(p-p') + \frac{i}{2\pi}\mathrm{P}\frac{1}{p-p'} ,$$

where P stands for principal part. This is evidently different from the usual constructions for self-adjoint operators.

Let us now pass on to the position operator $\widehat{x}$, defined on those square integrable functions on the half-line, $\psi \in \mathcal{H}_>$, such that $\int_0^\infty \mathrm{d}x\, x^2 |\psi(x)|^2 < +\infty$. This operator is self-adjoint on its domain. Additionally, it is true that $[\widehat{x}, \widehat{p}] = i\hbar$ on a dense domain. However, in the case at hand the position operator is bounded from below, and Pauli's theorem therefore applies. It follows that $\widehat{p}$ is not a self-adjoint operator and admits no self-adjoint extension in $\mathcal{H}_>$.

We shall now rephrase the statements of the previous two paragraphs in a different way. The domain of the operator adjoint to $\widehat{p}$, i.e., $\widehat{p}^\dagger$, according to von Neumann's formula, is $D(\widehat{p}) \oplus N(i) \oplus N(-i)$, where $N(\pm i)$ are the spaces of eigenvectors of $\widehat{p}^\dagger$ with eigenvalues $\pm i$, respectively. This comes about because, even though $\widehat{p}$, being symmetric, has no imaginary eigenvalues, its adjoint does have eigenvectors with imaginary eigenvalues. As a matter of fact, the functions $\exp(i\lambda x/\hbar)$, with λ a complex number with positive imaginary part, are all of them eigenvectors of $\widehat{p}^\dagger$ of eigenvalue λ. Notice that the dimension of $N(\lambda)$, the space of eigenvectors with eigenvalue λ, is the same for all λ with positive imaginary part. This is called the first deficiency index of the operator $\widehat{p}$, while the second one is the dimension of the space of eigenvectors for any given eigenvalue of negative imaginary part, in our case, 0. That is, the deficiency indices of $\widehat{p}$ are $(1, 0)$. But self-adjoint operators have deficiency indices $(0, 0)$, as do essentially self-adjoint operators, which, although not self-adjoint, have a unique self-adjoint extension, namely, their closure. If we were to build a self-adjoint extension of $\widehat{p}$ we would need to include somehow those elements of $D(\widehat{p}^\dagger)$ that are not in $D(\widehat{p})$, but in such a way that the imaginary parts compensate. When the deficiency indices are equal, this is achieved by the use of a unitary transformation from $N(\bar{\lambda})$ to $N(\lambda)$ (this is the von Neumann theory of self-adjoint extensions, whereby each self-adjoint extension in the Hilbert space of definition of an operator over a dense domain, with equal defect indices, is given by a unitary transformation between the deficiency subspaces). As the deficiency indices in our case are different, it is not possible to extend $\widehat{p}$ to include an action over the whole of the domain of $D(\widehat{p}^\dagger)$ such that the imaginary parts compensate. Since $\widehat{p}$ is a closed operator, symmetric over its domain, which is dense in $\mathcal{H}_>$, and its deficiency indices are unequal, one of them being 0, we say that it is a maximally symmetric operator, and the previous result is that it admits no self-adjoint extension over $\mathcal{H}_>$.

Retaking the complete set of generalized eigenfunctions $\psi_p(x)$, we can construct a POVM F. The concept has been introduced in the preceding section, and illustrated for the discrete case. In the present case, the POVM is a map from intervals in the real line to the positive operators over a Hilbert space that satisfies the properties presented above: the map for our case is given by

$$\langle \phi | F([a, b]) \psi \rangle = \int_a^b \mathrm{d}p \int_0^\infty \mathrm{d}x \int_0^\infty \mathrm{d}y\, \overline{\phi(x)} \psi_p(x) \overline{\psi_p(y)} \psi(y)\,,$$

over all $\phi, \psi \in \mathcal{H}_>$. This map sends intervals of the real line to positive operators acting on $\mathcal{H}_>$, which add together when the intervals are disjoint, and which add to the identity operator when summed over the real line, because of the completeness proved above. This differs from the usual decompositions for self-adjoint operators in that it does not fulfill the property that the positive operators be projectors, i.e., in our case $F([a,b])^2 \neq F([a,b])$, because of the lack of orthogonality shown above.

Now we have the POVM associated with $\hat{p}$, we can reconstruct the operator and the probability distribution of its values. The action of the operator is given by

$$(\hat{p}\varphi)(x) = \int_{-\infty}^{\infty} dp \int_0^{\infty} dy\, p\, \psi_p(x)\overline{\psi_p(y)}\varphi(y) \ .$$

The probability distribution for its possible values over a state φ is

$$\Pi_\varphi(p) = \left| \int_0^{\infty} dx\, \overline{\psi_p(x)}\varphi(x) \right|^2 \ .$$

Notice that this indeed satisfies all the requirements for it to be a probability.

Let us see in more detail the relationship between the probability density and the POVM. We have written above the expression for the action of $F([a,b])$. Symbolically, we can write $F(dp)$, and its expectation value on a state φ is

$$\langle \varphi | F(dp)\varphi \rangle = \int_0^{\infty} dx \int_0^{\infty} dy\, \overline{\varphi(x)}\, \frac{e^{ip(x-y)/\hbar}}{2\pi\hbar}\varphi(y)\, dp$$
$$= \Pi_\varphi(p)dp \ .$$

The expression just written defines in fact $\langle \phi | F(dp)\psi \rangle$, for all "decent" states $|\phi\rangle$ and $|\psi\rangle$, and therefore the whole POVM. This comes about as follows. Let $|\varphi_1\rangle = |\phi\rangle + |\psi\rangle$ and $|\varphi_2\rangle = |\phi\rangle + i|\psi\rangle$. On the one hand,

$$\Pi_{\varphi_1}(p)dp = \Pi_\phi(p)dp + \Pi_\psi(p)dp + \langle \phi | F(dp)\psi \rangle + \langle \psi | F(dp)\phi \rangle \ .$$

On the other,

$$\Pi_{\varphi_2}(p)dp = \Pi_\phi(p)dp - \Pi_\psi(p)dp + i\langle \phi | F(dp)\psi \rangle - i\langle \psi | F(dp)\phi \rangle \ ,$$

whence it follows

$$\langle \phi | F(dp)\psi \rangle = \frac{1}{2} \left(\Pi_{\varphi_1}(p) - i\Pi_{\varphi_2}(p) - (1-i)\Pi_\phi(p) - (1+i)\Pi_\psi(p) \right) dp \ .$$

Furthermore, the domain of the operator associated with the POVM is also defined by $\Pi_\psi(p)$: by integration by parts it can readily be seen that the domain of the operator $\hat{p}$ is precisely the set of states ψ for which the second moment of the probability distribution $\Pi_\psi(p)$ is finite, and there it coincides with $\|\hat{p}\psi\|^2$. Therefore, the probability density written above defines the POVM and the operator with it associated. Notice also that the probability density need not be a continuous function. The only requirement is that $\Pi_\psi(p)dp$ be a good measure on the real line.

Another important property of the POVM is covariance under displacements of momenta, which reflects the commutation relation $[\widehat{x}, \widehat{p}] = i\hbar$. Namely, it is readily computed that, for all real q,

$$\langle\phi|e^{iq\widehat{x}/\hbar}F([a,b])e^{-iq\widehat{x}/\hbar}\psi\rangle = \langle\phi|F([a+q,b+q])\psi\rangle .$$

In terms of the probability density, the statement is that

$$\Pi_{\psi_q}(p) = \Pi_\psi(p+q) ,$$

where $\psi_q = e^{-iq\widehat{x}/\hbar}\psi$ is the shifted state.

The probability density, written above for the case of pure states, can be easily generalized to mixed states. Let $\rho(y, x)$ be the matrix elements of the density matrix $\widehat{\rho}$ in the position representation. The probability density associated with this density matrix is then

$$\Pi_\rho(p) = \int_0^\infty dx \int_0^\infty dy \, \frac{e^{ip(x-y)/\hbar}}{2\pi\hbar}\rho(y, x) .$$

In what follows we will use only pure states in the discussion, bearing in mind that the generalization to mixed states is straightforward.

It would seem that something akin to the spectral theorem of self-adjoint operators has been achieved, and this is indeed the case. The difference, however, lies in that the expectation value of higher-order powers of $\widehat{p}$ on a state ψ, $\langle\psi, \widehat{p}^n\psi\rangle$, does not necessarily coincide with the corresponding moment of the distribution $\Pi_\psi(p)$, i.e., $\int_{-\infty}^\infty dp\, p^n \Pi_\psi(p)$, for $n \geq 3$. For the case at hand, for instance, this happens already for $n = 3$ if $\psi'(0) \neq 0$. As a matter of fact, in such a situation the expectation value for $\widehat{p}^3$ has an imaginary part. Nonetheless, the measures of momenta can be readily associated with $\Pi_\psi(p)$, which carries the relevant physical information. As such, there is no further need to complicate the situation by requiring ancilla or further detailed description of the measurement process; at least, not more detailed than what we normally require for the analysis of momentum in the full line.

10.3.2 Naimark's Dilation Theorem

Even so, we would normally like to understand the constructions above in terms of the more usual recipes for self-adjoint operators. We shall now make essential use of the uniqueness theorem for POVMs of maximally symmetric operators: given a maximally symmetric operator $\widehat{A}$ over a Hilbert space $\mathcal{H}$, there is a unique POVM, F_A (unique up to isomorphisms), such that its first operator moment coincides with the operator, and that the set of states over which the second moment exists is precisely the domain of $\widehat{A}$ [21, 27]. This means that if we are to construct by whichever means a POVM such that it fulfills these conditions, we will be obtaining again the same POVM.

Moreover, Naimark's dilation theorem tells us that any POVM associated with a symmetric operator defined on a dense subset of $\mathcal{H}$ can be constructed from a self-adjoint extension of the operator to a larger space $\tilde{\mathcal{H}}$, as follows: let E be the projection valued measure of the self-adjoint extension (i.e., a POVM that satisfies the further requirement that $E([a,b])^2 = E([a,b])$), and P the projection operator from $\tilde{\mathcal{H}}$ to $\mathcal{H}$. Then $F([a,b]) := PE([a,b])$ is a POVM associated with the symmetric operator.

As a consequence, if we are to build a self-adjoint extension of $\hat{p}$ in a larger space, we reproduce the unique POVM and the whole of the physical content of the operator from the usual analysis for the self-adjoint extension and a projection. Notice however that the possible extensions are infinite. Not so the POVM, and that makes the freedom of choice of extension even more interesting.

Back to the case of $\hat{p} = -i\hbar\partial_x$ defined on the half-line, we see that there is a simple possibility: to extend the operator to the full line, i.e., $-i\hbar\partial_x$ defined on a dense subset of $\mathcal{H} = L^2(\mathbf{R}, dx)$. Naturally enough, the action of this operator on the elements of $D(\hat{p})$ is the same as that of $\hat{p}$, so it is an extension. This is not the only self-adjoint extension on the whole real line, of course. As a matter of fact, the deficiency indices for the direct sum of the momentum operators on the positive and negative half-lines are $(1,1)$, thus signaling that a one-dimensional continuum of alternative self-adjoint extensions exist. They differ by the presence of a jump function located on $x = 0$. However, most natural is the momentum on the full line, defined over absolutely continuous functions, i.e., with no jump on $x = 0$. This is a self-adjoint operator, for which the standard spectral analysis is applicable.

More concretely, the projection valued measure E is given by the expression

$$\langle \phi | E([a,b]) \psi \rangle = \int_a^b dp \int_{-\infty}^{\infty} dx \int_{-\infty}^{\infty} dy \, \overline{\psi_p(y)} \psi_p(x) \overline{\phi(x)} \psi(y) \, ,$$

with $\psi_p(x) = \exp(ipx/\hbar)/\sqrt{2\pi\hbar}$, as before, but now defined over the whole real line. The projection P in our case is simply $(P\psi)(x) = \Theta(x)\psi(x)$, with Θ being Heaviside's step function.

The probability distribution for the momentum operator over the whole real line for a state ψ is of course the modulus squared of the wave function in the momentum representation, and its restriction to states that belong to $D(\hat{p})$ is none other than the probability distribution associated with the POVM.

We see then that the reason we could not do the standard analysis for the momentum operator on the half-line is that we are being, in a way, far too restrictive in the behavior near $x = 0$ of the states on which it can act. There is a reminder of the full line, seen for instance in the principal part that forbids orthogonality, or in the fact that the higher moments of the probability distribution do not in general agree with the expectation value of powers of the restricted momentum operator. If our states were such that the function ψ and all of its derivatives were zero at $x = 0$, there would be no problem with the higher moments, and we would get no imaginary part for the powers of $\widehat{p}$ over such states. However, such a strong restriction on the allowable wave functions would cut out many physically sensible cases.

Let us now write slightly more formally the extension procedure we have performed to reobtain and interpret the probability density: we have started with a maximally symmetric operator $\widehat{p} = -i\hbar\partial_x$ acting on the dense domain $D(\widehat{p}) \subset \mathcal{H}_> = L^2(\mathbf{R}^+, dx)$, with deficiency indices $(1, 0)$. We have then considered an extension in $L^2(\mathbf{R}, dx)$, making use of the isomorphism

$$L^2(\mathbf{R}, dx) = L^2(\mathbf{R}^+, dx) \oplus L^2(\mathbf{R}^-, dx) \ .$$

The extension has been the natural one, i.e., $-i\hbar\partial_x$ on the full line (although, as we repeatedly stated, this is not the only possible extension; another easy choice would have been some extension of $\widehat{p} \oplus (-\widehat{p})$, for instance). The standard spectral analysis for this operator produces the probability density, which, when restricted to $\mathcal{H}_>$, gives $\Pi_\psi(p)$, the probability density out of which the POVM for the operator we started from can be built.

10.4 The POVM of the Aharonov–Bohm Time Operator

Within the long-running discussion on the concept of time in quantum mechanics, and in particular with respect to the status of the time–energy uncertainty relations, Aharonov and Bohm introduced in an important paper [2], among other questions, "a clock" to measure time from the position and momentum of a freely moving test particle. The corresponding operator was obtained by a simple symmetrization of the classical expression my/p_y, where y and p_y are, respectively, the position and momentum of the test particle. By the same token, the operator obtained by symmetrizing the classical expression for the arrival time at $x = 0$ of a freely moving particle having position x and momentum p, $t = -mx/p$, is given by [5, 28]

$$\widehat{T}_{AB} := -\frac{m}{2}\left(\widehat{x}\widehat{p}^{-1} + \widehat{p}^{-1}\widehat{x}\right) \ ,$$

Note the minus sign in comparison to the clock time in [2]. In spite of the somewhat subtle difference (in concept and sign) with the original time operator introduced in [2] we shall refer to this operator as the Aharonov–Bohm (time-of-arrival) operator. Without regard for topological considerations, it is clear that it has the correct commutation relation with the free particle Hamiltonian on the line, $\widehat{H}_0 = \widehat{p}^2/2m$, namely

$$[\widehat{H}_0, \widehat{T}_{AB}] = i\hbar$$

(notice that in Heisenberg's picture this entails that $d\widehat{T}_{AB}(t)/dt = -1$, so it flows contrary to parametric time, which is the correct sign for a time of arrival). It is thus a good candidate for a time operator, with the plausible physical interpretation, given by the correspondence rule, that it is related to the time of arrival (notice that other quantization rules might possibly give a different result, although this is the operator obtained not just by the symmetrization rule, but also by applying Weyl, Rivier, or Born–Jordan quantizations [29]). However, it also follows, from Pauli's theorem, that it cannot be a self-adjoint operator acting on a dense space.

Let us then examine first the question of the domain of $\widehat{T}_{AB}$. In order to do that, it is useful at this point to consider the Hilbert space of the free particle in the momentum representation, $\mathcal{H}_p := L^2(\mathbf{R}, dp)$ (on this space we know how $\widehat{p}^{-1}$ acts, thanks to the spectral theorem). Formally, we then obtain

$$\widehat{T}_{AB} \to \frac{i\hbar m}{2}\left(\frac{1}{p^2} - \frac{2}{p}\frac{\partial}{\partial p}\right).$$

There are other alternative expressions, such as $-i\hbar m p^{-1/2}\partial_p p^{-1/2}$, which would be valid for $p > 0$, or for $p < 0$ by analytic continuation, see, e.g., [3]. At any rate, $\widehat{T}_{AB}$ understood as a differential operator presents a singular point at $p = 0$, and this is the source of all the difficulties that have appeared in the literature.

The differential operator written above can only be applied to absolutely continuous functions, but there are further requirements. One such is that $\widehat{T}_{AB}\psi$ belongs to $\mathcal{H}_p$, i.e., that it be square integrable. This poses a restriction due to the singularity of the operator at $p = 0$. On computation, one finds that the singularity is avoided if the function ψ has one of the following possible behaviours close to $p \to 0$: either $\psi(p) \sim p^{1/2}$, or $\psi(p)/p^{3/2} \to 0$. However, this is not enough to fix the domain of the operator, as Paul noticed long ago [11]. Given that, at least formally, $\widehat{T}_{AB}$ is symmetric, this should also be a requirement on its domain. Integration by parts, and demanding that $\langle\varphi|\widehat{T}_{AB}\varphi\rangle$ be equal to $\langle\widehat{T}_{AB}\varphi|\varphi\rangle$ for all φ in the domain of $\widehat{T}_{AB}$, leads us to exclude the first possibility, thus defining the domain of $\widehat{T}_{AB}$, $D(\widehat{T}_{AB})$, as the set of absolutely continuous square integrable functions of p on the real line, such that $\psi(p)/p^{3/2} \to 0$ as $p \to 0$ and $\|\widehat{T}_{AB}\psi\|^2$ is finite. As for the alternative expression $-i\hbar m p^{-1/2}\partial_p p^{-1/2}$, let us ask that $(\Theta(p) - i\Theta(-p))\psi(p)/\sqrt{|p|}$ be absolutely continuous for ψ to be in its domain. This, together with the further requirement of symmetry, leads us to a domain that coincides with that of $\widehat{T}_{AB}$. Since the respective actions also coincide over this domain, we see that they are but equivalent differential expressions for the operator, once the adequate domain is taken into account.

In order to apply von Neumann's formula, we have to check whether there are states ψ in $\mathcal{H}_p$ such that for all $\varphi \in D(\widehat{T}_{AB})$ the following expression holds:

$$\langle\psi|\left(\widehat{T}_{AB} + i\right)\varphi\rangle = 0,$$

since then ψ is an eigenvector of $\widehat{T}^{\dagger}_{AB}$ with eigenvalue i. Analogously, we also have to study the case with eigenvalue $-i$. By integration by parts, and application of the condition that all functions in $D(\widehat{T}_{AB})$ satisfy, namely, that $\varphi(p)/p^{3/2}$ tends to zero as p tends to zero, it is found that there are *two* independent eigenvectors with eigenvalue i, and none with eigenvalue $-i$. The relevant eigenvectors are

$$\psi_{\pm}(p) = \Theta(\pm p)\,\sqrt{\pm p}\,e^{-p^2/2m\hbar}\,.$$

Notice that, contrary to the expectations of some authors [30, 3, 4], both of them have to be taken into account: there are no requirements of derivability or continuity for the functions in the domain of the adjoint.

The deficiency indices are therefore $(2, 0)$, and we have a maximally symmetric operator. As we have seen in the previous section, this implies that no self-adjoint extension can exist, and we should redo for this case the analysis performed for the momentum operator on the half-line. However, the most convenient way of doing that is by passing to the energy representation, as was already pointed out by Allcock in his seminal work [31, 32, 33], and emphasized again by Kijowski [10, 34]. This change of representation is useful, from the mathematical point of view, because it implements a theorem [27] which states that a simple symmetric operator with deficiency indices $(n, 0)$ can be decomposed as a direct sum of n operators with deficiency indices $(1, 0)$. Since each of these is in fact isomorphic to the momentum operator on the half-line, we will be able to use directly the previous results.

The change of representation corresponds to the decomposition of the Hilbert space $\mathcal{H}_p$ into the subspaces of positive and negative momentum, i.e.,

$$L^2\left(\mathbf{R}, dp\right) = L^2\left(\mathbf{R}^+, dE\right) \oplus L^2\left(\mathbf{R}^+, dE\right)$$
$$= \mathcal{H}_+ \oplus \mathcal{H}_-\,,$$

where the first subspace is that of positive momenta, whereas the second corresponds to negative p. The explicit isomorphism is given by

$$\psi_{\pm}(E) = (m/2E)^{1/4}\psi(\pm\sqrt{2mE})\,,$$
$$\psi(p) = \left(\frac{|p|}{m}\right)^{1/2}\left[\Theta(p)\psi_+\left(\frac{p^2}{2m}\right) + \Theta(-p)\psi_-\left(\frac{p^2}{2m}\right)\right]\,,$$

where $\psi \in \mathcal{H}_p$, $\psi_{\pm} \in \mathcal{H}_{\pm}$, and the isomorphism relates $\psi \leftrightarrow (\psi_+, \psi_-)$. The factor $E^{-1/4}$ is due to the change in the measure from dp to dE and reciprocally for the factor $(|p|/m)^{1/2}$. The interesting point is that given this isomorphism, the time operator of Aharonov and Bohm takes the form $-i\hbar\partial_E$. The domain of the operator, $D(\widehat{T}_{AB})$, is sent by the isomorphism into the direct sums of square integrable absolutely continuous functions in each subspace, such that for each subspace we have the restriction that $\psi_{\alpha}(E)E^{-1/2} \to 0$ as $E \to 0$.

But this is exactly the case considered above, since the requirement $\psi_\alpha(0) = 0$ and the square integrability of $\psi'_\alpha(E)$ imply the restriction stated before. In other words, we have the isomorphism

$$\hat{T}_{AB} = (-i\hbar\partial_E) \oplus (-i\hbar\partial_E) = \hat{T}_+ \oplus \hat{T}_- \,,$$

where $\hat{T}_\pm$ are isomorphic to the momentum operator on the half-line.

Therefore, the constructions carried out in the previous section can be immediately translated to this situation, but taking into account that the energy spectrum is degenerate whereas the position spectrum is not. For instance, the complete nonorthogonal set of generalized eigenfunctions ψ_p is doubled here into a set with a continuous parameter t (the notation is intended to be suggestive) and a discrete one, with values $+$ or $-$, as follows:

$$\psi_+^{(t)}(E) = \left(\frac{1}{\sqrt{2\pi\hbar}} e^{iEt/\hbar}, 0 \right)$$

and

$$\psi_-^{(t)}(E) = \left(0, \frac{1}{\sqrt{2\pi\hbar}} e^{iEt/\hbar} \right) \,.$$

In what follows we will not make the explicit distinction between the element of the full Hilbert space and its component, if only one is zero.

These functions transform under the isomorphism to give the following expressions:

$$\tilde{\psi}_\alpha^{(t)}(p) = \Theta(\alpha p) \left(\frac{\alpha p}{2\pi m\hbar} \right)^{1/2} e^{ip^2 t/2m\hbar} \,. \tag{10.3}$$

As Dirac's notation will also prove useful in this context later, we shall also denote these functions as

$$\langle p|t, \alpha \rangle = \tilde{\psi}_\alpha^{(t)}(p) \,, \tag{10.4}$$

thus defining the state $|t, \alpha\rangle$.

It is straightforward to prove completeness, i.e., $\sum_\alpha \int_{-\infty}^\infty dt\, \tilde{\psi}_\alpha^{(t)}(p') \tilde{\psi}_\alpha^{(t)}(p) = \delta(p - p')$. Alternatively, $\sum_\alpha \int_{-\infty}^\infty dt\, \psi_\alpha^{(t)}(E') \psi_\alpha^{(t)}(E) = \delta(E - E')1$, where 1 is the two by two identity matrix ($\delta(E - E')1$ is the identity operator on the full Hilbert space $\mathcal{H}_+ \oplus \mathcal{H}_-$). In Dirac's notation,

$$\sum_\alpha \int_{-\infty}^\infty dt\, |t, \alpha\rangle\langle t, \alpha| = 1 \,.$$

Nonorthogonality is also a direct translation:

$$\langle t', \alpha'|t, \alpha \rangle = \int_0^\infty dE\, \psi_{\alpha'}^{(t')}(E) \psi_\alpha^{(t)}(E) = \int_{-\infty}^\infty dp\, \tilde{\psi}_{\alpha'}^{(t')}(p) \tilde{\psi}_\alpha^{(t)}(p) =$$
$$= \frac{1}{2}\delta_{\alpha\alpha'} \left(\delta(t - t') + \frac{i}{\pi} P \frac{1}{t - t'} \right) \,.$$

It now behooves us to compute the POVM or, alternatively, the probability distribution for measured values of the $\widehat{T}_{AB}$ operator from which it can be readily recovered. By direct translation, we can write the probability density as

$$\Pi^K_{(\psi_+,\psi_-)}(t) = \left| \int_0^\infty dE\, \frac{e^{-iEt/\hbar}}{\sqrt{2\pi\hbar}}\psi_+(E) \right|^2 +$$
$$\left| \int_0^\infty dE\, \frac{e^{-iEt/\hbar}}{\sqrt{2\pi\hbar}}\psi_-(E) \right|^2 ,$$

in the energy representation, or as

$$\Pi^K_\psi(t) = \left| \int_0^\infty dp\, \left(\frac{p}{2\pi m\hbar}\right)^{1/2} e^{-ip^2 t/2m\hbar}\psi(p) \right|^2 +$$
$$\left| \int_{-\infty}^0 dp\, \left(\frac{-p}{2\pi m\hbar}\right)^{1/2} e^{-ip^2 t/2m\hbar}\psi(p) \right|^2 ,$$

in the momentum representation. Using Dirac's notation, it would read

$$\Pi^K_\psi(t) = |\langle t, +|\psi\rangle|^2 + |\langle t, -|\psi\rangle|^2 .$$

This is, of course, the same as Kijowski's probability density, introduced in Sect. 10.2. The essential property of covariance under transformations generated by the Hamiltonian is also evident and a direct translation of the covariance property signaled in the previous section. Physically, it means that the probability of arriving at t for a given state is equal to the probability of arriving at $t - \tau$ for the same state evolved a time τ. This is the reflection on the probability density of the canonical commutation relation $[\widehat{H}, \widehat{T}_{AB}] = i\hbar$.

The domain of $\widehat{T}_{AB}$, now defined through the POVM, can be characterized as the set of elements of the Hilbert space for which $\int_{-\infty}^\infty dt\, \Pi^K_\psi(t)t^2$ is finite, and this quantity defines $\langle \widehat{T}_{AB}\psi|\widehat{T}_{AB}\psi\rangle$ (thus realizing the minimum variance demanded by Kijowski and Werner [10, 35]).

As before, we would like to understand all these constructions in terms of a generalized self-adjoint extension, by using the uniqueness of the POVM associated with a maximally symmetric operator, and Naimark's theorem. From the structure of the time operator of Aharonov and Bohm, namely $\widehat{T}_{AB} = (-i\hbar\partial_E) \oplus (-i\hbar\partial_E)$ on $L^2(\mathbf{R}^+, dE) \oplus L^2(\mathbf{R}^+, dE)$, it follows that a *natural* and simple extension (natural in the sense of following the natural extension in the analogy) is the operator $(-i\hbar\partial_E) \oplus (-i\hbar\partial_E)$ acting on $L^2(\mathbf{R}, dE) \oplus L^2(\mathbf{R}, dE)$, which is obviously self-adjoint. In other words, we have introduced negative energies, respecting the twofold degeneracy of the initial spectrum. On this space the (doubled) Fourier transform acts as a unitary transformation that provides us with the *time* representation of the states, and the probability density for the time of arrival over a pure state in this extended space is nothing but the modulus squared of the wave function in the time representation. Notice that the time representation corresponds to the doubled space $L^2(\mathbf{R}, dt) \oplus L^2(\mathbf{R}, dt)$. We reobtain the by now usual probability density for time of arrival, by restriction to the initial Hilbert space. The restrictions to the initial Hilbert space of the applications of the spectral theorem are therefore related to $\Pi_\psi^K(t)$, as stated before.

The problems that the vicinity of $p = 0$ (alternatively $E = 0$) pose for the analysis of the operator of Aharonov and Bohm are thus seen to be due to the (physically imposed) restriction to the space of positive energies. There is a reminder of the negative energies in aspects such as the nonorthogonality of the set of generalized eigenfunctions, similarly to what happens in the paradigmatic example of the previous section.

10.5 Other Time Operators

As the perusal of the table of contents of this volume will have made clear, there are many different time observables to consider. This entails the possibility that there be several other time operators to consider as well. Many of those have indeed been examined in Chap. 8, and it is not necessary to reproduce here a detailed analysis of all those. However, it is instructive to consider a number of those from the point of view of POVMs. In particular, Grot, Rovelli, and Tate [3] proposed a self-adjoint modification of Aharonov and Bohm's time-of-arrival operator that deserves some attention, and whose structure can be better understood after analyzing an operator with dimensions of time that appears in periodic systems.

10.5.1 Periodic Systems

Following Pegg [36], let us consider a finite or infinite dimensional system, with a Hamiltonian bounded from below, whose spectrum is purely discrete. Without lack of generality, we can assume that the ground state energy is zero. Assume furthermore that all the other eigenvalues of the Hamiltonian are commensurate. That is, all the eigenvalues E_i can be written as $E_i = 2\pi\hbar r_i/T$, with r_i a natural number, and T is the period. In this case, the following POVM will provide us with something akin to a time operator:

$$F(d\alpha) = \sum_{i,j} \frac{d\alpha}{T} e^{-2\pi i(r_i - r_j)\alpha/T} |E_i\rangle\langle E_j| .$$

It is clearly true that

$$\int_{\alpha_0}^{\alpha_0+T} \langle\psi|F(d\alpha)\psi\rangle = 1 \tag{10.5}$$

for all normalized states.

The operator constructed from the POVM is thus

$$\widehat{A} = \int_{\alpha_0}^{\alpha_0+T} F(d\alpha)\alpha = \frac{1}{T}\int_{\alpha_0}^{\alpha_0+T} d\alpha\,\alpha \sum_{i,j} e^{-2\pi i(r_i-r_j)\alpha/T}|E_i\rangle\langle E_j| \,. \quad (10.6)$$

This operator is not canonically conjugate to the Hamiltonian, as can be readily computed:

$$[\widehat{H},\widehat{A}] = i\hbar \sum_{i\neq j} e^{-2\pi i\alpha_0(r_i-r_j)/T}|E_i\rangle\langle E_j| = i\hbar\left(|\alpha_0\rangle\langle\alpha_0| - \widehat{1}\right)\,, \quad (10.7)$$

where we have used the definition

$$|\alpha\rangle = \sum_i e^{-2\pi i\alpha r_i/T}|E_i\rangle \,.$$

Observe that using this definition we can write $\widehat{A} = \frac{1}{T}\int_{\alpha_0}^{\alpha_0+T} d\alpha\,\alpha|\alpha\rangle\langle\alpha|$, and that the set $\{|\alpha\rangle\}_{\alpha\in[\alpha_0,\alpha_0+T)}$ gives us an overcomplete resolution of the identity,

$$\frac{1}{T}\int_{\alpha_0}^{\alpha_0+T} d\alpha\,|\alpha\rangle\langle\alpha| = \widehat{1}.$$

This lack of canonical commutation relations, however, comes about because of the periodicity of the system. The fact that the energy eigenvalues are commensurate leads to recurrence, and it is impossible to make out, in this context, in which particular recurrence we are, just from observing the system itself. In order to appreciate this point better, consider the derivative of the expectation value of the operator $\widehat{A}$ over a generic state $|\psi\rangle$, $\langle\widehat{A}\rangle$,

$$\frac{d\langle\widehat{A}\rangle}{dt} = 1 - |\langle\psi(t)|\alpha_0\rangle|^2 \,. \quad (10.8)$$

The last term can be understood as the probability density of measuring α_0 as the value of $\widehat{A}$ at time t modulo the period. Thus, the expectation value of $\widehat{A}$ follows time unless the overlap with α_0 becomes big, which forces the resetting of this system clock.

This example shows one way of spoiling the canonical commutation relation so as to obtain a time operator; clearly, this is not the only method.

In order to clarify the construction of Pegg, and to understand better some of the subtleties in interpreting POVMs physically, let us consider a system with two degrees of freedom. Let the energy eigenstates be written as $|0\rangle$ and $|1\rangle$. The Hamiltonian is thus $\widehat{H} = \hbar\omega|1\rangle\langle1|$. A state $|\alpha\rangle$ is given by $|\alpha\rangle = |0\rangle + \exp(-i\omega\alpha)|1\rangle$, which leads to $\widehat{A}$ as follows:

$$\widehat{A} = \frac{1}{T}\int_{\alpha_0}^{\alpha_0+T} \alpha|\alpha\rangle\langle\alpha| = (\alpha_0+\frac{\pi}{\omega})\widehat{1} + \frac{i}{\omega}\left(e^{-i\omega\alpha_0}|1\rangle\langle0| - e^{i\omega\alpha_0}|0\rangle\langle1|\right) \,. \quad (10.9)$$

None of the $|\alpha\rangle$ states is an eigenvector of $\widehat{A}$. The commutator of the Hamiltonian and $\widehat{A}$ is easily computed to be

$$[\widehat{H}, \widehat{A}] = i\hbar \left(e^{-iw\alpha_0}|1\rangle\langle 0| + e^{iw\alpha_0}|0\rangle\langle 1|\right) = i\hbar \left(|\alpha_0\rangle\langle\alpha_0| - \widehat{1}\right) . \qquad (10.10)$$

Let us consider the probability for the state to be found in state $\alpha\rangle$ if the state of the system is a generic normalized $(|\beta|^2 + |\gamma|^2)$ state $|\psi\rangle = \beta|0\rangle, +\gamma|1\rangle$: it is simply $|\langle\alpha|\psi\rangle|^2/2$ (because of normalization of $|\alpha\rangle$). Since we have a resolution of the identity in terms of the states $|\alpha\rangle$, we can actually define a different probability: the probability density for the measurement adapted to the POVM to produce the result α. This is given by the the probability density $P(\alpha)$

$$P(\alpha) = \frac{1}{T}|\langle\alpha|\psi\rangle|^2 = \frac{1}{T}\left(1 + \bar{\beta}\gamma e^{iw\alpha} + \beta\bar{\gamma}e^{-iw\alpha}\right) . \qquad (10.11)$$

Notice that the probabilities mentioned pertain to different measurements. In the first case, we are asking whether the state is α or not, with probabilities $|\langle\alpha|\psi\rangle|^2/2$ and $1 - |\langle\alpha|\psi\rangle|^2/2$; the operator being measured is $|\alpha\rangle\langle\alpha|/2$. Whereas in the second case we are asking with which probability will the outcome α be the result of measuring the operator $\widehat{A}$. The process of carrying out a number of measurements of $\widehat{A}$ adapted to the POVM provides us with the density $P(\alpha)$, and the expectation value and the second moment of this distribution are computed to be

$$\langle\alpha\rangle = \int_{\alpha_0}^{\alpha_0+T} d\alpha\, \alpha P(\alpha) = \alpha_0 + \frac{\pi}{w} - \frac{i}{w}\left(\gamma\bar{\beta}e^{iw\alpha_0} - \beta\bar{\gamma}e^{-iw\alpha_0}\right) \quad (10.12)$$

$$\Delta\alpha^2 = \int_{\alpha_0}^{\alpha_0+T} d\alpha\, (\alpha - \langle\alpha\rangle)^2\, P(\alpha) =$$

$$= \frac{\pi^2}{3w^2} + \frac{1}{w^2}\left(\gamma^2\bar{\beta}^2 e^{2iw\alpha_0} + 2\beta\bar{\gamma}e^{-iw\alpha_0} + 2\gamma\bar{\beta}e^{iw\alpha_0} + \qquad (10.13)\right.$$

$$\left. +\beta^2\bar{\gamma}^2 e^{-2iw\alpha_0} - 2|\gamma|^2|\beta|^2\right) . \qquad (10.14)$$

On the other hand, one can obtain

$$\langle\psi|\widehat{A}|\psi\rangle = \alpha_0 + \frac{\pi}{w} - \frac{i}{w}\left(\gamma\bar{\beta}e^{iw\alpha_0} - \beta\bar{\gamma}e^{-iw\alpha_0}\right) , \qquad (10.15)$$

$$(\Delta A)^2 = \langle\psi|\widehat{A}^2|\psi\rangle - \langle\psi|\widehat{A}|\psi\rangle^2 =$$

$$= \frac{1}{w^2}\left(1 + \gamma^2\bar{\beta}^2 e^{2iw\alpha_0} - 2|\gamma|^2|\beta|^2 + \beta^2\bar{\gamma}^2 e^{-2iw\alpha_0}\right) , \qquad (10.16)$$

which differs in the variance. This comes about because the experimental arrangement is dedicated to measuring $\widehat{A}$ and not $\widehat{A}^2$; in other words, the measurement is unsharp.

Notice that in this case a self-adjoint operator has been constructed, which is not canonically conjugate to the Hamiltonian. This should be compared to the construction by Grot, Rovelli, and Tate [3].

10.5.2 Grot, Rovelli, and Tate

Instead of the operator $\widehat{T}_{AB}$ of Aharonov and Bohm, which can be written in the momentum representation as $-i\hbar m p^{-1/2}\partial_p p^{-1/2}$, Grot, Rovelli, and Tate proposed the set of operators $\widehat{T}_\epsilon$, with momentum representation

$$\widehat{T}_\epsilon \to -i\hbar m \sqrt{f_\epsilon(p)}\,\partial_p\,\sqrt{f_\epsilon(p)}\,,$$

where $f_\epsilon(p)$ equals $1/p$ for $|p|$ bigger than a given ϵ, and is regular at the origin.

These operators are thus self-adjoint, after the adequate choice of domains and closure, and do not present the problems that Aharonov and Bohm's operator have. However, physically they are nowhere as useful as $\widehat{T}_{AB}$, because a state which should be arriving very close to a given time has its arrival distribution very much spread out from that maximum: Oppenheim et al. [4] pointed out that the particle, at the predicted time of arrival, is found far away from the point of arrival with probability $1/2$. This should be compared with the analysis of the generalized states $|t,\alpha\rangle$ in [28]. In fact, this quirk of the eigenstates of $\widehat{T}_\epsilon$ is due to the following expression:

$$[\widehat{H}_0, \widehat{T}_\epsilon] = i\hbar \widehat{p} f_\epsilon(\widehat{p})\,.$$

As in the case of Pegg's operator, Pauli's theorem is circumvented because the commutation relation is not canonical. This means that this operator does not fully flow with time, and this is the source of the difficulties mentioned before.

10.6 Arrival States

Let us consider again Aharonov and Bohm's time-of-arrival operator. It satisfies the canonical commutation relation only with the Hamiltonian of the free particle. Indeed, it has been purpose-built for such an instance, and it becomes meaningless in an interacting context. However, having been able to understand fully this observable in the free case, we are even more convinced that similar observables can be given meaning for all circumstances. A generic axiomatic approach à la Kijowski seems somewhat misplaced, since for the generic case it is not certain that a particle will arrive at a given point, there is no Galilei invariance, the generic stationary state will be radically different from the momentum eigenstates, etc. We are thus forced to think anew.

For this purpose, it is useful to recall the expression of Kijowski's distribution in terms of the states $|t,\alpha\rangle$, defined in (10.3) and (10.4):

$$\Pi_\psi^K(t) = \sum_{\alpha=\pm} |\langle t,\alpha|\psi\rangle|^2 . \tag{10.17}$$

The (nonnormalizable) states $|t,\alpha\rangle$ have been obtained as (generalized) eigenstates of Aharonov and Bohm's operator,[4] and thus are inextricably tied to the free particle. Notice in particular that the covariance property of Kijowski's distribution is transmuted into

$$e^{i\widehat{H}_0 s/\hbar}|t,\alpha\rangle = |t+s,\alpha\rangle ,$$

where again $\widehat{H}_0 = \widehat{p}^2/2m$ is the free particle Hamiltonian. It follows that we could rewrite expression (10.17) as

$$\Pi_\psi^K(t) = \sum_{\alpha=\pm} |\langle 0,\alpha|\psi(t)\rangle|^2 .$$

Further note that instead of the labels $\alpha = \pm$, which refer to the sign of the components of momenta that make up the states $|t,\alpha\rangle$, we might as well use labels $\beta = \{L,R\}$, standing for arrivals from the left (corresponding to positive momenta) or the right (negative momenta). Additionally, let us remark that in all the preceding discussion the point of arrival has been fixed at the origin of coordinates, while further generality might be achieved by simple translation. All this suggests that we give a special emphasis to the state $|0,\alpha\rangle = e^{-i\widehat{H}_0 t/\hbar}|t,\alpha\rangle$ and its (spatial) translates. To remark this emphasis we shall denote those states as $|v_{\beta,x}\rangle$, which we will refer to as "arrival states from the {left,right} at point x", or as "crossing states." The relevance of these states for our purposes lies in the first place in the fact that the (formal) operator

$$\sum_{\beta} |v_{\beta,x}\rangle\langle v_{\beta,x}|$$

is a quantization of the classical observable $J^L - J^R$, where $J^{L,R}$ is the flux of particles from the {left,right}. Secondly, the rewriting of Kijowski's distribution for the free particle, (10.17), in terms of the crossing states,

$$\Pi_\psi^K(t) = \sum_{\beta=L,R} \langle\psi(t)|v_{\beta,x}\rangle\langle v_{\beta,x}|\psi(t)\rangle , \tag{10.18}$$

brings forward a change of emphasis: whereas in (10.17) the time-of-arrival distribution is obtained from the overlap of the *initial* wave function with the states associated with arrival of free particles at the instant t, from the left ($\alpha = +$) or from the right ($\alpha = -$), in (10.18) it is computed as the overlap

---

[4] Normalizable states formed by their linear combination behave as expected for an arrival state, and its arrival becomes arbitrarily sharp in space–time as shown in [28].

of the *evolved* wave function with the *constant* states that measure arrivals (from the left, $\beta = L$, or from the right, $\beta = R$). As it stands, for the free particle case, this rewriting is utterly irrelevant. Not so, however, in a more general context.

The actual procedure of measuring times of arrival could be naively described as follows: set up a good particle detector at the point of arrival, prepare the state with whichever preparing procedure one desires, start running the clock whenever the preparing procedure has been carried out, and note down the instant the detector clicks. This entails a "waiting" disposition of the detector. On the other hand, an expression such as (10.17) pertains to a "predictive" measurement of the density of arrivals, which implies full knowledge of the evolution of the particle after a measurement. Imagine that we prepare a state and erroneously assume that the particle will evolve freely, while in fact there is a potential which randomly changes in time. The predictive measurement would undoubtedly fail, while the "waiting" procedure would describe adequately the different arrivals. This second perspective could be also termed "unconditional": the arrival or otherwise of a particle at point x is directly measured in physical space at every instant, using local definitions that are in no way conditioned by the different potentials in which the particle might be moving.

This point of view, insisting on the form (10.18) of Kijowski's distribution and its extension to situations different from the free particle, has been put forward in [20], inspired by Wigner's formalization of the time–energy uncertainty relation [37], and further developed in [38]. In this latter work, the realization that there are many different alternative quantizations of the classical observable $J^L - J^R$ led the authors to an analysis of several proposals for generalizations to the interacting case of Kijowski's distribution [39, 40, 41, 42], using the concept of "crossing state."

On the strength of the analysis carried out in [38], in which we found that for the class of crossing states studied there only the one described above had the correct behavior in quasiclassical situations, we suggest that the best "ideal"[5] description of the measurement of times-of-arrival we have so far consists of (10.18), applied also to non-free particles, i.e. with states $|\psi(t)\rangle$ evolved with arbitrary (one-dimensional) Hamiltonians.

The interpretation of expression (10.18) must also change when applied to the interacting case. In such a situation, it is no longer a probability

---

[5] We use the term "ideal" to denote quantities that are computed intrinsically to the system, i.e., without introducing further couplings to measurement apparatus or ancillary systems. Please note that otherwise this term is very often value-laden and has been used by various authors in many different ways. In particular, in the literature "ideal" detectors often denote other quantum systems used for measurement, whereas here we are asserting that at this level of abstraction there could exist a detector for which the measurement procedure can be described with variables intrinsic to the system. In this manner, the position operator, to give but an example, has an associated ideal measurement procedure.

density, but simply an arrival density. For instance, in the harmonic oscillator there might be many arrivals, or it could happen that there is hardly any transmission through a potential barrier, thus depleting the number of arrivals.

This procedure, and the interpretation of $\Pi_\psi(t)$ as an arrival density, suggests another extension to the ideal description of times of arrival to the many particle case. When we have a system with, say, two particles, our ideal detector (see footnote 5) might click twice at different times. As before, the number of actual arrivals is highly dependent on the potential and the preparation state. In the case of several free particles, the integral $\int_{-\infty}^{+\infty} dt\, \Pi_\psi(t)$ should be exactly the number of particles present. The ideal detector being considered would have an individual response to the arrival of each individual particle, and thus should be represented by a one particle operator in the formalism of second quantization.

Before addressing this construction in the next section, let us point out that the analysis we have just carried indicates that observable quantities need not be associated with POVMs in a simple manner. Consider the case of one single particle in a scattering situation. The density of arrivals does not normalize to one in general: the particle might be reflected, and no arrival detected behind the scattering barrier. This implies that the naive POVM would not be normalized to one, i.e. the sum of the components will not be the identity. Brunetti and Fredenhagen have proposed a way out for this case: suitably redefine the corresponding operator, obtained as the first operator moment of the density of arrivals, or else as

$$\widehat{T}_{x,\text{arrivals}} = \sum_{\beta=L,R} \int_{-\infty}^{+\infty} dt\, t\, U^\dagger(t) |v_{\beta,x}\rangle \langle v_{\beta,x}| U(t)\,,$$

with $U(t)$ the applicable evolution operator. This operator $\widehat{T}_{x,\text{arrivals}}$ has a kernel (a null subspace: the set of states which it sends to 0). Consider the pre-Hilbert space obtained by quotienting the whole Hilbert space by this subspace. Complete it to a Hilbert space in the usual manner. A new operator $\widehat{T}'_x$ is defined over this new space as follows: let $[\psi]$ be an element of the quotient Hilbert space with representative ψ in the original space. Then $\widehat{T}'_x[\psi] = [\widehat{T}_{x,\text{arrivals}}\psi]$. This operator will have an associated POVM adequately normalized.

If the possibility of many arrivals exists, a similar construction might be carried out, by sectoring the Hilbert space into subspaces according to the value of $\int_{-\infty}^{+\infty} dt\, \Pi_\psi(t)$. It is however far simpler to concentrate on the simple definition of the density of arrivals, without needing to have recourse to POVMs, except to justify that we remain in the realm of standard quantum mechanics.

In a similar manner, we need not be concerned about the POVM character or otherwise of an extension to the case of many particles, to which we now direct our attention.

10.7 Times of Arrival for Identical Particles

We have made more explicit the many particle distribution sketched above in [43], of which this subsection is an abbreviated version. Following the standard construction of the formalism of second quantization (see [44] for reference), adequate for one particle operators, we define creation and annihilation operators $\widehat{v}_\beta^\dagger(x)$ and $\widehat{v}_\beta(x)$, as those that connect the crossing state $|v_{\beta,x}\rangle$ and the vacuum state $|0\rangle$,

$$|v_{\beta,x}\rangle = \widehat{v}_\beta^\dagger(x)|0\rangle \quad \text{and} \quad \widehat{v}_\beta(x)|v_{\beta,x}\rangle = |0\rangle .$$

In terms of the creation and annihilation operators for plane waves, these operators are written as

$$\widehat{v}_\beta(x) = \int_{-\infty}^{+\infty} dp \langle v_{\beta,x}|p\rangle \widehat{a}_p = \int_{-\infty}^{+\infty} dp \left(\frac{\alpha p}{2\pi\hbar m}\right)^{1/2} \Theta(\alpha p)e^{ipx}\widehat{a}_p ; \quad (10.19)$$

$$\widehat{v}_\beta^\dagger(x) = \int_{-\infty}^{+\infty} dp \langle p|v_{\beta,x}\rangle \widehat{a}_p^\dagger = \int_{-\infty}^{+\infty} dp \left(\frac{\alpha p}{2\pi\hbar m}\right)^{1/2} \Theta(\alpha p)e^{-ipx}\widehat{a}_p^\dagger \quad (10.20)$$

(remember the correspondence $\beta = L \leftrightarrow \alpha = +$ and $\beta = R \leftrightarrow \alpha = -$).

Now, by inspection of expression (10.18) and following standard procedure, we *define* an arrival density at time t at point x as

$$\Pi_\psi(t,x) = \sum_{\beta=L,R} \langle \psi(t)|\widehat{v}_\beta^\dagger(x)\widehat{v}_\beta(x)|\psi(t)\rangle . \quad (10.21)$$

In this definition we have obviated all topological considerations of domains, closure and the like.

Actually, when dealing with more than one particle, there are other observables related with times of arrival, such as the time–time correlation function. The quantum–mechanical description and computation of those quantities is carried out, in the formalism of second quantization, as expectation values of products of operators in Heisenberg's or interaction representation. For this purpose it is convenient to define the arrival density operator at instant t and point x,

$$\widehat{\Pi}(t,x) = U^\dagger(t) \sum_{\beta=L,R} \widehat{v}_\beta^\dagger(x)\widehat{v}_\beta(x)U(t) =$$

$$= \int dp\,dq\,\frac{|pq|^{1/2}}{2\pi\hbar m}\Theta(pq)e^{i(q-p)x/\hbar}\widehat{a}_p^\dagger(t)\widehat{a}_q(t) , \quad (10.22)$$

where $\widehat{a}_p^\dagger(t)$ and $\widehat{a}_q(t)$ are the time evolved creation and annihilation operators for plane waves, with evolution operator $U(t)$, i.e., $\widehat{a}_p^\dagger(t) = U^\dagger(t)\widehat{a}_p^\dagger U(t)$ and similarly $\widehat{a}_p(t) = U^\dagger(t)\widehat{a}_p U(t)$. Notice that

$$\Pi_\psi(t,x) = \langle \psi|\widehat{\Pi}(t,x)|\psi\rangle ,$$

which is an alternative formulation of the covariance property of the density of arrivals. From these operators $\widehat{\Pi}(t,x)$ we can construct a time operator that would be a generalization of Aharonov and Bohm's, as follows:

$$\widehat{T}_x = \int_{-\infty}^{+\infty} dt\, t\, \widehat{\Pi}(t,x) = \int dt\, dp\, dq\, \frac{|pq|^{1/2}}{2\pi\hbar m}\Theta(pq)e^{i(q-p)x/\hbar}\, t\, \widehat{a}_p^\dagger(t)\widehat{a}_q(t) \ .$$

By construction this is simply a one particle operator, which coincides with $\widehat{T}_{AB}$ over states whose content is just one free particle. A particularly important property of the arrival density operator is that the density of arrivals $\Pi_\psi(t,x)$ over any state is covariant in time if the Hamiltonian is independent of time (as has been assumed all along). Even though the properties of covariance and positivity do not, by themselves, completely fix the density of times of arrival, they are minimal requirements, the lack of which would seriously impair any proposal.

For the sake of completeness, let us note down the flux operator for many particles:

$$\widehat{j}(0,x) = \frac{-i\hbar}{2m}\left(\widehat{\psi}^\dagger(x)\partial_x\widehat{\psi}(x) - \left(\partial_x\widehat{\psi}^\dagger(x)\right)\widehat{\psi}(x)\right) =$$
$$= \frac{1}{4\pi\hbar m}\int dp\, dq\, e^{i(q-p)x/\hbar}(p+q)\widehat{a}_p^\dagger\widehat{a}_q \ . \tag{10.23}$$

At other times one obtains $\widehat{j}(t,x)$ as

$$\widehat{j}(t,x) = U^\dagger(t)\widehat{j}(0,x)U(t) = \frac{1}{4\pi\hbar m}\int dp\, dq\, e^{i(q-p)x/\hbar}(p+q)\widehat{a}_p^\dagger(t)\widehat{a}_q(t) \ , \tag{10.24}$$

(again assuming that the Hamiltonian is independent of time).

A straightforward comparison of (10.22) and (10.24) reveals that in $\widehat{\Pi}$ we compute a geometric mean, to be contrasted to the arithmetic mean character of the flux operator. The quantities being averaged are the incoming and outgoing momenta, weighed with an oscillatory factor dependent on the position of arrivals. As a matter of fact, the comparison is not as straightforward, since the density of arrivals is positive, whereas this is not the case for the flux. We thus see that the arrival density operator can be understood as a quantization of the classical flux, but, more adequately, as a quantization of $J^L - J^R$, where $\{J^L, J^R\}$ are the fluxes of {left,right} going particles.

The numerical computation of densities of arrivals for symmetric and antisymmetric states of two particles (required for bosons and fermions, respectively) reveals the behavior one would normally expect [43]. The full bosonic or fermionic character of the particles considered, however, will only be apparent in an arrival–arrival correlation function; in particular in the two point function

$$\langle\psi| :\widehat{\Pi}(t,0)\widehat{\Pi}(0,0): |\psi\rangle \ , \tag{10.25}$$

where $:\ :$ stands for normal ordering (notice that $:\widehat{\Pi}(t,0)\widehat{\Pi}(0,0):$ is in fact the two particle part of $\widehat{\Pi}(t,0)\widehat{\Pi}(0,0)$).

10.8 How can Kijowski's distribution be measured?

Within the standard interpretation, the predicted values of observable quantities are only realized when the "proper" measurement is performed. It is important to keep this in mind to avoid conceptual pitfalls such as the idea that a zero probability density at a point is not compatible with Π_K different from zero there; in fact it is, since the corresponding operators do not commute. This has generated some debate [45, 46]. What a measurement for some abstract operator should really be in practice is not something that follows directly from the formalism, and in fact the absence of systematic recipes to achieve that goal was seen by Wigner as a major drawback of quantum theory [47]. In the particular case of the arrival time, the measurement model of Chap. 8 [48], provides in principle a set of operations to measure Kijowski's distribution: Assume first a sharply defined laser-illuminated region for $x \geq 0$ and ground state atoms incident from the left with positive momenta. The distribution of first (fluorescence) photons is distorted with respect to an ideal distribution, such as Π_K, because of the backaction of the laser which induces atomic reflection from the laser region at very low incident energies. By enhancing the relative weight of low-energy components in the initial state to compensate for these losses,[6] the first photon distribution coincides with Kijowski's [49].

If the laser-illuminated region has, more reallistically, a finite width, there will be undetected atoms both because of too low or too high incident energies with respect to the characteristic atom–laser interaction energy. Applying a similar modification of the initial state to compensate for the two effects and taking the limit of a point-like laser, Kijowski's distribution is again obtained [50], but now for states with both positive and negative incident momenta. An extension to the case with potential interaction has been considered in [50].

10.9 Conclusions and Outlook

The different constructions presented in this chapter provide a good setting for the time of arrival within the language of quantum mechanics, as the ideal concept to which experiments might converge, in exactly the same manner as operators such as the position operator provide the ideal concept to which experiments might converge. In fact each relevant experiment will probably have to be thought anew, and its most salient facts properly modeled. At any rate, we believe ourselves justified in thinking that the long prevalent view that there is no place for ideal time observables has been superseded,

---

[6] This is formally done by the operator normalization procedure of Brunetti and Fredenhagen [12] and requires the vanishing of wave packet component at zero energy. In practice, the initial state modification, up to a normalization constant, may be performed automatically, i.e., for arbitrary incident states, by "filtering" the incident state with an appropriately chosen potential interaction previous to the laser region.

and that we do have the (admittedly elementary) tools for dealing with the simplest ones without in any way distorting the standard framework of quantum mechanics. Actually, measuring the ideal distribution of Kijowski in an interesting case, i.e., when it differs significantly from the current density, is still an open experimental challenge, but at least the limits in which an operational procedure would lead to that goal have been found. In a complementary, opposite direction, we might also wonder what are present day time-of-flight experiments actually measuring: what is the ideal quantity that we get from them, possibly in some high accuracy limit? To answer this question it is necessary to use models and describe specific measurements [52, 53, 54, 55, 56]. In particular, the measurement model in [48, 51, 57] has shown that from the distribution of first spontaneous photons, and depending on the experimental setting (defined by different parameters such as laser intensity, incident energy, laser detuning, or atomic lifetime) and on the realization of deconvolutions to subtract expected measurement delays, different quantities may in principle be obtained apart from arrival time distributions, such as current densities, particle densities, kinetic energy densities and others [48, 58].

The theory of E. Galapon and coworkers [59, 60, 61, 62, 63], based on selfadjoint time-of-arrival operators for confined systems has also been shown to lead to Kijowski distribution as the length of the confining box increases to infinity [62]. Moreover the confined time-of-arrival eigenfunctions evolve to have point supports at the arrival point at their respective eigenvalues in the limit of arbitrarily large confining lengths. A direct operational interpretation of the discrete times of the confined case is still missing and is one of the challenges for future research. In any case, the discrete–continuum smooth transition found may be a useful tool to generate new theories of first time of arrival with interacting potentials, that can be later translated to the continuum and operationally interpreted or compared with existing operational proposals.

Acknowledgments

The authors wish to thank the fruitful discussions held with R. Sala, R. Leavens, S. Brouard, E. Galapon, G. C. Hegerfeldt, D. Seidel, V. Hannstein, J. A. Damborenea and many others (most of which are contributors to this volume). Support by CERION (Canadian European Research Initiative on Nanostructures), Ministerio de Ciencia y Tecnología (grants no. AEN99-0315 and BFM2000-0816-C03-03), The University of the Basque Country (grants no. UPV 063.310-EB187/98 and 9/UPV 00039.310-13507/2001), and the Basque Government is gratefully acknowledged.

References

1. W. Pauli: in *Encyclopedia of Physics*, ed. by S. Flugge (Springer, Berlin, 1958), Vol. V/1, p. 60
2. Y. Aharonov, D. Bohm: Phys. Rev. **122**, 1649 (1961)

3. N. Grot, C. Rovelli, R.S. Tate: Phys. Rev. A **54**, 4676 (1996). quant-ph/9603021
4. J.A. Oppenheim, B. Reznik, W.G. Unruh: Phys. Rev. A **59**, 1804 (1999). quant-ph/9807043
5. J.G. Muga, R. Sala Mayato, J.P. Palao: Superlattices Microstruct. **23**, 833 (1998). quant-ph/9801043
6. R. Giannitrapani: Int. J. Theor. Phys. **36**, 1575 (1997). quant-ph/9611015
7. V. Delgado, J.G. Muga: Phys. Rev. A **56**, 3425 (1997). quant-ph/9704010
8. M. Toller: Phys. Rev. A **59**, 960 (1999). quant-ph/9805030
9. J. León: J. Phys. A **30**, 4791 (1997). quant-ph/9608013
10. J. Kijowski: Rept. Math. Phys. **6**, 361 (1974)
11. V.H. Paul: Ann. Phys. (Leipzig) **9**, 252 (1962)
12. R. Brunetti, K. Fredenhagen: Phys. Rev. A **66**, 044101 (2002)
13. J. von Neumann: *Mathematical Foundations of Quantum Mechanics* (Princeton University Press, Princeton, NJ, 1955)
14. J.G. Muga, C.R. Leavens: Phys. Rep. **338**, 353 (2000)
15. B. Mielnik, G. Torres-Vega: Concepts of Physics **2**, 81 (2005)
16. J. Kijowski: Concepts of Physics **2**, 99 (2005)
17. L. Hardy: Quantum theory from five reasonable axioms (2001). quant-ph/0101012
18. C.A. Fuchs: Quantum Foundations in the Light of Quantum Information, in *Decoherence and its Implications in Quantum Computation and Information Transfer: Proceedings of the NATO Advanced Research Workshop, Mykonos Greece, June 25–30, 2000*, ed. by A. Gonis, P. E. A. Turchi (IOS Press, Amsterdam, 2001), pp. 38–82. quant-ph/0106166
19. P. Busch: Quantum states as effect valuations: giving operational content to von Neumann's no-hidden-variables theorem. Phys. Rev. Lett. 91, 120403 (2003). quant-ph/9909073
20. A.D. Baute, R. Sala Mayato, J.P. Palao, J.G. Muga, I.L. Egusquiza: Phys. Rev. A **61**, 022118 (2000). quant-ph/9904055
21. P. Busch, M. Grabowski, P.J. Lahti: *Operational Quantum Mechanics* (Springer, Berlin, 1995)
22. W.M. de Muynck, W.D. Baere, H. Martens: Found. Phys. **24**, 1589 (1994)
23. E.B. Davies: *Quantum Theory of Open Systems* (Academic Press, London, 1976)
24. G. Ludwig: *Foundations of Quantum Mechanics*, Vol. I (Springer, Berlin, 1983)
25. A.S. Holevo: *Probabilistic and Statistical Aspects of Quantum Theory* (North-Holland, Amsterdam, 1982)
26. B. Huttner, A. Muller, J.D. Gautier, H. Zbinder, N. Gisin: Phys. Rev. A **54**, 3783 (1996)
27. N.I. Akhiezer, I. M. Glazman: *Theory of Linear Operators in Hilbert Space* (Dover, New York, 1963)
28. J.G. Muga, C.R. Leavens, J.P. Palao: Phys. Rev. A **58**, 4336 (1998). quant-ph/9807066
29. R. Sala, J.P. Palao, J.G. Muga: Phys. Lett. **A231**, 304 (1997). quant-ph/9901042
30. M. Razavi: Can. J. Phys. **49**, 3075 (1971)
31. G.R. Allcock: Ann. Phys. (N.Y.) **53**, 253 (1969)
32. G.R. Allcock: Ann. Phys. (N.Y.) **53**, 286 (1969)
33. G.R. Allcock: Ann. Phys. (N.Y.) **53**, 311 (1969)
34. J. Kijowski: Phys. Rev. A **59**, 897 (1999)
35. R. Werner: J. Math. Phys. **27**, 793 (1986)

36. D.T. Pegg: Phys. Rev. A **58**, 4307 (1998)
37. E.P. Wigner: in *Aspects of Quantum Theory*, ed. by A. Salam, E.P. Wigner (Cambridge University Press, London, 1972)
38. A.D. Baute, I.L. Egusquiza, J.G. Muga: Phys. Rev. A **64**, 012501 (2001)
39. A.D. Baute, I.L. Egusquiza, J.G. Muga, R. Sala Mayato: Phys. Rev. A **61**, 052111 (2000). quant-ph/9911088
40. J. León, J. Julve, P. Pitanga, F.J. de Urríes: Time of arrival through a quantum barrier (1999). quant-ph/9903060
41. J. León, J. Julve, P. Pitanga, F.J. de Urríes: Phys. Rev. A **61**, 062101 (2000). quant-ph/0002011
42. J. León: in *Extensions of Quantum Physics*, ed. by A. Horzela, E. Kapuscik, (Apeiron, Montreal, 2002), p.29. quant-ph/0008025
43. A.D. Baute, I.L. Egusquiza, J.G. Muga: Phys. Rev. A **65**, 032114 (2002)
44. G. Baym: *Lectures on Quantum Mechanics* (W. A. Benjamin, Reading, MA, 1974)
45. C.R. Leavens: Phys. Lett. A **303**, 154 (2002); **338**, 19 (2005); **345**, 251 (2005)
46. I.L. Egusquiza, J.G. Muga, B. Navarro, A. Ruschhaupt: Phys. Lett. A **313**, 498 (2003)
47. E.P. Wigner: in *Contemporary Research in the Foundations and Philosophy of Quantum Theory*, ed. by C. A. Hooker (Reidel, Dordrecht, 1973), p. 369
48. J.A. Damborenea, I.L. Egusquiza, G.C. Hegerfeldt, J.G. Muga: Phys. Rev. A **66**, 052104 (2002)
49. G.C. Hegerfeldt, D. Seidel, J.G. Muga: Phys. Rev. A **68**, 022111 (2003)
50. G.C. Hegerfeldt, D. Seidel, J.G. Muga, B. Navarro: Phys. Rev. A **70**, 012110 (2004)
51. J.A. Damborenea, I.L. Egusquiza, G.C. Hegerfeldt, J.G. Muga: J. Phys. B **36**, 2657 (2003)
52. M.M. Ali, A.S. Majumdar, D. Home, S. Sengupta: Phys. Rev. A **68**, 042105 (2003); A.K. Pan, M.M. Ali, D. Home: Phys. Lett. A **352**, 296 (2006)
53. R.S. Bondurant: Phys. Rev. A **69**, 062104 (2004)
54. J. Ruseckas, B. Kaulakys: Phys. Rev. A **66**, 052106 (2002)
55. T. Okubo, T. Odagaki: Phys. Rev. E **73**, 026128 (2006)
56. M. Debicki, A. Gozdz: Int. J. Mod. Phys. E **15**, 437 (2006)
57. V. Hannstein, G.C. Hegerfeldt, J.G. Muga: J. Phys. B **38**, 409 (2005)
58. J.G. Muga, D. Seidel, G. Hegerfeldt: J. Chem. Phys. **122**, 154106 (2005)
59. E.A. Galapon: Proc. R. Soc. A **458**, 451 (2002). quant-ph/9908033
60. E.A. Galapon: Proc. R. Soc. A **487**, 2671 (2002).
61. E.A. Galapon, R. Caballar, R.T. Bahague: Phys. Rev. Lett. **93**, 180406 (2004)
62. E.A. Galapon, F. Delgado, J.G. Muga, I. Egusquiza: Phys. Rev. A **72**, 042107 (2005)
63. E.A. Galapon, R.F. Caballar, R.T. Bahague: Phys. Rev. A **72**, 062107 (2005)

11

Experimental Issues in Quantum–Mechanical Time Measurement

Aephraim M. Steinberg

Department of Physics, University of Toronto, Toronto, ON, Canada M5S 1A7
aephraim@physics.utoronto.ca

11.1 Time Operators Versus Real Measurements

Time has a unique status in quantum mechanics. In general, to develop a quantum–mechanical description of some observable, one writes down an operator whose eigenvalues represent the possible outcomes of the measurement; the operator is chosen to yield the classical results in the correspondence-principle limit. Generally, this latter limit is unambiguous.

In the case of time measurements, there are in fact two problems with this prescription. First, as is well known, it is not possible to construct a Hermitian time operator canonically conjugate to the Hamiltonian (see the Introduction for some of the subtleties, however, connected with this argument of Pauli's). Second, though less well appreciated, there is no such thing as a time measurement in classical physics either. Just as in quantum physics, time plays the role of a free parameter classically, while observables such as position and momentum evolve as functions of time. When we measure "the time," what are we doing? We cannot measure the time of a particle; the particle exists at all times, after all. We may look at a clock and measure the position of one of its hands, and infer a statement about the time based on some (hopefully good) assumptions as to how that position evolves temporally. If we measure correlations between this clock-hand position and some other observable, such as the state of a particle detector, we may interpret this in terms of detection times, etc. But of course, this time must be referred to some origin. Therefore, we really measure durations—whether this is the duration of an ion's flight in a time-of-flight mass spectrometer, or the length of time between a clock being being synchronized to NIST and a detector observing neutrinos from a distant supernova, it is always a duration.

In practice, there are many different experimental techniques for measuring quantities with dimensions of time, but all involve measuring an observable with some other dimensions and then relying on some calibration. Although these techniques have various degrees of directness, and some rely on internal degrees of freedom while others rely on external ones, the problem is generally

A. M. Steinberg: *Experimental Issues in Quantum–Mechanical Time Measurement*,
Lect. Notes Phys. **734**, 333–353 (2008)
DOI 10.1007/978-3-540-73473-4_11 ⓒ Springer-Verlag Berlin Heidelberg 2008

not discussed at the classical level, because the different techniques always yield the same results. Time is what a clock measures. We all have a long-standing working definition of what this means, and when a new technique is developed for measuring time, it is judged on whether or not it agrees with existing clocks. Fundamentally, this is all we mean by the accuracy of a time measurement; if it agrees with the steady evolution of the Earth around the Sun and the moon around the Earth, and all the other processes we take to proceed at a constant rate in nature, then it serves its purpose, because we have no other direct access to the free parameter t.

A common paradigm for measuring delay times in practice is as follows. One event serves as a "trigger." This event (particle detection, for instance) causes a circuit to fire, setting off a linear voltage ramp. A second event sends a "stop" signal, halting the rise of the voltage; the final voltage is thus proportional to the time delay between the "start" and "stop" signals. This voltage is recorded, as a measure of the time.[1] Standard "time-to-amplitude convertors" function in this fashion; for that matter, an oscilloscope operates the same way, with the linear voltage ramp hooked up to control the horizontal sweep on a display, so that a measurement of the position of a spot serves as an indirect measurement of the time of an event relative to some trigger. Many practical time measurements are even less direct. Optical time-of-flight measurements may rely on interference between two waves with a given relative delay; our understanding of interference and prior knowledge of the optical frequency allows us to infer the delay time from the interference pattern. In the field of ultrafast (femtosecond) lasers, extremely rapid nonlinear processes are used to generate second-harmonic radiation if and only if two beams coexist in an appropriate medium at the same time. To measure the duration and shape of a laser pulse, one beam is alternately delayed and advanced by translation of a mirror, and the intensity of second-harmonic light is monitored. Based on our knowledge of the speed of light propagation in air and of the nonlinear optical process, we infer temporal characteristics. In condensed-matter physics and in nuclear magnetic resonance, many important measurements involve determinations of the time it takes for a certain population or coherence to decay. Such measurements frequently rely on a "pump–probe" paradigm, in which the delay between two pulses is varied, and the size of some resulting effect (such as a pulse echo) is monitored. From the behavior of the directly measured observable as a function of the adjustable delay time, one infers the timescale of the microscopic process. (Even the "directly" adjustable delay time must, of course, be measured using the sort of indirect technique described early.) In these latter cases, as in situations where the particle's own

---

[1] N. of E.: There are two commonly found but different meanings for "time delay." The "delay" described here does not coincide with the the time delay of scattering theory (cf. Chap. 2). The latter refers to the difference between the passage times of wave packets with and without interaction.

motion towards some detector is used to keep track of its time of flight, one can think of the particle as carrying its own "clock."

We all have certain intuitions about how such measurements behave, and a quantum theory of observation ought to be constructed in order to reproduce so far as possible the expected characteristics of the classical measurements. One classical assumption is that the quantity being measured is independent of the details of the measurement technique—note, however, that this is far from obvious given the range of widely disparate experimental scenarios alluded to above. Measuring the time delay between a particle or a pulse peak entering and leaving some medium may be classically equivalent to measuring the magnitude of some effect the particle has on the medium or the medium has on the particle, but the quantities being measured are in fact different physical observables. There is no guarantee that when the corresponding observables are measured in the quantum limit, they will exhibit the same numerical equality. In particular, quantum mechanics does not share the same notion of "trajectory" as classical physics; for this reason, no straightforward inference may be made about the time spent in a given region based solely on observations of entry and exit times. The inescapable conclusion is that while one can try to construct quantum–mechanical descriptions of specific time measurements, it may not always be possible to construct a *unique* description of time measurement which will correctly determine the outcome of different classes of experiment. Many different quantities with the dimensions of time can be defined theoretically, but in the end, whether or not they should be considered time measurements can only be determined by comparing them with the types of experimental observation which are in fact used to operationally define time even at the classical level.

11.2 Arrival-Time Measurements

11.2.1 Techniques for Measuring Arrival Time

In general, experimental determination of arrival time [1] relies on some reference. For relatively slow processes, the reference may be an accepted external standard, such as a properly synchronized clock. When this is not feasible, arrival-time measurements are in fact always determinations of the *delay* between some predetermined "trigger" event (particle emission, detection of the particle prior to its traversal of the region of interest, or detection of some other particle) and final detection of the particle. Rather than a von Neumann interaction of the form $\mathcal{H} = g(t)P \cdot O$, where P is the canonical momentum of the "pointer" or measuring device, and O is the operator to be measured, the device here interacts with the particle at two different times. Furthermore, the coupling Hamiltonian does not directly depend on the arrival time of the particle; the observable O more usually represents a function of particle position, e.g., whether or not the particle is in the final region. Time enters instead through the free evolution of the measuring device.

If the delay of an electric or optical pulse is to be determined on an oscilloscope, the actual measurement is of field amplitude or intensity as a function of time, and the propagation time is simply inferred from the spatial separation of points on the scope display. As described earlier, this processing may also be performed automatically, as in a time-to-amplitude convertor, in which case the overall time measurement is complete when a specific final voltage is measured. Naturally, these techniques are limited not only by the reliability of the pointer's time evolution, but also by one's practical ability to trigger on the "start" and "stop" events. For measurements of very short times, there is often no detector with sufficient time resolution. One technique used in such cases is the "streak camera" method, in which the voltage ramp is not terminated when the particle is detected, but is instead used directly to alter where an image of the particle is formed. The particle, while in some interaction region, experiences a time-dependent force, and when its position is subsequently measured, this provides an indirect observation of the magnitude of this force (and thus the time) at which the particle was in the interaction region.

These arrival-time measurements generally reflect the "group delay" $d\phi/d\omega$ of a wavepacket peak,[2] although they are often triggered by a feature other than the peak, for instance, the moment at which the waveform crosses a preset amplitude, or a prearranged fraction of its peak height. Of course, if one knows the transmission function $T(\omega)$ and its phase $\phi(\omega) \equiv \arg T(\omega)$, one can infer the group delay via differentiation. For limited bandwidths and sufficiently well-behaved transmission functions, it is a fair approximation that the entire pulse travels at the group velocity, but this naturally breaks down once dispersion comes into play, necessitating a more careful definition of the quantity to be measured.

Not infrequently, delay measurements are accomplished by directly measuring the phase, and assuming the validity of the stationary-phase approximation. For instance, one of the first experimental studies of the delay time for tunneling [2, 3, 4, 5], in the guise of frustrated total internal reflection [6, 7, 8], was carried out by measuring the phase shift as a function of frequency [9, 10]. The observation that the phase shift varied little as a function of frequency appeared to confirm the prediction that the delay time in tunneling could be anomalously small. Similarly, some early work in the microwave regime, on tunneling through waveguides beyond cutoff, relied on similar techniques [11]. In the latter experiments, a numerical Fourier transform was performed to calculate directly what a transmitted waveform would have resembled if a broadband incident pulse had been used. More recently, measurements of tunneling time via electron interferometry have also been discussed [12]. It is important to note that since absolute phase cannot be measured, these measurements invariably involve comparisons, electronic or interferometric, between the delayed signal and some reference. At optical frequencies, this

---

[2] This is a "delay" with respect to the passage of the free motion system.

reference must generally be derived from the same source as the pulse being studied; at lower frequencies, it can be possible to use independent sources with well-known absolute frequency.

Interference techniques may also make a more-or-less direct determination of the group delay, for a broadband source. "White-light" interference fringes occur for broadband sources, provided that the path-length difference in the interferometer is not larger than the coherence time of the light. This is because at zero path-length difference, all frequency components simultaneously exhibit constructive interference, but once the distance $\Delta L > c/\Delta\omega$ (where ΔL is the path-length difference between two arms of the interferometer; c is the speed of propagation; and $\Delta\omega$ is the bandwidth), some components experience constructive interference while others experience destructive interference. The envelope of the interference pattern thus maps out the Fourier transform of the spectrum, which may (in the case of a transform-limited pulse) be equivalent to the temporal profile (or wave function) of the pulse itself. The peak of the interference pattern occurs when frequencies centered around the peak frequency all experience the same phase shift, i.e., when $\frac{d}{d\omega}[\phi(\omega) - \omega t] = 0$. Adjusting external delays (i.e., the path-length difference) to find the peak of the interference pattern after an obstruction has been placed in one arm of the interferometer therefore allows a direct measurement of the group delay time $\delta\tau_g \equiv d\phi(\omega)/d\omega$ relative to the transmission of the unobstructed path. These techniques are sometimes appreciated for their single-particle nature, although in linear propagation there is no important difference between the predictions of a classical wave theory and those of quantum electrodynamics.

When the intensity (or probability density) varies on too short a timescale to be determined by direct measurement, it may still be possible to use intensity-dependent effects to determine by how much a pulse peak has been delayed. In laser physics, a common technique for doing this with femtosecond resolution is known as nonlinear autocorrelation or cross-correlation. In a medium with a nonlinear polarizability $P_{NL} = \chi^{(2)} E_1 E_2$, radiation may be emitted at the sum of the frequencies of fields E_1 and E_2 (second-harmonic generation, in the limit of degenerate frequencies). The intensity of this radiation is proportional to the instantaneous product of the two intensities $|E_1|^2$ and $|E_2|^2$, and if in a pulsed experiment, the total energy radiated into the second harmonic is measured, this provides a measure of the intensity–intensity overlap $\int dt |E_1(t)|^2 |E_2(t)|^2$ of the two fields. If one pulse is delayed, it is possible to adjust the path length of the other pulse until this overlap is once more maximized, thus providing an indirect measure of the arrival time relative to the time of appearance of the unobstructed pulse. This technique was used by Spielmann et al. to study the time delay for femtosecond laser pulses to traverse an optical tunneling barrier [13].

Higher-order correlation functions (as in intensity–intensity, or Hanbury–Brown–Twiss, interferometry [14]) may also be measured interferometrically. For instance, there is a nonlinear optical effect discovered by Hong, Ou, and Mandel [15] which can be thought of roughly as a time-reversed variant of the nonlinear autocorrelation technique. This effect relies on spontaneous parametric downconversion, a process in which a crystal with a $\chi^{(2)}$ nonlinearity absorbs a pump photon at ω_0 and emits in its place a pair of photons at frequencies spread symmetrically about $\omega_0/2$, energy conservation being assured by the anticorrelation of the frequencies of the two photons. The photons are emitted simultaneously to within their coherence lengths, which are typically on the order of 10 or 20 fs. If the two photon wave packets meet simultaneously at opposite sides of a 50/50 beam splitter, a quantum interference effect related to Bose statistics causes them to exit the beam splitter along the same (randomly chosen) direction; detectors placed at the two exit ports of the beam splitter will never register photons simultaneously. On the other hand, if the two photons arrive at different times, each will make an independent choice at the beam splitter, leading to coincidence counts in half of the cases. Thus by changing the path length of one photon's trip until the coincidence rate is minimized, one can ensure that the photons are meeting simultaneously at the beam splitter [15, 16, 17, 18]. If an obstruction such as a tunnel barrier is placed in one arm of the two-photon interferometer, the coincidence dip recorded as a function of external path length will shift, and this shift is a measure of the delay time for traversing the barrier. It is interesting to note that these experiments are typically performed with a continuous-wave argon laser as the pump, so the state of the light is in fact stationary in time. It is only the *correlations* between the photons which have the very fast (15 fs) time dependence. Once a photon is detected, it is possible to say that its twin has "collapsed" into a 15-fs wave packet, but prior to that time, the system is better seen as a superposition of 15-fs wavepackets with centers at every possible position. Nevertheless, these measurements (performed with detectors thousands of times too slow to directly resolve femtosecond-scale delays) reproduce the expected group delay, measuring the extra time taken by *any* of these hypothetical wave packets to traverse the obstruction.

11.2.2 Superluminal Arrival Times

The question of operational definitions of experimentally accessible times gained a sense of urgency due to the controversy over the process of tunneling alluded to above. Despite its absence from the overwhelming majority of textbook descriptions of tunneling (one early exception being [19]), the tunneling-time problem has a long and illustrious history. At the heart of the problem is the fact that the classical expression for the kinetic energy of a particle inside a tunnel barrier is negative, so that a semiclassical estimate of its velocity becomes imaginary.[3] This makes it impossible to make the naïve first approximation that the duration of a tunneling event is the barrier width divided by the velocity $\sqrt{2E/m}$. Already, we see the first hint that the classical

---

[3] Such a local definition of kinetic energy becomes problematic in quantum mechanics; see for instance [20, 21, 22].

equivalence of a broad range of definitions for traversal time cannot persist, for this one yields an imaginary number, while most measurement techniques will be certain to yield positive values.

Within a few years of the first predictions of tunneling, discussions appeared of the time spent by a particle in a "forbidden" region, and of the use of the stationary phase approximation to calculate properties of tunneling wave packets [23, 2]. Wigner [3] and Eisenbud [24] studied the relationship between scattering phase shifts and the delay time, making explicit the connection between these quantities and the principle of causality. As discussed in Sect. 2.4.4, Wigner observed that "the 'retardation' cannot assume arbitrarily large negative values, in classical theory it could not be less than $-2a$," where a is the radius of the scattering potential; in other words, a classical particle cannot leave the scattering center before it arrives. He notes that "It will be seen that the wave nature of the particles does permit some infringement of [this inequality]." It is primarily with this very infringement that we are concerned here. Does wave mechanics truly allow particles to exit a barrier before they enter it, and in particular, do such effects violate relativistic causality? One could reasonably suspect that the nonrelativistic nature of the Schrödinger equation is at fault here, but more careful analyses using the Dirac equation show that such superluminal transmission (which occurs in cases where all relevant energy scales are far less than the electron rest mass in the first place) persists [25]. The conflict is made even more clear by turning to optical analogs of tunneling, as the same problems arise with Maxwell's (fully relativistic) equations, and since one begins in the relativistic regime, it is relatively easy to achieve conditions under which the group delay is predicted to be superluminal.

Of course, superluminal and even negative group velocities were already known to occur in electromagnetism, and had been reconciled with causality by Sommerfeld and Brillouin [26]. Their work showed that no real signal could propagate faster than the vacuum velocity of light c in any medium obeying the Kramers–Kronig relations, even in regions of anomalous dispersion. In these regions, the absorption and the strong frequency dispersion cause the stationary-phase approximation to break down, as an incident pulse is distorted beyond recognition and no single transmitted peak may be observed. Conventional wisdom has it that such a breakdown occurs in every limit where the group velocity exceeds c. Nevertheless, as early as 1970, Garrett and McCumber showed theoretically that for short enough interaction lengths, absorbing media could indeed transmit undistorted (but attenuated) Gaussian pulses at superluminal, infinite, or even negative group velocities [27]. This prediction was experimentally verified using the technique of nonlinear autocorrelation by Chu and Wong [28]. More recently, anomalous optical delay times as large as 62 ns (large enough to be observed by direct photodetection and storage of an oscilloscope trace) were observed by Wang, Kuzmich, and Dogariu [29] in conditions of near transparency, following a prediction by Chiao and others [30, 31]. This work has been followed up by several groups [32, 33], and various attempts have been made to distinguish the speed of information from that of a pulse peak, confirming that the former is indeed causal [32]. With the advent of negative index "metamaterials," superluminal propagation has also been studied in microwave transmission-like measurements [34]. In parallel, some novel proposals for unattenuated superluminal *reflection* have been made [35, 36], although subtleties remain, related to the proper treatment of the propagation of pulses in two or three dimensions [37]. An excellent recent review on anomalous velocities in optical propagation is [38].

Even before this work on anomalous dispersion, the question of superluminal wave-packet transmission in tunneling was put on a firmer footing by [39]. Hartman was not satisfied by MacColl's 1932 observation that there is "no appreciable" delay in tunneling, and he was concerned about the effects of preferential transmission of higher energy components in a wave packet. In a rigorous treatment of the tunneling of wave packets through a rectangular barrier, he indeed found that for very thick barriers, such distortion occurred that no peak could be identified which might appear at the group delay time. For thin barriers, his results were in agreement with the stationary phase prediction, but there was no conflict with causality. Roughly speaking, the prediction is that for thicknesses smaller than one decay length of the evanescent wave ($d < 1/\kappa$), a transmitted particle of energy much less than the barrier height ($E \ll V_0$; $k \ll k_0$) will appear to have traveled at its initial velocity of $\hbar k/m$. This delay is related to the fact that phase is accumulated as the evanescent ($e^{-\kappa x}$) and antievanescent ($e^{+\kappa x}$) waves change in relative size, as the two have different (but constant) phases. For thicker barriers ($\kappa d \gg 1$), there is no phase change across most of the barrier, since the wave function is dominated by real exponential decay. The so called "phase time" (an unfortunate term, as it is related to the group velocity and not the phase velocity, but one which has become conventional and will therefore be adopted for the sake of this book) is defined as the group delay plus the free motion term, and therefore saturates at the finite value $2m/\hbar k\kappa$, the time it would take the free incident particle to traverse two exponential decay lengths $1/\kappa$. Hartman confirmed that for intermediate barrier thicknesses, larger than $1/\kappa$ but small enough that the pulse was not significantly distorted, this saturation effect did indeed occur. As the distance traversed continues to grow, but the time required to traverse it remains roughly constant, it is clear that one eventually reaches a regime where the apparent propagation speed exceeds c.

The quantum–mechanical tunneling problem is mathematically equivalent to a number of situations in electromagnetism, including frustrated total internal reflection, transmission through a waveguide beyond cutoff, and transmission through a "photonic bandgap" (in one dimension, also known more prosaically as a dielectric mirror). Most examples of massive-particle tunneling occur on time and distance scales which have made direct arrival-time measurements infeasible, and so much of the work on tunneling times has taken place in the microwave and optical domains. Measurements of phase shifts [10, 9, 11], quantum interferograms [40], pulse peaks and rising edges [41, 42], and nonlinear autocorrelation [13] all agree with one another and with the phase time prediction, despite the superluminal nature of this quantity. Similar results have since been obtained for other surprising examples of propagation, in diffraction [43, 44], Bessel beams [45], and anomalous dispersion in gain media [29].

While this state of affairs may seem surprising, it has been pointed out, notably by Büttiker and Landauer, that no physical law guarantees that an incoming peak turns into an outgoing peak [46]. Certainly there need not be any causal connection between an incoming peak and an outgoing one, as demonstrated in a striking electronics experiment [47]. As explained in [8] and references therein, no *information* can be sent faster than light using these effects. To sum the arguments up briefly, the wave exiting the barrier can be expressed by integrating a *perfectly causal* response function:

$$\Psi(x = d, t) = \int_0^\infty d\tau f(\tau)\Psi(x = 0, t - d/c - \tau) \,. \qquad (11.1)$$

The behavior of $\Psi(x = 0)$ at times $t_0 > t - d/c$ has no impact whatever on the behavior of $\Psi(x = d, t)$. Remarkably, the effect of a tunnel barrier is to perform an extrapolation into the future (a Taylor expansion [48, 49]) by combining paths with slightly different delays 180° out of phase with one another [50, 51]. Whether for a Gaussian wave packet or a piece by Mozart, the outcome is a convincing reproduction of the incident wave form; some workers therefore argue that signal propagation for such frequency-band-limited waves indeed occurs faster than c [52]. Other workers take the more cautious view that information velocity should refer to the arrival of "new" information, and have experimented with modified pulses containing "abrupt" signals [32], demonstrating experimentally that when these changes are sufficiently sharp, they indeed propagate causally. However one chooses to think about superluminal group delays, it is clear that the phase time is an accurate prediction for the type of measurement it is designed to describe; it does indeed indicate the time delay for a given wave or probability distribution to reach its maximum value, and thus the *most likely* delay time observable via direct detection of individual particles. The various measurement schemes which are sensitive to this property all agree in the quantum regime as they do in the classical one. In the cases of superluminal transmission without attenuation [30, 31, 29, 32, 33], it is interesting to ask about the *energy* velocity, which by standard definitions also appears to exceed c. Several interesting attempts have been made [53, 54, 55] to clarify this situation by devising a more complete definition of energy velocity. Some recent reviews [56, 57, 58] offer a variety of perspectives on superluminal tunneling and related issues. The remaining question is whether or not this delay is the *only* reasonable way of defining the traversal time for a tunneling particle.

11.3 Dwell or Interaction Time Measurements

11.3.1 Theoretical Proposals

Büttiker and Landauer were prominent among those arguing that in cases where the delay time was anomalously small, it must not reflect the actual duration for which the particle was in the region of interest. Of course, there is nothing special about the particular barrier for which the phase time becomes smaller than d/c, or the one for which it becomes negative; if in these cases, the "dwell time" spent by the particle in the region differs from the measured time of flight inferred from wave-packet peaks, then it is reasonable to suppose that these times are in general different quantities. It so happens in the classical regime that the time spent by a particle in a given region is equal to the difference between its exit and entrance times; quantum mechanically, this relation is true only as an approximation valid in certain parameter ranges, and it must in fact be recognized that at least two different physical quantities have been given the same name because of the accuracy of this approximation in our familiar experience.

The "dwell" or "sojourn" time τ_d seems the most natural answer to the question "how much time does a particle spend in the barrier region?" It can be defined alternately for the time-dependent or the time-independent case. In the former, its natural statement is as the time integral of the instantaneous probability that the particle is inside the barrier (assumed to extend from $-d/2$ to $d/2$):

$$\tau_d \,(\text{time-dependent}) \equiv \int_{-\infty}^{\infty} dt \int_{-d/2}^{d/2} dx |\Psi(x,t)|^2 \ . \tag{11.2}$$

In the latter case, it is simply the probability density within the barrier, divided by the incident flux J_I:

$$\tau_d \,(\text{time-independent}) \equiv \frac{1}{J_I} \int_{-d/2}^{d/2} dx |\Psi(x)|^2 \ . \tag{11.3}$$

In the limit of a monochromatic wave packet, these two formulas yield the same result, although for packets of finite extent, corrections may be important [59]. A variety of expressions for the dwell time is discussed in Chap. 2. The importance of definitions in the quantum regime cannot be overexaggerated. In the classical limit, τ_d (the time spent within the potential step) and the phase time τ^{Ph} (the time between arrival at the leading edge of the step and departure from the trailing edge) are of course identical, and equal to $d/v = md/\hbar k$, as can easily be verified analytically. There is only one sensible quantity to term the "traversal time" in this case, but this fact does not follow directly from the structure of the two quantum–mechanical definitions.

The dwell time may appear unsatisfactory as a candidate for a traversal time for several reasons. Foremost, it is a characteristic of an entire wave function, comprising both transmitted and reflected portions. One might well expect that transmitted and reflected particles could spend differing amounts of time in the barrier. (Without a doubt, one would expect them to spend different amounts of time on the *far* side of the barrier—a finite amount for the transmitted particles and none for the reflected ones—whereas the formulations of (11.2) and (11.3) leave no room to introduce this distinction.) Its definition is so natural that many researchers have argued that it must at least reflect the *weighted average* of transmission and reflection times, $\tau_d = |T|^2\tau_T + |R|^2\tau_R$ (with T and R the transmission and reflection amplitudes, respectively), but even this assertion has been hotly disputed [60, 4, 61, 62, 63, 64, 5].

The second seeming problem with the dwell time is one it shares with the phase time. It is not guaranteed to be greater than the barrier thickness upon the speed of light, d/c. In fact, in the low-energy limit $k \rightarrow 0$ of a rectangular-barrier tunneling problem with barrier height $V_0 \equiv \hbar^2 k_0^2/2m$, the wave is almost entirely reflected by the first interface, and $|\Psi|^2$ is negligible in the barrier, leading τ_d to vanish as $2mk/\hbar\kappa k_0^2$.

Once more, one is led to approach the problem of defining a time by considering potential experimental schemes for its measurement. What do we mean when we ask about the time a particle *spent* in a region, as distinct from observations about its most likely appearance at the far side? We are naturally considering some hypothetical interaction between the particle and the barrier, or some other form of time-dependent interaction. In the 1982 paper which is widely viewed as having rekindled the tunneling-time fire, Büttiker and Landauer proposed a *Gedanken experiment* which would allow one to infer the duration of the tunneling process. Consider a particle tunneling through a rectangular barrier. Now modulate the height of the barrier by a small amount, at some relatively low-frequency Ω. Clearly, the transmission is lowest when the barrier is highest, and vice versa. But now imagine that Ω becomes greater and greater, until $\Omega \gg 1/\tau_t$, that is, until the barrier goes through more than one oscillation during the "duration" τ_t of the tunneling event. Naturally, the modulation of the transmitted wave will be washed out. Büttiker and Landauer therefore solved the problem of the oscillating barrier, and looked for this critical frequency Ω_c. They then postulated that the traversal time was $\tau^{BL} \equiv 1/\Omega_c$. When the calculation was performed in the opaque limit ($\kappa d \gg 1$), they found the following result:

$$\tau^{BL} = md/\hbar\kappa . \tag{11.4}$$

This is a striking result. Recalling that the local wavevector inside the barrier is $i\kappa$, we see that this is exactly the time we would expect from a semiclassical or WKB (Wentzel-Kramers-Brillouin) approach ($md/\sqrt{2mE}$)—aside from the fact that we here find a real number, despite the imaginary value of the wave vector. Due to the similarity of the formulas, nevertheless, the Büttiker–Landauer time is also frequently referred to as the "semiclassical time." (Far above the barrier, both τ^{Ph} and τ_d in fact approach the semiclassical time

$\tau_s \equiv md/\hbar|k|$.) Since this time is proportional to d, it rarely becomes smaller than d/c; in fact, it would only do so for $m/\hbar\kappa > c$, which is the relativistic limit, where the Schrödinger equation should not be expected to be valid. (In reality, for geometries more complicated than the rectangular barrier, it has been noted that this time may vanish identically, leading once more to causality problems [65, 7, 40]; in addition, [66] have pointed out examples in which no direct, general relation exists between the critical frequency Ω_c and the duration of the tunneling process, thus casting doubt on the interpretation of (11.4) as the tunneling time.)

While above the barrier, the semiclassical time closely resembles the phase time (missing only the oscillations due to multiple reflections at the barrier edges, which become insignificant in the WKB limit), but it looks nothing at all like τ^{Ph} below the barrier, diverging when $E = V_0$ (where $V_0 \equiv \hbar^2 k_0^2/2m$ is the height of the barrier) and falling in the opaque limit ($\kappa d \gg 1$) to $md/\hbar k_0$ as opposed to diverging like $\tau^{Ph} \to 2m/\hbar k k_0$. The phase time diverges for $k \to 0$, but is independent of d; the Büttiker–Landauer time is well-behaved as $k \to 0$, and is proportional to d.

Büttiker went on to consider another "clock," to see if different types of perturbations would bring to light the same timescale. Expanding on work due to [67] and [68], he considered an electron tunneling through a barrier to which a small magnetic field $\mathbf{B} = B_0\hat{\mathbf{z}}$ is confined. Suppose the electron's spin is initially pointing along $\hat{\mathbf{x}}$. The magnetic field causes it to precess in the x–y plane at the Larmor frequency $\omega_L = 2\mu_B B_0/\hbar$, where μ_B is the Bohr magneton. If one measures the polarization of the transmitted electron, one will find it to have precessed through some angle θ_y, and nothing could be more natural than to ascribe this to precession at ω_L for the duration τ_y of the tunneling event, leading to the "Larmor time" $\tau_y \equiv \theta_y/\omega_L$. This internal clock is certainly closely related to many standard experimental measurement techniques, and is of course reminiscent of nuclear magnetic resonance studies of spin precession and associated decay times. It can be shown that this time $\tau_y = -\hbar\frac{\partial}{\partial V_0}\arg(T)$, and is thus closely related to the phase time $\tau^{Ph} = dm/\hbar k + \hbar\frac{\partial}{\partial E}\arg(T)$. The two times show superluminal behavior at low energies.

Büttiker's insight was that this early expression for the Larmor time made the implicit assumption that by taking the $B_0 \to 0$ limit, one could neglect the tendency of the electron to align itself with respect to the magnetic field. In reality, due to the interaction Hamiltonian $\mathcal{H}_{int} = +2\mu_B B_0 S_z$, a spin-up electron sees an effective potential with a higher barrier than that seen by a spin-down electron, and therefore has a lower transmission probability. As the $\hat{\mathbf{x}}$-polarized electrons are equal superpositions of $S_z = \pm 1/2$, this preferential transmission will tend to rotate the polarization out of the x–y plane towards the $-z$-axis, so that the transmitted electron beam is slightly spin-polarized antiparallel to the applied $\mathbf{B}$ field. Büttiker showed that both this out-of-plane rotation and the in-plane precession were first order in B_0, and furthermore, that the former dominated the latter in the opaque limit. Defining a second Larmor time related to the polar rotation according to $\tau_z \equiv \theta_z/\omega_L = -\hbar \frac{\partial}{\partial V_0}\ln|T|$, he found this timescale to reproduce the $md/\hbar\kappa$ behavior he and Landauer had already calculated by considering the modulated barrier. Suggesting that the true interaction time should take into account the full three-dimensional rotation of the electron's spin, he proposed that the interaction time was $\tau_x \equiv \sqrt{\tau_y^2 + \tau_z^2}$. This time agrees with the oscillating-barrier result τ^{BL} in both the low- and high-energy limits.

A fair number of other methods for defining the tunneling time have been proposed on purely theoretical grounds, without direct connection to any specific experimental scheme. For example, a Feynman-path approach in which the durations of all relevant Feynman paths were averaged with the weighting factor $\exp\{iS[x(t)]/\hbar\}$ yielded the "complex time" $\tau_c = \tau_y - i\tau_z$ [60, 62, 69, 70, 71, 64]. It is easy to observe that the magnitude of this time is Büttiker's Larmor time, while its real and imaginary parts are (for the rectangular barrier) the dwell time and minus the semiclassical time, respectively. (An earlier approach [72, 73] yielded a similar complex time, whose real part was the group delay, rather than the dwell time.) Nevertheless, it is only by comparison with the outcomes of actual or potential measurements that one can judge the suitability of a theoretical proposal for the tunneling time.

11.3.2 Some Experimental Examples

The interest in tunneling times was largely driven by the condensed matter community, and some effort went into attempting to carry out time measurements in electronic systems. As has already been observed, the appropriate techniques did not exist for measuring electronic wave-packet times, but it was possible to measure various types of interaction times for tunneling electrons [74].

While it did not prove feasible to carry out directly either the Larmor or the Büttiker–Landauer proposal, Guéret et al. measured a related effect and interpreted it in terms of barrier traversal time. They studied tunneling of electrons through GaAs/GaAlAs heterostructures. Although the spin precession of the tunneling electrons could not be studied directly, they argued that an applied magnetic field also leads to a Lorentz force, deflecting the electron as it tunnels, thereby increasing the distance it needs to travel, and consequently exponentially suppressing the tunneling. By performing a series of static measurements on the tunneling current as a function of applied magnetic field, they were able to infer the energy dependence of the tunneling current, and hence Büttiker's τ_z. They found good agreement with the theoretical predictions, and some took this as support for the semiclassical time. Of course, this measurement has a similar status to the time-independent determination of a series of phase shifts for the sake of calculating the group delay by numerical differentiation; it does not test the validity of the definition, and makes no direct measurement of an interaction time.

An ingenious experiment was carried out by Esteve et al. [75] in rough analogy with the Büttiker–Landauer proposal. In this experiment, a current-biased Josephson junction undergoes macroscopic quantum tunneling between different local minima of the Josephson phase. When the system (not an individual particle!) tunnels, a voltage is produced. By terminating the circuit at a resistor following a delay line, the researchers argued that reflections would be introduced, and might alter the tunneling rate. If, however, the delay line was sufficiently long, the reflections would arrive after the tunneling event was complete. They therefore studied the mean rate of tunneling as a function of delay line length, and determined a crossover time beyond which the resistor appeared to have no further effect on the tunneling decay. Once more, real measurements of time often involve not the direct determination of some observable, but rather careful analysis of a large set of time-independent measurements to estimate a parameter on which the overall system behavior depends. And as in the case of the magnetic deflection of tunneling electrons, such "interaction times" seemed to be in good agreement with the semiclassical times, rather than the phase time.

More recently, some optical experiments have succeeded in measuring quantities related to the Larmor or semiclassical times, as well as pulse–peak arrival times. Deutsch and Golub [76] performed an experiment to measure the Larmor tunneling time for photons. Their experiment utilizes an analogy between the spin of an electron and the spin of a photon, whose polarization state can be described by a point on the Poincaré sphere given by the Stokes parameters $\mathbf{S}$. The equation of motion for the Stokes parameters for a beam of light propagating along the x-axis through a medium with an anisotropic refractive index is given by $d\mathbf{S}/dx = \mathbf{\Omega} \times \mathbf{S}$, where $\mathbf{\Omega}$ is the precession rate of the tip of the $\mathbf{S}$ vector on the Poincaré sphere arising from the anisotropic index of refraction. This equation is formally identical to the one describing the precession of the tip of the electron spin vector $\boldsymbol{\sigma}$ on the Bloch sphere arising from an applied magnetic field $d\boldsymbol{\sigma}/dt = \mathbf{\Omega}_{\mathrm{L}} \times \boldsymbol{\sigma}$, when the optical precession rate $\mathbf{\Omega}$ is identified (apart from a proportionality constant) with the rate of Larmor precession $\mathbf{\Omega}_{\mathrm{L}}$. This analogy between electron and photon spin precession led Deutsch and Golub to suggest an optical implementation of the Larmor clock measurement of the tunneling time of Baz' and Rybachenko (latter corrected and generalized by Büttiker). The basic idea is to replace electrons with photons, and to replace a uniform magnetic field confined to the electron tunnel barrier region with a uniform birefringent medium confined to the corresponding optical tunnel barrier. Thus instead of utilizing the precession of the electron spin as an internal clock to measure the Larmor tunneling time, they utilized the precession of the $\mathbf{S}$ vector of the photon as an internal clock. In their experiment, they used frustrated total internal reflection between two glass prisms as the tunnel barrier. The gap between the prisms, which served as the tunnel barrier, was filled with a birefringent fluid (a liquid crystal). They then measured the Stokes parameters of the transmitted light. In this way, they too confirmed the Larmor time predictions.

In 1997, Balcou and Dutriaux [77] measured two quantities with dimensions of a time in a single experiment on frustrated total internal reflection. In this process, the angle of incidence plays the role of incident energy; and angles closer to the critical angle are transmitted with higher probability than angles further beyond critical. Since any bounded beam incident on the barrier has a range of incident angles, it is found that the transmitted beam is deflected relative to the incident one, in the same way that a tunneling electron will on average have higher energy than the typical incident electron. By measuring this angle, the researchers concluded, one can determine the Larmor time (specifically, Büttiker's τ_z). Simultaneously, the different angular components pick up different phase shifts, which manifests itself as a transverse shift of the *center* of the transmitted beam (as in the Goos–Hänchen effect [78]). This shift was interpreted as an indirect measure of the actual time of flight, and is clearly mathematically similar to both τ_g and τ_y. Here, then, was a clear example where no unique definition of the tunneling time was possible, for two simultaneous measurements on a system yielded two results, each in good agreement with the appropriate theory. The ambiguity is not a theoretical problem, but merely a recognition that there is a lack of defined terms. We cannot, as in the classical regime, simply measure "the" time of a process; we must decide precisely what measurement we are interested with, and use the appropriate theoretical and experimental tools for that measurement.

11.4 Weak Measurements

11.4.1 Interpretational Aspects

A new light can be cast on dwell-time measurements by using the "weak measurement" formalism of Aharonov et al. [79, 80]; see Chap. 13. In quantum mechanics, it is straightforward to define an operator Θ_B which is 1 if the particle is in the barrier region and 0 otherwise. Such a projection operator is Hermitian, and may correspond to a physical observable. Its expectation value simply measures the integrated probability density over the region of interest—it is this expectation value divided by the incident flux which yields the dwell time, as defined earlier. The central problem, once more, is the absence of well-defined histories (or trajectories) in standard quantum theory: as remarked above, the dwell time measures a property of a wave function with both transmitted and reflected portions, and does not display a unique decomposition into portions corresponding to these individual scattering channels (see Chap. 2 and [81]).

Some workers calculate the expectation value not for the initial state but rather for the final state [82, 83, 84]. This answers questions about transmitted particles known to be incident from the left no better than does the usual dwell time; instead of discarding information about late times it discards information about early times. Approaches relying on projector algebra in general have been analyzed in [81] and [85]. Other related approaches follow phase space trajectories [82], Bohm trajectories [86, 87, 88, 89, 90, 91, 92], or Feynman paths [60, 62, 71, 64, 93, 69, 70]. No consensus has been reached as to the validity and the relationship of these various approaches. Once more, in order to choose among various theoretical definitions, we must return to considering the type of measurement actually contemplated.

Weak measurement theory does precisely this, by showing one how to analyze "conditional measurements" in quantum mechanics: that is, how to predict outcomes of measurements not for entire ensembles, but for *sub*ensembles determined both by state preparation and by a subsequent postselection. In the case which concerns us, the state is prepared with a particle incident from the left, and selected to have a particle emerging on the right at late times. Due to the time reversibility of the wave equation, results of intervening measurements depend both on the initial and the final state. This formalism relies only on standard quantum theory, and yields a result which is completely general for any measurement arising from a von Neumann-style measurement interaction, in the limit where the interaction strength is kept low enough to avoid irreversibly disturbing the quantum evolution. This low strength implies great measurement uncertainty on any individual shot, but an average may be calculated for a large number of data runs. We have shown [94, 95] how to apply this formalism to tunneling, and the time we find is identical to the complex time of Sokolovski, Baskin, and Connor, $\tau_c = \tau_y - i\tau_z$. But thanks to the "weak measurement" formalism, it becomes clear what the physical significance of the real and imaginary parts is: the real part (the in-plane Larmor time) quantifies how strongly the tunneling particle will affect a clock with which it interacts; this is the portion which corresponds to a classical measurement outcome. The imaginary part, on the other hand, describes the amount of back action the measuring apparatus will exert on the particle (the sensitivity of the tunneling probability to small perturbations, in other words, as in Büttiker's out-of-plane Larmor rotation). While the former effect remains constant as the measurement is made weaker and weaker, the back action may be made arbitrarily small by resorting to extremely "gentle" (and consequently uncertain) measurements. Among other attractive properties, these conditional times automatically satisfy the relationship $\tau_d = |T|^2\tau_T + |R|^2\tau_R$.

The generality of the times obtained in this way suggests that it may be possible to apply them to a broad variety of problems, at least approximately, even in cases where exact solution would be intractable. It has already been shown that not only are the Larmor times a clear subset of these "conditional times," but that the counterintuitive effects of absorption on light propagating through layered media can be qualitatively understood by application of these complex times [95]. The equality of the Büttiker–Landauer time and $-\mathrm{Im}\tau_c$ makes sense given that the oscillating barrier approach in fact studies the sensitivity to perturbations in the barrier height. The direct connection to measurement outcomes lifts the ambiguity present in other "projector approaches" and the Feynman-path formalism. Finally, it is possible using these methods to calculate conditional probability distributions for transmitted or reflected particle positions as a function of time, and directly investigate questions about whether tunneling particles spend significant lengths of time in the center of the barrier, whether only the leading edge of the wave is transmitted, etc. Since these probability distributions may have large values on both sides of the barrier simultaneously, and independent "weak measurements" can be shown to add linearly (unlike "strong" measurements of noncommuting observables), it is interesting to speculate about whether a statistical demonstration that during tunneling, a particle is "in two places at once" might be possible. Work continues on all of these issues. The connection of superluminality, weak values, and "superoscillations" has been followed up in [96, 97, 98]. Some work has also been carried out to analyze how to take a step beyond these expectation-value-like tunneling times and calculate higher moments, or entire distributions [99, 100, 101].

11.4.2 Potential Experimental Tests

We believe that in order to study these quantum subensembles, laser-cooled atoms offer a unique tool. They can routinely be cooled into the quantum regime, where their de Broglie wavelengths are on the order of microns, and their time evolution takes place in the millisecond regime. They can be directly imaged, and if they are made to impinge on a laser-induced tunnel barrier, transmitted and reflected clouds should be spatially resolvable. With various internal degrees of freedom (hyperfine structure as well as Zeeman sublevels), they offer a great deal of flexibility for studying the various interaction times and nonlocality-related issues. In addition, extensions to dissipative interactions and questions related to irreversible measurements and the quantum–classical boundary are easy to envision [102].

We are working on an atom–optics experiment which will let us directly test these questions. We start with a Bose condensate of Rubidium-87 atoms, at a temperature of about 200 nK. We use a tightly focussed beam of intense light detuned far to the blue of the D2 line to create a dipole-force potential for the atoms [103, 104, 105]. We can readily create repulsive potentials with maxima much higher than the energy of the Rubidium atoms. Acousto-optical modulation of the beam allows us to shape these potentials with nearly total freedom, such that we can have the atoms impinge on a thin plane of repulsive light, whose width would be on the order of the cold atoms' de Broglie wavelength. This is because the beam may be focussed down to a spot several microns across. This focus may be rapidly displaced [106, 107] by using acousto-optic modulators. As the atomic motion is in the millimeter per second range, the atoms respond only to the time-averaged intensity, which can be arranged to have a nearly arbitrary profile.

Ultracold atoms thus provide a unique system in which to study tunneling. By using optical pumping, stimulated Raman transitions, and other such probes, we will be able to go beyond simple wave packet studies to investigate the interactions of tunneling atoms while in the forbidden region itself [108, 109]. This will allow us to study the direct analog of the Larmor times, along with extensions to particles of higher spin. It should also provide an arena in which to study the effect of dissipation and decoherence on quantum measurement, as well as to make measurements at two locations at the same time in order to further investigate the strange features of quantum tunneling [110, 111, 100, 101]. In this way, we hope to shed new light on this fascinating phenomenon, but also on nonlocality in quantum mechanics more generally.

11.5 Conclusion

As we have seen several times now, the problem of time measurement in quantum mechanics is difficult primarily because time measurement is a subtle and indirect business even in the classical world. The fact that some of our familiar intuitions about these indirect measurements should break down in the quantum limit ought not surprise us terribly. In particular, it is observed time and again that tunneling (for instance) may be described by two quite different sorts of timescale. One is related to the variation of transmission phase shifts—common examples are the phase time τ^{Ph} and the τ_y Larmor time—while the other is related to the variation of transmission *probabilities*—the Larmor time τ_z is the standard example of this sort of "interaction time." While certain measurements which appear reliable at the classical level agree with τ_z in the tunneling regime, other equally valid classical measurements tend to reproduce the phase time. Perhaps more interesting is to take the correspondence-principle limit in the other direction. If a Larmor time measurement is carried out on a particle undergoing classical propagation, what do we expect? The spin precesses as usual, and it will be found that $\tau_g = \tau_y$ is the unique timescale for traversal. In this limit, transmission *probability* has no dependence on energy, and therefore τ_z vanishes. This definition of the "interaction time" measures an effect which does not exist at the classical level (related not to the magnitude of the particle's influence on some hypothetical measuring device, but rather to the sensitivity of the particle to back action from the measurement apparatus). It is therefore difficult to argue that such a definition is the appropriate extension of classical concepts to the quantum regime.

Certainly, nothing can be more clear than that there are various useful definitions of timescales for quantum events, which generally seem to be divisible into the two rough categories described above. We should not be hampered by the existence of only a single classical timescale to search perpetually for a *single* "tunneling time" in quantum mechanics. Instead, we must accept that theory can only answer well-posed questions about specific experimental arrangements, and that at least two different timescales (or the real and imaginary parts of a single complex time) are necessary to fully describe the behavior of a tunneling particle. This does not mean that we are forced to abandon any program of seeking universality. There are already strong hints that this complex time is indeed universal, in that quite a few very different experimental schemes for studying traversal times appear to yield either its real or imaginary part, depending simply on which aspect of tunneling is probed. As always, the definition of appropriate observables will have to continue to be informed by the interplay between theory and experiment.

References

1. J.G. Muga, C.R. Leavens: Phys. Rep. **338**, 353 (2000)
2. L.A. MacColl: Phys. Rev. **40**, 621 (1932)
3. E.P. Wigner: Phys. Rev. **98**, 145 (1955)
4. E.H. Hauge, J.A. Stovneng: Rev. Mod. Phys. **61**, 917 (1989)
5. R. Landauer, Th. Martin: Rev. Mod. Phys. **66**, 217 (1994)
6. R.Y. Chiao, P.G. Kwiat, A.M. Steinberg: Physica B **175**, 257 (1991)
7. Th. Martin, R. Landauer: Phys. Rev. A **45**, 2611 (1992)
8. R.Y. Chiao, A.M. Steinberg: in *Progress in Optics* Vol. XXXVII, ed. by E. Wolf (Elsevier, Amsterdam, 1997), pp. 347–406
9. C.K. Carniglia, L. Mandel: Phys. Rev. D **3**, 280 (1971)
10. C.K. Carniglia, L. Mandel: J. Opt. Soc. Am. **61**, 1035 (1971)
11. A. Enders, G. Nimtz: J. Phys. I France **2**, 1693 (1992)
12. J.C. Martinez, E. Polatdemir: Appl. Phys. Lett. **84**, 1320 (2004)
13. C. Spielmann, R. Szipöcs, A. Stingl, F. Krausz: Phys. Rev. Lett. **73**, 2308 (1994)
14. R. Hanbury-Brown, R.Q. Twiss: Proc. Roy. Soc. A (London) **248**, 199 (1958)
15. C.K. Hong, Z.Y. Ou, L. Mandel: Phys. Rev. Lett. **59**, 2044 (1987)
16. A.M. Steinberg, P.G. Kwiat and R.Y. Chiao: Phys. Rev. A **45**, 6659 (1992)
17. J. Jeffers, S.M. Barnett: Phys. Rev. A **47**, 3291 (1993)
18. J.H. Shapiro, K.-X. Sun: J. Opt. Soc. Am. B **11**, 1130 (1994)
19. D. Bohm: *Quantum Theory* (Prentice Hall, Englewood Cliffs, NJ, 1951)
20. C.C. Real, J.G. Muga, S. Brouard: Am J. Phys. **65**, 157 (1997)
21. J.G. Muga, J.P. Palao, R. Sala: Phys. Lett. A **238**, 90 (1998)
22. Y. Aharonov, S. Popescu, D. Rohrlich, L. Vaidman: Phys. Rev. A **48**, 4084 (1993)
23. E.U. Condon: Rev. Mod. Phys. **3**, 43 (1931)
24. L. Eisenbud: Ph.D. Thesis, Princeton University (1948)
25. C.R. Leavens, G.C. Aers: Phys. Rev. B **40**, 5387 (1989)
26. L. Brillouin: *Wave Propagation and Group Velocity* (Academic Press, New York, 1960)
27. C.G.B. Garrett, D.E. McCumber: Phys. Rev. A **1**, 305 (1970)
28. S. Chu, S. Wong: Phys. Rev. Lett. **48**, 738 (1982)
29. L.J. Wang, A. Kuzmich, A. Dogariu: Nature **406**, 277 (2000)
30. R.Y. Chiao: Phys. Rev. A. **48**, R34 (1993)
31. A.M. Steinberg, R.Y. Chiao: Phys. Rev. A **49**, 2071 (1994)
32. M.D. Stenner, D.J. Gauthier, M.A. Neifeld: Nature **425**, 695 (2003)
33. G.M. Gehring, A. Schweinsberg, C. Barsi, N. Kostinski, R.W. Boyd: Science **312**, 895 (2006)

34. O.F. Siddiqui, S.J. Erickson, G.V. Eleftheriades, M. Mojahedi: IEEE Trans. Microwave Theory Tech. **52**, 1449 (2004)
35. P. Tournois: IEEE J. Quant. Electron. **33**, 519 (1997)
36. P. Tournois: Opt. Lett. **30**, 815 (2005)
37. K.J. Resch, J.S. Lundeen, A.M. Steinberg: IEEE J. Quant. Electron. **37**, 794 (2001)
38. R.W. Boyd, D.J. Gauthier: Prog. Opt. **43**, 497, E. Wolf ed., Elsevier (2002)
39. T.E. Hartman: J. App. Phys. **33**, 3427 (1962)
40. A.M. Steinberg, P.G. Kwiat, R.Y. Chiao: Phys. Rev. Lett. **71**, 708 (1993)
41. A. Ranfagni, D. Mugnai, P. Fabeni, G.P. Pazzi: Appl. Phys. Lett. **58**, 774 (1991)
42. A. Enders, G. Nimtz: J. Phys. I France **3**, 1089 (1993)
43. A. Ranfagni, P. Fabeni, G.P. Pazzi, D. Mugnai: Phys. Rev. E **48**, 1453 (1993)
44. A. Ranfagni, D. Mugnai: Phys. Rev. E **54**, 5692 (1996)
45. D. Mugnai, A. Ranfagni, R. Ruggeri: Phys. Rev. Lett. **84**, 4830 (2000)
46. M. Büttiker, R. Landauer: Phys. Rev. Lett. **49**, 1739 (1982)
47. M.W. Mitchell, R.Y. Chiao: Phys. Lett. A **230**, 133 (1997)
48. Y.Japha, G. Kurizki: Phys. Rev. A **53**, 586 (1996)
49. G. Diener: Phys. Lett. A **223**, 327 (1996)
50. A.M. Steinberg, P.G. Kwiat, R.Y. Chiao: in *Perspectives in Neutrinos, Atomic Physics and Gravitation, XXVIIIe Rencontre de Moriond*, ed. by J. Trân Thanh Vân et al. (Editions Frontières, Gif-sur-Yvette, France, 1993)
51. A.M. Steinberg: La Recherche **281**, 46 (11/1995)
52. G. Nimtz, W. Heitmann: Prog. Quant. Electron. **21**, 81 (1997)
53. G. Diener: Phys. Lett. A **235**, 118 (1997)
54. J. Peatross, S.A. Glasgow, M. Ware: Phys. Rev. Lett. **84**, 2370 (2000)
55. M. Ware, S.A. Glasgow, J. Peatross: Opt. Exp. **9**, 519 (2001)
56. C.A.A. de Carvalho, H.M. Nussenzveig: Phys. Rev. **364**, 83 (2002)
57. P.C.W. Davies: Am. J. Phys. **73**, 23 (2005)
58. H.G. Winful: Phys. Rep. **436**, 1 (2006)
59. E.H. Hauge, J.P. Falck, T.A. Fjeldly: Phys. Rev. B **36**, 4203 (1987)
60. D. Sokolovski, L.M. Baskin: Phys. Rev. A **36**, 4604 (1987)
61. M. Büttiker: 'Traversal, Reflection and Dwell Time for Quantum Tunneling', in *Electronic Properties of Multilayers and low Dimensional Semiconductors*, ed. by J. M. Chamberlain, L. Eaves, J. C. Portal (Plenum, New York, 1990), pp. 297–315
62. D. Sokolovski, J.N.L. Connor: Phys. Rev. A **42**, 6512 (1990)
63. V.S. Olkhovsky, E. Recami: Phys. Rep. **214**, 339 (1992)
64. D. Sokolovski, J.N.L. Connor: Phys. Rev. A **47**, 4677 (1993)
65. M. Büttiker, R. Landauer: Phys. Scr. **32**, (1985)
66. J.A. Stovneng, E.H. Hauge: J. Stat. Phys. **57**, 841 (1989)
67. A.I. Baz': Sov. J. Nucl. Phys. **5**, 161 (1967)
68. V.F. Rybachenko: Sov. J. Nucl. Phys. **5**, 635 (1967)
69. H.A. Fertig: Phys. Rev. Lett. **65**, 2321 (1990)
70. H.A. Fertig: Phys. Rev. B **47**, 1346 (1993)
71. P. Hänggi: in *Lectures on Path Integration*, ed. by H.A. Cerdeira et al. (World Scientific, Singapore, 1993)
72. E. Pollak, W.H. Miller: Phys. Rev. Lett. **53**, 115 (1984)
73. J.P. Falck, E.H. Hauge: Phys. Rev. B **38**, 3287 (1988)

74. R. Landauer: Nature **341**, 567 (1989)
75. D. Esteve, J.M. Martinis, C. Urbina, E. Turlot, M.H. Devoret: Phys. Scr. **T29**, 121 (1989)
76. M. Deutsch, J.E. Golub: Phys. Rev. A **53**, 434 (1996)
77. Ph. Balcou, L. Dutriaux: Phys. Rev. Lett. **78**, 851 (1997)
78. A. Ghatak, S. Banerjee: Appl. Opt. **28**, 1960 (1989)
79. Y. Aharonov, D.Z. Albert, L. Vaidman: Phys. Rev. Lett. **60**, 1351 (1988)
80. Y. Aharonov, L. Vaidman: Phys. Rev. A **41**, 11 (1990)
81. S. Brouard, R. Sala, J.G. Muga: Phys. Rev. A **49**, 4312 (1994)
82. J.G. Muga, S. Brouard, R. Sala: Phys. Lett. A **167**, 24 (1992)
83. B.A. van Tiggelen, A. Tip, A. Lagendijk: J. Phys. A **26**, 1731 (1993)
84. S. Brouard, R. Sala, J.G. Muga: Europhys. Lett. **22**, 159 (1993)
85. C.R. Leavens: Found. Phys. **25**, 229 (1995)
86. C. Dewdney, B.J. Hiley: Found. Phys. **12**, 27 (1982)
87. C.R. Leavens: Solid State Commun. **74**, 923 (1990)
88. C.R. Leavens, G.C. Aers, Solid State Commun. **78**, 1015 (1991)
89. C.R. Leavens, G. C. Aers: in *Scanning Tunneling Microscopy III*, ed. by R. Wiesendanger and H.-J. Güntherodt (Springer-Verlag, Berlin, 1993)
90. C.R. Leavens: Phys. Lett. A **178**, 27 (1993)
91. C.R. Leavens, G. Iannaccone, W.R. McKinnon: Phys. Lett. A **208**, 17 (1995)
92. C.R. Leavens, W.R. McKinnon: Phys. Rev. A **51**, 2748 (1995)
93. D. Sokolovski, J.N.L. Connor: Solid State Commun. **89**, 475 (1994)
94. A.M. Steinberg: Phys. Rev. Lett. **74**, 2405 (1995)
95. A.M. Steinberg: Phys. Rev. A **52**, 32 (1995)
96. Y. Aharonov, N. Erez, B. Reznik: Phys. Rev. A **65**, 052124 (2002)
97. Y. Aharonov, N. Erez, B. Reznik: J. Mod. Opt. **50**, 1139 (2003)
98. D. Sokolovski, A.Z. Msezane, V.R. Shaginyan: Phys. Rev. A **71**, 064103 (2005)
99. G. Iannaccone: in *Adriatico Research Conference on Tunnelling and its Implications* (ICTP, Trieste, 1996); preprint quant-ph/9611018
100. K.J. Resch, A.M. Steinberg: Phys. Rev. Lett. **92**, 130402 (2004)
101. K.J. Resch, J.S. Lundeen, A.M. Steinberg: Phys. Lett. A **324**, 125 (2004)
102. A.M. Steinberg: Superlattices Microstruct. **23**, 823 (1998)
103. S.L. Rolston, C. Gerz, K. Helmerson, P.S. Jessen, P.D. Lett, W.D. Phillips, R.J. Spreeuw, C.I. Westbrook: 'Trapping atoms with optical potentials'. In: Proc. *SPIE, Vol. 1726, Shanghai International Symposium on Quantum Optics*, ed. by D.-H. Wang, Z. Wang, (1992), pp. 205–211
104. J.D. Miller, R.A. Cline, D.J. Heinzen: Phys. Rev. A **47**, R4567 (1993)
105. N. Davidson, H.J. Lee, C.S. Adams, M. Kasevich, S. Chu: Phys. Rev. Lett. **74**, 1311 (1995)
106. A. Steinberg, R. Thompson, V. Bagnoud, K. Helmerson, W. Phillips: in *15th International Conference on Atomic Physics*, abstract booklet, University of Amsterdam, The Netherlands (1996)
107. P. Rudy, R. Ejnisman, A. Rahman, S. Lett, N.P. Bigelow: 1997 OSA Tech. Dig. **12**, 67 (1997)
108. A.M. Steinberg: J. Kor. Phys. Soc. **35**, 122 (1999)
109. D. Boosé, F. Bardou: Europhys. Lett. **53**, 1 (2001)
110. A.M. Steinberg, S. Myrskog, H.S. Moon, H.A. Kim, J. Fox, J.B. Kim, Ann. Phys. (Leipzig) **7**, 593 (1998)
111. A.M. Steinberg: in *Causality and Locality in Modern Physics*, ed. by G. Hunter et al. (Kluwer, Dordrecht, 1998), p. 431

12

Microwave Experiments on Tunneling Time

Daniela Mugnai and Anedio Ranfagni

"Nello Carrara" Institute of Applied Physics, CNR Florence Research Area,
Via Madonna del Piano 10, 50019 Sesto Fiorentino, Italy

12.1 An Overview of Theoretical Models of Tunneling in the Electromagnetic Framework

The tunneling process represents one of the most nonclassical predictions in quantum mechanics. Within this process, one of the questions most debated is that of the "traversal time," that is, the time spent during the passage through a barrier. The question of how much time tunneling takes is certainly not new, but it is one that has not yet been resolved.

As is known, considerable difficulties are encountered when performing a direct measurement of the tunneling time due to the very short time involved: its scale is given by the ratio between a few wavelengths of the wave function and a velocity comparable with that of the light in vacuum. With a solid-state device the tunneling time involved may be, typically, of the order of femtoseconds. In Josephson junctions, this time seems to increase up to the order of 10^2 ps. However, only a few experimental results are currently available; further investigations would, therefore, be worthwhile, also in connection with novel theoretical aspects.

Tunneling also occurs in optics, so it is possible to establish a close analogy between particle motion and electromagnetic wave propagation, which can be of help in understanding this intriguing problem. The implications of this topic, which are ultimately related to the questions connected with particle–wave dualism, have recently been connected also to relativistic problems, owing to the possibility of observing superluminal motions. In other words, the question of whether quantum tunneling is faster than light has now been brought to the fore [1, 2, 3].

For optical tunneling in the visible region, the magnitude of the tunneling time is still of the order of femtoseconds, but a decisive increase in this time can be obtained by increasing the wavelength up to microwaves. In this way, the timescale is magnified up to nanoseconds, and measurements can easily be performed [4].

D. Mugnai and A. Ranfagni: *Microwave Experiments on Tunneling Time*, Lect. Notes Phys.
734, 355–397 (2008)
DOI 10.1007/978-3-540-73473-4_12   © Springer-Verlag Berlin Heidelberg 2008

An experimental device suitable for simulating quantum tunneling consists of an undersized waveguide in which evanescent modes take place for frequencies below the cutoff one (see Sect. 12.2). By analogy with the quantum tunneling of a particle through a barrier, the waveguide can be regarded as a one-dimensional barrier for electromagnetic waves.

The analogy, however, goes beyond what is outlined here, since quantum tunneling and electromagnetic waves are described by closely related wave equations. In fact, the time-independent Schrödinger equation for the motion of a particle of energy E in a potential V_0 is formally identical to the Helmholtz equation for the propagation of a scalar field (electric or magnetic component of the wave). The only difference lies in the dispersive relation, which reflects the different time dependence in the Schrödinger equation with respect to the d'Alembert equation. Once the dispersion relations are properly taken into account, the results of quantum mechanics can be adopted for waveguides provided that the substitution $(\hbar/m) \rightarrow (c^2/2\pi\nu)$ is made (m is the mass of the particle, ν is the frequency of the wave, and c is the light velocity). Even if the two wave equations describe the evolution of two quite different quantities—the quantum wave function and the electromagnetic field—this does not prevent a test being made of quantum-mechanical models, which are based on the evolution of wave packets, provided that the above substitution is made. There is, however, a limitation in this analogy since in contrast to tunneling particle, an electromagnetic pulse consists of many photons and can be probed in a noninvasive way.

The fact that the results of such a simulation are best described by quantum-mechanical models, suitably translated into a classical electromagnetic framework, constitutes the proof that quantum tunneling can actually be simulated by these kinds of experiments. A model we are concerned with is the phase time, that is, the group delay as calculated by the stationary phase method. The predictions of this procedure can, within certain limits, be paradoxically small, implying a barrier-traversal velocity greater than the speed of light c (Hartman effect)[5]. However, by working in the time domain, that is, by measuring directly the delay time, it is rather difficult to perform measurements sufficiently below the cutoff frequency because of the severe limitations caused by the attenuation of the signal. Thus, with this kind of device it was not possible to obtain clear experimental evidence of the existence of superluminal behavior. On the contrary, by performing measurements of the phase shift as a function of frequency, it is possible to reach regions appreciably below the cutoff; the tunneling time can then be derived by means of a Fourier-transform analysis. The results obtained confirm that the effective group velocity for evanescent waves can exceed c [6].

Dispersion is always present, in the case of a waveguide, making the signal analysis more complicated, especially in tunneling situations where the forerunners strongly influence the time of arrival (as explained in Sect. 12.1.3).

We wish to recall that both in the presence and in the absence of dispersion, the upper limit of the signal velocity in classically allowed motions is represented by the speed of light, as demonstrated for wave propagation in dispersive media [7]. However, such arguments do not appear clearly applicable to classically forbidden situations such as tunneling processes or, more generally, in the presence of evanescent waves (see below).

Tunneling time in superluminal cases could be considered as a practical case of a weak value observable within the framework of the weak measurement theory, where mean values which would be strictly forbidden for any complete ensemble can be obtained for a small subensemble [8].

In dealing with the microwave simulation of tunneling, a theoretical interpretation was modeled on the basis of a path-integral solution of the telegrapher's equation, analytically continued to imaginary time [9]. The salient features of this model can be summarized as follows. It is known that, in the absence of dissipation, the semiclassical delay time is simply given by $\tau_s = L/|v_{gr}|$, where L is the length of the waveguide and v_{gr} is the group velocity, which for the TE_{01} mode is $v_{gr} = c\sqrt{1 - \omega_0^2/\omega^2}$, $\omega = 2\pi\nu$ the angular frequency and $\omega_0 = 2\pi\nu_0$ the cutoff angular frequency. By taking into account the fact that dissipation produces a shift in the cutoff from $\nu_0 = c/2b$ (b is the width of the rectangular waveguide) to an effective cutoff $\tilde{\nu}_0 = (\nu_0^2 + \delta^2)^{1/2}$, where $\delta = \tilde{a}/\lambda$ ($\tilde{a}$ is related to the dissipative parameter a entering the telegrapher's equation and λ the free-space wavelength) (Fig. 12.1). The question is posed as to what extent model of this kind can account for the experimental results. What clearly emerges from the comparison of the experimental results obtained with microwave simulation (see Sect. 2.1) with the existing theoretical models is the good agreement with the corresponding theoretical curves as deduced from quantum–mechanical models [10]. In particular, delay-time data are in agreement with the phase-time model $\tau_\phi = \partial\Delta\phi/\partial\omega$, (which represents the real part of the delay), while data for τ_z (which represents the imaginary part) are in agreement with the relative theoretical curve $\tau_z = \partial(\ln T^{1/2})/\partial\omega$, the complex transmission amplitude of the barrier being $T^{1/2}\exp(i\Delta\phi)$ [11].

As for the prediction of the above modified semiclassical model-based, as said, on path-integral solutions of the telegrapher's equation—the relative curve of the delay vs frequency appears to be in rather good agreement with the absolute value of the complex traversal time $(\tau_\phi^2 + \tau_z^2)^{1/2}$, rather than with each (real or imaginary) component. This is nothing but the Büttiker model based on a Larmor clock [12] in agreement with the more sophisticated path-integral treatments of the tunneling time [13, 14]. Noteworthy is the extreme brevity of τ_ϕ (the real part of the delay), throughout below the cutoff ν_0, which is confirmed by the experimental results. Superluminal effects can be observed below a given frequency ν_c where the curve of τ_ϕ crosses the constant value L/c (see the inset in Fig. 12.1). Below $\tilde{\nu}_0$, the Büttiker model is practically coincident with τ_z (the imaginary part of the delay) as well as with the modified semiclassical model. So, on this basis, we can conclude that the path-integral treatment of the telegrapher's equation makes the semiclassical model a plausible one for a qualitative interpretation of the tunneling time.

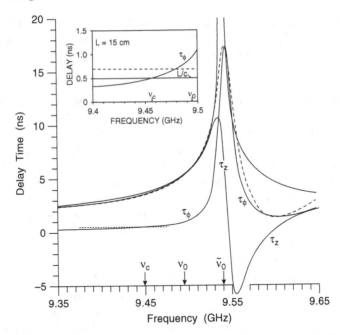

Fig. 12.1. Curves of delay time vs frequency for a length of the undersized waveguide $L = 15$ cm, relative to different models. The *heavy line* represents the modified semiclassical model resulting from the telegrapher's equation with $\tilde{a} = 0.1$ c. This implies a shift in the cutoff frequency from $\nu_0 = 9.494$ GHz to $\tilde{\nu}_0 \simeq 9.54$ GHz. The *dashed line* represents the Büttiker model. In the inset the limit of superluminal behavior is shown by *continuous line*; the *dashed line* represents the delay in a waveguide of the same length with no narrowing

Within this framework, it was shown that in tunneling processes the effective velocity turns out to be increased by an (imaginary) dissipation and, well below the cutoff frequency, can actually overcome the light velocity c. This conclusion was also confirmed by an extension of the analysis made in order to compare the traversal time results, relative to a beat-envelope signal, with the ones deduced from the distribution function of the "randomized time" and its analytical continuation in imaginary time. What clearly emerges from this analysis is the inverted role of the dissipative parameter, which in tunneling acts as an "accelerator" of the motion [15, 16].

This apparent paradox can be explained by considering the fact that, when we say dissipation in tunneling, we are not dealing with a true dissipation (as when we consider the wave equation in allowed processes) but rather with an imaginary quantity that was introduced in order to obtain the analytical continuation of the wave equation.[1] By following the same procedure adopted

---

[1] A true dissipation, introduced ad hoc in a tunneling experiment, produces just the opposite effect, that is, a slowing down of the motion [17].

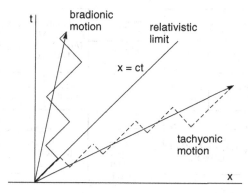

Fig. 12.2. Trajectory of a stochastic motion for a classically allowed case (*continuous line*) and a forbidden—or tunneling—case (*dashed line*). In the first case, we have a reversal of the motion in the x space (bradionic motion), while in the second case the reversal happens in the t (imaginary) time (tachyonic motion)

by Kac [18] in order to individuate a stochastic model related to the telegrapher's equation, a stochastic model related to tunneling processes was found [19]. The basic assumption for this model is the interchange of the space x with the time t (Fig. 12.2); in this way, we obtain a new equation of motion suitable for describing superluminal behavior. In addition, a close correspondence between quantum relativistic motion and wave propagation in the presence of dissipation is established. In particular, it is shown that the telegrapher's equation is connected to the Klein–Gordon equation of a bradyonic type when the transformation [20, 21]

$$\varphi(x,t) = u(x,t)\exp\left(-\frac{imc^2t}{\hbar}\right) \tag{12.1}$$

is made, where $\varphi(x,t)$ is the solution of the Klein–Gordon equation and $u(x,t)$ is the solution of the telegrapher's equation. Analogously, a connection with the Klein–Gordon equation of a tachyonic type can be established when the transformation is

$$\varphi(x,t) = u(x,t)\exp\left(\mp\frac{mc^2t}{\hbar}\right) . \tag{12.2}$$

A schematic representation of these connections is shown in Fig. 12.3. The attenuation of the wave function due to the exponential function in (12.2) represents an important practical limitation for the observation of superluminal behaviors (for further details about the stochastic model see Sect. 1.2).

Let us now consider the problem of the signal velocity analysis. According to Fox, Kuper, and Lipson [22], the front edge of a superluminal wave packet will never exceed light speed, even if the group velocity is greater than c. Similar conclusions have also been drawn in more recent works [23, 24]. Therefore,

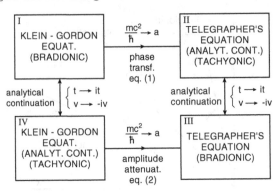

Fig. 12.3. Block diagram showing the connection between quantum relativistic and telegrapher's equations

even if it is well established that the upper limit of a signal velocity is represented by the light speed in vacuum, it is not fully understood whether these conclusions also hold true for tunneling processes and/or propagation in the presence of evanescent waves. By following Brillouin [7], the propagation of a pulse in the x-direction can be described by a contour integral in the complex plane of ω [19, 25]

$$\psi(x,t) = \frac{1}{2\pi}\mathrm{Re}\int_\gamma \frac{\exp[-i(\omega t - kx)]}{\omega - \omega_i}\, d\omega \qquad (12.3)$$

where ω_i is the frequency of the incoming signal. The wave number for a waveguide is $k = \sqrt{\omega^2 - \omega_0^2}/c$. For $\omega \longrightarrow \infty$, we have $k \longrightarrow \omega/c$, and the exponential function in the integral becomes $\exp[-i\omega(t - x/c)]$. By deforming the contour of the integration into an infinite semicircle, this implies that the integral is zero for $t < x/c$ and that the first forerunner of the signal cannot arrive before a time given by $t_0 = x/c$. In the case of evanescent waves, the exponential function becomes $\exp(-i\omega t - \kappa x)$ where $\kappa = \sqrt{\omega_0^2 - \omega^2}/c$. The integral which gives the signal is again zero if the domain of integration extends to infinity. However, if we limit this range (the range in which we have evanescent waves), the integral is different from zero. We wish to point out that, because of the dependence of the result on $\omega_0 t_0 = 2\pi x/\lambda_0$, the contribution for $t/t_0 < 1$ is strongly attenuated by increasing the distance x. Therefore, for sufficiently large distances—say of a few cutoff wavelenghts λ_0—the superluminal contribution becomes quite negligible, and we again obtain the usual result that nothing arrives before t_0 (Fig. 12.4) [19, 26].

Thus, by limiting the range of integration in the ω domain, that is, by considering the finite spectral extension of the pulse, we can actually obtain that "something" arrives before the usual forerunner for $0 \le t \le x/c$ and for short distances. We are aware that a finite spectral extension does not represent a "true signal" which, on the contrary, requires an infinite spectrum. By limiting the spectral extension, the signal is profoundly modified, since

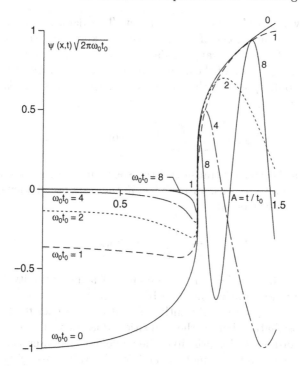

Fig. 12.4. Behavior of the signal shape $\psi(x,t)$ as a function of the normalized time $A = t/t_0 (t_0 = x/c)$ for ω_i less than but comparable to the cutoff (angular) frequency ω_0. Note that the contribution for $A \leq 1$ strongly depends on the value $\omega_0 t_0$, that is, on the distance x from the launcher, and tends to disappear when x is of the order of a few wavelengths

its spatial and temporal extension (initially supposed to be finite) becomes infinite. In this way $\psi(x,t)$ does not represent a true signal and, although its profile is traveling with a group velocity $v_{gr} > c$, there is no contradiction with relativity. In the case of tunneling, the choice of limiting the frequency domain is supported by the fact that evanescent waves have a finite spectral extension. We wish to note, however, that any practical signal necessarily has a finite spectral extension. The relative delay time, sometimes referred to as technical signal or technical information delay [27], is not necessarily coincident with the front edge delay which requires a consideration of the limit $\omega \longrightarrow \infty$. This is a delicate point since infinite frequencies, as well as infinite time duration, do not exist in physical phenomena. However, in practical applications, such as telecommunications, and radar, the propagation velocity involved is beyond dispute. The case of tunneling for the role played by evanescent waves is a different matter. These waves are not properly propagating; however, there is no doubt that in the experiments to which we refer "something" is propagated, even if the spectral extension of the "signal" is well confined within the ω-domain of evanescent waves.

Let us try to focalize better this point. The deformation induced in a signal due to the finiteness of the spectrum is known as the Gibb's effect, which consists of the appearance of damped oscillations before and after the transition. The extension of the oscillations depends on the cutting frequency $\bar{\omega}$, and tends to disappear as the cutting frequency tends towards infinity. As formulas, we have

$$f(t) = \frac{1}{2\pi} \int_{-\infty}^{\infty} F(\omega) \exp(i\omega t) \, d\omega \,, \tag{12.4}$$

where $F(\omega)$ is the Fourier transform (the spectrum) of $f(t)$. In the presence of a frequency limitation (as, for instance, in the response of a low-pass filter), the pulse can be written as

$$f_{\bar{\omega}}(t) = \frac{1}{2\pi} \int_{-\bar{\omega}}^{\bar{\omega}} F(\omega) \exp(i\omega t) \, d\omega \,, \tag{12.5}$$

and shows the Gibb's effect in the presence of a discontinuity (the signal). The behavior is well known and can be experimentally verified especially if the output of the signal is sufficiently delayed with respect to the input one, so that the oscillation before the transition does not violate the causality principle. According to this principle, the value of the answer to an excitation $f(t)$ to a given instant t cannot depend on the value that $f(t)$ takes after t. What does happen if this principle, according to the prediction of (12.5), is not satisfied? Simply, the initial "acausal" oscillation is not observed, as shown in Fig. 12.5.

This means that the pulses used in microwave tunneling simulations (pulse duration ~ 1 μs, pulse separation ~ 1 ms) for measuring delays of the order of nanoseconds should be considered as causal signals (even if, because of the frequency limitation, they do not have the features of a perfect signal). Thus, by tending the pulse separation to infinity (and, therefore, the repetition frequency to zero), the behavior should remain the same and superluminal effects in the group velocity should continue to be observed. In fact, a signal can be defined as a pulse which supplies information. In order to have this condition it is necessary to have a wave packet (or pulse) which must be also a "single event".

An effective experimental check of this point has been performed by confirming the superluminal behavior also for single microwave pulses [28]. In [28], the signal velocity is identified with that of the first forerunner which, by following Sommerfeld, cannot overcame the light speed in vacuum. However, as said before, a different conclusion can be reached by limiting the spectral extension. Moreover, a different analysis of the forerunner (based on the Steven's procedure) showed results very similar to those related to the phase-time model which predicts superluminal behavior (see Sect. 12.1.3). An estimate of the timescale of the effect can be performed by considering that the exponent in (12.2), that is, $mc^2 t/\hbar$, has to be of the order of some units

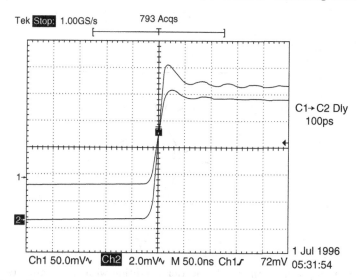

Fig. 12.5. Experimental observation of the Gibb's effect in delay-time measurements. The initial "acausal" oscillation, before the transition, is not observed while it appears only the one after the transition

in order to have a nonnegligible amplitude of $\varphi(x,t)$. For relativistic electrons with $mc^2 \simeq 0.5$ MeV, we have a timescale of the order of 10^{-21} s. However, this quantity is replaced by $\omega_0 t$ in the optical analogy, and the timescale becomes more accessible. So, for microwave experiments with $\omega_0 \approx 10^{10}$ s^{-1}, the resulting time delay is in the range of nanoseconds, while for photon tunneling experiments $\hbar\omega_0$ is of the order of few electronvolts and the time delay is of the order of femtoseconds. Both these timescales have been experimentally confirmed [6, 11, 29, 30, 31].

Another kind of system in which optical tunneling takes place consists of a grating followed by a paraffin prism: if the period of the grating is lower than the wavelength of the incident wave, all the diffracted waves, except the zero-order one, are evanescent ones which are transformed into ordinary waves by refraction on the paraffin prism (see Sect. 12.3).

A third kind of system which can be utilized for studying optical tunneling consists of two prisms (with a refractive index greater than the refractive index of air) facing each other and separated by an air gap. When the total reflection takes place in the first prism, evanescent waves originate in the gap (see Sect. 12.4). For this kind of experiment, a pure electromagnetic model can be formulated.

12.1.1 Frustrated Total Reflection: An Electromagnetic Model

In order to analyze the experimental situation mentioned above, let us consider a system formed by two half-spaces (two prisms in the experimental

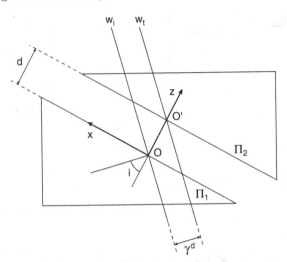

Fig. 12.6. Two prisms in the total reflection condition with the coordinate system adopted in the theoretical analysis

setup), limited by plane-parallel boundaries, filled with a homogeneous and nondispersive medium of refractive index n, and separated by a vacuum gap, as shown in Fig. 12.6.

A plane electromagnetic wave impinges on the gap with an incidence angle larger than the limit angle $i_0 = \sin^{-1}(1/n)$. By using standard methods, it is possible to evaluate the transmitted field $\mathbf{E}^t$ (and $\mathbf{H}^t$ too) and its deformation with respect to the impinging wave. By comparing them, it is possible to evaluate the time taken by the wave to "travel" through the gap.

By choosing an $\mathbf{i}$, $\mathbf{j}$, $\mathbf{k}$ (coordinates x, y, z) Cartesian reference system, as shown in Fig. 12.6, from the boundary conditions (namely, the continuity conditions of the tangential components of the fields at the surfaces of discontinuity of the refractive index), we have

- In the half-space $z < 0$, the field is the superposition of two waves, the impinging wave $\mathbf{E}^i$, $\mathbf{H}^i$ and a reflected wave $\mathbf{E}^r$, $\mathbf{H}^r$.
- In the gap, the field is the superposition of two evanescent waves, one attenuated in the positive direction of $\mathbf{k}$ and the other attenuated in the reverse direction $-\mathbf{k}$. The former will be denoted as "progressive" wave $\mathbf{E}^+$, $\mathbf{H}^+$, the latter as "regressive" wave $\mathbf{E}^-$, $\mathbf{H}^-$.
- In the half-space $z > d$, the field is composed of a single transmitted wave, $\mathbf{E}^t$, $\mathbf{H}^t$, propagating in the same direction as the impinging wave.

In the case of a monochromatic impinging wave, the incident E_y^i, reflected E_y^r, transmitted E_y^t, progressive E_y^+ and regressive E_y^- fields can be written as [32]

$$E_y^i = E_0 \exp\left[i\frac{\omega}{c}n(\alpha x + \gamma z)\right]\exp(-i\omega t) \tag{12.6}$$

$$E_y^r = E_0\,\rho\,\exp\left[i\frac{\omega}{c}n(\alpha x - \gamma z)\right]\exp(-i\omega t) \tag{12.7}$$

$$E_y^t = E_0\,\tau\,\exp\left[i\frac{\omega}{c}n(\alpha x + \gamma(z-d))\right]\exp(-i\omega t) \tag{12.8}$$

$$E_y^+ = E_0\,p\,\exp\left[i\frac{\omega}{c}(n\alpha x + i\Gamma z)\right]\exp(-i\omega t) \tag{12.9}$$

$$E_y^- = E_0\,r\,\exp\left[i\frac{\omega}{c}(n\alpha x - i\Gamma z)\right]\exp(-i\omega t) \tag{12.10}$$

where ρ denotes the amplitude reflection coefficient and τ the transmission coefficient of the gap. From the continuity conditions for the tangential component of the electric fields across Π_1 and Π_2, we can derive the (complex) transmission coefficient

$$\tau(\omega) = \frac{4in\gamma\Gamma}{e_1(\Gamma + in\gamma)^2 - e_2(\Gamma - in\gamma)^2} \tag{12.11}$$

where

$$\gamma = \sqrt{1 - \alpha^2}, \quad \Gamma = \sqrt{n^2 - 1 - n^2\gamma^2} \quad e_1 = \exp(-\omega\Gamma d/c) \quad e_2 = 1/e_1 \tag{12.12}$$

It appears from (12.11) that the gap behaves like a low-pass filter, since for $|\omega| \to \infty$, $\tau \to 0$.

The expression $\tau(\omega)$ allows us to evaluate the traversal time of the gap. We consider the impinging wave front w_i through O (at $x = 0$, $z = 0$) and the transmitted wave front w_t passing through O' ($x = 0$ and $z = d$), as in Fig. 12.6. According to (12.7) and (12.8), the phase ϕ_i of the former is $\phi_i = -\omega t$; that of the latter is $\phi_t = \arg(\tau) - \omega t$. It follows that a given value of the phase is at O at time $t = 0$, and at O' at time $\arg(\tau)/\omega$.

The conclusion is that the phase takes a time

$$t_{\mathrm{ph}} = \arg(\tau)/\omega \tag{12.13}$$

in passing from O to O', that is, to travel distance γd (note that even for $\omega = 0$ the phase delay has a finite value, $t_{\mathrm{ph}} = \arctan[(n^2\gamma^2 - \Gamma^2)d/2n\gamma c]$).

For nonmonochromatic waves, the group delay is

$$t_{\mathrm{gr}} = t_{\mathrm{ph}} + \nu\frac{\partial t_{\mathrm{ph}}}{\partial\nu} = \frac{\partial}{\partial\omega}[\arg(\tau(\omega))]\,. \tag{12.14}$$

From (12.11) we have

$$\arg(\tau(\omega)) = \arctan\left[\frac{n^2\gamma^2 - \Gamma^2}{2n\Gamma\gamma}\tanh\left(\frac{\omega\Gamma d}{c}\right)\right], \tag{12.15}$$

hence t_{gr} can be obtained as

$$t_{\text{gr}} = \frac{2n\Gamma\gamma\, t_d\,(n^2\gamma^2 - \Gamma^2)}{(2n\Gamma\gamma)^2 \cosh^2(t_d\omega) + (n^2\gamma^2 - \Gamma^2)^2 \sinh^2(t_d\omega)}, \qquad t_d = \frac{\Gamma d}{c}.$$

$$(12.16)$$

In Fig. 12.7 we show results of t_{ph} as a function of Γ for some values of the frequency ν ($\omega = 2\pi\nu$), for $d = 2$ cm and $n = 1.49$.

Results of t_{gr}, as a function of Γ, are reported in Fig. 12.8 for the same parameter values as in Fig. 12.7.

The results obtained show a clear superluminal behavior both in phase and group delays. This behavior is dependent on the frequency, as well as on Γ: in particular, for $\Gamma = n\gamma$, that is, for $\gamma^2 = (n^2 - 1)/2\,n^2$, the phase and group delays go to zero. For higher values of Γ, they became negative.

12.1.2 More About the Stochastic Model

The stochastic motion of a particle moving along a line with constant velocity v and suffering collision that can reverse its motion is described by an equation of motion equal to the telegrapher's one [18]. In the absence of dissipation, the said motion does not suffer reversal, and the delay time for a displacement L is given by L/v. On the contrary, in the presence of dissipation, the time is described by a randomized quantity, the distribution $g(r, t)$ of which is a two-variable time function. The first time, r, is a fictitious time that describes the delay which a particle would take if the motion was without reversals.

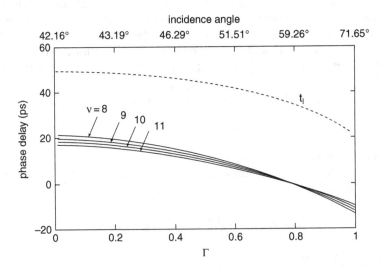

Fig. 12.7. Phase delay (*continuous line*) between the wavefronts w_i and w_t, as deduced from (12.13), as a function of Γ (or of incidence angle) for $n = 1.49, d = 2$ cm and for some values of the frequency ν ranging from 8 GHz (*upper curve*) to 11 GHz (lower curve). The *dashed line* represents the time t_l of a wave traveling distance γd at the light speed

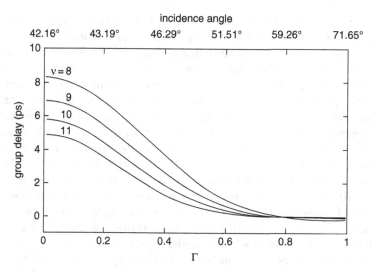

Fig. 12.8. Group delay, as deduced from (12.16), for the same parameter values as in Fig. 12.7

The second time, t, is the true time, that is, the delay that a particle suffering reversals spends in going from one given point to another.

By using the time distribution $g(r,t)$, we can derive the average time $\bar{r}$, which results in [15]

$$\bar{r}(t) = \frac{1}{2a}\left(1 - e^{-2at}\right) , \tag{12.17}$$

where a is the friction coefficient that enters the telegrapher's equation. For classically forbidden processes, the average time t can be evaluated as

$$\langle t \rangle = \frac{\int_{-\infty}^{\infty} it\, g(ir, it)\, d(it)}{\int_{-\infty}^{\infty} g(ir, it)\, d(it)} \tag{12.18}$$

where $g(ir, it)$ is the distribution function of a stochastic process analytically continued to the imaginary time [33]. When developed, (12.18) produces the following result:

$$\langle t \rangle = \frac{r[2\sin ar - ar] + ir[2\cos ar - 1 + 2a^2 r^2/3]}{(2\cos ar - 1) + 2i(ar - \sin ar)}$$
$$\equiv \operatorname{Re}\langle t \rangle + i\operatorname{Im}\langle t \rangle , \tag{12.19}$$

where $r = L/v$ is the semiclassical time and v is the velocity in the forbidden region (barrier). Thus, the average tunneling time is a complex quantity which, for L/v tending to zero, can be expressed as

$$\langle t \rangle \simeq a\left(\frac{L}{v}\right)^2 + i\frac{L}{v} \tag{12.20}$$

where L/v represents the imaginary part and parameter a is responsible for the real part, which is zero if $a = 0$ and depends quadratically on L/v. However, if a is zero, must we believe that a real part of the tunneling time does not exist? This is certainly not the case, since any dephasing through the boundaries of the barrier can contribute to real time (phase-time model). An alternative approach for evaluating this duration is represented by Feynman's transition elements [34]. By applying these, we can obtain a contribution to the tunneling time of the $(L/v) \exp(-S_0/\hbar)$ type, that is, a result according to which the real-time duration is provided by the semiclassical time (L/v) attenuated by the factor $\exp(-S_0/\hbar)$, with S_0 as the classical action. In typical cases, with S_0 of the order of a few units of $\hbar$, this factor is of the order of $10^{-2} - 10^{-3}$, and a contribution of a few picoseconds is compatible with semiclassical times of the order of 10^2 ps, as in Josephson junctions [35]. In this work, it has been concluded that this contribution should be added to the one evaluated by the stochastic model which, in this case, turned out to be on the same scale of picoseconds. On the contrary, we have to believe that the two contributions—namely the one given by the stochastic model and the one evaluated by the transition elements—are really the same result. In fact, even within the framework of the transition elements, we arrived at a result of the type [36]

$$\langle t \rangle \simeq \frac{imc^2}{\hbar} \left(\frac{L}{v}\right)^2 \tag{12.21}$$

which, by identifying $imc^2/\hbar$ with a [19], becomes

$$\langle t \rangle_R \simeq a \left(\frac{L}{v}\right)^2 \tag{12.22}$$

exactly as the real part in (12.20). This result tends to put the stochastic model in a different light, one that is not limited by the presence of dissipation (which is always present in macroscopic systems), but that is also capable of interpreting situations in which dissipation is absent or negligible.

Another and perhaps more important result, one that is peculiar to the stochastic approach, was obtained when the range of the variable L/v was not limited to small values, but was extended, depending on the parameter a, to large values of L/v. In this way, we found that real time as a function of L/v shows the unexpected presence of a peak (Fig. 12.9), situated approximately at a value of L, so that the semiclassical time L/v is nearly equal to the time a^{-1}. Behind this peak, the curve continues with a nearly quadratic law.

This behavior, that is, the presence of a peak in the real part of the traversal time, supports the hypothesis of a kind of resonance, which is also confirmed by the curve of the imaginary part of the time. In fact, this curve shows a typical shape (Fig. 12.10), with a zero in correspondence with the position of the above-mentioned peak which is a characteristic of resonances. The origin of this resonance is not completely clear: we note that its occurrence, for $L/v \approx a^{-1}$, could be interpreted by rewriting this condition as $aL \approx v$, and

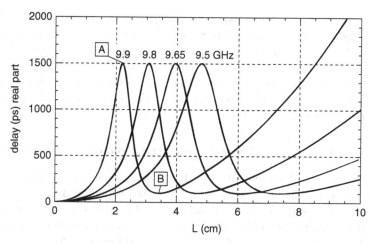

Fig. 12.9. Real part of the traversal time as a function of the position L inside the barrier (infinitely long) computed for several values of the frequency ν below the cutoff one at 10 GHz and for a fixed value of the dissipative constant $a = 2$ ns^{-1}. The position of the peak (whose amplitude is equal to $3/a$ ns) is nearly coincident with $L = v/a$, so by lowering a it tends to increase, while decreases increasing a. So, the possibility to detect this peak depends on a suitable choice of the parameter values. A and B indicate the stationary points in the curve

hypothesizing that it occurs—producing a strong increase in real time—when the semiclassical (imaginary) velocity v becomes comparable with the quantity aL. The latter has velocity dimensions and acts in the sense of slowing down the motion, thus increasing the traversal time of the barrier.

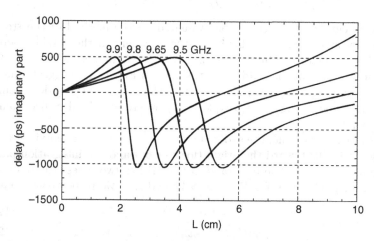

Fig. 12.10. Imaginary part of the traversal time as a function of the position L inside the barrier for the same parameter values as in Fig. 9

More important than the theoretical interpretation is the experimental test of this behavior. Two experiments were performed in the microwave range (wavelength, $\lambda \simeq 3$ cm). One measured the penetration time in a barrier consisting of a rectangular waveguide excited below the cutoff frequency (see Sect. 12.2.2); the other measured the lateral shift of a microwave beam in the case of frustrated total reflection (Sect. 12.4). For a more detailed theoretical interpretation, see [37] and [38].

12.1.3 The Procedure of Stevens and the Role of the Forerunner

We analyze here a development of a theoretical model due to Stevens [39]. The essence can be summarized as follows. According to Brillouin [7], the signal velocity in a dispersive medium can be determined by evaluating the arrival of the main front of a uniform pulse. The latter is initially described by a function of the type $\exp(-i\omega_i t)$ for $t \geq 0$ and zero for $t < 0$; for a propagation along the x axis, it can be expressed by the contour integral in the complex plane of ω (similar to (12.3)) as

$$\psi(x,t) = \frac{1}{2\pi i} \int_\gamma \frac{\exp[-i(\omega t - k_\omega x)]}{\omega - \omega_i} \, d\omega \,, \tag{12.23}$$

where γ is a closed path including ω_i. For a nonrelativistic particle of mass m, the wave number is expressed as $k_\omega = \hbar^{-1}[2m(E - V_0)]^{1/2}$ for $E = \hbar\omega$ either for $E > V(x) \equiv V_0$ or $E < V(x) \equiv V_0$. In order to evaluate the above integral, it is suitable to introduce the new variables $z^2 = \omega - V_0/\hbar$ and $\xi = x(2m/\hbar)^{1/2}$. Thus, by substituting in (12.23), we obtain

$$\psi(\xi,t) = \frac{\exp(-iV_0 t/\hbar)}{2\pi i} \int_\Gamma \frac{\exp[i(z\xi - z^2 t)]}{z^2 + V_0\hbar^{-1} - \omega_i} 2z \, dz \,. \tag{12.24}$$

The integral in (12.24) can be evaluated by the method of the steepest descent, by analyzing the behavior of the exponent in the neighborhood of the saddle point of the function $F(z) = z\xi - z^2 t$, that is, where $z = \xi/2t$. The conformation of the saddle point, shown in Fig. 12.11, suggests that the appropriate integration path Γ, at least in the region of the saddle point, is the straight line $z = (\xi/2t) + \rho \exp(-i\pi/4)$, where ρ is the distance from the saddle point.

With increasing time, the saddle point moves along the real axis of z from infinity (at $t = 0$) toward the origin (for $t \to \infty$) and the integration path Γ analogously behaves. The integrand in (12.24) has two poles situated on the real and imaginary axes for $E = \hbar\omega_i > V_0$ and $E < V_0$, respectively, that is,

$$z = \pm(\omega_i - V_0\hbar^{-1})^{1/2}, \quad E > V_0 \tag{12.25}$$

$$z = \pm i(V_0\hbar^{-1} - \omega_i)^{1/2}, \quad E < V_0 \tag{12.26}$$

For $E > V_0$, the main contribution to the integral (12.24) arises as soon as the integration path arrives at one pole of (12.25) and is given by

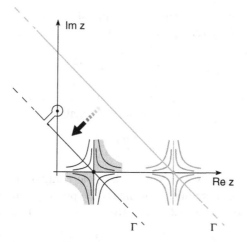

Fig. 12.11. Integration path Γ in the complex plane z at two different instants. The integration path is coincident with the steepest descent line of the saddle point. In the dashed area, the real part of the exponent of (12.24) is greater than zero, in the other one, the real part is less than zero. When the path arrives at the pole, its contribution must be considered

$\exp[i\xi(\omega_i - V_0\hbar^{-1})^{1/2}]\exp(-i\omega_i t)$. This quantity is definitely larger than the saddle point contribution constituting a small forerunner of the signal before the integration path has reached the pole. The principal part of the front of the pulse travels with a uniform velocity

$$v = \frac{x}{t} = \left(\frac{\hbar}{2m}\right)^{1/2}\frac{\xi}{t} = 2(\omega_i - V_0\hbar^{-1})^{1/2}\left(\frac{\hbar}{2m}\right)^{1/2} = \frac{\hbar k}{m}, \qquad (12.27)$$

in agreement with the semiclassical analysis, which gives the traversal time as

$$\tau = \frac{m}{\hbar}\frac{L}{|k|}. \qquad (12.28)$$

For $E < V_0$, the situation is more complicated. When the integration path arrives at one pole, now situated on the imaginary axis (12.26), the pole contribution is now given by

$$\exp[-\xi(V_0\hbar^{-1} - \omega_i)^{1/2}]\exp(-i\omega_i t). \qquad (12.29)$$

This quantity describes the arrival of an important part of the pulse attenuated in amplitude by the factor

$$\exp[-\xi(V_0\hbar^{-1} - \omega_i)^{1/2}] = \exp\{-[2m(V_0 - E)]^{1/2}x/\hbar\} = e^{-|k|x}, \qquad (12.30)$$

which for $x = L$ gives the expected result. More important, the time of arrival of this contribution, as determined by the crossing of the integration path over the real axis (which has the same value of that of the crossing over the imaginary axis, see Fig. 12.11), is given by

$$\frac{\xi}{2t} = (V_0\hbar^{-1} - \omega_i)^{1/2} .$$
(12.31)

In this way we find that also in the case of tunneling an important part of the front of the pulse travels with the uniform velocity

$$v = \frac{\hbar|k|}{m} .$$
(12.32)

However, in tunneling cases in which the pole contribution is given by the quantity (12.29), the contribution of the saddle point is definitely not smaller. In fact, it can be seen that (12.24), in the saddle point approximation, gives

$$\psi(\xi, t) = \left(\frac{2}{\pi\xi}\right)^{1/2} \exp\left\{i\left[\left(\frac{\xi^2}{4t^2} - \frac{V_0}{\hbar}\right)t - \frac{3}{4}\pi\right]\right\}$$
$$\times \left[\left(\frac{\xi}{2t}\right)^{3/2}\left(\frac{\xi^2}{4t^2} + V_0\hbar^{-1} - \omega_i\right)^{-1}\right] .$$
(12.33)

The modulus of ψ decreases to zero either for $t \to 0$ or for $t \to \infty$, and is mainly characterized by the last factor in squared brackets, which reaches its maximum near the coordinate value z_0 given by (12.31). At $z = z_0$, (12.24) becomes

$$\psi(x, t)_{z=z_0} = \frac{1}{\sqrt{2\pi|k|x}} \exp\left[-i\left(\omega_i t + \frac{3\pi}{4}\right)\right] .$$
(12.34)

and, by comparison with (12.30), it turns out that the contribution of the saddle point at z_0 is never negligible but rather comparable to or greater than the one corresponding to the arrival of the attenuated pulse. More precisely, on the basis of (12.33), we find that in the forbidden region ($\hbar\omega_i < V_0$), the contribution of the saddle point tends to anticipate the arrival of the signal with respect to the prediction of (12.28), while in the allowed region ($\hbar\omega_i > V_0$) the influence of the saddle point contribution should be negligible.

Let us now consider the total solution of integral (12.24) by including in the analysis the saddle point, the pole and the line-integral contributions [40, 41]. By putting $\Omega = \hbar^{-1}(E - V_0)$ we have that the two poles, as given by (12.25) and (12.26), are situated at $\pm\Omega^{1/2}$.

For $\Omega > 0$ (allowed region), the main contribution to the integral arises as soon as the integration path arrives at one pole (at a time given by $2t/\xi = \Omega^{-1/2}$) with a constant oscillating expression of unitary amplitude $\exp[i(\xi\Omega^{1/2} - \omega_i t)]$. Previously, during the approach to the pole, the integration gives an oscillating part the amplitude of which increases to about $1/2$ the final amplitude; afterwards, while leaving the pole, we have a contribution—in opposition of phase—the amplitude of which decreases from $1/2$ the final one to very small values [7]. The superposition of the several contributions gives

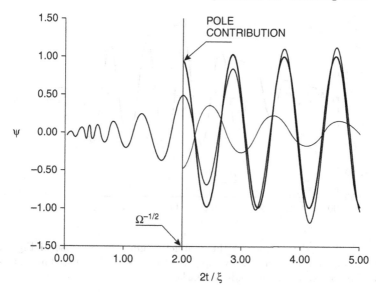

Fig. 12.12. Representation of the wave function after a barrier length $\xi = x(2m/\hbar)^{1/2} = 16$ and $\Omega = 0.25$ (classically allowed region) as a function of time. The pole contribution is at $2t/\xi = \Omega^{-1/2}$. This contribution is of unitary amplitude and frequency $\omega_i = 0.9$ and combined with the line contribution gives rise to a continuous oscillating function of increasing amplitude, which tends asymptotically to unity

rise to a continuous oscillating function the amplitude of which increases considerably around point $\Omega^{-1/2}$, as shown in Fig. 12.12.

The contribution due to the saddle point contributes to the forerunner of the signal. The latter is given by the whole line integral, which mainly consists of the saddle point contribution in the tunneling region ($\Omega < 0$). For the considered case of Fig. 12.12, inclusion of the saddle slightly modifies the shape of the complete signal whose time of arrival, measured at the midpoint of the maximum amplitude, can be also evaluated by simply considering the total envelope (Fig. 12.13).

For $\Omega < 0$ (forbidden region), the situation is similar, but the results are strongly influenced by the forerunners. When the integration path arrives at one pole, now situated on the imaginary axis of z, the pole contribution describes the arrival of a part of the pulse attenuated in amplitude by the factor $\exp(-|k|x)$. In this way, a significant contribution is made to the signal that travels with a uniform velocity $v = \hbar|k|/m$. However, in this case, the contribution of the saddle point, which constitutes part of the forerunner, is not negligible but rather comparable to or greater than the one corresponding to the arrival of the attenuated pulse. Moreover, whenever it is not negligible, the latter is typically so deformed that at the instant of arrival of the pole contribution, the signal shows nothing peculiar that would suitably recognize

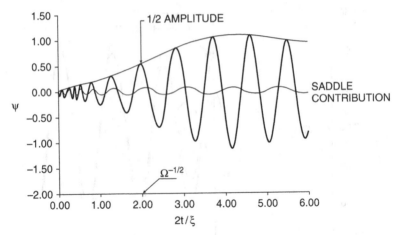

Fig. 12.13. Time dependence of the complete signal for the case of Fig. 12.12. The saddle contribution and the envelope are also shown. The time of arrival, taken halfway of the maximum amplitude, is marked

its presence. An example of a computed signal alone and then combined with the saddle contribution is shown in Fig. 12.14.

As before, the time of arrival of the complete signal can be determined simply by considering the envelope and taking the time required to reach 1/2 of the maximum amplitude. For a classically forbidden region, this time is quite near the one predicted by only considering the saddle point (which, sufficiently below the barrier, tends to coincide with the forerunner). In Fig. 12.15, several computations of the arrival time of the complete signal have been performed as a function of Ω, for different values of the barrier length. We note that the resulting behaviors are in qualitative agreement with the predictions of the other models, especially the phasetime one in the tunneling region. More interestingly, these theoretical predictions appear to be confirmed by the results of delay-time obtained by a microwave simulation below the cutoff (see Sect. 12.2).

12.2 Sub cutoff Microwave Propagation in Waveguide

12.2.1 Delay Time Through a Rectangular Potential Barrier

We report here the results of an experiment in the microwave range in which a step-narrowing in the waveguide simulates a quantum-mechanical rectangular potential barrier, as shown in Fig. 12.16.

This experiment represents a contribution toward an understanding of a complicated problem mainly for the facility of performing a microwave experiment, which magnifies the timescale up to nanoseconds. The measurements

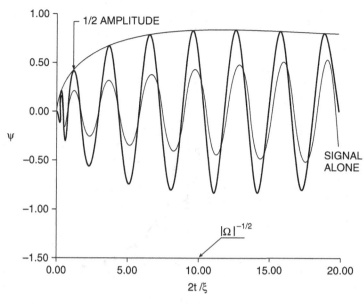

Fig. 12.14. Time dependence of the complete signal for $\xi = 4$ and $\Omega = -0.01$ (classically forbidden region) and $\omega_i = 1$. The signal alone, without saddle contribution, is also shown. The time of arrival, taken halfway of the maximum amplitude and according to the semiclassical analysis, is marked

can be compared with theoretical predictions as deduced from the existing models of quantum tunneling since the transposition is relatively easy, as explained in Sect. 12.1.

Let us consider the X-band microwave circuit of Fig. 12.17 in which there is a step-narrowing represented by a portion of waveguide in P band of length L, section $a' \times b'$ (7.9×15.8 mm^2), less than the normal size in X band $a \times b$ (10.16×22.86 mm^2).

The signals S_1 and S_2 were detected before and after the narrowing, respectively, and sent to an oscilloscope where the delay time is directly measured. The microwave signal—like a step function—was supplied by a klystron modulated by a pin modulator the fall time of which (<10 ns) was suitable to measure delay times down to <1 nanosecond. The delay time was measured as the distance between the falls at half-height of the signal after and before the narrowing, and was detected until the deformation of the signal between the input and the output was modest and the measurement was still reliable.

Measurements of the delay time performed as a function of the frequency, for $L = 15$ cm, are reported in Fig. 12.18 together with the theoretical predictions relative to the quantum-mechanical models (see the following section).

We note that, although none of the theoretical curves is far from the experimental results, above the cutoff, the experimental results tend to reproduce both the τ_{BL} and τ_ϕ models. Below the cutoff, however, the phasetime model

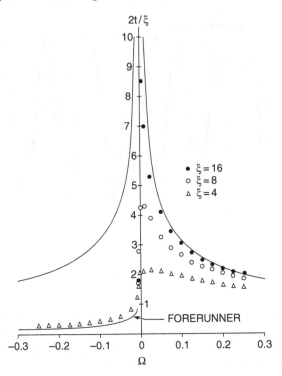

Fig. 12.15. Tunneling time duration determined by evaluating the arrival time of the complete signal, for some values of the barrier length. In the classically allowed region ($\Omega > 0$), the results are more or less comparable with respect to the semi-classical model (continuous line), while in the tunneling region ($\Omega < 0$) they differ deeply from it having much smaller tunneling times

τ_ϕ appears to be most adequate. Measurements with $L = 20$, 10, and 5 cm were also performed using the same experimental apparatus. In the first case ($L = 20$ cm) the peak of the delay moves toward the cutoff and the value of the maximum increases, as physically expected. However, for this distance problems arose due to the great attenuation of the pulse; the delay was detectable only in a small range below the cutoff, thus making it difficult to discriminate among the models. For $L = 10$ and 5 cm, the peak of delay was displaced toward the higher frequencies as theoretically expected, and below the cutoff the data well followed the phasetime model in both cases. On the other hand, the measurements were less reliable with shorter narrowing lengths because of spurious effects due to the discontinuities where the guide section changed abruptly from $a \times b$ to $a' \times b'$ and higher-order evanescent modes may have played a role. These discontinuities act as parallel inductive shunts and influence the phase behavior [42] even if their effect in the frequency dependence is quite negligible. In fact, the variation in the height of the guide introduced in the equivalent circuit a negligible capacitive reactance

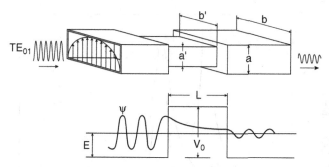

Fig. 12.16. TE_{01} mode propagating in the X-band waveguide with a narrowing of length L which creates the barrier. The quantum-mechanical analog is shown

$X_C \approx 13,000\,\Omega$ at 9.5 GHz where the characteristic impedance of the X-band guide was $\sim 500\,\Omega$. On the contrary, the width variation caused a considerable inductive reactance $X_L \approx 2000\,\Omega$. However, in order to eliminate or reduce the effect of discontinuities at the connection between the X-band guide and

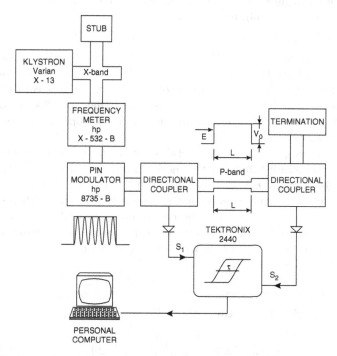

Fig. 12.17. Experimental setup employed to measure the delay time with a pulse like a step function. The pulse is created by modulating at 1000 Hz, with a square wave, the continuous wave generated by a klystron. The delay time is measured from the separation of the pulses S_1 and S_2 detected before and after the narrowing

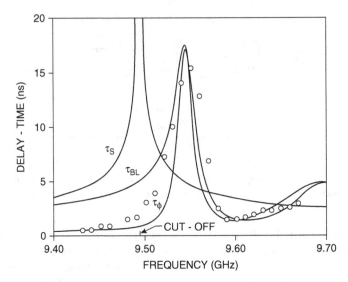

Fig. 12.18. Delay time (*open circles*) as a function of the frequency for $L = 15$ cm. The continuous lines are the theoretical curves of τ_{BL}, τ_s, and τ_ϕ as given by (12.40), (12.45), and (12.46), respectively

the P-band narrowing, an X-P adapter which allows a "soft" passage from the X to P band was then employed. For $L = 15$ cm, the delay-time measurements performed with the above-mentioned adapters almost coincided with those performed without adapters, thus showing that the effect of discontinuities, even if producing some disturbance, did not appreciably modify the transmitted signal.

In order to compare the experimental results with the quantum-mechanical models, let us consider that, for the mode $TE_{0,1}$, the refractive index in the waveguide is given by $n = [1 - (\lambda/2b)^2]^{1/2}$ (λ is the wavelength in the free space) and that the guide represents a dispersive medium with $n < 1$ for $\lambda < 2b$. When $\lambda > 2b$, the index becomes imaginary and evanescent waves originate. The experimental apparatus corresponds to a rectangular potential barrier, as sketched in Fig. 12.16, with the substitution

$$k = \frac{2\pi}{\lambda}n \rightarrow \frac{1}{\hbar}[2m(E - V_0)]^{1/2} . \tag{12.35}$$

More precisely, the "momentum" of (12.35) is expressed as

$$k = \frac{2\pi}{\lambda}\frac{v_{\mathrm{gr,g}}}{c} = 2\pi \left(\frac{\nu^2}{c^2} - \frac{1}{(2b')^2}\right)^{1/2} , \tag{12.36}$$

where $v_{\mathrm{gr,g}} = c[1 - (\lambda/2b')^2]^{1/2}$ is the group velocity in the narrowed guide, c is the light velocity, and ν is the frequency. The "momentum" for the section of width b is

$$k_1 = 2\pi \left(\frac{\nu^2}{c^2} - \frac{1}{(2b)^2} \right)^{1/2} \tag{12.37}$$

and the quantity

$$k_0 = 2\pi \left(\frac{1}{(2b')^2} - \frac{1}{(2b)^2} \right)^{1/2} = \left(k_1^2 - k^2 \right)^{1/2} \tag{12.38}$$

corresponds to a potential step of height V_0.

The delay time can be derived simply by modifying the corresponding quantum-mechanical results. This is easily done by considering that the dispersion relation, which for quantum waves in a region of constant potential is $\omega = (\hbar k^2/2m) + (V_0/\hbar)$, becomes as (12.36) and (12.37) for electromagnetic waves in a wave guide. By differentiating, the "group velocity," which for quantum waves is $d\omega/dk = \hbar k/m$, becomes $d\omega/dk = c^2 k/\omega$ for the waveguide. Thus we can adopt the formulas of quantum mechanics for waveguides by the substitution

$$\frac{\hbar}{m} \rightarrow \frac{c^2}{2\pi\nu} . \tag{12.39}$$

Accordingly, the Büttiker and Landauer traversal time [12], defined as the time τ_x through a rectangular barrier expressed by means of the energy dependence of the complex transmission amplitude that includes both phase changes and amplitude, can be written as

$$\tau_{BL} \equiv \tau_x = (\tau_y^2 + \tau_z^2)^{1/2}$$

$$\simeq \frac{1}{2\pi} \left[\left(\frac{\partial}{\partial\nu} \Delta\phi \right)^2 + \left(\frac{\partial(\ln T^{1/2})}{\partial\nu} \right)^2 \right]^{1/2} , \tag{12.40}$$

where the transmission coefficient T and the phase change $\Delta\phi$ in transmission through the barrier are given by

$$T = \left\{ 1 + \left[\frac{(k_1^2 - k^2)^2}{4k_1^2 k^2} \right] \sin^2(kL) \right\}^{-1} \tag{12.41}$$

$$\tan(\Delta\phi) = \frac{k^2 + k_1^2}{2kk_1} \tan(kL) . \tag{12.42}$$

According to relation (12.39) and (12.36–12.38), τ_y and τ_z are given by

$$\tau_y = \frac{2\pi\nu}{c^2 k} \frac{\partial\Delta\phi}{\partial k} = \frac{2\pi\nu}{c^2} \frac{k_1}{k} \left[\frac{2kL(k_1^2 + k^2) - k_0^2 \sin(2kL)}{4k^2 k_1^2 + k_0^4 \sin^2(kL)} \right] , \tag{12.43}$$

$$\tau_z = \frac{2\pi\nu}{c^2 k} \frac{\partial}{\partial k} \ln T^{1/2} = \frac{2\pi\nu}{c^2} \frac{k_0^2}{k^2} \left[\frac{(k_1^2 + k^2) \sin^2(kL) - (k_0^2 kL/2) \sin(2kL)}{4k^2 k_1^2 + k_0^4 \sin^2(kL)} \right] . \tag{12.44}$$

Note that for $k_1 \ll k, k_0$ and opaque barriers $(kL \gg 1)$, (12.44) tends to the simple semiclassical result

$$\tau_s = \frac{2\pi\nu}{c^2} \frac{L}{|k|} . \tag{12.45}$$

On the other hand, in the cutoff region where $k_1 \simeq k_0$, τ_y (named in the literature as dwell time) as given by (12.43) is practically coincident with the phase-time delay τ_ϕ given by

$$\tau_\phi = \frac{2\pi\nu}{c^2 k_1} \frac{\partial \Delta\phi}{\partial k_1} = \frac{2\pi\nu}{c^2} \frac{1}{k\,k_1} \left[\frac{-k_0^4 \sin(2kL) + 2k_1 k^2(k_1^2 + k^2)L}{k_0^4 \sin^2(kL) + 4k^2 k_1^2} \right] . \tag{12.46}$$

In all the above equations (12.41–12.46), k, k_1, and k_0 are given by (12.36–12.38) even for frequencies lower than the cutoff ($\nu_0 = c/(2b') = 9.494$ GHz) where k becomes imaginary.

Equations (12.40), (12.45) and (12.46) are reported in Fig. 12.18, together with the experimental results. As previously anticipated, we can see that the phase-time model appears to be the most adequate one for reproducing the experimental results, while, in other cases, this model appears inadequate (see Sects. 12.2.2 and 12.4)

Measures of delay time and amplitude below the cutoff were performed also in a different way, by utilizing a two frequencies method. The configuration of the experimental setup is shown in Fig. 12.19.

The beat envelope of the two signals at two very near frequencies ν_1 and ν_2 is detected. Because of the nearly quadratic characteristic of the detectors, the ratio of the components of the amplitude of the beat V_2 and V_1 (measured after and before the narrowing, respectively) is proportional to the product of the square root of the transmission coefficient at the two frequencies $T_1^{1/2} T_2^{1/2}$ [44]. Thus, for a given frequency ν_1, by varying ν_2 and by determining with a fitting procedure the slope of the curve $\ln T_2^{1/2}$ as a function of ν in the neighborhood of ν_1, we get τ_z by means of the second term in (12.40). The results obtained for $L = 15$ cm are reported in Fig. 12.20.

As expected, one can observe that the data derived from the beat-envelope amplitudes reasonably follow the theoretical curve of τ_z while the experimental data of beat delay better follow the curve of τ_ϕ. In the same figure, the semiclassical model, obtained in the framework of telegrapher's equation (see Sect. 12.1) is also reported.

12.2.2 Penetration Time Inside a Barrier of Infinite Length

We describe here an experiment, at microwave scale, which simulates a quantum-mechanical rectangular barrier of infinite length [36]. In this experiment, the signal is constituted by squared pulses, which modulate the carrier and is measured simultaneously before the barrier edge and inside the barrier, in a variable position l (see the inset in Fig. 12.21). The waveguide in the

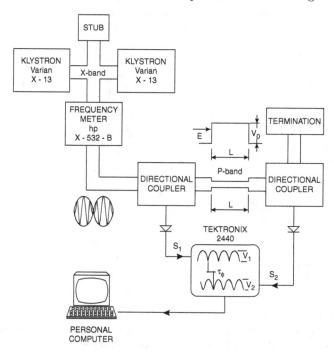

Fig. 12.19. Experimental setup employed to measure the delay time and the beat-envelope amplitude V_1 and V_2 with two slightly different frequencies generated by two klystrons. From V_1 and V_2 the time τ_z is derived

P-band was filled by a dielectric, Teflon, so that the frequency of the carrier was above the cutoff. The two signals were sent to a dual channel oscilloscope suitable for measuring the temporal delay between the two signals with sufficient accuracy. Reliable measurements required the use of frequencies near the cutoff one ($\nu_0 = 9.494$ GHz) and a penetration depth in the barrier of a few centimeters, in order to have an acceptable attenuation of the waves.

The measurements were performed at several values of the frequency carrier that ranged between 9.46 and 9.28 GHz. They were then compared with the curve of the real part of the traversal time as given by the stochastic model ((12.19) in Sect. 12.1.2). This was computed for a given frequency ν, which determined the velocity through the relation $v \equiv |v_{\mathrm{gr,g}}| = c\sqrt{(\widetilde{\nu}_0/\nu)^2 - 1}$, with $\widetilde{\nu}_0$ as the effective cutoff frequency given by $\widetilde{\nu}_0 = \sqrt{\nu_0{}^2 + \widetilde{a}^2\nu^2/c^2}$, by selecting the value for $\widetilde{a}$ which best described the experimental data.

In extreme cases ($\nu = 9.46$ and 9.28 GHz), the results obtained showed a clear monotonic (nearly quadratic) behavior, increasing with l, the description of which by the stochastic model required values for a of the order of 1 ns^{-1} or less. More interesting were the cases at intermediate frequencies (9.39, 9.33 GHz), which exhibited more complicated behavior, with the presence of a

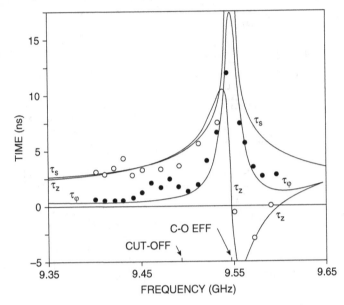

Fig. 12.20. Experimental results of τ_z vs frequency (*open circles*) as deduced from beat envelope amplitude data for $L = 15$ cm, compared with the relative theoretical curve and beat-delay data (*full circles*) compared with the curve of τ_ϕ. The heavy continuous line (τ_s) represents the modified semiclassical model resulting from the solution of the telegrapher equation with $\tilde{a} = 0.1c$ which produces a shift of the cutoff frequency from $\nu_0 \simeq 9.49$ GHz to $\tilde{\nu}_0 \simeq 9.54$ GHz

more or less pronounced peak as predicted by the theoretical model in some cases (see Fig. 12.21).

In Fig. 12.21 we report the results obtained for $\nu = 9.33$ GHz. These showed a clear increase in the delay up to values of l around 2.5 cm. However, for larger values of l, the delay tended to decrease or, at least, to saturate. The resulting peak was, however, not as accentuated as predicted by the above theory, but rather was damped as if a damping coefficient were present in the resonance (a phenomenon analogous to the lowering of the coefficient Q of a resonant circuit or cavity [43]). By taking into account that in tunneling cases the dissipative parameter a represents an imaginary dissipation, that is, not dissipative but rather reactive, it is plausible to introduce a truly dissipative effect by including a suitable imaginary part in parameter a, so that $a \to a+ib$. In this way, we obtained more damped peaks in the curves of the real part of the time, as shown in Fig. 12.21, where the lower curve is obtained for $a = 2.25 - i0.5$ ns^{-1}.

It seems, therefore, that by means of this experiment we have obtained a sufficiently clear demonstration of the validity of the stochastic model for tunneling, even independently of the existence of the peak in the curve of the delay time. This peak, however, strongly supports the theoretical model,

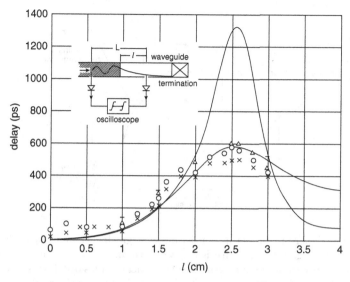

Fig. 12.21. Penetration time results (*small crosses, triangles* and *circles* refer to different series of measurements) as obtained with the experimental setup shown in the inset. The carrier frequency was $\nu = 9.33$ GHz, the cutoff frequency $\nu_0 = 9.494$ GHz. The upper curve refers to $a = 2.25$ ns^{-1}; the lower curve is obtained by including an imaginary part, namely $a = 2.25 - i0.5$ ns^{-1}

even if its implications are rather surprising. In fact, the traversal time of the forbidden region tends to decrease for certain values of l, while the distance increases. This unusual behavior deserves further investigation before any definitive conclusion can be safely drawn. It is worth noting, however, that in the region of the resonance peak the punctual velocity, as given by the inverse of the derivative, supplies velocities which are infinite in the two stationary points (A and B in Fig. 12.9, Sect. 12.1.2) and negative in the intermediate interval.

12.3 Delay-Time Measurements in a Diffraction Experiment

Here we describe an experiment of optical tunneling performed in order to measure the delay time for evanescent waves generated by diffraction. In the experiment, the phase-delay time was directly measured and group delay was then deduced from the previous results. Microwaves with a wavelength of about 3 cm were used in connection with a metal-strip grating, the period of which was chosen so that all the diffracted waves, except the zero-order one, were surface (evanescent) waves attenuated along the direction perpendicular to the grating. One of the first-order waves was transformed into an ordinary

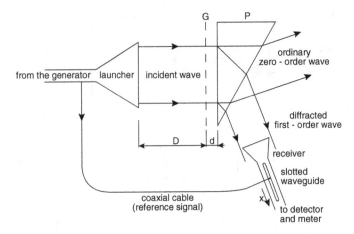

Fig. 12.22. Experimental setup consisting of a grating G and a paraffin prism P separated by a gap whose width is d. Two horn antennas (the launcher and the receiver) allow accurate phase measurements by means of a slotted waveguide where the reference signal is combined with the signal picked up by the receiver

wave by refraction on a paraffin prism; it was then revealed by means of a receiver. The experimental setup is shown in Fig. 12.22.

Besides the grating and the prism, it includes two horn antennas, one as launcher and the other as receiver. The latter is followed by a slotted waveguide in which the signal picked up is combined with a reference signal derived from the generator. In this way, we could make accurate phase measurements by detecting the probe position corresponding to a minimum (which exactly indicates the opposition of phases of the two waves) of the amplitude of the resulting signal [44].

The results are shown in Fig. 12.23. Here, the probe position x is reported as a function of the distance d between the grating and the prism. The period of the grating is $p = 3$ cm and the measurements were made at a frequency $\nu = 9.24$ GHz, below the cutoff frequency $\nu_0 = c/p = 10$ GHz.

The measurements were performed while keeping all components at fixed position; only the gap distance d was varied. Since $D_0 \equiv d + D = 52$ cm was taken to be constant, the distance D between the grating and the launcher also changed.

By performing a set of measurements like the one shown in Fig. 12.23, each for a different frequency, we could derive the delay as a function of the frequency. The results are shown in Fig. 12.24, together with the theoretical curves. In the absence of a grating, a prism, etc., the total propagation time of the wave for the indicated distance would be D_0/c; thus, the time attributable to the traversal of the gap would be $d/c = (D_0/c) - (D/c)$. Consider that for large D/λ the velocity prior to reaching the grating is, in an excellent approximation, just c. However, as the gap varies away from zero, the relative

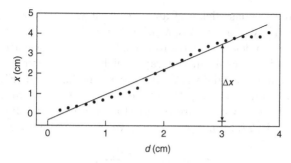

Fig. 12.23. Probe position x, relative to a constant phase value, as a function of gap width d for $\nu = 9.24$ GHz. The probe position is approximately linear in the gap size and it is only the average slope of the curve that is ultimately used in (12.49). For this reason our results are not sensitive to slight deviations from straight line behavior (the waviness in the curve) nor to the zero position of the x variable. In practice, we used the variation Δx for a gap width $d = 3$ cm

phase of the reference signal and the wave that actually passes through the grating, etc., change and the probe is moved until they match again.

A probe displacement Δx means that the wave through the prism has gained some time (with our sign convention). Therefore, by using the phase velocity $v_{\mathrm{ph,g}}$ within the slotted waveguide (where the probe is located), this yields

$$\tau_{\mathrm{ph}}(\nu) = \frac{d}{c} - \frac{\Delta x(\nu)}{v_{\mathrm{ph,g}}} , \qquad (12.47)$$

for the time in the gap (as a function of frequency), where $v_{\mathrm{ph,g}} = c/\sqrt{1 - (\lambda/2b)^2}$, λ is the free-space wavelength and $b = 22.86$ mm is the width of the waveguide. The quantity Δx is positive when the x position increases as it goes away from the receiver horn (see Fig. 12.22).

By substituting $v_{\mathrm{ph,g}}$ in (12.47), the phase delay can be rewritten as

$$\tau_{\mathrm{ph}}(\nu) = \frac{d}{c} \left[1 - \frac{\Delta x(\nu)}{\Delta x'(\nu)} \right] , \qquad (12.48)$$

where, by denoting the wavelength in the waveguide with λ_{g},

$$\Delta x'(\lambda) = d \frac{\lambda_{\mathrm{g}}}{\lambda} = \frac{d}{\sqrt{1 - (\lambda/2b)^2}} \qquad (12.49)$$

is the variation of the probe position in the waveguide corresponding to the variation d in free space. This means that if Δx turns out to be equal (or nearly equal) to $\Delta x'$, the phase delay in the gap is equal (or nearly equal) to zero. In the example of Fig. 12.23, when referring to a frequency below the cutoff, the phase delay for a gap of 3 cm turns out to be ~ 10 ps, while the phase delay for the same distance in free space is 100 ps. This shows—as expected—that the phase delay of an evanescent wave is very much shorter than the phase delay of a free propagation.

As for the group delay, we have to consider that in free space it coincides with the phase delay, while in the gap, which is a dispersive system, it is different. By fitting each data set with a straight line as in Fig. 12.23, we can determine, through (12.48), the phase delay for selected values of d. The results obtained for $d = 3$ cm are reported (full circles) as a function of the frequency in Fig. 12.24. From these values, we can evaluate the group delay as

$$\tau_{\mathrm{gr}} = \tau_{\mathrm{ph}} + \nu \frac{\Delta \tau_{\mathrm{ph}}}{\Delta \nu} . \tag{12.50}$$

The group delays are shown in Fig. 12.24 by means of open circles with fiducial bars. The fiducial bars have been estimated by the χ^2 criterion as $\sigma \approx \left(\sum_{i=1}^{N} \Delta_i^2 / N \right)^{1/2}$ where Δ_i are the differences (residuals) between the

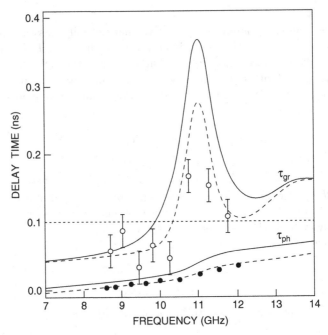

Fig. 12.24. Delay-time results and associated theoretical curves for a gap width of $d = 3$ cm for which the corresponding time for light velocity propagation would be 0.1 ns (*dotted line*). Solid circles are the experimental phase-time delays obtained from phase measurements. The two lower curves are the fitting of the experimental data (*dashed line*) and the phase delay (*continuous line*) as predicted by the theoretical model $\tau_{\mathrm{ph}} = \Delta\phi/\omega$. *Open circles* with fiducial bars represent group-delay results derived from phase-delay data. The two upper curves are the group-delay model (*continuous line*) and the group delay deduced from the fitting curve below (*dashed line*)

theoretical and the N experimental values of τ_{ph}. It is interesting to note that the group delay understood to be the ratio of the variation of $\Delta\phi$ over the frequency variation $\Delta\omega$, is exactly what we could obtain by operating simultaneously with two waves at frequencies ω and $\omega + \Delta\omega$ and by measuring the delay of the beat (a way of obtaining the group delay) in both the allowed and forbidden regions. In this way, our procedure for determining the group delay is directly connected to a real physical situation.

Since calculation of the group delay involves the derivative of the measured quantities, it is reasonable to calculate first a smoothing of the data, and to deduce then the group delay from that. Ideally, if a theoretical description of the experiment which depended on one or two parameters were available, the measured data could be used to establish the values of those parameters. Then for the group velocity the derivative of the phase velocity could be taken as given by a theoretical function dependent on the measured parameters (and the error bars would reflect uncertainty in the values of the parameters). In contrast, a straightforward calculation of the derivative by considering the differences of the experimental values would exaggerate the normal variation of the experimental output. Unfortunately, an exact theoretical calculation of the phase shift would be difficult, mainly because the prism is in the near field of the grating and the wavelength is close to the critical value for the grating. Moreover, the slits themselves are of intermediate geometry, being neither infinite slits nor circles; each of them carries different phase factors.

It is convenient to proceed along both lines. We used a phenomenological model to motivate reasonable curve fitting and then use the fitted curve to calculate the derivative. However, we also took the data and performed the most naive kind of derivative calculation. As will be seen below, the latter did not place τ_{gr} quite so deeply into the superluminal regime, although over a significant range even this leaves no doubt that the delays are less than would be obtained from velocity c.

The phenomenological model is based on the theoretical description of tunneling, namely [45]

$$\tan(\Delta\phi) = \frac{nk^2 - \kappa^2}{(1+n)k\kappa}\tanh(\kappa d) \tag{12.51}$$

where $k = 2\pi\nu/c$, $\kappa = (2\pi/c)\sqrt{\nu_0^2 - \nu^2}$, and $n = 1.49$ is the refractive index of the paraffin. In addition to the above phase change calculated, there is a contribution due to the passage from forbidden to allowed propagation. These contributions have been calculated in a variety of situations, although not for the intermediate type geometry (nonrectangular, finite slits) of this experiment. We therefore assumed that an additional phase change occurred, that it went smoothly from zero to its full value (like a hyperbolic tangent). We let the actual value of the phase shift be one of the parameters for the curve fitting.

The resulting data fit (with net phase change close to $2\pi/5$) is shown in the figure. It is the associated curve the derivative of which is used in the calculation of group velocity. As indicated, the figure also displays (as circles) a calculation of the group-delay that does not depend on any curve-fitting assumptions. For both the minimalist and the "informed" calculations and for frequencies below the cutoff, the group delay fell convincingly below that associated with the velocity of light.

When the delay measurements were made by using pulse modulation (like a "step" function in which the transition has a duration of about 10 ns and the spectral width $\sim$ 100 MHz), the experimental setup of Fig. 12.22 turned out to be not very suitable when the measurements were made by varying d and D (the sum $d + D$ kept constant). In fact, standing wave effects, which gave rise to the small undulation shown in Fig. 12.23, became amplified in the group-delay data since they are related to the derivative of the phase delay.

In addition, there was a modification in the results of delay measurements because of the so-called speed-up effect. This was due to the variation of the transmission coefficient in the frequency interval corresponding to the spectral width of the pulse. In fact, the barrier acted as a high-pass filter enhancing the transmission of the high-frequency components of the signal. This effect could be evaluated by noting that the transmitted pulse turned out to be shifted toward the high-momentum values by an amount given by [46]

$$\frac{\Delta k}{k} = \frac{(\Delta s_k)^2}{2} \frac{c^2}{\omega} \frac{\partial}{\partial \omega} \ln T^{1/2} \,, \tag{12.52}$$

where Δs_k is the spectral width in momentum space. Since $(\partial/\partial\omega) \ln T^{1/2}$ is τ_z, that is, the imaginary component of the delay, and assuming $\Delta k/k \approx \Delta t/t$ (t being the duration of the complete travel in the experiment), we obtained an enhancement given by

$$\delta\tau = \left[\frac{t}{2} \left(\frac{\Delta\omega}{\omega} \right)^2 \omega \right] \tau_z \,. \tag{12.53}$$

For the considered frequencies we estimated the factor in parentheses to be of the order of 10^{-2} so that the measured delay should be shortened by about 1% of the imaginary component. This, in turn, could be assumed to be nearly coincident with the semiclassical time given by

$$\tau_s = \frac{2\pi\nu}{c^2} \frac{L}{\kappa_2} \,, \tag{12.54}$$

where the quantity κ_2 is obtained from amplitude measurements. Some results are reported in Table 12.1, together with the corresponding τ_s which are in rather good agreement with the theoretical value of κ. Therefore, even if appreciable, this interesting effect does not represent an important deviation from the expected results.

More reliable results of group delay were also obtained by measuring, using a lock-in amplification technique, the phase delay of a sinusoidal modulation

Table 12.1. Attenuation constant κ_2 and semiclassical traversal time τ_s as deduced from amplitude measurements. The variation $\Delta\tau_\phi$ of the real delay time τ_ϕ with respect to the free-motion time L/c, as measured for $L = 3$ cm by modulating at 10 MHz, is compared with the value deduced from phase measurements of the carrier (best fit procedure). The reported error is consistent with the resolution of the lock-in amplifier

ν (GHz)	κ_2 (cm^{-1})	τ_s (ps)	$\Delta\tau_\phi = \tau_\phi - L/c$ lock-in measurements. (ps)	$\Delta\tau_\phi$ phase measurements. (ps)
9.01	—	—	-97 ± 55	~ -48
9.30	0.84	235	—	~ -46
9.42	0.71	279	—	~ -43
9.82	0.575	357	—	~ -38
11.0	—	—	186 ± 55	~ 171

which directly supplied the group delay or, more exactly, the variation $\Delta\tau_\phi$ with respect to the free motion time L/c (see Table 12.1).

The modulation frequency ν_m was fixed at 10 MHz so that the spectral width of the signal was only 20 MHz. The sensitivity of this measurement was not high, since a delay of 100 ps corresponds to a dephasing of only 0.36°. Each result was obtained by a best fit of the data, relative to measurements of delay time vs the gap width d, and was affected by a large error as reported in Table 12.1 for only two frequency values. This was due to the complexity of the procedure. Nevertheless, we can conclude that the results obtained in this way are in agreement with those derived from the phase-delay measurements reported in the last column of Table 12.1.

12.4 Tunneling Time in Frustrated Total Reflection

Let us now consider two microwave experiments that deal with tunneling time in the presence of frustrated total reflection.

The first experiment consists in measuring the delay in a gap of a few centimeters between two paraffin prisms when total reflection takes place in the first prism and evanescent waves originate in the gap. The experimental setup is shown in Fig. 12.25.

In addition to two paraffin prisms (refractive index $n = 1.49$, angles: 30°, 60°, 90°), it consists of two horn antennas, with relative detectors, one as launcher and the other as receiver. For angles of incidence $i \simeq 60°$ greater than the critical angle $i_0 = \sin^{-1}(1/n) \simeq 42°$, total reflection takes place, and evanescent waves, moving along the x-direction and amplitude attenuating in z-direction, originate in the gap. In this way, the gap acts like a quantum-mechanical potential barrier, and the traversal time can be identified with

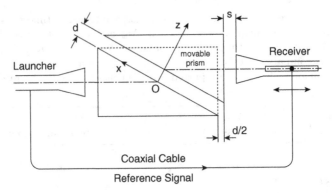

Fig. 12.25. Experimental setup consisting of two paraffin prisms separated by a gap, the width of which is d. Two horn antennas, the launcher and the receiver, allow accurate phase measurements by means of a slotted waveguide, in which the reference signal is combined with the signal picked up by the receiver. The reference axes for the theoretical analysis are also indicated

the tunneling time [47, 48]. For frequencies around 10 GHz (the wavelength $\lambda \simeq 3$ cm), we first performed phase-delay measurements from which the group delay could be deduced. For this purpose, after the receiver horn, we placed a slotted waveguide in which the signal picked up was combined with a reference signal derived from the generator. Phase measurements, as a function of the gap width d, were carried out by detecting a probe position corresponding to a minimum of the amplitude of the resulting signal, according to a procedure described in Sect. 12.3. We repeated this kind of measurement for several frequency values in the $\nu =$ 9–10 GHz range for gap widths from zero to $d = 2$ cm (the strong attenuation prevented our obtaining reliable results for larger values) [49].

The phase delays τ_{ph} obtained were very small, limited within the 4−6 ps range, and slightly dependent on the frequency, as shown in Fig. 12.26. The group delay could be deduced from the phase delay according to the relation $\tau_{gr} = \tau_{ph} + \nu(d\tau_{ph}/d\nu)$. However, the relative errors in the τ_{ph} data made an evaluation of the derivative of τ_{ph} uncertain. On the other hand, according to the theoretical analysis reported in Sect. 12.1.1, the term containing the phase-delay variation turned out to be nearly equal to the phase delay and, consequently, the group delay tended to zero in this case.

Confirmation of this result was obtained by direct group-delay measurements, which were carried out with pulse modulation of the carrier, rather than with monochromatic microwaves as in the phase-delay measurements. In view of the smallness of the quantity being measured, it is very important to check that the pulse shape (width, rise and fall-time) does not appreciably change when passing through the barrier; otherwise, the group-delay loses its meaning. Moreover, the geometry of the experimental setup had to be carefully controlled. Reliable results were obtained with pulse modulation, ~ 80

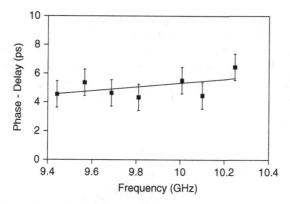

Fig. 12.26. Phase-delay results for several frequency values, obtained by varying the gap width from zero to $d = 2$ cm and maintaining constant s, the position of the receiver horn relative to the movable prism

ns width, rise and fall time of few nanoseconds, frequency repetition of $\sim$100 kHz, by comparing the input and the output pulses with a two-channel digital real-time oscilloscope, the time resolution of which was down to a few picoseconds. After almost compensating for the delay due to the travel before and after the gap, we measured a delay of 134 ± 8 ps when a gap of ~ 2 cm was filled with a paraffin slab. The delay decreased to 87 ± 7 ps when the gap was empty. We thus obtained a net advance in time of 47 ± 15 ps. Since the increase of the path ray in the paraffin was ~ 1 cm (see Fig. 12.25), a value which implies a delay of ~ 50 ps, we concluded that the group delay in the gap was of the order of a few picoseconds, thus demonstrating a clear superluminal behavior, in both the phase and group velocities.

In terms of electromagnetic waves, an analysis of what occurred in the experiment described above was made by considering a system formed by two half-spaces, limited by plane-parallel boundaries, at $z = 0$ and $z = d$, respectively, filled with a homogeneous and non dispersive medium of refractive index n, and separated by a vacuum gap. As described in Sect. 12.1.1, from a comparison of the impinging and transmitted pulses, it is possible to evaluate the time taken by the pulse to travel across the gap. Here, we limit ourselves to reporting the most significant results. If the impinging wave is a δ-like function of the time, picked on the wave front that at $t = 0$ passes through the origin O in Fig. 12.6, the transmitted pulse at $z = d$ can be simply expressed analytically, for $\Gamma \ll \gamma$, as [32, 49]

$$E_t \simeq \frac{2c}{n\gamma d} \frac{\exp(-\frac{2ct}{n\gamma d})}{1 + \exp(-\frac{\pi ct}{\Gamma d})} \tag{12.55}$$

where $\Gamma = \sqrt{n^2\alpha^2 - 1}$ and $(\omega/c)\Gamma$ is the attenuation constant of the evanescent wave along the z-coordinate in the gap. Equation (12.55) tells us that our system behaves approximately like a low-pass filter [50], the time constant of which is $\tau = n\gamma d/2c$. It is thus expected that the delay of a signal crossing the gap should be of the order of τ. More precisely, the delay coincides considerably with the time-constant difference $\tau - \tau'$, where $\tau' = \Gamma d/\pi c$. Since τ and τ' are typically of the same order of magnitude (tens of picoseconds), we obtain very small delays. In Fig. 12.27 a general view of the propagation of a pulse impinging on the gap is shown, with its reflected part and its transmitted part through the gap.

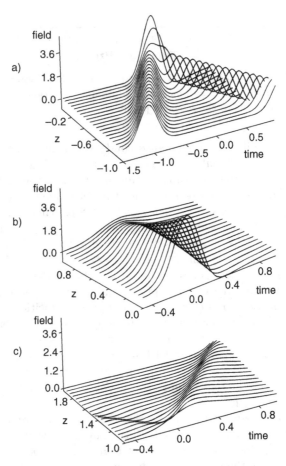

Fig. 12.27. General view of the propagation of a Gaussian pulse (half-width in time is 0.2) for an angle of incidence on the gap $i \simeq 43°$, $\gamma = 0.73$, $\Gamma = 0.16$, taking the light velocity to be $c = 1$. The impinging and reflected pulses, are shown in (**a**), the field in the gap, in (**b**), the transmitted pulse, in (**c**). We note that, while in (**a**) and (**c**) the pulse velocity is nearly equal to the unity, in (**b**) the velocity is approximately five times greater. This value is demonstrated by the small delay (~ 0.2) at the end of the gap. We note, however, that with these parameter values (especially $c = 1$), the deformation of the pulse is not negligible

What emerges from this three-dimensional representation is that, while the velocity is unitary ($\Delta t \approx 1$ for $\Delta z = 1$ corresponding to a velocity $c/n\gamma \approx 1$ before and after the barrier), the velocity during the traversal of the barrier is indubitably higher. This is because on this scale the delay for travelling a unitary gap width ($\Delta z = 1$) is much smaller (say, $\Delta t \approx 0.2$, which corresponds to a velocity $\sim 5c$).

More accurate results can be obtained by properly integrating (12.55), in its exact expression, over time. This can make it possible to obtain the shape of the output in the case that the input signal is, for example, a step function or a ramp. The results are reported in Fig. 12.28, and show that the delay varies from 36 ps for $\Gamma = 0.1$ (that is in proximity of the limit angle of incidence $i_0 \simeq 42°$) to nearly zero for $\Gamma = 0.8$, which corresponds to the experimental situation in which we obtained a delay of few picoseconds.

A detailed analysis of the propagation inside the forbidden region is reported in [51] and [52]. A different approach, based on the path integral method, is given in [53] and [54].

The second experiment consisted of measuring the shift of the microwave beam outgoing from the launcher and traversing the gap of a few centimeters between the two paraffin prisms, still in total reflection condition. This experiment is an extension to the microwave range of an analogous experiment already made in the optical range [48]. With reference to the inset in Fig. 12.29 (incidence and critical angles $\sim 60°$ and $42°$, respectively) we have quantity D—which is a measure of the traversal time—as twice displacement δ. The

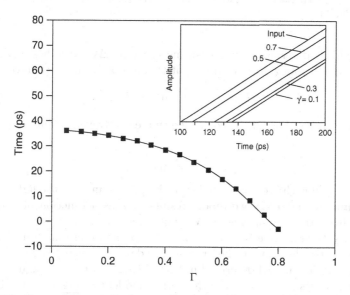

Fig. 12.28. Group-delay results (*squares*) computed by properly integrating (12.55) for several values of Γ and $d = 2$ cm. The inset shows the determination of the delay in the case that the input signal (as well as the output) is a ramp

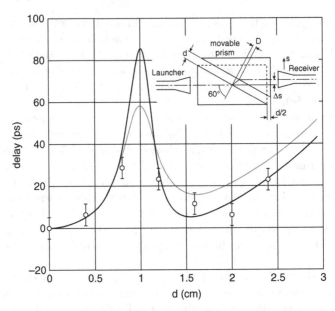

Fig. 12.29. Traversal time results as a function of the gap width d between the two paraffin prisms. The experimental setup is shown in the inset. The microwave beam at 9.33 GHz, in conditions of frustrated total internal reflection, exhibits a shift Δs, in the transmitted part through the gap, which is related to the traversal time. The curves correspond to the real part of the tunneling time as given by the theoretical model of Sect. 12.1.2 for $a = 35$ ns^{-1} (*continuous line*), and $a = (35 - i2.5)$ ns^{-1}(*shaded line*)

latter could be determined by measuring the shift δ of the beam while the gap was varied from zero to d. If we put this into a formula, we have

$$D = 2\delta = 2(d \sin i - \Delta s). \tag{12.56}$$

Once the shift δ is known, we could determine the traversal time as [48]

$$\tau = \frac{nD}{c \sin i} \tag{12.57}$$

where $n = 1.49$ is the refractive index of the prisms and c is the light velocity in vacuum. In Fig. 12.29 we report results of τ as a function of d as they resulted from the measurements. We note that, in spite of a nonnegligible uncertainty (mainly due to the uncertainty in the determination of Δs) in determining the duration of the process, the data exhibit a peak as predicted by the theoretical model described in Sect. 12.1.2. In the same figure we also report the curves as given by the theoretical model (see (12.18) and (12.19)), where the velocity is given, in this case, by $v = c/\sqrt{(\sin i)^2 n^2 - 1}$. The value of a which best fit the experimental data is between 30 and 35 ns^{-1}. A small imaginary part, say $b = -2.5$ ns^{-1}, was also admissible. While confirming the data previously obtained, these results seem to give further confirmation of the validity of the stochastic model in relation to tunneling processes.

12.5 Concluding Remarks

On the basis of the experiments reported here, we can conclude that the optical tunneling (at microwave scale) is a useful method for investigating tunneling processes. Since the wave equations which govern the waves evolution are formally equal to those related to quantum-mechanical particles, optical tunneling can be usefully utilized for simulating quantum tunneling. It therefore represents a good observatory for investigating processes that, in quantum mechanics, are difficult to observe directly.

However, there are some important differences. The first is due to the fact that measurements are always invasive in quantum-mechanical systems, while in optical tunneling, when the system is macroscopic (as in the microwave experiments here presented), measurements are noninvasive.

Another difference is related to the velocity of the processes. Within the electromagnetic framework, the wave equations are relativistic, while quantum particles seldom need the relativistic formalism. Indeed, they are usually described by the Schrödinger equation, rather than by the Klein–Gordon or Dirac ones.

A third difference is connected to the different timescale in which the superluminal effects can be observed. This timescale is confined within a range that is accessible for microwave scale and photon tunneling, while it appears to be rather inaccessible even for light relativistic particles, such as electrons, for which the time scale is of the order of 10^{-21} s.

The fast behavior observed in optical tunneling is attributed to group velocity. The question as to whether the superluminality can be extended to signal velocity is still open, and needs further investigations.

References

1. R. Landauer: Nature **365**, 692 (1993)
2. J.A. Stövneng, E.H. Hauge: Phys. World **23**, November (1993)
3. J. Brown: *New Sci* **1971**, 26 (1995)
4. A. Ranfagni, D. Mugnai, P. Fabeni, G.P. Pazzi: Phys. Scripta **42**, 508 (1990)
5. T.E. Hartman: J. Appl. Phys. **33**, 3427 (1962)
6. A. Enders, G. Nimtz: J. Phys. (Paris) I **2**, 1693 (1992)
7. L. Brillouin: *Wave Propagation and Group Velocity* (Academic Press, New York, 1960)
8. Y. Aharonov, L. Vaidman: Phys. Rev. A **41**, 11 (1990)
9. D. Mugnai, A. Ranfagni, R. Ruggeri, A. Agresti: Phys. Rev. Lett. **68**, 259 (1992)
10. A. Ranfagni, D. Mugnai, P. Fabeni, G.P. Pazzi: Appl. Phys. Lett. **58**, 774 (1991)

11. A. Ranfagni, D. Mugnai, P. Fabeni, G.P. Pazzi, G. Naletto, C. Sozzi: Physica B **175**, 283 (1991)
12. M. Büttiker: Phys. Rev. B **27**, 6178 (1983)
13. D. Sokolovski, J. N. L. Connor: Phys. Rev. A **42**, 6512 (1990)
14. D. Sokolovski, S. Brouard, J. N. L. Connor: Phys. Rev. A **50**, 1240 (1994)
15. D. Mugnai, A. Ranfagni, R. Ruggeri, A. Agresti: Phys. Rev. E **49**, 1771 (1994)
16. D. Mugnai, A. Ranfagni, R. Ruggeri, A. Agresti: Phys. Rev. E **50**, 790 (1994)
17. G. Nimtz, H. Spieker, H. M. Brodowsky: J. Phys. (Paris) I **4**, 1379 (1994)
18. M. Kac: Rocky Mountain J. Math. **4**, 497 (1974)
19. A. Ranfagni, D. Mugnai: Phys. Rev. E. **52**, 1128 (1995)
20. B. Gaveau, T. Jacobson, M. Kac, L. S. Schulman: Phys. Rev. Lett. **53**, 419 (1984)
21. A. Ranfagni, D. Mugnai, A. Agresti: Phys. Lett. A**175**, 334 (1993)
22. R. Fox, C. G. Kuper, S. G. Lipson: Proc. R. Soc. Lond. A **316**, 515 (1970)
23. K. Hass, P. Busch: Phys. Lett. A **185**, 9 (1994)
24. P. Moretti, A. Agresti: Il Nuovo Cimento B **110**, 905 (1995)
25. D. Mugnai, A. Ranfagni, R. Ruggeri, A. Agresti, E. Recami: Phys. Lett. A **209**, 227 (1995)
26. Another approach to the problem can be found in J.G. Muga, M. Büttiker: Phys. Rev. A **62**, 023808 (2000)
27. W. Heitmann, G. Nimtz: Phys. Lett. A **196**, 154 (1994)
28. M. Mojhedi, E. Schamiloglu, F. Hegeler, K.J. Malloy: Phys. Rev. E **62**, 5758 (2000)
29. A. Enders, G. Nimtz: Phys. Rev. E **48**, 632 (1993); J. Phys. (Paris) I **3**, 1089 (1993)
30. A.M. Steinberg, P.G. Kwiat, R.Y. Chiao: Phys. Rev. Lett. **71**, 708 (1993)
31. Ch. Spielmann, R. Szipöcs, A. Stingl, F. Krausz: Phys. Rev. Lett. **73**, 2308 (1994)
32. D. Mugnai, A. Ranfagni, L. Ronchi: Atti della Fondazione Giorgio Ronchi vol. **LIII**, 777 (1998)
33. A. Ranfagni, R. Ruggeri, A. Agresti: Found. Phys. **28**, 515 (1998)
34. R. Feynman, A. Hibbs: *Quantum Mechanics and Path Integrals* (McGraw-Hill, New York, 1965), Chap. 7
35. P. Moretti, D. Mugnai, A. Ranfagni, M. Cetica: Phys. Rev. A **60**, 5087 (1999)
36. A. Ranfagni, R. Ruggeri, C, Susini, A. Agresti, P. Sandri: Phys. Rev. E **63**, 025102 (2001)
37. A. Ranfagni, R. Ruggeri, D. Mugnai, A. Agresti, C. Ranfagni, P. Sandri: Phys. Rev. E **67**, 066611 (2003)
38. A. Ranfagni, D. Mugnai, M. A. Vitali: Phys. Rev. E **69**, 057603 (2004)
39. K.W.H. Stevens: Eur. J. Phys. **1**, 98 (1980); J. Phys. C: Solid State Phys. **16**, 3649 (1983)
40. A. Ranfagni, D, Mugnai, A. Agresti: Phys. Lett. **158**, 161 (1991)
41. S. Brouard, J.G. Muga: Phys. Rev. A **54**, 3055 (1996)
42. N. Marcuvitz: *Waveguide Handbook* (McGraw-Hill, New York, 1951), Chap. 5
43. F. Terman: *Electronic and Radio Engineering* (McGraw-Hill, New York, 1955), Chap. 3
44. F.E. Terman, J.M. Pettit: *Electronic Measurements* (McGraw-Hill, New York, 1952), p. 428
45. D. Mugnai, A. Ranfagni, L.S. Schulman: Phys. Rev. E **55**, 3593 (1997)

46. M.S. Marinov, B. Segev, in: *Tunneling and its Implications, ICTP-Adriatico Research Conference at Trieste, Italy, July 30-August 2, 1996*, ed. by D. Mugnai, A. Ranfagni, L.S. Schulman (World Scientific, Singapore, 1997)
47. S. Bosanac: Phys. Rev. A **28**, 577 (1983)
48. P. Balcou, L. Dutriaux: Phys. Rev. Lett. **78**, 851 (1997)
49. D. Mugnai, A. Ranfagni, L. Ronchi: Phys. Lett. A **247**, 281 (1998)
50. J. Millman, H. Taub: *Pulse and Digital circuits* (McGraw-Hill, New York, 1956), Chap. 2
51. D. Mugnai: Optics Commun. **175**, 309 (2000)
52. D. Mugnai: Optics Commun. **188**, 17 (2001), and **207**, 95 (2002)
53. A. Agresti, P. Sandri, C. Ranfagni, A. Ranfagni, R. Ruggeri: Phys. Rev. E **66**, 067606 (2002)
54. A. Ranfagni, R. Ruggeri, P. Sandri, A. Agresti: Phys. Rev. E **65**, 037601 (2002)

The Two-State Vector Formalism: An Updated Review

Yakir Aharonov[1,2] and Lev Vaidman[1]

[1] School of Physics and Astronomy, Raymond and Beverly Sackler Faculty of Exact Sciences, Tel Aviv University, Tel-Aviv 69978, Israel
yakir@post.tau.ac.il
vaidman@post.tau.ac.il

[2] Department of Physics and Department of Computational and Data Sciences College of Science, George Mason University, Fairfax, VA 22030, USA

13.1 Introduction

The two-state vector formalism of quantum mechanics is a time-symmetrized approach to standard quantum theory particularly helpful for the analysis of experiments performed on pre- and post-selected ensembles. It allows to see numerous peculiar effects which naturally arise in this approach. In particular, the concepts of "weak measurements" (standard measurements with weakening of the interaction) and "weak values" (the outcomes of weak measurements) reveal a very unusual but consistent picture. Recently, more and more effects are viewed as manifestations of weak measurements and more and more weak measurement experiments have been performed. The polemic about the validity of the approach and the meaning of its concepts never stopped. The number of papers written on the subject almost doubled since publication of the first version of the review. The current review does not explain in details the new results, but it puts the development of the approach in the proper context and provides citations for further reading.

13.2 Descriptions of Quantum Systems

13.2.1 The Quantum State

In the standard quantum mechanics, a system at a given time t is described completely by a quantum state

$$|\Psi\rangle \,, \tag{13.1}$$

defined by the results of measurements performed on the system in the past relative to the time t. (It might be that the system at time t is not described

Y. Aharonov and L. Vaidman: *The Two-State Vector Formalism: An Updated Review*, Lect. Notes Phys. **734**, 399–447 (2008)
DOI 10.1007/978-3-540-73473-4_13 © Springer-Verlag Berlin Heidelberg 2008

by a pure quantum state, but by a mixed state (density matrix). However, we can always assume that there is a composite system including this system which is in a pure state.) The status of a quantum state is controversial: there are many papers on reality of a quantum state and numerous interpretations of this "reality." However, it is noncontroversial to say that the quantum state yields maximal information about how this system can affect other systems (in particular, measuring devices) interacting with it at time t. Of course, the results of all measurements in the past, or just the results of the last complete measurement, also have this information, but these results include other facts too, so the quantum state is the most concise information about how the quantum system can affect other systems at time t.

The concept of a quantum state is time-asymmetric: it is defined by the results of measurements in the *past*. This fact by itself is not enough for the asymmetry: in classical physics, the state of a system at time t defined by the results of the complete set of measurements in the past is not different from the state defined by the complete measurements in the future. This is because for a classical system the results of measurements in the future are defined by the results of measurements in the past (and vice versa). In quantum mechanics this is not so: the results of measurements in the future are only partially constrained by the results of measurements in the past. Thus, the concept of a quantum state is genuinely time-asymmetric. The question arises: does the asymmetry of a quantum state reflects the time asymmetry of quantum mechanics, or it can be removed by reformulation of quantum mechanics in a time-symmetric manner?

13.2.2 The Two-state Vector

The two-state vector formalism of quantum mechanics (TSVF) originated in a seminal work of Aharonov, Bergmann, and Lebowitz (ABL) [1] removes this asymmetry. It provides a time-symmetric formulation of quantum mechanics. A system at a given time t is described completely by a *two-state vector*

$$\langle \Phi | \, | \Psi \rangle , \tag{13.2}$$

which consists of a quantum state $|\Psi\rangle$ defined by the results of measurements performed on the system in the past relative to the time t and of a backward evolving quantum state $\langle\Phi|$ defined by the results of measurements performed on this system after the time t. Again, the status of the two-state vector might be interpreted in different ways, but a noncontroversial fact is that it yields maximal information about how this system can affect other systems (in particular, measuring devices) interacting with it at time t.

The description of the system with the two-state vector (13.2) is clearly different from the description with a single quantum state (13.1), but in both cases we claim that "it yields maximal information about how this system can affect other systems (in particular, measuring devices) interacting with it at time t." Does it mean that the TSVF has different predictions than the standard quantum approach? No, the two formalisms describe the same theory with the same predictions. The difference is that the standard approach is time asymmetric and it is assumed that only the results of the measurements in the past exist. With this constraint, $|\Psi\rangle$ indeed contains maximal information about the system at time t. The rational for this approach is that if the results of the future measurements relative to the time t exist too, then "now" is after time t and we cannot return back in time to perform measurements at t. Therefore, taking into account results of future measurements is not useful. In contrast, the TSVF approach is time symmetric. There is no preference to the results of measurements in the past relative to the results of measurements in the future: both are taken into account. Then, there is more information about the system at time t. The maximal information (without constraints) is contained in the two-state vector $\langle\Phi|\ |\Psi\rangle$.

If the TSVF has the same predictions as standard quantum mechanics, what is the reason to consider it? And what about the argument that when the results of future measurements are known it is already too late to make measurements at time t? How might the two-state vector be useful? The answer to the first question is that it is important to understand the time symmetry of nature (described by quantum mechanics). The time asymmetry of the standard approach might be solely due to the usage of time-asymmetric concepts. The answer to the second question is that there are many situations in which we want to know how a system affected other systems in the past. The TSVF proved to be particularly useful after introduction of *weak measurements* [2, 3, 4] which allowed to see that systems described by some two-state vectors can affect other system at time t in a very peculiar way. This has led to the discovery of numerous bizarre effects [5, 6, 7, 8]. It is very difficult to understand these effects in the framework of standard quantum mechanics; some of them can be explained via a miraculous interference phenomenon known as *super-oscillations* [9, 10].

13.2.3 How to Create Quantum Systems Corresponding to Various Complete Descriptions?

The maximal complete description of a quantum system at time t is a two-state vector (13.2). We will name the system which has such a description as *pre- and postselected*. (Again, it might be that at time t the system is not described by a "pure" two-state vector. However, we can assume that there is a composite system including this system which is described by a two-state vector.) In some circumstances, the system might have only a partial description. For example, if time t is "present" and the results of the future measurements do not exist yet, then at that time, the system is described only by a usual forward evolving quantum state (13.1): the *preselected* system. Later, when the results of the future measurements will be obtained, the description will be completed to the form (13.2). It is also possible to arrange a situation in which, until some measurements in the future, the complete description of the system at time t is the backward evolving quantum state $\langle\Phi|$: the *postselected* system. We will now explain how all these situations can be achieved.

Single Forward-Evolving Quantum State

In order to have *now* a system the complete description of which at time t is a single quantum state (13.1), there should be a complete measurement in the past of time t and no measurement on the system after time t, see Fig. 13.1 a. The system in the state $|\Psi\rangle$ is obtained when a measurement of an observable A at time t_1 is performed, $t_1 < t$, obtaining a specific outcome $A = a$ such that the created state $|a\rangle$ performs unitary evolution between t_1 and t governed by the Hamiltonian H,

$$U(t_1, t) = e^{-i \int_{t_1}^{t} H dt}, \tag{13.3}$$

to the desired state:

$$|\Psi\rangle = U(t_1, t) |a\rangle. \tag{13.4}$$

The time "now," t_{now} should either be equal to the time t, or it should be known that during the time period $[t, t_{now}]$ no measurements have been performed on the system. The state $|\Psi\rangle$ remains to be the complete description of the system at time t until the future measurements on the system will be performed yielding additional information.

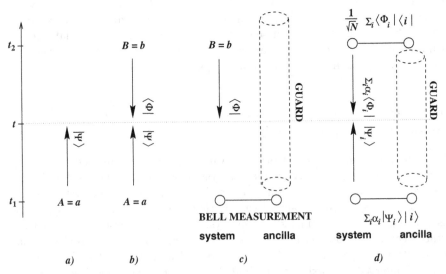

Fig. 13.1. Description of quantum systems: (**a**) pre-selected, (**b**) pre- and postselected, (**c**) postselected, and (**d**) generalized pre- and postselected

The Two-State Vector

In order to have *now* a system the complete description of which at time t is a two-state vector (13.2), there should be a complete measurement in the past of time t and a complete measurement after the time t, see Fig. 13.1b. In addition to the measurement $A = a$ at time t_1, there should be a complete measurement at t_2, $t_2 > t$, obtaining a specific outcome $B = b$ such that the backward time evolution from t_2 to t leads to the desired state

$$\langle \Phi| \; = \; \langle b| \, U^\dagger(t, t_2) \,. \qquad (13.5)$$

The time "now", t_{now} is clearly larger than t_2. The two-state vector $\langle \Phi| \, |\Psi\rangle$ is the complete description of the system at time t starting from the time t_2 and forever.

A Single Backward-Evolving Quantum State

We have presented above a description of quantum systems by a single forward-evolving quantum state (13.1) and by a two-state vector (13.2). It is natural to ask: Are there systems described by a single backward-evolving quantum state? The notation for such a state is

$$\langle \Phi| \,. \qquad (13.6)$$

A measurement of B at time t_2, even in the case it yields the desired outcome $B = b$, is not enough. The difference between preparation of (13.1) and (13.2) is that at present, t, the future of a quantum system does not exist (the future measurements have not been performed yet), but the past of a quantum system exists: it seems that even if *we* do not know it, there is a quantum state of the system evolving towards the future defined by the results of measurements in the past. Therefore, in order to obtain a quantum system described by a backward-evolving quantum state (13.2), in addition to the postselection measurement performed after time t, we have to *erase* the past.

How to erase the past of a quantum system? A complete measurement before the time t certainly partially erases the information which the system had before the measurement, but it also creates the new information: the result of this measurement. It creates another quantum state evolving forward in time, and this is, really, what we need to erase. We have to achieve the situation in which no information arrives from the past. It seems impossible given the assumption that all the past is known. However, if we perform a measurement on a composite system containing our system and an auxiliary system, *an ancilla*, then it can be done, see Fig. 13.1c. Performing a Bell-type measurement results in one of a completely correlated states of the system and the ancilla (the Einstein-Podolsky-Rosen (EPR)-type state). In such a state, each system has equal probability to be found in any state. However, the measurement on one system fixes the state of the other, so, in addition to the Bell-type measurement we need to "guard" the ancilla such that no measurement could be performed on it until now. Again, the complete description of a quantum system by a single (this time backward-evolving) quantum state can be achieved only for a period of time until the measurements on the ancilla would fix the forward-evolving quantum state for the system.

The backward evolving state is a premise not only of the two-state vector formalism, but also of "retrodictive" quantum mechanics [11, 12, 13, 14, 15], which deals with the analysis of quantum systems based on a quantum measurement performed in the future relative to the time in question. It is also relevant to "consistent histories" and "decoherent histories" approaches [16, 17].

13.2.4 The Generalized Two-State Vector

The descriptions we described above correspond to an "ideal" case. We have assumed that complete measurements have been performed on the system in the past, or in the future or both. The philosophical question is this: can we assume that going sufficiently far away to the past, far away to the future and far in the sense of considering composite systems larger and larger, at the end there always be a complete description in the form of a two-state vector. Usually we do put constraints on how far we go (at least regarding the future and the size of the system). In constructing the situation in which a system is described by a backward-evolving quantum state only, we already limited ourselves to a particular system instead of being satisfied by the correct claim that our system is a part of a composite system (which includes also the ancilla) which does have forward-evolving quantum state. As in the standard approach, limiting our analysis to a particular system leads to descriptions with *mixed* states. There are situations in which the forward-evolving state is a mixed state (the system is correlated to an ancilla) and the backward-evolving state is another mixed state (the system correlated to another ancilla). Although the generalization to the mixed states is straightforward, it is not obvious what is its most convenient form. For a powerful, but somewhat cumbersome formalism, see [18]. However, there is a particular case which is not too difficult to describe. It corresponds to another "pure" two-state vector description: *generalized two-state vector*.

Generalized two-state vector [4] is the name for the superposition of two-state vectors

$$\sum_i \alpha_i \langle \Phi_i | \, |\Psi_i \rangle \, . \tag{13.7}$$

In general, the sets $\{|\Psi_i\rangle\}$, $\{\langle \Phi_i|\}$ need not be orthogonal. Then, the normalization should be chosen consistently, although it is not very important since in main applications of this concept the normalization does not affect anything.

For simplicity, we will consider the case of zero free Hamiltonian for the system and for the ancilla. In order to obtain the generalized two-state vector (13.7) we have to prepare at t_1 the system and the ancilla in a correlated state $\sum_i \alpha_i |\Psi_i\rangle |i\rangle$, where $\{|i\rangle\}$ is a set of orthonormal states of the ancilla. Then we have to "guard" the ancilla such that there will be no measurements or any other interactions performed on the ancilla until the postselection measurement of a projection on the correlated state $1/\sqrt{N} \sum_i |\Phi_i\rangle|i\rangle$, see Fig. 13.1d. If we obtain the desired outcome, then the system is described at time t by the generalized two-state vector (13.7).

13.3 Ideal Quantum Measurements

13.3.1 Von Neumann Measurements

In this section I shall discuss how a quantum system characterized by a certain description interacts with other systems. Some particular types of interactions are named *measurements* and the effect of these interactions characterized as the results of these measurements. The basic concept is an *ideal quantum measurement* of an observable C. This operation is defined for preselected quantum systems in the following way:

> If the state of a quantum system before the measurement was an eigenstate of C with an eigenvalue c_n then the outcome of the measurement is c_n and the quantum state of the system is not changed.

The standard implementation of the ideal quantum measurement is modeled by the von Neumann Hamiltonian [19]:

$$H = g(t)PC , \qquad (13.8)$$

where P is the momentum conjugate to the pointer variable Q, and the normalized coupling function $g(t)$ specifies the time of the measurement interaction. The outcome of the measurement is the shift of the pointer variable during the interaction. In an ideal measurement the function $g(t)$ is nonzero only during a very short period of time, and the free Hamiltonian during this period of time can be neglected.

13.3.2 The Aharonov–Bergmann–Lebowitz Rule

For a quantum system described by the two-state vector (13.2), the probability for an outcome c_n of an ideal measurement of an observable C is given by [1, 4]

$$\text{Prob}(c_n) = \frac{|\langle\Phi|\mathbf{P}_{C=c_n}|\Psi\rangle|^2}{\sum_j |\langle\Phi|\mathbf{P}_{C=c_j}|\Psi\rangle|^2} . \qquad (13.9)$$

For a quantum system described by a *generalized two-state vector* (13.7) the probability for an outcome c_n is given by [4]

$$\text{Prob}(c_n) = \frac{|\sum_i \alpha_i \langle\Phi_i|\mathbf{P}_{C=c_n}|\Psi_i\rangle|^2}{\sum_j |\sum_i \alpha_i \langle\Phi_i|\mathbf{P}_{C=c_j}|\Psi_i\rangle|^2} . \qquad (13.10)$$

Another important generalization of the formula (13.9) is for the case in which the postselection measurement is not complete and therefore it does not specify a single postselection state $\langle \Phi |$. Such an example was recently considered by Cohen [20] in an (unsuccessful [21]) attempt to find constraints to the applicability of the ABL formula. In this case, the postselection measurement is a projection on a *degenerate* eigenvalue of an observable $B = b$. The modified ABL formula is [21]:

$$\text{Prob}(c_n) = \frac{\|\mathbf{P}_{B=b}\mathbf{P}_{C=c_n}|\Psi\rangle\|^2}{\sum_j \|\mathbf{P}_{B=b}\mathbf{P}_{C=c_j}|\Psi\rangle\|^2} . \tag{13.11}$$

This form of the ABL formula allows to connect it to the standard formalism of quantum theory in which there is no post-selection. In the limiting case when the projection operator $\mathbf{P}_{B=b}$ is just the unity operator $\mathbf{I}$, we obtain the usual expression:

$$\text{Prob}(c_n) = \|\mathbf{P}_{C=c_n}|\Psi\rangle\|^2 . \tag{13.12}$$

13.3.3 Three-Boxes Example

Consider a particle which can be located in one out of three boxes. We denote the state of the particle when it is in box i by $|i\rangle$. At time t_1 the particle is prepared in the state

$$|\Psi\rangle = \frac{1}{\sqrt{3}}(|1\rangle + |2\rangle + |3\rangle) . \tag{13.13}$$

At time t_2 the particle is found to be in the state

$$|\Phi\rangle = \frac{1}{\sqrt{3}}(|1\rangle + |2\rangle - |3\rangle) . \tag{13.14}$$

We assume that in the time interval $[t_1, t_2]$ the Hamiltonian is zero. Therefore, at time t, $t_1 < t < t_2$, the particle is described by the two-state vector

$$\langle \Phi | \, |\Psi\rangle = \frac{1}{3}((\langle 1| + \langle 2| - \langle 3|) \, (|1\rangle + |2\rangle + |3\rangle) . \tag{13.15}$$

Probably the most peculiar fact about this single particle is that it can be found with certainty in two boxes [4]. Indeed, if at time t we open box 1, we are certain to find the particle in box 1; and if we open box 2 instead, we are certain to find the particle in box 2. These results can be obtained by straightforward application of the ABL formula (13.9). Opening box i corresponds to measuring the projection operator $\mathbf{P}_i = |i\rangle\langle i|$. The corresponding operators appearing in (13.9) are

$$\mathbf{P}_{\mathbf{P}_i=1} = |i\rangle\langle i|, \qquad \mathbf{P}_{\mathbf{P}_i=0} = \sum_{j \neq i} |j\rangle\langle j| \tag{13.16}$$

Therefore, the calculation of the probability to find the particle in box 1 yields:

$$\text{Prob}(\mathbf{P}_1 = 1) = \frac{|\langle\Phi|1\rangle\langle1|\Psi\rangle|^2}{|\langle\Phi|1\rangle\langle1|\Psi\rangle|^2 + |\langle\Phi|2\rangle\langle2|\Psi\rangle + \langle\Phi|3\rangle\langle3|\Psi\rangle|^2} = \frac{|\frac{1}{3}|^2}{|\frac{1}{3}|^2 + |0|^2} = 1 .$$
(13.17)

Similarly, we obtain $\text{Prob}(\mathbf{P}_2 = 1) = 1$. Note, that if we open both box 1 and box 2, we might not see the particle at all.

This example can be generalized to the case of a large number of boxes N. A single particle described by a two-state vector

$$\frac{1}{N}(\langle1| + \langle2| + ... - \sqrt{N-2}\langle N|) (|1\rangle + |2\rangle + ... + \sqrt{N-2}|N\rangle) .$$
(13.18)

This single particle is, in some sense, simultaneously in $N-1$ boxes: whatever box is opened (except the last one) we are certain to find the particle there.

Recently, we found that the particle is simultaneously in several boxes even in a more robust sense [22]. We cannot find it simultaneously in all boxes if we look at all of them, but a single photon can! We found that a photon will scatter from our pre- and postselected particle, as if there were particles in all boxes.

The analysis of the three-boxes example has interesting features also in the framework of the consistent histories approach [23, 24, 25]. On the other hand, it generated significant controversies. The legitimacy of counterfactual statements were contested, see discussion in Sect 5.4, the Kastner criticism [26] and Vaidman's reply [27], and it was claimed by Kirkpatrick [28] that the three-boxes example does not exhibit genuine quantum paradoxical feature because it has a classical counterpart. Very recently Ravon and Vaidman [29] showed that Kirkpatrick's proposal fails to mimic quantum behavior and that the three-box example is one of not too many classical tasks which can be done better using quantum tools. (We could not see a refutation of this statement in Kirkpatrick's reply [30].) This is the paradoxical feature of the three-box experiment which was overlooked by Leavens et al. [31] who considered variations of the three-box experiment with modified pre- and postselected states.

Recently, a setup equivalent to the three-box example was presented as a novel *counterfactual computation* method [32]. The analysis of this proposal in the framework of the two-state vector formalism [33] shows that one cannot claim that the computer yields the result of computation without actually performing the computation and therefore, the proposal fails to provide counterfactual computation for all possible outcomes as it was originally claimed.

13.3.4 The Failure of the Product Rule

An important difference between pre- and postselected systems and preselected systems only is that the *product rule* does not hold [34]. The product rule, which does hold for preselected quantum systems is that if $A = a$ and

$B = b$ with certainty, then it is certain that $AB = ab$. In the three-boxes case we know with certainty that $\mathbf{P}_1 = 1$, $\mathbf{P}_2 = 1$. However, $\mathbf{P}_1\mathbf{P}_2 = 0$.

Another example of this kind in a which measurement in one place affects the outcome of a measurement in another place is a pre- and postselected pair of separate spin-$\frac{1}{2}$ particles [35]. The particles are prepared, at time t_1, in a singlet state. At time t_2 measurements of σ_{1x} and σ_{2y} are performed and certain results are obtained, say $\sigma_{1x} = 1$ and $\sigma_{2y} = 1$, i.e., the pair is described at time t, $t_1 < t < t_2$, by the two-state vector

$$\frac{1}{\sqrt{2}} \langle \uparrow_x | \, \langle \uparrow_y | \, (| \uparrow_z \rangle | \downarrow_z \rangle - | \downarrow_z \rangle | \uparrow_z \rangle) \,. \tag{13.19}$$

If at time t a measurement of σ_{1y} is performed (and if this is the only measurement performed between t_1 and t_2), then the outcome of the measurement is known with certainty: $\sigma_{1y}(t) = -1$. If, instead, only a measurement of σ_{2x} is performed at time t, the result of the measurement is also certain: $\sigma_{2x}(t) = -1$. The operators σ_{1y} and σ_{2x} obviously commute, but nevertheless, measuring $\sigma_{2x}(t)$ clearly disturbs the outcome of the measurement of $\sigma_{1y}(t)$: it is not certain anymore.

Measuring the product $\sigma_{1y}\sigma_{2x}$, is, in principle, different from the measurement of both σ_{1y} and σ_{2x} separately. In our example, the outcome of the measurement of the product *is* certain, the ABL formula (13.9) yields $\sigma_{1y}\sigma_{2x} = -1$. Nevertheless, it does not equal the product of the results which must come out of the measurements of σ_{1y} and σ_{2x} when every one of them is performed without the other.

Note measurability of the product $\sigma_{1y}\sigma_{2x}$ using only local interactions. Indeed, we may write the product as a modular sum, $\sigma_{1y}\sigma_{2x} = (\sigma_{1y} + \sigma_{2x})\mathrm{mod}4 - 1$. It has been shown [36] that nonlocal operators such as $(\sigma_{1y} + \sigma_{2x})\mathrm{mod}4$ can be measured using solely local interactions.

Hardy [37] analyzed another very spectacular example in which an electron and a positron are found with certainty if searched for in a particular place, but, nevertheless, if both are searched simultaneously, there is certainty *not* to find them together. Again, the failure of the product rule explains this counterintuitive situation and the far reaching conclusions of Hardy's paper seem not to be warranted [34].

The two spin-$\frac{1}{2}$ particles example with a small modification of omitting the measurement at time t_2 performed on a second particle, but instead, "guarding" it starting from time t_1 against any measurement, is a demonstration of obtaining a quantum system described only by a backward-evolving quantum state $\langle \uparrow_x |$. The probability distribution for outcomes of spin-component measurements performed at time t is identical to that of a particle in a preselected state $| \uparrow_x \rangle$. Note that for quantum systems which are postselected only, the product rule does hold.

Recently [38] it has been shown that pre- and postselection allows another related peculiar feature: "a posteriori" realization of super-correlations maximally violating the CHSH bound, which have been termed as Popescu–Rohrlich boxes [39].

13.3.5 Ideal Measurements Performed on a System Described by Generalized Two-State Vector

Another modification, replacing the measurements at t_2 on two particles by measurement of a nonlocal variable such as a Bell operator on both particles and guarding the second particle between t_1 and t_2 produces a *generalized two-state vector* for the first particle. Such particles might have a peculiar feature that the outcome of spin component measurements is certain in a continuum of directions. This is a surprising result because the preselected particle might have definite value of spin component at most in one direction and the particle described by two-state vector will have definite results of spin component measurements in two directions: one defined by preselection and one defined by postselection (the directions might coincide). For example [4], the particle described by a generalized two-state vector

$$\cos\chi\langle\uparrow_z \mid \mid \uparrow_z\rangle - \sin\chi\langle\downarrow_z \mid \mid \downarrow_z\rangle , \qquad \chi \in \left(0, \frac{\pi}{2}\right) , \qquad (13.20)$$

will yield the outcome $\sigma_\eta = 1$ for the cone of directions $\hat{\eta}$ making angle θ with the z axes such that $\theta = 4\arctan\sqrt{\tan\chi}$. This can be verified directly using the formula (13.10), but we will bring another argument for this result below.

The generalized two-state vector is obtained when there is a particular result of the nonlocal measurement at time t_2. It is interesting that we can construct a particular measurement at time t_2 such that whatever the outcome will be there will be a cone of directions in which the spin has a definite value. These cones intersect in general in four lines. It can be arranged that they will "touch" on, say x-axis and intersect in y- and z-axes. Then, in all cases we will be able to ascertain the value of σ_x, σ_y, and σ_z of a single particle [5].

The problem was also analyzed in the framework of the standard approach [40, 41] and after coining the name "The Mean King Problem" continued to be a topic of an extensive analysis. It has been generalized to the spin-1 particle [42] and to a higher dimentional case [43, 44]. The research continues until today [45, 46, 47, 48, 49, 50]. Moreover, today's technology converted from gedanken quantum game to a real experiment. Schulz et al. [51] performed this experiment with polarized photons (instead of spin-$\frac{1}{2}$ particles).

13.4 Weak Measurements

13.4.1 Introduction

The most interesting phenomena which can be seen in the framework of the TSVF are related to *weak measurements* [3]. A weak measurement is a standard measuring procedure (described by the Hamiltonian (13.8)) with weakened coupling. In an ideal measurement, the initial position of the pointer

Q is well localized around zero and therefore the conjugate momentum P has a very large uncertainty which leads to a very large uncertainty in the Hamiltonian of the measurement (13.8). In a weak measurement, the initial state of the measuring device is such that P is localized around zero with small uncertainty. This leads, of course, to a large uncertainty in Q and therefore the measurement becomes imprecise. However, by performing the weak measurement on an ensemble of N identical systems we improve the precision by a factor $\sqrt{N}$ and in some special cases we can obtain good precision even in a measurement performed on a single system [2].

The idea of weak measurements is to make the coupling with the measuring device sufficiently weak so that the change of the quantum state due to the measurements can be neglected. In fact, we require that the two-state vector is not significantly disturbed, i.e., neither the usual, forward-evolving quantum state, nor the backward-evolving quantum state is changed significantly. Then, the outcome of the measurement should be affected by both states. Indeed, the outcome of a weak measurement of a variable C performed on a system described by the two-state vector $\langle \Phi | \, | \Psi \rangle$ is the *weak value* of C:

$$C_w \equiv \frac{\langle \Phi | C | \Psi \rangle}{\langle \Phi | \Psi \rangle} . \tag{13.21}$$

Strictly speaking, the readings of the pointer of the measuring device will cluster around $\text{Re}(C_w)$. In order to find $\text{Im}(C_w)$ one should measure the shift in P [3].

The weak value for a system described by a generalized two-state vector (13.7) is [4]:

$$C_w = \frac{\sum_i \alpha_i \langle \Phi_i | C | \Psi_i \rangle}{\sum_i \alpha_i \langle \Phi_i | \Psi_i \rangle} . \tag{13.22}$$

Next, let us give the expression for the weak value when the postselection measurement is not complete. Consider a system preselected in the state $|\Psi\rangle$ and postselected by the measurement of a degenerate eigenvalue b of a variable B. The weak value of C in this case is:

$$C_w = \frac{\langle \Psi | \mathbf{P}_{B=b} C | \Psi \rangle}{\langle \Psi | \mathbf{P}_{B=b} | \Psi \rangle} . \tag{13.23}$$

This formula allows us to find the outcome of a weak measurement performed on a preselected (only) system. Replacing $\mathbf{P}_{B=b}$ by the unity operator yields the result that the weak value of a preselected system in the state $|\Psi\rangle$ is the expectation value:

$$C_w = \langle \Psi | C | \Psi \rangle . \tag{13.24}$$

Let us show how the weak values emerge as the outcomes of weak measurements. We will limit ourselves to two cases: first, the weak value of the preselected state only (13.24) and then, the weak value of the system described by the two-state vector (13.21).

In the weak measurement, as in the standard von Neumann measurement, the Hamiltonian of the interaction with the measuring device is given by (13.8). The weakness of the interaction is achieved by preparing the initial state of the measuring device in such a way that the conjugate momentum of the pointer variable, P, is small, and thus the interaction Hamiltonian (13.8) is small. The initial state of the pointer variable is modeled by a Gaussian centered at zero:

$$\Psi_{in}^{MD}(Q) = (\Delta^2\pi)^{-1/4}e^{-Q^2/2\Delta^2} . \tag{13.25}$$

The pointer is in the "zero" position before the measurement, i.e., its initial probability distribution is

$$\text{Prob}(Q) = (\Delta^2\pi)^{-1/2}e^{-Q^2/\Delta^2} . \tag{13.26}$$

If the initial state of the system is a superposition $|\Psi\rangle = \Sigma\alpha_i|c_i\rangle$, then after the interaction (13.8) the state of the system and the measuring device is:

$$(\Delta^2\pi)^{-1/4}\Sigma\alpha_i|c_i\rangle e^{-(Q-c_i)^2/2\Delta^2} . \tag{13.27}$$

The probability distribution of the pointer variable corresponding to the state (13.27) is:

$$\text{Prob}(Q) = (\Delta^2\pi)^{-1/2}\Sigma|\alpha_i|^2e^{-(Q-c_i)^2/\Delta^2} . \tag{13.28}$$

In case of the ideal measurement, this is a weighted sum of the initial probability distribution localized around various eigenvalues. Therefore, the reading of the pointer variable in the end of the measurement almost always yields the value close to one of the eigenvalues. The limit of weak measurement corresponds to $\Delta \gg c_i$ for all eigenvalues c_i. Then, we can perform the Taylor expansion of the sum (13.28) around $Q = 0$ up to the first order and rewrite the probability distribution of the pointer in the following way:

$$\text{Prob}(Q) = (\Delta^2\pi)^{-1/2}\Sigma|\alpha_i|^2e^{-(Q-c_i)^2/\Delta^2} =$$
$$(\Delta^2\pi)^{-1/2}\Sigma|\alpha_i|^2(1 - (Q-c_i)^2/\Delta^2) = (\Delta^2\pi)^{-1/2}e^{-(Q-\Sigma|\alpha_i|^2c_i)^2/\Delta^2} \tag{13.29}$$

But this is exactly the initial distribution shifted by the value $\Sigma|\alpha_i|^2c_i$. This is the outcome of the measurement, in this case the weak value is the expectation value:

$$C_w = \Sigma|\alpha_i|^2c_i = \langle\Psi|C|\Psi\rangle . \tag{13.30}$$

The weak value is obtained from statistical analysis of the readings of the measuring devices of the measurements on an ensemble of identical quantum systems. But it is different conceptually from the standard definition of expectation value which is a mathematical concept defined from the statistical analysis of the *ideal* measurements of the variable C all of which yield one of the eigenvalues c_i.

Now let us turn to the system described by the two-state vector (13.2). As usual, the free Hamiltonian is assumed to be zero so it can be obtained by preselection of $|\Psi\rangle$ at t_1 and postselection of $|\Phi\rangle$ at t_2. The (weak) measurement interaction of the form (13.8) takes place at time t, $t_1 < t < t_2$. The state of the measuring device after this sequence of measurements is given (up to normalization) by

$$\Psi^{MD}(Q) = \langle \Phi | e^{-iPC} | \Psi \rangle e^{-Q^2/2\Delta^2} . \tag{13.31}$$

After simple algebraic manipulations we can rewrite it (in the P-representation) as

$$\tilde{\Psi}^{MD}(P) = \langle \Phi | \Psi \rangle \, e^{-iC_w P} \, e^{-\Delta^2 P^2/2} \tag{13.32}$$

$$+ \langle \Phi | \Psi \rangle \sum_{n=2}^{\infty} \frac{(iP)^n}{n!} [(C^n)_w - (C_w)^n] e^{-\Delta^2 P^2/2} .$$

If Δ is sufficiently large, we can neglect the second term of (13.32) when we Fourier transform back to the Q-representation. Large Δ corresponds to weak measurement in the sense that the interaction Hamiltonian (13.8) is small. Thus, in the limit of weak measurement, the final state of the measuring device (in the Q-representation) is

$$\Psi^{MD}(Q) = (\Delta^2 \pi)^{-1/4} e^{-(Q-C_w)^2/2\Delta^2} . \tag{13.33}$$

This state represents a measuring device pointing to the weak value (13.21).

Weak measurements on pre- and postselected ensembles yield, instead of eigenvalues, a value which might lie far outside the range of the eigenvalues. Although we have shown this result for a specific von Neumann model of measurements, the result is completely general: any coupling of a pre- and postselected system to a variable C, provided the coupling is sufficiently weak, results in effective coupling to C_w. This weak coupling between a single system and the measuring device will not, in most cases, lead to a distinguishable shift of the pointer variable, but collecting the results of measurements on an ensemble of pre- and postselected systems will yield the weak values of a measured variable to any desired precision.

When the strength of the coupling to the measuring device goes to zero, the outcomes of the measurement invariably yield the weak value. To be more precise, a measurement yields the real part of the weak value. Indeed, the weak value is, in general, a complex number, but its imaginary part will contribute only a (position dependent) phase to the wave function of the measuring device in the position representation of the pointer. Therefore, the imaginary part will not affect the probability distribution of the pointer position which is what we see in a usual measurement. However, the imaginary part of the weak value also has physical meaning. It is equal to the shift of the Gaussian wave function of the measuring device in the momentum representation. Thus, measuring the shift of the momentum of the pointer will yield the imaginary part of the weak value.

The research of weak measurements continues until today. Recently, Botero [52] noted that in some cases the pointer of the weak measurements in some cases has narrower distribution after the weak measurement interaction than it has before. Note also recent different ways of the analysis of the weak measurement effect [53, 54, 55, 56, 57, 58, 59, 60].

13.4.2 Examples: Measurements of Spin Components

Let us consider a simple Stern–Gerlach experiment: measurement of a spin component of a spin-$\frac{1}{2}$ particle. We shall consider a particle prepared in the initial state spin "up" in the $\hat{x}$ direction and postselected to be "up" in the $\hat{y}$ direction. At the intermediate time we measure, weakly, the spin component in the $\hat{\xi}$ direction which is bisector of $\hat{x}$ and $\hat{y}$, i.e., $\sigma_\xi = (\sigma_x + \sigma_y)/\sqrt{2}$. Thus $|\Psi\rangle = |\uparrow_x\rangle$, $|\Phi\rangle = |\uparrow_y\rangle$, and the weak value of σ_ξ in this case is

$$(\sigma_\xi)_w = \frac{\langle\uparrow_y|\sigma_\xi|\uparrow_x\rangle}{\langle\uparrow_y|\uparrow_x\rangle} = \frac{1}{\sqrt{2}}\frac{\langle\uparrow_y|(\sigma_x+\sigma_y)|\uparrow_x\rangle}{\langle\uparrow_y|\uparrow_x\rangle} = \sqrt{2}\,. \tag{13.34}$$

This value is, of course, "forbidden" in the standard interpretation where a spin component can obtain the (eigen)values ± 1 only.

An effective Hamiltonian for measuring σ_ξ is

$$H = g(t)P\sigma_\xi\,. \tag{13.35}$$

Writing the initial state of the particle in the σ_ξ representation, and assuming the initial state (13.25) for the measuring device, we obtain that after the measuring interaction the quantum state of the system and the pointer of the measuring device is

$$\cos(\pi/8)|\uparrow_\xi\rangle e^{-(Q-1)^2/2\Delta^2} + i\sin(\pi/8)|\downarrow_\xi\rangle e^{-(Q+1)^2/2\Delta^2}\,. \tag{13.36}$$

The probability distribution of the pointer position, if it is observed now without postselection, is the sum of the distributions for each spin value. It is, up to normalization,

$$\text{Prob}(Q) = \cos^2(\pi/8)e^{-(Q-1)^2/\Delta^2} + \sin^2(\pi/8)e^{-(Q+1)^2/\Delta^2}\,. \tag{13.37}$$

In the usual strong measurement, $\Delta \ll 1$. In this case, as shown on Fig. 13.2a, the probability distribution of the pointer is localized around -1 and $+1$ and it is strongly correlated to the values of the spin, $\sigma_z = \pm 1$.

Weak measurements correspond to a Δ which is much larger than the range of the eigenvalues, i.e., $\Delta \gg 1$. Figure 13.2b shows that the pointer distribution has a large uncertainty, but it is peaked between the eigenvalues, more precisely, at the expectation value $\langle\uparrow_x|\sigma_\xi|\uparrow_x\rangle = 1/\sqrt{2}$. An outcome of an individual measurement usually will not be close to this number, but it can be found from an ensemble of such measurements, see Fig. 13.2c. Note, that we have not yet considered the postselection.

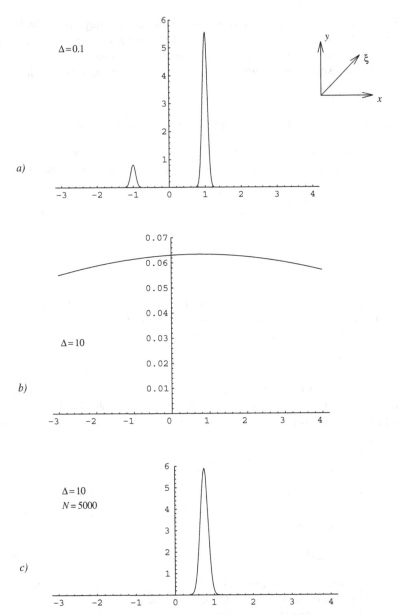

Fig. 13.2. Spin component measurement without post-selection: Probability distribution of the pointer variable for measurement of σ_ξ when the particle is preselected in the state $|\uparrow_x\rangle$. **(a)** Strong measurement, $\Delta = 0.1$. **(b)** Weak measurement, $\Delta = 10$. **(c)** Weak measurement on the ensemble of 5000 particles. The original width of the peak, 10, is reduced to $10/\sqrt{5000} \simeq 0.14$. In the strong measurement **(a)** the pointer is localized around the eigenvalues ± 1, while in the weak measurements **(b)** and **(c)** the peak is located in the expectation value $\langle\uparrow_x|\sigma_\xi|\uparrow_x\rangle = 1/\sqrt{2}$

In order to simplify the analysis of measurements on the pre- and postselected ensemble, let us assume that we first make the postselection of the spin
of the particle and only then look at the pointer of the device that weakly
measures σ_ξ. We must get the same result as if we first look at the outcome
of the weak measurement, make the postselection, and discard all readings
of the weak measurement corresponding to the cases in which the result is
not $\sigma_y = 1$. The postselected state of the particle in the σ_ξ representation is
$\langle \uparrow_y | = \cos(\pi/8)\langle \uparrow_\xi | - i \sin(\pi/8)\langle \downarrow_\xi |$. The state of the measuring device after
the postselection of the spin state is obtained by projection of (13.36) onto
the postselected spin state:

$$\Phi(Q) = \mathcal{N}\left(\cos^2(\pi/8)e^{-(Q-1)^2/2\Delta^2} - \sin^2(\pi/8)e^{-(Q+1)^2/2\Delta^2}\right), \quad (13.38)$$

where $\mathcal{N}$ is a normalization factor. The probability distribution of the pointer
variable is given by

$$\mathrm{Prob}(Q) = \mathcal{N}^2\left(\cos^2(\pi/8)e^{-(Q-1)^2/2\Delta^2} - \sin^2(\pi/8)e^{-(Q+1)^2/2\Delta^2}\right)^2. \quad (13.39)$$

If the measuring interaction is strong, $\Delta \ll 1$, then the distribution is
localized around the eigenvalues ± 1 (mostly around 1 since the pre- and postselected probability to find $\sigma_\xi = 1$ is more than 85%), see Fig. 13.3a and b.
But when the strength of the coupling is weakened, i.e., Δ is increased, the
distribution gradually changes to a single broad peak around $\sqrt{2}$, the weak
value, see Fig. 13.3c–e.

The width of the peak is large and therefore each individual reading of the
pointer usually will be far from $\sqrt{2}$. The physical meaning of the weak value
can, in this case, be associated only with an ensemble of pre- and postselected
particles. The accuracy of defining the center of the distribution goes as $1/\sqrt{N}$,
so increasing N, the number of particles in the ensemble, we can find the weak
value with any desired precision, see Fig. 13.3f.

In our example, the weak value of the spin component is $\sqrt{2}$, which is only
slightly more than the maximal eigenvalue, 1. By appropriate choice of the
pre- and postselected states we can get pre- and postselected ensembles with
arbitrarily large weak value of a spin component. One of our first proposals
[6] was to obtain $(\sigma_\xi)_w = 100$. In this case the postselected state is nearly
orthogonal to the preselected state and, therefore, the probability to obtain
appropriate postselection becomes very small. While in the case of $(\sigma_\xi)_w = \sqrt{2}$
the pre- and postselected ensemble was about half of the preselected ensemble,
in the case of $(\sigma_\xi)_w = 100$ the postselected ensemble will be smaller than the
original ensemble by the factor of $\sim 10^{-4}$.

13.4.3 Weak Measurements Which Are not Really Weak

We have shown that weak measurements can yield very surprising values
which are far from the range of the eigenvalues. However, the uncertainty of
a single weak measurement (i.e., performed on a single system) in the above

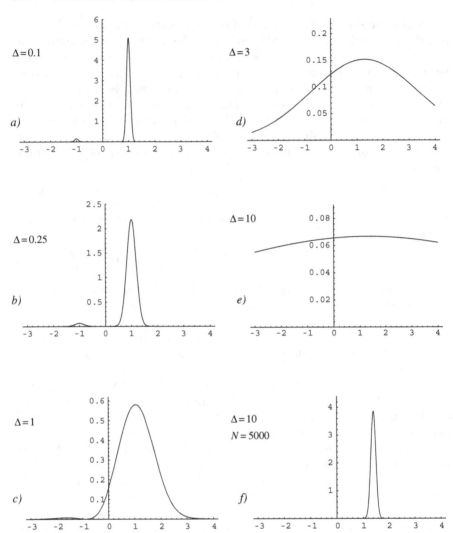

Fig. 13.3. Measurement on pre- and postselected ensemble: Probability distribution of the pointer variable for measurement of σ_ξ when the particle is preselected in the state $|{\uparrow_x}\rangle$ and postselected in the state $|{\uparrow_y}\rangle$. The strength of the measurement is parameterized by the width of the distribution Δ. (a) $\Delta = 0.1$; (b) $\Delta = 0.25$; (c) $\Delta = 1$; (d) $\Delta = 3$; (e) $\Delta = 10$. (f) Weak measurement on the ensemble of 5000 particles; the original width of the peak, $\Delta = 10$, is reduced to $10/\sqrt{5000} \simeq 0.14$. In the strong measurements (a)–(b) the pointer is localized around the eigenvalues ± 1, while in the weak measurements (d)–(f) the peak of the distribution is located in the weak value $(\sigma_\xi)_w = \langle{\uparrow_y}|\sigma_\xi|{\uparrow_x}\rangle/\langle{\uparrow_y}|{\uparrow_x}\rangle = \sqrt{2}$. The outcomes of the weak measurement on the ensemble of 5000 pre- and postselected particles, (f), are clearly outside the range of the eigenvalues, (-1,1)

example is larger than the deviation from the range of the eigenvalues. Each single measurement separately yields almost no information and the weak value arises only from the statistical average on the ensemble. The weakness and the uncertainty of the measurement goes together. Weak measurement corresponds to small value of P in the Hamiltonian (13.8) and, therefore, the uncertainty in P has to be small. This requires large Δ, the uncertainty of the pointer variable. Of course, we can construct measurement with large uncertainty which is not weak at all, for example, by preparing the measuring device in a mixed state instead of a Gaussian, but no precise measurement with weak coupling is possible. So, usually, a weak measurement on a single system will not yield the weak value with a good precision. However, there are special cases when it is not so. Usual strength measurement on a single pre- and postselected system can yield "unusual" (very different from the eigenvalues) weak value with a good precision. Good precision means that the uncertainty is much smaller than the deviation from the range of the eigenvalues.

Our example above was not such a case. The weak value $(\sigma_\xi)_w = \sqrt{2}$ is larger than the highest eigenvalue, 1, only by ~ 0.4, while the uncertainty, 1, is not sufficiently large for obtaining the peak of the distribution near the weak value, see Fig. 13.3c. Let us modify our experiment in such a way that a single experiment will yield meaningful surprising result. We consider a system of N spin-$\frac{1}{2}$ particles all prepared in the state $|\uparrow_x\rangle$ and postselected in the state $|\uparrow_y\rangle$, i.e., $|\Psi\rangle = \prod_{i=1}^{N} |\uparrow_x\rangle_i$ and $\langle\Phi| = \prod_{i=1}^{N}\langle\uparrow_y|_i$. The variable which is measured at the intermediate time is $C \equiv (\sum_{i=1}^{N}(\sigma_i)_\xi)/N$. The operator C has $N+1$ eigenvalues equally spaced between -1 and $+1$, but the weak value of C is

$$C_w = \frac{\prod_{k=1}^{N}\langle\uparrow_y|_k \sum_{i=1}^{N}((\sigma_i)_x + (\sigma_i)_y) \prod_{j=1}^{N}|\uparrow_x\rangle_j}{\sqrt{2}\,N(\langle\uparrow_y|\uparrow_x\rangle)^N} = \sqrt{2}\,. \qquad (13.40)$$

The interaction Hamiltonian is

$$H = \frac{g(t)}{N}P\sum_{i=1}^{N}(\sigma_i)_\xi\,. \qquad (13.41)$$

The initial state of the measuring device defines the precision of the measurement. When we take it to be the Gaussian (6), it is characterized by the width Δ. For a meaningful experiment we have to take Δ small. Small Δ corresponds to large uncertain P, but now, the strength of the coupling to each individual spin is reduced by the factor $1/N$. Therefore, for large N, both the forward-evolving state and the backward-evolving state are essentially not changed by the coupling to the measuring device. Thus, this single measurement yields the weak value. In [7] it is proven that if the measured observable is an average on a large set of systems, $C = (\sum_i^N C_i)/N$, then we can always construct a single, good precision measurement of the weak value. Here let us present just numerical calculations of the probability distribution of the measuring device for N pre- and postselected spin-$\frac{1}{2}$ particles. The state of the pointer after the postselection for this case is

$$\mathcal{N}\sum_{i=0}^{N}\frac{(-1)^{i}}{(i!(N-i)!)}\left(\cos^{2}(\pi/8)\right)^{N-i}\left(\sin^{2}(\pi/8)\right)^{i}e^{-(Q-\frac{(2N-i)}{N})^{2}/2\Delta^{2}}\,.\quad(13.42)$$

The probability distribution for the pointer variable Q is

$$prob(Q)=\mathcal{N}^{2}\left(\sum_{i=0}^{N}\frac{(-1)^{i}}{(i!(N-i)!)}\left(\cos^{2}(\pi/8)\right)^{N-i}\left(\sin^{2}(\pi/8)\right)^{i}e^{-(Q-\frac{(2N-i)}{N})^{2}/2\Delta^{2}}\right)^{2}.$$

$$(13.43)$$

The results for $N = 20$ and different values of Δ are presented in Fig. 13.4. We see that for $\Delta = 0.25$ and larger, the obtained results are very good: the final probability distribution of the pointer is peaked at the weak value, $\left((\sum_{i=1}^{N}(\sigma_{i})_{\xi})/N\right)_{w} = \sqrt{2}$. This distribution is very close to that of a measuring device measuring operator O on a system in an eigenstate $|O=\sqrt{2}\rangle$. For N large, the relative uncertainty can be decreased almost by a factor $1/\sqrt{N}$ without changing the fact that the peak of the distribution points to the weak value.

Although our set of particles preselected in one state and postselected in another state is considered as one system, it looks like an ensemble. In quantum theory, measurement of the sum does not necessarily yield the same result as the sum of the results of the separate measurements, so conceptually our measurement on the set of particles differs from the measurement on an ensemble of pre- and postselected particles. However, in our example of weak measurements, the results are the same.

A less ambiguous case is the example considered in the first work on weak measurements [2]. In this work a single system of a large spin N is considered. The system is preselected in the state $|\Psi\rangle = |S_{x}=N\rangle$ and postselected in the state $|\Phi\rangle = |S_{y}=N\rangle$. At an intermediate time the spin component S_{ξ} is weakly measured and again the "forbidden" value $\sqrt{2}N$ is obtained. The uncertainty has to be only slightly larger than $\sqrt{N}$. The probability distribution of the results is centered around $\sqrt{2}N$, and for large N it lies clearly outside the range of the eigenvalues, $(-N, N)$. Unruh [61] made computer calculations of the distribution of the pointer variable for this case and got results which are very similar to what is presented in Fig.13.4.

An even more dramatic example is a measurement of the kinetic energy of a tunneling particle [8]. We consider a particle preselected in a bound state of a potential well which has negative potential near the origin and vanishing potential far from the origin; $|\Psi\rangle = |E=E_{0}\rangle$. Shortly later, the particle is postselected to be far from the well, inside a classically forbidden tunneling region; this state can be characterized by vanishing potential $|\Phi\rangle = |U=0\rangle$. At an intermediate time, a measurement of the kinetic energy is performed. The weak value of the kinetic energy in this case is

$$K_{w} = \frac{\langle U=0|K|E=E_{0}\rangle}{\langle U=0|E=E_{0}\rangle} = \frac{\langle U=0|E - U|E=E_{0}\rangle}{\langle U=0|E=E_{0}\rangle} = E_{0}\,.\quad(13.44)$$

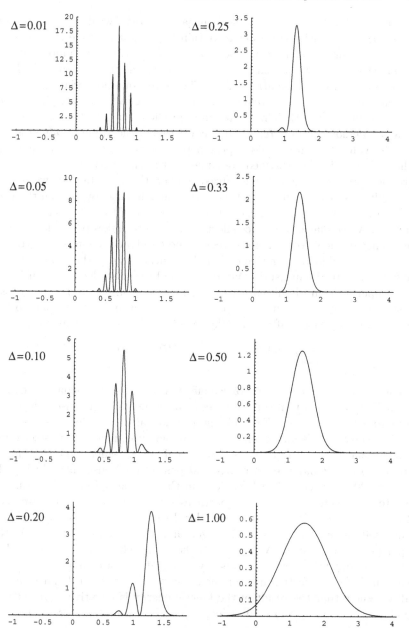

Fig. 13.4. Measurement on a single system: Probability distribution of the pointer variable for the measurement of $A = (\sum_{i=1}^{20} (\sigma_i)_\xi)/20$ when the system of 20 spin-$\frac{1}{2}$ particles is preselected in the state $|\Psi_1\rangle = \prod_{i=1}^{20} |\uparrow_x\rangle_i$ and postselected in the state $|\Psi_2\rangle = \prod_{i=1}^{20} |\uparrow_y\rangle_i$. While in the very strong measurements, $\Delta = 0.01$–0.05, the peaks of the distribution located at the eigenvalues, starting from $\Delta = 0.25$ there is essentially a single peak at the location of the weak value, $A_w = \sqrt{2}$

The energy of the bound state, E_0, is negative, so the weak value of the kinetic energy is negative. In order to obtain this negative value the coupling to the measuring device need not be too weak. In fact, for any finite strength of the measurement we can choose the postselected state sufficiently far from the well to ensure the negative value. Therefore, for appropriate postselection, the usual *strong* measurement of a positive definite operator invariably yields a negative result! This weak value predicted by the two-state vector formalism demonstrates a remarkable consistency: the value obtained is exactly the value that we would expect a particle to have when the particle is characterized in the intermediate times by the two wave functions, one in a ground state, and the other localized outside the well. Indeed, we obtain this result precisely when we postselect the particle far enough from the well that it could not have been kicked there as a result of the intermediate measurement. A peculiar interference effect of the pointer takes place: destructive interference in the whole "allowed" region and constructive interference of the tails in the "forbidden" negative region. The initial state of the measuring device $\Phi(Q)$, due to the measuring interaction and the postselection, transforms into a superposition of shifted wave functions. The shifts are by the (possibly small) eigenvalues, but the superposition is approximately equal to the original wave function shifted by a (large and/or forbidden) weak value:

$$\sum_n \alpha_n \Psi^{MD}(Q - c_n) \simeq \Psi^{MD}(Q - C_w) \,. \tag{13.45}$$

These surprising, even paradoxical effects are really only gedanken experiments. The reason is that, unlike weak measurements on an ensemble, these are extremely rare events. For yielding an unusual weak value, a single preselected system needs an extremely improbable outcome of the postselection measurement. Let us compare this with a weak measurement on an ensemble. In order to get N particles in a pre- and postselected ensemble which yield $(\sigma_\xi)_w = 100$, we need $\sim N10^4$ particles in the preselected ensemble. But, in order to get a single system of N particles yielding $(S_\xi)_w = 100N$, we need $\sim 10^{4N}$ systems of N preselected particles. In fact, the probability to obtain an unusual value by error is much larger than the probability to obtain the proper postselected state. What makes these rare effects interesting is that there is a strong (although only one-way) correlation: for example, every time we find in the postselection measurement the particle sufficiently far from the well, we know that the result of the kinetic energy is negative, and not just negative: it is equal to the weak value, $K_w = E_0$, with a good precision.

13.4.4 Relations Between Weak and Strong Measurements

In general, weak and strong measurements do not yield the same outcomes. The outcomes of strong measurements are always the eigenvalues while the outcomes of weak measurements, the weak values, might be very different from the eigenvalues. However, there are two important relations between them [4].

(i) If the description of a quantum system is such that a particular eigenvalue of a variable is obtained with certainty in case it is measured strongly, then the weak value of this variable is equal to this eigenvalue. This is correct in all cases, i.e., if the system described by a corresponding single (forward or backward evolving) eigenstate, or if it is described by a two-state vector, or even if it is described by a generalized two-state vector.

(ii) The inverse of this theorem is true for dichotomic variables such as projection operators of spin components of spin-$\frac{1}{2}$ particles. The proofs of both statements are given in [4].

Let us apply the theorem (i) for the example of three boxes when we have a large number of particles all pre- and postselected in the two-state vector (13.15). The *actual* story is as follows: A macroscopic number N of particles (gas) were all prepared at t_1 in a superposition of being in three separated boxes (13.13). At later time t_2 all the particles were found in another superperposition (13.14) (this is an extremely rare event). In between, at time t, *weak measurements* of a number of particles in each box, which are, essentially, usual measurements of pressure in each box, have been performed. The readings of the measuring devices for the pressure in the boxes 1, 2, and 3 were

$$p_1 = p \,,$$
$$p_2 = p \,, \tag{13.46}$$
$$p_3 = -p \,,$$

where p is the pressure which is expected to be in a box with N particles.

We are pretty certain that this "actual" story never took place because the probability for the successful postselection is of the order of 3^{-N}; for a macroscopic number N it is too small for any real chance to see it happening. However, given that the postselection does happen, we are safe to claim that the results (13.46) are correct, i.e., the measurements of pressure at the intermediate time with very high probability have shown these results.

Indeed, the system of all particles at time t (signified by index i) is described by the two-state vector

$$\langle \Phi | \, |\Psi\rangle = \frac{1}{3^N} \prod_{i=1}^{i=N} (\langle 1|_i + \langle 2|_i - \langle 3|_i) \prod_{i=1}^{i=N} (|1\rangle_i + |2\rangle_i + |3\rangle_i) \,. \tag{13.47}$$

Then, intermediate measurements yield, for each particle, probability 1 for the the following outcomes of measurements:

$$\mathbf{P}_1 = 1 \,,$$
$$\mathbf{P}_2 = 1 \,, \tag{13.48}$$
$$\mathbf{P}_1 + \mathbf{P}_2 + \mathbf{P}_3 = 1 \,,$$

where $\mathbf{P}_1$ is the projection operator on the state of the particle in box 1, etc. Thus, from (13.48) and theorem (i) it follows:

$$(\mathbf{P}_1)_w = 1 \, ,$$
$$(\mathbf{P}_2)_w = 1 \, , \qquad\qquad (13.49)$$
$$(\mathbf{P}_1 + \mathbf{P}_2 + \mathbf{P}_3)_w = 1 \, .$$

Since for any variables, $(X+Y)_w = X_w + Y_w$ we can deduce that $(\mathbf{P}_3)_w = -1$.

Similarly, for the "number operators" such as $\mathcal{N}_1 \equiv \Sigma_{i=1}^N \mathbf{P}_1^{(i)}$, where $\mathbf{P}_1^{(i)}$ is the projection operator on the box 1 for a particle i, we obtain:

$$(\mathcal{N}_1)_w = N \, ,$$
$$(\mathcal{N}_2)_w = N \, , \qquad\qquad (13.50)$$
$$(\mathcal{N}_3)_w = -N \, .$$

In this rare situation the "weak measurement" need not be very weak: a usual measurement of pressure is a weak measurement of the number operator. Thus, the time-symmetrized formalism yields the surprising result (13.46): the result of the pressure measurement in box 3 is negative! It equals minus the pressure measured in the boxes 1 and 2.

Of course, the negative pressure was not measured in a real laboratory (it requires an extremely improbable postselection), but a nonrobust weak measurement for three-box experiment has been performed in a laboratory [62].

Another example of relation between strong and weak measurements is Hardy's paradox [37]. The analysis of strong measurements appears in [34] and the weak measurements are analyzed in detail in [63]. See also discussions of a realistic experimental proposals [64, 65, 66, 67].

An application of the inverse theorem yields an alternative proof of the results regarding strong measurements of spin components of a spin-$\frac{1}{2}$ particle described by the generalized two-state vector (13.20). Indeed, the linearity property of weak measurements yields a "geometrical picture" for weak values of spin components of a spin-$\frac{1}{2}$ particle. The operators σ_x, σ_y, and σ_z are a complete set of spin operators and they yield a geometry in the familiar three-dimensional space. Each generalized two-state vector of a spin-$\frac{1}{2}$ particle corresponds to a vector in this three-dimensional space with components equal to the weak values of σ_x, σ_y, and σ_z. We call it "weak vector." The weak value of a spin component in an arbitrary direction, then, is given by the projection of the weak vector on this direction. If the weak vector is real and its value larger than 1, then there is a cone of directions the projection on which is equal 1. This yields an alternative proof that in some situations there is a continuum of directions forming a cone in which the result of a spin-component measurements are known with certainty, see Sect. 13.3.5.

13.4.5 Experimental Realizations of Weak Measurements

Realistic weak measurements (on an ensemble) involve preparation of a large preselection ensemble, coupling to the measuring devices of each element of the ensemble, postselection measurement which, in all interesting cases, selects only a small fraction of the original ensemble, selection of corresponding measuring devices, and statistical analysis of their outcomes. In order to obtain good precision, this selected ensemble of the measuring devices has to be sufficiently large. Although there are significant technological developments in "marking" particles running in an experiment, clearly the most effective solution is that the particles themselves serve as measuring devices. The information about the measured variable is stored, after the weak measuring interaction, in their other degree of freedom. In this case, the postselection of the required final state of the particles automatically yields the selection of the corresponding measuring devices. The requirement for the postselection measurement is, then, that there is no coupling between the variable in which the result of the weak measurement is stored and the postselection device.

An example of such a case is the Stern–Gerlach experiment where the shift in the momentum of a particle, translated into a spatial shift, yields the outcome of the spin measurement. Postselection measurement of a spin component in a certain direction can be implemented by another (this time strong) Stern–Gerlach coupling which splits the beam of the particles. The beam corresponding to the desired value of the spin is then analyzed for the result of the weak measurement. The requirement of nondisturbance of the results of the weak measurement by postselection can be fulfilled by arranging the shifts due to the two Stern–Gerlach devices to be orthogonal to each other. The details are spelled out in [6].

An analysis of a realistic experiment which can yield large weak value Q_w appears in [68]. Duck, Stevenson, and Sudarshan [69] proposed a slightly different optical realization which uses a birefringent plate instead of a prism. In this case the measured information is stored directly in the spatial shift of the beam without being generated by the shift in the momentum. Ritchie, Story, and Hulet [70] adopted this scheme and performed the first successful experiment measuring the weak value of the polarization operator. Their results are in very good agreement with theoretical predictions. They obtained weak values which are very far from the range of the eigenvalues, $(-1, 1)$, their highest reported result is $Q_w = 100$. The discrepancy between calculated and observed weak value was 1%. The RMS deviation from the mean of 16 trials was 4.7%. The width of the probability distribution was $\Delta = 1000$ and the number of pre- and postselected photons was $N \sim 10^8$, so the theoretical and experimental uncertainties were of the same order of magnitude. Their other run, for which they showed experimental data on graphs (which fitted very nicely theoretical graphs), has the following characteristics: $Q_w = 31.6$, discrepancy with calculated value 4%, the RMS deviation 16%, $\Delta = 100$, $N \sim 10^5$. A similar optical experiment has been successfully performed several years ago [71].

Recently, optical weak measurement experiments moved to the field of fiber optics [72, 73, 74]. Another step prevents now any sceptic to argue that the unusual outcomes of weak measurement are a classical effect because macroscopic number of photons are involved in these experiments. The weak measurement of photon polarization have been performed with single particles [75]. Note also a more controversial issue of measurement of "time of arrival" [76] for which weak measurement technique were also applied [77, 78, 79].

Already at 1990 [3] we gave an example of a gedanken experiment in which pre- and postselection lead to a superluminal propagation of light. Steinberg and Chiao [80, 81] connected this to superluminal effect observed for tunneling particles. The issue was analyzed recently by Aharonov et al. [82] and Sokolovsky et al. [83]. Rohrlich and Aharonov [84] also predicted that there is really a physical meaning for this superluminal propagation: we should expect Cherenkov radiation in such experiment.

Note also proposals for *weak nonlocal* measurements [85, 86]. In these works it was pointed out that observation of correlations between outcomes of local weak measurements can yield values of nonlocal variables. However, these methods are very inefficient, and the methods of efficient nonlocal measurements [36] require conditions which contradict conditions of weak measurements, so we doubt that there will be efficient weak nonlocal measurement proposals suitable for realization in a laboratory.

13.5 The Quantum Time-Translation Machine

13.5.1 Introduction

To avoid possible misinterpretations due to the name "time machine," let us explain from the outset what our machine [7] can do and how it differs from the familiar concept of "time machine." Our device is not for time travel. All that it can accomplish is to change the rate of time flow for a closed quantum system. Classically, one can slow down the time flow of a system relative to an external observer, e.g., by fast travel. Our quantum time machine is able to change the rate of time flow of a system for a given period by an arbitrary, even *negative*, factor. Therefore, our machine, contrary to any classical device, is capable of moving the system to its "past." In that case, at the moment the machine completes its operation the system is in a state in which it was some time *before the beginning* of the operation of the time machine. Our machine can also move the system to the future, i.e., at the end of the operation of the time machine the system is in a state corresponding to some later time of the undisturbed evolution.

A central role in the operation of our time machine is played by a peculiar mathematical identity which we discuss in Sect. 13.5.2. In order to obtain different time evolutions of the system we use the gravitational time dilation effect which is discussed in Sect. 13.5.3. In Sect. 13.5.4 we describe the design and the operation of our time machine. The success of the operation of our time machine depends on obtaining a specific outcome in the postselection quantum measurement. The probability of the successful postselection measurement is analyzed in Sect. 13.5.5. The concluding discussion of the limitations and the advantages of our time machine appear in Sect. 13.5.6.

13.5.2 A Peculiar Mathematical Identity

The peculiar interference effect of weak measurements (13.45), that a particular superposition of identical Gaussians shifted by small values yields the Gaussian shifted by a large value occurs not just for Gaussians, but for a large class of functions. Consider now that the system is described by such a wave function and the shifts are due to the time evolutions for various periods of time. Then, this effect can be a basis of a (gedanken) time machine. A specific superposition of time evolutions for short periods of time δt_n yields a time evolution for a large period of time Δt

$$\sum_{n=0}^{N} \alpha_n U(\delta t_n)|\Psi\rangle \sim U(\Delta t)|\Psi\rangle . \tag{13.51}$$

This approximate equality holds (with the same δt_n and Δt) for a large class of states $|\Psi\rangle$ of the quantum system, and in some cases even for all states of the system.

In order to obtain different time evolutions $U(\delta t_n)$ we use the gravitational time dilation effect. For finding the appropriate δt_n and α_n we will rely on

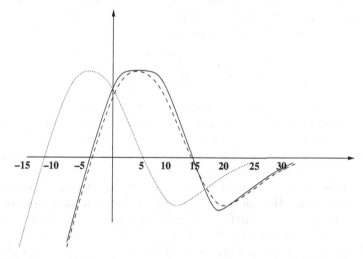

Fig. 13.5. Demonstration of an approximate equality given by (13.53): The sum of a function shifted by the 14 values c_n between 0 and 1 and multiplied by the coefficients α_n (c_n and α_n are given by (13.52) with $N = 13$, $\eta = 10$) yields approximately the same function shifted by the value 10. The *dotted line* shows $f(t)$; the *dashed line* shows $f(t - 10)$, the RHS of (13.53); and the *solid line* shows the sum, the LHS of (13.53)

the identity (13.45) for a particular weak measurement. We choose

$$c_n = n/N, \qquad \alpha_n = \frac{N!}{(N-n)!n!}\eta^n(1-\eta)^{N-n} \,, \qquad (13.52)$$

where $n = 0, 1, ..., N$. Note, that the coefficients α_n are terms in the binomial expansion of $[\eta+(1-\eta)]^N$ and, in particular, $\sum_{n=0}^{N} \alpha_n = 1$. The corresponding "weak value" in this case is η and for a large class of functions (the functions with Fourier transform bounded by an exponential) we have an approximate equality

$$\sum_{n=0}^{N} \alpha_n f(t - c_n) \simeq f(t - \eta) \,. \qquad (13.53)$$

The proof can be found in [87]. Here we only demonstrate it on a numerical example, Fig. 13.5. Even for a relatively small number of terms in the sum (14 in our example), the method works remarkably well. The shifts from 0 to 1 yield the shift by 10. The distortion of the shifted function is not very large. By increasing the number of terms in the sum, the distortion of the shifted function can be made arbitrarily small.

13.5.3 Classical Time Machines

A well-known example of a time machine is a rocket which takes a system to a fast journey. If the rocket is moving with velocity V and the duration of the journey (in the laboratory frame) is T, then we obtain the time shift (relative to the situation without the fast journey):

$$\delta t = T \left(1 - \sqrt{1 - \frac{V^2}{c^2}} \right) \,. \qquad (13.54)$$

For typical laboratory velocities this effect is rather small, but it has been observed experimentally in precision measurements in satellites and, of course, the effect is observed on decaying particles in accelerators. In such a "time machine," however, the system necessarily experiences external force, and we consider this a conceptual disadvantage.

In our time machine we use, instead of the time dilation of special relativity, the gravitational time dilation. The relation between the proper time of the system placed in a gravitational potential ϕ and the time of the external observer ($\phi = 0$) is given by $d\tau = dt\sqrt{1 + 2\phi/c^2}$. We produce the gravitational potential by surrounding our system with a spherical shell of mass M and radius R. The gravitational potential inside the shell is $\phi = -GM/R$. Therefore, the time shift due to the massive shell surrounding our system, i.e., the difference between the time period T of the external observer at a large distance from the shell and the period of the time evolution of the system (the proper time), is

$$\delta t = T \left(1 - \sqrt{1 - \frac{2GM}{c^2 R}} \right) . \tag{13.55}$$

This effect, for any man-made massive shell, is too small to be observed by today's instruments. However, the conceptual advantage of this method is that we do not "touch" our system. Even the gravitational field due to the massive spherical shell vanishes inside the shell.

The classical time machine can only *slow down* the time evolution of the system. For any reasonable mass and radius of the shell, the change of the rate of the time flow is extremely small. In the next section we shall describe our quantum time machine which amplifies the effect of the classical gravitational time machine (for a spherical shell of the same mass), and makes it possible to speed up the time flow for an evolution of a system, as well as to change its direction.

13.5.4 Quantum Gravitational Time Machine

In our machine we use the gravitational time dilation and a quantum interference phenomenon which, due to the peculiar mathematical property discussed in Sect. 13.5.2, amplifies the time translation. We produce the superposition of states shifted in time by small values δt_n (due to spherical shells of different radii) given by the left-hand side of (13.51). Thus, we obtain a time shift by a possibly large, positive or negative, time interval Δt.

The wave function of a quantum system $\Psi(q, t)$, considered as a function of time, usually has a Fourier transform which decreases rapidly for large frequencies. Therefore, the sum of the wave functions shifted by small periods of time $\delta t_n = \delta t c_n$, and multiplied by the coefficients α_n, with c_n and α_n given by (13.52), is approximately equal to the wave function shifted by the large time $\Delta t = \delta t \eta$. Since the equality (13.53) is correct with the same coefficients for all functions with rapidly decreasing Fourier transforms, we obtain for each q, and therefore for the whole wave function,

$$\sum_{n=0}^{N} \alpha_n \Psi(q, t - \delta t_n) \simeq \Psi(q, t - \Delta t) . \tag{13.56}$$

Thus, a device which changes the state of the system from $\Psi(q, t)$ to the state given by the left-hand side of (13.56) generates the time shift of Δt. Let us now present a design for such a device and explain how it operates.

Our machine consists of the following parts: a massive spherical shell, a mechanical device—"the mover"—with a quantum operating system, and a measuring device which can prepare and verify states of this quantum operating system.

The massive shell of mass M surrounds our system and its radius R can have any of the values $R_0, R_1, ..., R_N$. Initially, $R = R_0$.

The mover changes the radius of the spherical shell at time $t = 0$, waits for an (external) time T, and then moves it back to its original state, i.e., to the radius R_0.

The quantum operating system (QOS) of the mover controls the radius to which the shell is moved for the period of time T. The Hamiltonian of the QOS has $N + 1$ nondegenerate eigenstates $|n\rangle$, $n = 0, 1, ..., N$. If the state of the QOS is $|n\rangle$, then the mover changes the radius of the shell to the value R_n.

The measuring device preselects and postselects the state of the QOS. It prepares the QOS before the time $t = 0$ in the initial state

$$|\Psi_{in}\rangle_{QOS} = \mathcal{N} \sum_{n=0}^{N} \alpha_n |n\rangle , \tag{13.57}$$

with the normalization factor

$$\mathcal{N} = \frac{1}{\sqrt{\sum_{n=0}^{N} |\alpha_n|^2}} . \tag{13.58}$$

After the mover completes its operation, i.e., after the time $t = T$, we perform another measurement on the QOS. One of the nondegenerate eigenstates of this measurement is the specific "final state"

$$|\Psi_f\rangle_{QOS} = \frac{1}{\sqrt{N+1}} \sum_{n=0}^{N} |n\rangle . \tag{13.59}$$

Our machine works only if the postselection measurement yields the state (13.59). Unfortunately, this is a very rare event. We shall discuss the probability of obtaining the appropriate outcome in the next section.

Assume that the postselection measurement is successful, i.e., that we do obtain the final state (13.59). We will next show that in this case, assuming an appropriate choice of the radii R_n, our "time machine" shifts the wave function of the system by the time interval Δt. The time shift is defined relative to the situation in which the machine has not operated, i.e., the radius of the shell was not changed from the initial value R_0. In order to obtain the desired time shift $\Delta t = \delta t \eta$ we chose the radii R_n such that

$$\delta t_n \equiv \frac{n \delta t}{N} = T \left(\sqrt{1 - \frac{2GM}{c^2 R_0}} - \sqrt{1 - \frac{2GM}{c^2 R_n}} \right) . \tag{13.60}$$

The maximal time shift in the different terms of the superposition (the left-hand side of (13.51)) is $\delta t_N = \delta t$. The parameter η is the measure of the "quantum amplification" relative to the maximal (classical) time shift δt. If the radius R_0 of the shell is large enough that the time dilation due to the shell in its initial configuration can be neglected, (13.60) simplifies to

$$\delta t_n = T\left(1 - \sqrt{1 - \frac{2GM}{c^2 R_n}}\right).$$

(13.61)

Let us assume then that we have arranged the radii according to (13.61) and we have prepared the quantum operating system of the mover in the state (13.57). Then, just prior to the operation of the time machine the overall state is the direct product of the corresponding states of the system, the shell, and the mover,

$$\mathcal{N}|\Psi(q,0)\rangle|R_0\rangle \sum_{n=0}^{N} \alpha_n n\rangle,$$

(13.62)

where $|R_0\rangle$ signifies that the shell, together with the mechanical part of the mover, is at the radius R_0. Although these are clearly macroscopic bodies. we assume that we can treat them quantum–mechanically. We also make an idealized assumption that these bodies do not interact with the environment, i.e., no element of the environment becomes correlated to the radius of the shell.

Once the mover has operated, changing the radius of the spherical shell, the overall state becomes

$$\mathcal{N}|\Psi(q,0)\rangle \sum_{n=0}^{N} \alpha_n |R_n\rangle|n\rangle.$$

(13.63)

For different radii R_n, we have different gravitational potentials inside the shell and, therefore, different relations between the flow of the proper time of the system and the flow of the external time. Thus, after the external time T has elapsed, just before the mover takes the radii R_n back to the value R_0, the overall state is

$$\mathcal{N} \sum_{n=0}^{N} \alpha_n |\Psi(q, T - \delta t_n)\rangle|R_n\rangle|n\rangle.$$

(13.64)

Note that now the system, the shell, and the QOS are correlated: the system is not in a pure quantum state. After the mover completes its operation, the overall state becomes

$$\mathcal{N} \sum_{n=0}^{N} \alpha_n |\Psi(q, T - \delta t_n)\rangle|R_0\rangle|n\rangle.$$

(13.65)

There is still a correlation between the system and the QOS.

The last stage is the postselection measurement performed on the QOS. It puts the QOS and, consequently, our quantum system, in a pure state. After the successful postselection measurement, the overall state is

$$\left(\sum_{n=0}^{N} \alpha_n |\Psi(q, T - \delta t_n)\rangle\right)|R_0\rangle\left(\frac{1}{\sqrt{N+1}}\sum_{n=0}^{N}|n\rangle\right).$$

(13.66)

We have shown that the wave function of the quantum system $\Psi(q,t)$ is changed by the operation of the time machine into $\sum_{n=0}^{N} \alpha_n |\Psi(q, T-\delta t_n)\rangle$. Up to the precision of the approximate equality (13.53) (which can be arbitrarily improved by increasing the number of terms N in the sum), this wave function is indeed $|\Psi(q, T - \Delta t)\rangle$! Note that for $\Delta t > T$, the state of the system at the moment the time machine has completed its operation is the state in which the system was before the beginning of the operation of the time machine.

13.5.5 The Probability of the Success of the Quantum Time Machine

The main conceptual weakness of our time machine is that usually it does not work. Successful postselection measurements corresponding to large time shifts are extremely rare. Let us estimate the probability of the successful postselection measurement in our example. The probability is given by the square of the norm of the vector obtained by projecting the state (13.66) on the subspace defined by state (13.59) of the QOS:

$$\text{Prob} = ||\frac{\mathcal{N}}{\sqrt{N+1}}(\sum_{n=0}^{N} \alpha_n |\Psi(q, T - \delta t_n)\rangle)|R_0\rangle||^2 . \tag{13.67}$$

In order to obtain a time shift without significant distortion, the wave functions shifted by different times δt_n have to be such that the scalar products between them can be approximated by 1. Taking then the explicit form of α_n from (13.52), we evaluate the probability (13.67), obtaining

$$\text{Prob} \simeq \frac{\mathcal{N}^2}{N} . \tag{13.68}$$

The normalization factor $\mathcal{N}$ given by (13.58) decreases very rapidly for large N. Even if we use a more efficient choice of the initial and the final states of the QOS (see [3]) for the amplification, $\eta > 1$, the probability decreases with N as $1/(2\eta - 1)^N$.

The small probability of the successful operation of our time machine is, in fact, unavoidable. At the time just before the postselection measurement, the system is in a mixture of states correlated to the orthogonal states of the QOS (see (13.65)). The probability of finding the system at that time in the state $|\Psi(q, T - \Delta t)\rangle$, for Δt which differs significantly from the time periods δt_n, is usually extremely small. This is the probability to find the system, by a measurement performed "now," in the state in which it was supposed to be at some other time. For any real situation this probability is tiny but not equal precisely to zero, since all systems with bounded energies have wave functions with nonvanishing tails. The successful operation of our time machine is a particular way of "finding" the state of the quantum system shifted by the period of time $\Delta = \eta\delta$. Therefore, the probability for success cannot be larger than the probability of finding the shifted wave function by direct measurement.

One can wonder what has been achieved by all this rather complicated procedure if we can obtain the wave function of the system shifted by the time period Δt simply by performing a quantum verification measurement at the time T of the state $|\Psi(q, T - \Delta t)\rangle$. There is a very small chance for the success of this verification measurement, but using our procedure the chance is even smaller. What our machine can do, and we are not aware of any other method which can achieve this, is to shift the wave function in time *without knowing* the wave function. If we obtain the desired result of the postselection measurement (the postselection measurement performed on *the measuring device*), we know that the wave function of the system, whatever it is, is shifted by the time Δt. Not only is the knowledge of the wave function of the system inessential for our method, but even the very nature of the physical system whose wave function is shifted by our time machine need not be known. The only requirement is that the energy distribution of the system decreases rapidly enough. If the expectation value of the energy can be estimated, then we can improve dramatically the probability of the success of our procedure. The level of difficulty of the time shift without distortion depends on the magnitude of the energy dispersion ΔE and not on the expectation value of energy $\langle E \rangle$. For quantitative analysis of this requirement see [87].

The operation of our time machine can be considered as *a superposition of time evolutions* [7] for different periods of time δt_n. This name is especially appropriate if the Hamiltonian of the system is bounded, since in this case the approximate equality (13.51) is correct for all states $|\Psi\rangle$.

13.5.6 Time Translation to the Past and to the Future

Let us spell out again what our machine does. Assume that the time evolution of the state of the system is given by $|\Psi(t)\rangle$. By this we mean that this is the evolution *before* the operation of the time machine and this is also the evolution later, provided we *do not* operate the time machine. The state $|\Psi(t)\rangle$ describes the actual past states of the system and the counterfactual future states of the system, i.e., the states which will be in the case we do not disturb the evolution of the system by the operation of our time machine. Define "now," $t = 0$, to be the time at which we begin the operation of the time machine. The time interval of the operation of the time machine is T. Moving the system to the *past* means moving it to the state in which the system actually was at some time $t < 0$. Moving the system to the future means moving it to the state in which it would have wound up after undisturbed evolution at some future time $t > T$. Evidently, the classical time machine does neither of these, since all it can achieve is that at time T the system is in the state corresponding to the time t, $0 < t < T$.

When we speed up or slow down the rate of the time evolution, the system passes through all states of its undisturbed evolution only once. More bizarre is the situation when we reverse the direction of the time flow, thus ending up, after completing the operation of the time machine, in the state in which the system was before $t = 0$. In this case the system passes three times through some states during its evolution.

For our time machine to operate properly, it is essential that the system is isolated from the external world. In the case of the time translation to the state of the past, the system has to be isolated not only during the time of the operation of the time machine, but also during the whole period of intended time translation. If the system is to be moved to the state in which it was at the time t, $t < 0$, then it has to be isolated from the time t until the end of operation of the time machine. This seems to be a limitation of our time machine. It leads, however, to an interesting possibility. We can send a system to its *counterfactual past*, i.e., to the past in which it was supposed to be if it were isolated (or if it were in any environment chosen by us).

Consider an excited atom which we isolate in the vacuum at time $t = 0$ inside our time machine. And assume that our time machine made a successful time translation to a negative time t, such that $|t|$ is larger than the lifetime of the excited atomic state. Since the atom, now, is not in the environment it was in the past, we do not move the atom to its actual state in the past. Instead, we move the atom to the state of its counterfactual past. By this we mean the state of the isolated atom which, under its normal evolution in the vacuum during the time period $|t|$ winds up in the excited state. In fact, this is the state of the atom together with an incoming radiation field. The radiation field is exactly such that it will be absorbed by the atom. Although our procedure is very complicated and only very rarely successful, still, it is probably the easiest way to prepare the precise incoming electromagnetic wave which excites a single atom with probability one.

13.5.7 Experimental Realization of the Quantum Time-Translation Machine?!

Suter [88] has claimed to perform an experimental realization of the quantum time-translation machine using a classical Mach–Zehnder interferometer. The experimental setup of Suter, however, does not fall even close to the definition of the time machine. In his setup we know what is the system and what is its initial state. What he shows is that if we send a single mode of a radiation field through a birefringent retardation device which yields different retardations for two orthogonal polarizations, then placing the preselection polarization filter and the postselection polarization filter will lead to a much larger effect than can be achieved by preselection alone. Thus, it might seem like speeding up the time evolution, but this procedure fails all tests of universality. Different modes of radiation field speed up differently, an arbitrary wave packet is usually distorted, and for other systems (other particles) the device is not supposed to work at all.

Thus, the first basic requirement that the time machine has to work for various systems is not fulfilled from the beginning. And it cannot be easily modified since the "external" variable (which is supposed to be a part of the time machine) is the property of the system itself—the polarization of the radiation field. The next necessary requirement, that it works for a large class of the initial states of the system, cannot be fulfilled too. Indeed, he considers a superposition of only two time evolutions. This superposition can be identical to a longer evolution for a particular state, but not for a large class of states. As it has been shown [7, 87] a superposition of a large number of time evolutions is necessary for this purpose.

Suter, together with R. Ernst and M. Ernst, performed in the past another experiment which they called "An experimental realization of a quantum time-translation machine" [89]. In this experiment a very different system was used: the effect was demonstrated on the heteronuclear coupling between two nuclear spins. But the experimental setup was also applicable only to a specific system and only for a certain state. Therefore, the same criticism is applicable and, therefore, one should not call it an implementation of the time-translation machine.

Although the experiments of Suter are not implementations of the quantum time machine, still, they are interesting as *weak measurements*. The experiment of Suter with a birefringent retardation device can be considered as a weak measurement of a polarization operator. In fact, this is a variation of the experiments which were proposed [68] and performed [70] previously. The "weakness condition" of these two experiments follows from the localization of the beam (which was sent through a narrow slit). The "weak" regime of the experiment of Suter is achieved by taking the retardation small. The second experiment of Suter can be considered as the first weak measurement of a nuclear spin component.

13.6 Time Symmetry

13.6.1 Forward- and Backward-Evolving Quantum States

Before discussing the time symmetry of the pre- and postselected systems which are usually discussed in the framework of the two-state vector formalism, we will consider the question of differences between possibilities for manipulating forward-evolving quantum states (13.1) and backward-evolving state (13.6) which has been recently analyzed [90]. It is particularly important in the light of recent argument of Shimony [91] against equal status of forward- and backward-evolving quantum states.

A notable difference between forward- and backward-evolving states has to do with the *creation* of a particular quantum state at a particular time. In order to create the quantum state $|A = a\rangle$ evolving forward in time, we measure A before this time. We cannot be sure to obtain $A = a$, but if we obtain a different result $A = a'$ we can always perform a unitary operation and thus create at time t the state $|A = a\rangle$. On the other hand, in order to create the backward-evolving quantum state $\langle A = a|$, we measure A after time t. If we do not obtain the outcome $A = a$, we cannot repair the situation, since the correcting transformation has to be performed at a time when we do not yet know which correction is required. Therefore, a backward-evolving quantum state at a particular time can be created only with some probability, while a forward-evolving quantum state can be created with certainty. (Only if the forward-evolving quantum state is identical to the backward-evolving state we want to create at time t, and only if we know that no one touches the system at time t, can the backward-evolving state be created with certainty, since then the outcome $A = a$ occurs with certainty. But this is not an interesting case.)

The formalism of quantum theory is time reversal invariant. It does not have an intrinsic arrow of time. The difference with regard to the creation of backward and forward evolving quantum state follows from the "memory's" arrow of time. We can base our decision of what to do at a particular time only on events in the past, since future events are unknown to us. The memory time arrow is responsible for the difference in our ability to manipulate forward- and backward-evolving quantum states. However, the difference is only in relation to creation of the quantum state. As we will see below there are no differences with measurements in the sense of "finding out" what is the state at a particular time.

The ideal (von Neumann) measurement procedure applies both to forward evolving quantum states and to backward-evolving quantum states. In both cases, the outcome of the measurement is known after the time of the measurement. All that is known about what can be measured in an ideal (nondemolition) measurement of a forward-evolving quantum state can be applied also to a backward-evolving quantum state. There are constraints on the measurability of nonlocal variables, i.e., variables of composite systems with parts separated in space. When we consider instantaneous nondemolition measurements (i.e., measurements in which, in a particular Lorentz frame during an arbitrarily short time, local records appear which, when taken together, specify the eigenvalue of the nonlocal variable), we have classes of measurable and unmeasurable variables. For example, the Bell operator variable is measurable, while some other variables [92], including certain variables with product state eigenstates [93, 94], cannot be measured.

The procedure for measuring nonlocal variables involves entangled ancillary particles and local measurements, and can get quite complicated. Fortunately, there is no need to go into detail in order to show the similarity of the results for forward- and backward-evolving quantum states. The operational meaning of the statement that a particular variable A is measurable is that in a sequence of three consecutive measurements of A—the first taking a long time and possibly including bringing separate parts of the system to the same location and then returning them, the second being short and nonlocal, and the third, like the first, consisting of bringing together the parts of the system—all outcomes have to be the same. But this is a time symmetric statement; if it is true, it means that the variable A is measurable both for forward- and backward-evolving quantum states.

We need also to obtain the correct probabilities in the case that different variables are measured at different times. For a forward-evolving quantum state it follows directly from the linearity of quantum mechanics. For a backward-evolving quantum state, the simplest argument is the consistency between the probability of the final measurement, which is now $B = b$, given the result of the intermediate measurement $A = a$, and the result of the intermediate measurement given the result of the final measurement. We assume that the past is erased. The expression for the former is $|\langle A = a|B = b\rangle|^2$. For consistency, the expression for the latter must be the same, but this is what we need to prove.

In exactly the same way we can show that the same procedure for *teleportation* of a forward-evolving quantum state [95] yields also teleportation of a backward-evolving quantum state. As the forward-evolving quantum state is teleported to a space–time point in the future light cone, the backward-evolving quantum state is teleported to a point in the backward light cone. Indeed, the operational meaning of teleportation is that the outcome of a measurement in one place is invariably equal to the outcome of the same measurement in the other place. Thus, the procedure for teleportation of the forward-evolving state to a point in the future light cone invariably yields teleportation of the backward-evolving quantum state to the backward light cone.

The impossibility of teleportation of the backward-evolving quantum state outside the backward light cone follows from the fact that it will lead to teleportation of the forward-evolving quantum state outside the forward light cone, and this is impossible since it obviously breaks causality.

Another result which has been proved using causality argument is the *no cloning theorem* for backward-evolving quantum quantum states [90]. So, also in this respect there is no difference between forward- and backward-evolving quantum states.

The argument used above does not answer the question of whether it is possible to measure nonlocal variables in a *demolition measurement*. Demolition measurements destroy (for the future) the state and may be the quantum systems itself. Thus, obviously, a demolition measurement of a nonlocal variable of a quantum state evolving forward in time does not measure this variable for a quantum state evolving backward in time. Any nonlocal variable of a composite system can be measured with demolition for a quantum state evolving forward in time [96]. Recently, it has been shown [97] also that any nonlocal variable can be measured for a quantum state evolving backward in time. Moreover, the procedure is simpler and requires fewer entanglement resources.

The difference follows from the fact that we can change the direction of time evolution of a backward-evolving state along with complex conjugation of the quantum wave (flipping a spin). Indeed, all we need is to prepare an EPR state of our system and an ancilla. Guarding the system and the ancilla ensures that the forward-evolving quantum state of the ancilla is the flipped state of the system. For a spin wave function we obtain $\alpha\langle\uparrow| + \beta\langle\downarrow| \rightarrow -\beta^*|\uparrow\rangle + \alpha^*|\downarrow\rangle$. For a continuous variable wave function $\Psi(q)$ we need the original EPR state $|q - \tilde{q} = 0, \ p + \tilde{p} = 0\rangle$. Then, the backward-evolving quantum state of the particle will transform into a complex conjugate state of the ancilla $\Psi(q) \rightarrow \Psi^*(\tilde{q})$.

If the particle and the ancilla are located in different locations, then such an operation is a combination of time reversal and teleportation of a backward-evolving quantum state of a continuous variable [98].

We cannot flip and change the direction of time evolution of a quantum state evolving forward in time. To this end we would have to perform a Bell measurement on the system and the ancilla and to get a particular result (singlet). However, we cannot ensure this outcome, nor can we correct the situation otherwise. Moreover, it is easily proven that no other method will work either. If one could have a machine which turns the time direction (and flips) a forward-evolving quantum state, then one could prepare at will any state that evolves toward the past, thus signaling to the past and contradicting causality.

Let us consider now a pre- and postselected system. It is meaningless to ask whether we can perform a nondemolition measurement on a system described by a two-state vector. Indeed, the vector describing the system should not be changed *after* the measurement, but there is no such time: for a forward-evolving state, "after" means later, whereas for a backward-evolving state, "after" means before. It is meaningful to ask whether we can perform a *demolition* measurement on a system described by a two-state vector. The answer is positive [97], even for composite systems with separated parts.

Next, is it possible to teleport a two-state vector? Although we can teleport both forward- and backward-evolving quantum states, we cannot teleport the two-state vector. The reason is that the forward evolving state can be teleported only to the future light cone, while the backward-evolving state can be teleported only to the backward light cone. Thus, there is no space–time point to which both states can be teleported.

Finally, the answer to the question of whether it is possible to clone a two-state vector is negative, since neither forward-evolving nor backward-evolving quantum states can be cloned.

13.6.2 Time-Symmetric Aspects of Pre- and Postselected Systems

When a quantum system is described by the two-state vector (13.2) or the generalized two-state vector (13.7), the backward-evolving states enter on equal footing with the forward-evolving states. Note that the asymmetry in the procedure for obtaining the state (13.7) is not essential: we can start preparing $1/\sqrt{N} \sum_i |\Phi_i\rangle|i\rangle$ instead.

We will analyze now the symmetry under the interchange $\langle\Phi|\ |\Psi\rangle \leftrightarrow \langle\Psi|\ |\Phi\rangle$. This will be considered as a symmetry under reversal of the direction of the arrow of time. It is important to note that in general this interchange is not equivalent to the interchange of the measurements creating the two-state vector $A = a$ and $B = b$. An example showing the nonequivalence can be found in [99]. However, in order to simplify the discussion, we will assume that the free Hamiltonian is zero, and therefore $|\Psi\rangle = |A = a\rangle$ and $\langle\Phi| = \langle B = b|$. In this case, of course, the reversal of time arrow is identical to the interchange of the measurements at t_1 and t_2. If the free Hamiltonian is not zero, then an appropriate modification should be made [100].

The ABL rule for the probabilities of the outcomes of ideal measurements (13.9) is also explicitly time-symmetric: First, both $\langle\Phi|$ and $|\Psi\rangle$ enter the equation on equal footing. Second, the probability (13.9) is unchanged under the interchange $\langle\Phi|\ |\Psi\rangle \leftrightarrow \langle\Psi|\ |\Phi\rangle$.

The ABL rule for a quantum system described by a generalized two-state vector (13.7) is time-symmetric as well: $\langle\Phi_i|$ and $|\Psi_i\rangle$ enter the equation on equal footing. The manifestation of the symmetry of this formula under the reversal of the arrow of time includes complex conjugation of the coefficients. The probability (13.10) is unchanged under the interchange $\sum_i \alpha_i \langle\Phi_i|\ |\Psi_i\rangle \leftrightarrow \sum_i \alpha_i^* \langle\Psi_i|\ |\Phi_i\rangle$.

The outcomes of weak measurements, the weak values, are also symmetric under the interchange $\langle\Phi|\ |\Psi\rangle \leftrightarrow \langle\Psi|\ |\Phi\rangle$ provided we perform complex conjugation of the weak value together with the interchange. This is similar to complex conjugation of the Schrödinger wave function under the time reversal. Thus, also for weak measurements there is the time reversal symmetry: both $\langle\Phi|$ and $|\Psi\rangle$ enter the formula of the weak value on the same footing and there is symmetry under the interchange of the pre- and postselected states. The time symmetry holds for weak values of generalized two-state vectors (13.22): i.e., the interchange $\sum_i \alpha_i \langle\Phi_i|\ |\Psi_i\rangle \leftrightarrow \sum_i \alpha_i^* \langle\Psi_i|\ |\Phi_i\rangle$ leads to $C_w \leftrightarrow C_w^*$.

13.6.3 The Time Asymmetry

The symmetry is also suggested in using the language of "preselected" state and "postselected" state. In order to obtain the two-state vector (13.2) we need to preselect $A = a$ at t_1 and postselect $B = b$ at t_2. Both measurements might not yield the desired outcomes, so we need several systems out of which we pre- and postselect the one which is described by the two-state vector (13.2). However, the symmetry is not complete and the language might be somewhat misleading. It is true that we can only (post)select $B = b$ at t_2, but we can *prepare* instead of preselect $A = a$ at t_1. For preparation of $|a\rangle$ a single system is enough. If the measurement of A yields a different outcome a' we can perform a fast unitary operation which will change $|A = a'\rangle$ to $|A = a\rangle$ and then the time evolution to time t will bring the system to the state $|\Psi\rangle$. This procedure is impossible for creation of the backward-evolving state $\langle\Phi|$. Indeed, if the outcome of the measurement of B does not yield b, we cannot read it and then make an appropriate unitary operation *before* t_2 in order to get the state $\langle \Phi |$ at time t. We need several systems to post-select the desired result (unless by chance the first system has the desired outcome).

Although the formalism includes situations with descriptions by solely forward-evolving quantum state and by solely backward-evolving quantum states, here also there is a conceptual difference. For obtaining backward-evolving state it was necessary to have a guarded ancilla in order to erase the quantum state evolving from the past. Of course, there is no need for this complication in obtaining forward-evolving quantum state. The difference is due to fixed "memory" arrow of time: we know the past and we do not know the future. This asymmetry is also connected to the concept of a measurement. It is asymmetric because, by definition, we do not know the measured value before the measurement and we do know it after the measurement.

13.6.4 If *Measurements* are Time-Asymmetric, How the Outcomes of Measurements are Time-Symmetric?

Taking this asymmetry of the concept of measurement into account, how one can understand the time symmetry of the formulae for the probability of an intermediate measurements (13.9), (13.10) and for the formulae of weak values (13.21), (13.22)?

This is because these formulae deal with the results of the measurements which, in contrast with the concept of measurement itself, are free from the time asymmetry of a measurement. The results of measurements represent the way the system affects other systems (in this case measuring devices) and these effects, obviously, do not exhibit the time asymmetry of our memory. The time asymmetry of measurement is due to the fact that the pointer variable of the measuring device is showing "zero" mark before the measurement and not after the measurement. But the result of the measurement is represented by the *shift* of the pointer position. (If originally the pointer showed "zero" it is also represented by the final position of the pointer.) This shift is independent of the initial position of the pointer and therefore it is not sensitive to the time asymmetry caused by asymmetrical fixing of the initial (and not final) position of the pointer. The relations described in the formulae of the two-state vector formalism are related to these shifts and, therefore, the time symmetry of the formulae follows from the underlying time symmetry of the quantum theory. The shifts of the pointer variable in weak measurements were considered as "weak-measurements elements of reality" [101] where "elements of reality were identified with "definite shifts." This approach was inspired by the EPR elements of reality which are definite outcomes of ideal measurements, i.e., definite shifts in ideal measurement procedures. The next section discusses a controversy related to ideal measurements.

13.6.5 Counterfactual Interpretation of the ABL Rule

Several authors criticized the TSVF because of the alleged conflict between counterfactual interpretations of the ABL rule and predictions of quantum theory [102, 20, 103, 104]. The form of all these inconsistency proofs is as follows: The probability of an outcome $C = c_n$ of a quantum measurement performed on a preselected system, given correctly by (13.12), is considered. In order to allow the analysis using the ABL formula, a measurement at a later time, t_2, with two possible outcomes, which we denote by "1_f" and "2_f," is introduced. The suggested application of the ABL rule is expressed in the formula for the probability of the result $C = c_n$

$$\text{Prob}(C = c_n) = \text{Prob}(1_f)\, \text{Prob}(C = c_n\ ; 1_f) \qquad (13.69)$$
$$+ \text{Prob}(2_f)\, \text{Prob}(C = c_n\ ; 2_f)\,,$$

where $\text{Prob}(C = c_n\ ; 1_f)$ and $\text{Prob}(C = c_n\ ; 2_f)$ are the conditional probabilities given by the ABL formula, (13.9), and $\text{Prob}(1_f)$ and $\text{Prob}(2_f)$ are the probabilities of the results of the final measurement. In the proofs, the authors show that (13.69) is not valid and conclude that the ABL formula is not applicable to this example and therefore it is not applicable in general.

One us (L.V.) has argued [105, 106, 21] that the error in calculating equality (13.69) does not arise from the conditional probabilities given by the ABL formula, but from the calculation of the probabilities $\text{Prob}(1_f)$ and $\text{Prob}(2_f)$ of the final measurement. In all three alleged proofs, the probabilities $\text{Prob}(1_f)$ and $\text{Prob}(2_f)$ were calculated on the assumption that no measurement took place at time t. Clearly, one cannot make this assumption here since then the discussion about the probability of the result of the measurement at time t is meaningless. Thus, it is not surprising that the value of the probability $\text{Prob}(C = c_n)$ obtained in this way comes out different from the value predicted by the quantum theory. Straightforward calculations show that the formula (13.69) with the probabilities $\text{Prob}(1_f)$ and $\text{Prob}(2_f)$ calculated on the condition that the intermediate measurement has been performed leads to the result predicted by the standard formalism of quantum theory.

The analysis of counterfactual statements considers both *actual* and *counterfactual* worlds. The statement is considered to be true if it is true in counterfactual worlds "closest" to the actual world. In the context of the ABL formula, in the actual world the preselection and the postselection has been successfully performed, but the measurement of C has not (necessarily) been performed. On the other hand, in counterfactual worlds the measurement of C has been performed. The problem is to find counterfactual worlds "closest" to the actual world in which the measurement of C has been performed. The fallacy in all the inconsistency proofs is that their authors have considered counterfactual worlds in which C has not been measured.

Even if we disregard this fallacy there is still a difficulty in defining the "closest" worlds in the framework of the TSVF. In standard quantum theory it is possible to use the most natural definition of the "closest" world. Since the future is considered to be irrelevant for measurements at present time t, only the period of time before t is considered. Then the definition is:

(i) Closest counterfactual worlds are the worlds in which the system is described by the same quantum state as in the actual world.

In the framework of the TSVF, however, this definition is not acceptable. In the time-symmetric approach the period of time before and after t is considered. The measurement of C constrains the possible states immediately after t to the eigenstates of C. Therefore, if in the actual world the state immediately after t is not an eigenstate of C, no counterfactual world with the same state exists. Moreover, there is the same problem with the backward- evolving quantum state (the concept which does not exist in the standard approach) in the period of time before t. This difficulty can be solved by adopting the following definition of the closest world [106]:

(ii) Closest counterfactual worlds are the worlds in which the results of all measurements performed on the system (except the measurement at time t) are the same as in the actual world.

For the preselected only situation, this definition is equivalent to (i), but it is also applicable to the symmetric pre- and postselected situation. The definition allows to construct *time-symmetric counterfactuals* in spite of common claims that such concept is inconsistent [107].

An important example of counterfactuals in quantum theory are "elements of reality" which are inspired by the EPR elements of reality. The modification of the definition of elements of reality applicable to the framework of the TSVF [34] is:

(iii) If we can *infer* with certainty that the result of measuring at time t of an observable C is c, then, at time t, there exists an element of reality $C = c$.

The word "infer" is neutral relative to past and future. The inference about results at time t is based on the results of measurements on the system performed both before and after time t. Note that there are situations (e.g., the three-boxes example) in which we can "infer" some facts that cannot be obtained by neither "prediction" based on the past results nor "retrodiction" based on the future results separately.

The theorem (i) of Sect. 13.4.4 now can be formulated in a simple way: If $A = a$ is an element of reality then $A_w = a$ is the weak-measurement of reality. The theorem (ii) of Sect. 13.4.4 can be formulated as follows. If A is a dichotomic variable, a is an eigenvalue of A, and if $A_w = a$ is a weak-measurement element of reality, then $A = a$ is an element of reality.

The discussion about the meaning of time symmetric counterfactuals continues until today. Kastner changed her view on such counterfactuals from "inconsistent" to "trivial" [108]. See Vaidman's reply [109] and other very recent contributions on this issue [110, 111, 112].

13.7 Protective Measurements

Several years ago we proposed a concept of *protective measurements* [113, 114, 115] which provides an argument strengthening the consideration of a quantum state as a "reality" of some kind. We have shown that "protected" quantum states can be observed just on a single quantum system. On the other hand, if a single quantum state is "the reality" how "the two-state vector" can be "the reality"?

13.7.1 Protective Measurement of a Single Quantum State

In order to measure the quantum state of single system one has to measure expectation values of various observables. In general, the weak (expectation) value cannot be measured on a single system. However, it can be done if the quantum state is *protected* [113, 114]. The appropriate measurement interaction is again described the Hamiltonian (13.8), but instead of an impulsive interaction the adiabatic limit of slow and weak interaction is considered: $g(t) = 1/T$ for most of the interaction time T and $g(t)$ goes to zero gradually before and after the period T.

In this case the interaction Hamiltonian does not dominate the time evolution during the measurement, moreover, it can be considered as a perturbation. The free Hamiltonian H_0 dominates the evolution. In order to protect a quantum state this Hamiltonian must have the state to be a nondegenerate energy eigenstate. For $g(t)$ smooth enough we then obtain an adiabatic process in which the system cannot make a transition from one energy eigenstate to another, and, in the limit $T \to \infty$, the interaction Hamiltonian changes the energy eigenstate by an infinitesimal amount. If the initial state of the system is an eigenstate $|E_i\rangle$ of H_0 then for any given value of P, the energy of the eigenstate shifts by an infinitesimal amount given by the first-order perturbation theory: $\delta E = \langle E_i|H_{int}|E_i\rangle = \langle E_i|A|E_i\rangle P/T$. The corresponding time evolution $e^{-iP\langle E_i|A|E_i\rangle}$ shifts the pointer by the expectation value of A in the state $|E_i\rangle$. Thus, the probability distribution of the pointer variable, $e^{-(Q-a_i)^2/\Delta^2}$ remains unchanged in its shape, and is shifted by the expectation value $\langle A \rangle_i = \langle E_i|A|E_i\rangle$.

If the initial state of the system is a superposition of several nondegenerate energy eigenstates $|\Psi_1\rangle = \Sigma\alpha_i|E_i\rangle$, then a particular outcome $\langle A \rangle_i \equiv \langle E_i|A|E_i\rangle$ appears at random, with the probability $|\alpha_i|^2$ [61]. (Subsequent adiabatic measurements of the same observable A invariably yield the expectation value in the same eigenstate $|E_i\rangle$.)

13.7.2 Protective Measurement of a Two-State Vector

At first sight, it seems that protection of a two-state vector is impossible. Indeed, if we add a potential that makes one state a nondegenerate eigenstate, then the other state, if it is different, cannot be an eigenstate too. (The states of the two-state vector cannot be orthogonal.) But, nevertheless, protection of the two-state vector is possible [116].

The procedure for protection of a two-state vector of a given system is accomplished by coupling the system to another pre- and postselected system. The protection procedure takes advantage of the fact that weak values might acquire complex values. Thus, the effective Hamiltonian of the protection might not be Hermitian. Non-Hermitian Hamiltonians act in different ways on quantum states evolving forward and backwards in time. This allows simultaneous protection of two different states (evolving in opposite time directions).

Let us consider an example [116] of a two-state vector of a spin-$\frac{1}{2}$ particle, $\langle \uparrow_y || \uparrow_x \rangle$. The protection procedure uses an external pre- and postselected system S of a large spin N that is coupled to our spin via the interaction

$$H_{prot} = -\lambda \mathbf{S} \cdot \sigma \, . \tag{13.70}$$

The external system is preselected in the state $|S_x=N\rangle$ and postselected in the state $\langle S_y=N|$, that is, it is described by the two-state vector $\langle S_y=N||S_x=N\rangle$. The coupling constant λ is chosen in such a way that the interaction with our spin-$\frac{1}{2}$ particle cannot change significantly the two-state vector of the protective system S, and the spin-$\frac{1}{2}$ particle "feels" the effective Hamiltonian in which S is replaced by its weak value,

$$\mathbf{S}_w = \frac{\langle S_y = N|(S_x, S_y, S_z)|S_x = N\rangle}{\langle S_y = N|S_x = N\rangle} = (N, N, iN) \, . \tag{13.71}$$

Thus, the effective protective Hamiltonian is

$$H_{eff} = -\lambda N(\sigma_x + \sigma_y + i\sigma_z) \, . \tag{13.72}$$

The state $|\uparrow_x\rangle$ is an eigenstates of this (non-Hermitian) Hamiltonian (with eigenvalue $-\lambda N$). For backward-evolving states the effective Hamiltonian is the hermitian conjugate of (13.72) and it has different (nondegenerate) eigenstate with this eigenvalue; the eigenstate is $\langle \uparrow_y |$.

In order to prove that the Hamiltonian (13.70) indeed provides the protection, we have to show that the two-state vector $\langle \uparrow_y || \uparrow_x \rangle$ will remain essentially unchanged during the measurement. We consider measurement which is performed during the period of time, between pre- and postselection which we choose to be equal one. The Hamiltonian

$$H = -\lambda \mathbf{S} \cdot \sigma + P\sigma_\xi \tag{13.73}$$

can be replaced by the effective Hamiltonian

$$H_{eff} = -\lambda N(\sigma_x + \sigma_y + i\sigma_z) + P\sigma_\xi \, . \tag{13.74}$$

Indeed, the system with the spin S can be considered as N spin-1/2 particles all preselected in $|\uparrow_x\rangle$ state and postselected in $|\uparrow_y\rangle$ state. The strength of the coupling to each spin-1/2 particle is $\lambda \ll 1$, therefore during the time of the measurement their states cannot be changed significantly. Thus, the forward-evolving state $|S_x{=}N\rangle$ and the backward-evolving state $\langle S_y{=}N|$ do not change significantly during the measuring process. The effective coupling to such system is the coupling to its weak values.

Good precision of the measurement of the spin component requires large uncertainty in P, but we can arrange the experiment in such a way that $P \ll \lambda N$. Then the second term in the Hamiltonian (13.74) will not change significantly the eigenvectors. The two-state vector $\langle \uparrow_y || \uparrow_x \rangle$ will remain essentially unchanged during the measurement, and therefore the measuring device on this single particle will yield $(\sigma_\xi)_w = \frac{\langle \uparrow_y | \sigma_\xi | \uparrow_x \rangle}{\langle \uparrow_y | \uparrow_x \rangle}$.

The Hamiltonian (13.73), with an external system described by the two-state vector $\langle S_y = N || S_x = N \rangle$, provides protection for the two-state vector $\langle \uparrow_y || \uparrow_x \rangle$. It is not difficult to demonstrate that any two-state vector obtained by pre- and postselection of the spin-$\frac{1}{2}$ particle can be protected by the Hamiltonian (13.73). A general form of the two-state vector is $\langle \uparrow_\beta || \uparrow_\alpha \rangle$ where $\hat{\alpha}$ and $\hat{\beta}$ denote some directions. It can be verified by a straightforward calculation that the two-state vector $\langle \uparrow_\beta || \uparrow_\alpha \rangle$ is protected when the two-state vector of the protective device is $\langle S_\beta = N || S_\alpha = N \rangle$.

At least formally we can generalize this method to make a protective measurement of an arbitrary two-state vector $\langle \Psi_2 || \Psi_1 \rangle$ of an arbitrary system. Let us decompose the post-selected state $|\Psi_2\rangle = a|\Psi_1\rangle + b|\Psi_\perp\rangle$. Now we can define "model spin" states: $|\Psi_1\rangle \equiv |\tilde{\uparrow}_z\rangle$ and $|\Psi_\perp\rangle \equiv |\tilde{\downarrow}_z\rangle$. On the basis of the two orthogonal states we can obtain all other "model spin" states. For example, $|\tilde{\uparrow}_x\rangle = 1/\sqrt{2}\,(|\tilde{\uparrow}_z\rangle + |\tilde{\downarrow}_z\rangle)$, and then we can define the "spin model" operator $\tilde{\sigma}$. Now, the protection Hamiltonian, in complete analogy with the spin-$\frac{1}{2}$ particle case is

$$H_{prot} = -\lambda \mathbf{S} \cdot \tilde{\sigma} \, . \tag{13.75}$$

In order to protect the state $\langle \Psi_2 || \Psi_1 \rangle$, the pre-selected state of the external system has to be $|S_z{=}N\rangle$ and the postselected state has to be $\langle S_\chi{=}N|$ where the direction $\hat{\chi}$ is defined by the "spin model" representation of the state $|\Psi_2\rangle$,

$$|\tilde{\uparrow}_\chi\rangle \equiv |\Psi_2\rangle = \langle \Psi_1 | \Psi_2 \rangle |\tilde{\uparrow}_z\rangle + \langle \Psi_\perp | \Psi_2 \rangle |\tilde{\downarrow}_z\rangle \, . \tag{13.76}$$

However, this scheme usually leads to unphysical interaction and is good only as a gedanken experiment in the framework of nonrelativistic quantum theory where we assume that any Hermitian Hamiltonian is possible.

13.8 The TSVF and the Many-Worlds Interpretation of Quantum Theory

The TSVF fits very well into the many-worlds interpretation (MWI) [117], the preferred interpretation of quantum theory of one of us (L.V.) [118]. The counterfactual worlds corresponding to different outcomes of quantum measurements have in the MWI an especially clear meaning: these are subjectively actual different worlds. In each world, the observers of the quantum measurement call their world the actual one, but, if they believe in the MWI they have no paradoxes about ontology of the other worlds. The apparent paradox that a weak value at a given time might change from an expectation value to a weak value corresponding to a particular postselection is solved in a natural way: in a world with pre-selection only (before the postselection) the weak value is the expectation value; then this world splits into several worlds according to results of the postselection measurement and in each of these worlds the weak value will be that corresponding to the particular postselection. The time-symmetric concepts of "elements of reality," "weak-measurements elements of reality" are consistent and meaningful in the context of a particular world. Otherwise, at time t, before the "future" measurements have been performed, the only meaningful concepts are the concepts of the standard, time-asymmetric approach.

One of us (Y.A.) is not ready to adopt the far reaching consequences of the MWI. He proposes another solution [119]. It takes the TSVF even more seriously than it was presented in this paper. Even at present, before the "future" measurements, the backward evolving quantum state (or its complex conjugate evolving forward in time) exists! It exists in the same way as the quantum state evolving from the past exists. This state corresponds to particular outcomes of all measurements in the future. An element of arbitrariness: "Why this particular outcome and not some other?" might discourage, but the alternative (without the many-worlds)—the collapse of the quantum wave—is clearly worse than that.

Acknowledgments

It is a pleasure to thank Fred Alan Wolf for correcting an error of the previous version of the review and Juan Gonzalo Muga for helpful suggestions. This work has been supported in part by the European Commission under the Integrated Project Qubit Applications (QAP) funded by the IST directorate as Contract Number 015848 and by grant 990/06 of the Israel Science Foundation.

References

1. Y. Aharonov, P.G. Bergmann, J.L. Lebowitz: Phys. Rev. B **134**, 1410 (1964)
2. Y. Aharonov, D. Albert, A. Casher, L. Vaidman: Phys. Lett. A **124**, 199 (1987)
3. Y. Aharonov, L. Vaidman: Phys. Rev. A **41**, 11 (1990)
4. Y. Aharonov, L. Vaidman: J. Phys. A **24**, 2315 (1991)
5. L. Vaidman, Y. Aharonov, D. Albert: Phys. Rev. Lett. **58**, 1385 (1987)

6. Y. Aharonov, D. Albert, L. Vaidman: Phys. Rev. Lett. **60**, 1351 (1988)
7. Y. Aharonov, J. Anandan, S. Popescu, L. Vaidman: Phys. Rev. Lett. **64**, 2965 (1990)
8. Y. Aharonov, S. Popescu, D. Rohrlich, L. Vaidman: Phys. Rev. A **48**, 4084 (1993)
9. M.B. Berry: J. Phys. A **27**, L391 (1994)
10. M.B. Berry, S. Popescu: J. Phys. A **39**, 6965 (2006)
11. S.M. Barnett, D.T. Pegg, J. Jeffers, O. Jedrkiewicz, R. Loudon: Phys. Rev. A **62**, 022313 (2000)
12. S.M. Barnett, D.T. Pegg, J. Jeffers, O. Jedrkiewicz: J. Phys. B **33**, 3047 (2000)
13. S.M. Barnett, D.T. Pegg, J. Jeffers: J. Mod. Opt. **47**, 1779 (2000)
14. D.T. Pegg, S.M. Barnett, J. Jeffers: Phys. Rev. A **66**, 022106 (2002)
15. D.T. Pegg, K.L. Pregnell: J. Mod. Opt. **51**, 1613 (2004)
16. R.B. Griffiths: J. Stat. Phys. **36**, 219 (1984)
17. M. Gell-Mann, J.B. Hartle: in *Physical Origins of Time Asymmetry*, ed. by J. Halliwell et al. (Cambridge University Press, Cambridge, 1994)
18. B. Reznik, Y. Aharonov: Phys. Rev. A **52**, 2538 (1995)
19. J. von Neumann: *Mathematical Foundations of Quantum Theory* (Princeton University Press, Princeton, NJ, 1983)
20. O. Cohen: Phys. Rev. A **51**, 4373 (1995)
21. L. Vaidman: Phys. Rev. A **57**, 2251 (1998)
22. Y. Aharonov, L. Vaidman: Phys. Rev. A **67**, 042107 (2003)
23. R.B. Griffiths: Phys. Rev. A **54**, 2759 (1996); **57**, 1604 (1998)
24. A. Kent: Phys. Rev. Lett. **78**, 2874 (1997)
25. R.B. Griffiths: *Consistent Quantum Theory* (Cambridge University Press, Cambridge, 2003)
26. R.E. Kastner: Found. Phys. **29**, 851 (1999)
27. L. Vaidman: Found. Phys. **29**, 865 (1999)
28. K.A. Kirkpatrick: J. Phys. A: Math. Theor. **36**, 4891 (2003)
29. T. Ravon, L. Vaidman: J. Phys. A: Math. Theor. **40**, 2873 (2007)
30. K.A. Kirkpatrick: J. Phys. A: Math. Theor. **40**, 2883 (2007)
31. C.R. Leavens, I. Puerto Giménez, D. Alonso, R. Sala Mayato: Phys. Lett. A **359**, 416 (2006)
32. O. Hosten ,M.T. Rakher, J.T. Barreiro, N.A. Peters, P.G. Kwiat: Nature **439**, 949 (2006)
33. L. Vaidman: The Impossibility of the Counterfactual Computation for all Possible Outcomes. Phys. Rev. Lett. **98**, 160403(2006). quant-ph/0610174
34. L. Vaidman: Phys. Rev. Lett. **70**, 3369 (1993)
35. L. Vaidman: ' "Elements of Reality" and the Failure of the Product Rule'. In *Symposium on the Foundations of Modern Physics*, ed. by P.J. Lahti, P. Bush, P. Mittelstaedt (World Scientific, Cologne, 1993) pp. 406–417
36. Y. Aharonov, D. Albert, L. Vaidman: Phys. Rev. D **34**, 1805 (1986)
37. L. Hardy: Phys. Rev. Lett. **68**, 2981 (1992)
38. S. Marcovitch, B. Reznik, L. Vaidman: Phys. Rev. A **75**, 022102 (2007)
39. S. Popescu, D. Rohrlich: Found. Phys. **24**, 379 (1994)
40. S. Benmenahem: Phys. Rev. A **39**, 1621 (1989)
41. D. Mermin: Phys. Rev. Lett. **74**, 831 (1995)
42. Y. Aharonov, B.G. Englert: Zeit. Natur. A **56** 16 (2001)
43. B.G. Englert, Y. Aharonov: Phys. Lett. A **284**, 1 (2001)

44. P.K. Aravind: Z. Natur. A **58**, 85 (2003)
45. P.K. Aravind: Z. Natur. A **58**, 682 (2003)
46. M. Horibe, A. Hayashi, T. Hashimoto: Phys. Rev. A **71**, 032337 (2005)
47. A. Hayashi, M. Horibe, T. Hashimoto: Phys. Rev. A **71**, 052331 (2005)
48. J.P. Paz, A.J. Roncaglia, M.A. Saraceno: Phys. Rev. A **72**, 012309 (2005)
49. G. Kimura, H. Tanaka, M. Ozawa: Phys. Rev. A **73**, 050301 (2006)
50. T. Durt: Int. J. Mod. Phys. B **20**, 1742 (2006)
51. O. Schulz, R. Steinhbl, M. Weber, B. Englert, C. Kurtsiefer, H. Weinfurter: Phys. Rev. Lett. **90**, 177901 (2003)
52. A. Botero: Ph.D. Thesis, The University of Texas at Austin (2003). quant-ph/0306082
53. Y. Aharonov, A. Botero: Phys. Rev. A **72**, 052111 (2005)
54. A. Tanaka: Phys. Lett. A **297**, 307 (2002)
55. H.M. Wiseman: Phys. Rev. A **65**, 032111 (2002)
56. A.D. Parks: J. Phys. A **36**, 7185 (2003)
57. L.M. Johansen: Phys. Lett. A **322** 298 (2004)
58. L.M. Johansen: Phys. Rev. Lett. **93**, 120402 (2004)
59. L.M. Johansen, A. Luis: Phys. Rev. A **70**, 052115 (2004)
60. O. Oreshkov, T. Brun: Phys. Rev. Lett. **95**, 110409 (2005)
61. W. Unruh: Ann. N.Y. Acad. Sci. **755**, 560 (1995)
62. K.J. Resch, J.S. Lundeen, A.M. Steinberg: Phys. Lett. A **324**, 125 (2004)
63. Y. Aharonov, A. Botero, S. Popescu, B. Reznik, J. Tollaksen: Phys. Lett. A **301**, 130 (2002)
64. K. Molmer: Phys. Lett. A **292**, 151 (2001)
65. S.E. Ahnert, M.C. Payne: Phys. Rev. A **70**, 042102 (2004)
66. J.S. Lundeen, K.J. Resch, A.M. Steinberg: Phys. Rev. A **72**, 016101 (2005)
67. S.E. Ahnert, M.C. Payne: Phys. Rev. A **72**, 016102 (2005)
68. J.M. Knight, L. Vaidman: Phys. Lett. A **143**, 357 (1990)
69. M. Duck, P.M. Stevenson, E.C.G. Sudarshan: Phys. Rev. D **40**, 2112 (1989)
70. N.W.M. Ritchie, J.G. Story, R.G. Hulet: Phys. Rev. Lett. **66**, 1107 (1991)
71. A.D. Parks, D.W. Cullin, D.C. Stoudt: Proc. R. Soc. Lond A **454**, 2997 (1998)
72. N. Brunner, A. Acin, D. Collins, N. Gisin, V. Scarani: Phys. Rev. Lett. **93**, 180402 (2003)
73. N. Brunner, V. Scarani, M. Wegmuller, M. Legre, N. Gisin: Phys. Rev. Lett. **93**, 203902 (2004)
74. D.R. Solli, C.F. McCormick, R.Y. Chiao, S. Popescu, J.M. Hickmann: Phys. Rev. Lett. **92**, 043601 (2004)
75. G.L. Pryde, J.L. O'Brien, A.G. White, T.C. Ralph, H.M. Wiseman: Phys. Rev. Lett. **94**, 220405 (2005)
76. Y. Aharonov, J. Oppenheim, S. Popescu, B. Reznik, W.G. Unruh: Phys. Rev. A **57**, 4130 (1998)
77. J. Ruseckas, B. Kaulakys: Phys. Rev. A **66**, 052106 (2002)
78. S.E. Ahnert, M.C. Payne: Phys. Rev. A **69**, 042103 (2004)
79. Q. Wang, F.W. Sun, Y.S. Zhang, Jian-Li, Y.F. Huang, G.C. Guo: Phys. Rev. A **73**, 023814 (2006)
80. A.M. Steinberg: Phys. Rev. Lett. **74**, 2405 (1995)
81. R.Y. Chiao, A.M. Steinberg: Prog. Opt. **37**, 345 (1997)
82. Y. Aharonov, N. Erez, B. Reznik: Phys. Rev. A **65**, 052124 (2002)
83. D. Sokolovski, A.Z. Msezane, V.R. Shaginyan: Phys. Rev. A **71**, 064103 (2005)

84. D. Rohrlich, Y. Aharonov: Phys. Rev. A **66**, 042102 (2002)
85. K.J. Resch, A.M. Steinberg: Phys. Rev. Lett. **92**, 130402 (2004)
86. J.S. Lundeen, K.J. Resch: Phys. Lett. A **334**, 337 (2005)
87. L. Vaidman: Found. Phys. **21**, 947 (1991)
88. D. Suter: Phys. Rev. A **51**, 45 (1995)
89. D. Suter, M. Ernst, R.R. Ernst: Molec. Phys. **78**, 95 (1993)
90. L. Vaidman: J. Phys. A: Math. Theor. **40**, 3275 (2007)
91. A. Shimony: Found. Lett. **35**, 215 (2005)
92. S. Popescu, L. Vaidman: Phys. Rev. A **49**, 4331 (1994)
93. C.H. Bennett, D.P. DiVincenzo, C.A. Fuchs, T. Mor, E. Rains, P.W. Shor, J.A. Smolin, W.K. Wootters: Phys. Rev. A **59**, 1070 (1999)
94. B. Groisman, L. Vaidman: J. Phys. A: Math. Gen. **34**, 6881 (2001)
95. C.H. Bennett, G. Brassard, C. Crépeau, R. Jozsa, A. Peres, W.K. Wootters: Phys. Rev. Lett. **70**, 1895 (1993)
96. L. Vaidman: Phys. Rev. Lett. **90**, 010402 (2003)
97. L. Vaidman, I. Nevo: Int. J. Mod. Phys. B **20**, 1528 (2006)
98. L. Vaidman: Phys. Rev. A **49**, 1473 (1994)
99. A. Shimony: Erken. **45**, 337 (1997)
100. L. Vaidman: Fortschr. Phys. **46**, 729 (1998)
101. L. Vaidman: Found. Phys. **26**, 895 (1996)
102. W.D. Sharp, N. Shanks: Phil. Sci. **60**, 488 (1993)
103. D.J. Miller: Phys. Lett. A **222**, 31 (1996)
104. R. E. Kastner: Stud. Hist. Phil. Mod. Phys. B **30**, 237 (1999)
105. L. Vaidman: Stud. Hist. Phil. Mod. Phys. **30**, 373 (1999)
106. L. Vaidman: Found. Phys. **29**, 755 (1999)
107. D. Lewis: Nous **13**, 455 (1979); reprinted in *Philosophical Papers Vol II* (Oxford University Press, Oxford 1986) p. 32
108. R.E. Kastner: Phil. Sci. **70**, 145 (2003)
109. L. Vaidman: preprint. http://philsci-archive.pitt.edu/archive/00001108/
110. R.E. Kastner: Stud. Hist. Philos. Mod. Phys. B **35**, 57 (2004)
111. D.J. Miller: Found. Phys. Lett. **19**, 321 (2006)
112. M.S. Leifer, R.W. Spekkens: Phys. Rev. Lett. **95**, 200405 (2005)
113. Y. Aharonov, L. Vaidman: Phys. Lett. A **178**, 38 (1993)
114. Y. Aharonov, J. Anandan, L. Vaidman: Phys. Rev. A **47**, 4616 (1993)
115. Y. Aharonov, J. Anandan, L. Vaidman: Found. Phys. **26**, 117 (1996)
116. Y. Aharonov, L. Vaidman: 'Protective Measurements of Two-State Vectors'. In *Potentiality, Entanglement and Passion-at-a-Distance*, ed. by R.S. Cohen et al. (Kluwer Academic Publishers, Dordrecht, 1997) pp. 1–8
117. H. Everett: Rev. Mod. Phys. **29**, 454 (1957)
118. L. Vaidman: 'Many-Worlds Interpretation of Quantum Mechanics', in *The Stanford Encyclopedia of Philosophy (Summer 2002 Edition)*, ed. by E. N. Zalta <http://plato.stanford.edu/archives/sum2002/entries/qm-anyworlds/>.
119. Y. Aharonov, E. Y. Gruss: Two-time interpretation of quantum mechanics (2005). quant-ph/0507269

Index

Schrödinger equation, 356